SUBATOMIC PHYSICS

An Introduction to Nuclear and
Particle Physics, and Astrophysics

SUBATOMIC PHYSICS

An Introduction to Nuclear and Particle Physics, and Astrophysics

John Demetrius Vergados
University of Ioannina, Greece

Charalampos Moustakidis
Aristotle University of Thessaloniki, Greece

World Scientific

NEW JERSEY • LONDON • SINGAPORE • BEIJING • SHANGHAI • HONG KONG • TAIPEI • CHENNAI • TOKYO

Published by

World Scientific Publishing Co. Pte. Ltd.
5 Toh Tuck Link, Singapore 596224
USA office: 27 Warren Street, Suite 401-402, Hackensack, NJ 07601
UK office: 57 Shelton Street, Covent Garden, London WC2H 9HE

British Library Cataloguing-in-Publication Data
A catalogue record for this book is available from the British Library.

SUBATOMIC PHYSICS
An Introduction to Nuclear and Particle Physics and Astrophysics

ISBN 978-981-122-979-4 (hardcover)
ISBN 978-981-122-980-0 (ebook for institutions)
ISBN 978-981-122-981-7 (ebook for individuals)

For any available supplementary material, please visit
https://www.worldscientific.com/worldscibooks/10.1142/12085#t=suppl

Desk Editor: Nur Syarfeena Binte Mohd Fauzi

To the Students

Who helped in preparing these Lecture Notes

0.1 PREFACE

This book contains material of a set of Lectures on Nuclear and Particle Physics, which one of the authors (JDV) delivered to the students of KAIST University, Daejeon, during the Fall Semester of 2014.

Chapters 6 and 8 were written, while on a visit to the University of S. Carolina during the spring term of 2015, under the USC Provost's "Visiting Scholars Grant Program".

This material has been expanded and improved by the first author (CCM) based on his teaching at the Aristotelian University of Thessaloniki.

The net outcome is material to cover one undergraduate course on Nuclear and Particle Physics as well as Astrophysics together. The main idea is that there are many advantages in teaching all these subjects together, in addition to the situation that there may be cases of only few students interested in anyone of them separately. From a theoretical point of view it is expected that, understanding the similarities between atoms, nuclei and other hadrons and applying analogs from one to the other, have been very effective in research and, thus, they have lead to the development of these fields. From an experimental point of view the nuclear physicist may learn to search for interesting new phenomena motivated by advances in particle physics. The high energy physicist had better understand nuclear physics, if he wants to construct new devices, like detectors etc., appropriate for observing new high energy phenomena. Furthermore we should mention that an understanding of the areas of astrophysics and the physics of the cosmos, demands a strong coupling between them and both nuclear and particle physics. Our approach can, thus, be viewed as an investment, expected to yield high dividends.

The content of the book is intended as a menu from which the teacher can pick material according to his taste and interests as well as those of his students. We have tried to insert cross references to make a specific selection by the student from this menu as easily digested as possible. To this end we give reference to other existing, perhaps more specialized books. Occasionally we will make references to a more advanced companion book "The standard Model and Beyond" by one of us (JDV) published by the same house.

We do not presuppose that the student has any specialized knowledge beyond quantum physics, meaning non relativistic quantum mechanics. Some knowledge of relativistic kinematics may be useful, but we review this material anyway. We felt that it is more important for the student to have an overview of the subject, rather than having him loaded with proofs and technical details. The latter, if needed, can be taken from more advanced courses and materials later. We believe there is some intuitive understanding, which cannot be derived by starting with technically advanced material, but with an overview early on instead. Space limitations have, however, forced us to eliminate certain useful material and it is understood that some other authors might have made a different menu, which could have satisfied the student equally well.

Ch. C. Moustakidis

J. D. Vergados

May 2020

Contents

Chapter 1

Brief elementary particle history

1.1 A panoramic view of the world

The purpose of physics is to understand the world around us. This world at first sight appears very complex (a panoramic view is shown in Fig. 1.1 and Table 1.1). In view of the large range in the values of the quantities shown, one may think that this task is hopeless. Yet, after heroic human efforts over thousands of years and, especially, during the last hundred years, great progress has been made in that direction.

Table 1.1: Some representative sizes seen in nature.

	Order of magnitude of some typical objects	
1.	Size of the universe	: $L \approx 10^{30}$ cm
2.	Size of a typical galaxy cluster (over 10,000 members)	: $L \approx 10^{28}$ cm
3.	Size of a typical galaxy	: $L \approx 10^{23}$ cm
4.	Sun-Earth distance	: $L \approx 10^{12}$ cm
5.	Human size	: $L \approx 10^{2}$ cm
6.	Molecular size	: $L \approx 10^{-6}$ cm
7.	Atomic scale	: $L \approx 10^{-8}$ cm
8.	Size of the nucleus	: $L \approx 10^{-12}$ cm
9.	Size of the constituents of the nucleus (nucleons)	: $L \approx 10^{-13}$ cm
10.	Size of the electron	: $L \approx 10^{-16}$ cm

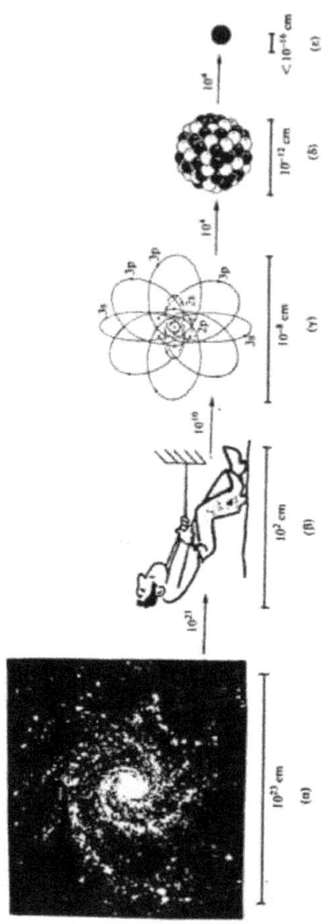

Fig. 1.1: The scale of 4 basic physical systems: (a) A galaxy like our own, (b) a human being like you, (c) a typical atom like Ar with A = 40, Z = 18, (d) a heavy nucleus like uranium, U, and (e) an electron. The number above the arrow indicates the scale factor involved in going to the next. As one can see the electron has a size smaller than 10^{-16} cm. It is considered to be an elementary point like particle.

1.2 Are there elementary constituents of matter?

What do we know about the composition of matter? It is a very old question, but the answer to it is not known a priori. Matter is continuous and indivisible according to the Greek philosophers Plato (427–348) and

Aristotle (384–322). It comes in discrete elementary parts, atoms, according to Democritus (460–437).

This controversy, however, is not a philosophical one and it must be settled by experiment, which is not trivial. How small are the scales one can observe?

The first observations were mainly performed with our eyes. Light falling on various objects reaches our eyes after it is scattered by these objects. The spectrum of the radiation emitted by our sun is continuous. Most of the emitted energy, however, lies in the bands of wavelengths between 4000 and 8000 A^0 or between 0.4 and 0.8 μm. No wonder that the natural selection of the theory of evolution lead to the creation of eyes, which have detection sensitivity to see in that range, which is the visible range. We can, of course, sense the infrared as heat through our skin. Anyway our early ancestors needed eyes to see the dangers coming from wild animals. They had no need to develop eyes to see radio waves coming, e.g., from radio stars. Some insects, like bees etc., developed eyes to see in the ultraviolet. In all cases we can see objects which have dimensions larger the employed wavelength. We cannot see smaller objects, no matter how good the magnification lens is.

We know to-day that we cannot see inside the atom. The ultraviolet region did not provide any characteristic signature for this. A new window was opened with the discovery by Röntgen of X-rays in 1895. The nature of this radiation and the mechanism of its production was not known, but it was later shown that it was an electromagnetic radiation with wavelengths between 0.1 and 10 A^0 (10^{-11}–10^{-9} m). Anyway, this peculiar radiation could be scattered by regular structures, e.g. between the layers of a crystal, which are at larger distances and give an image of the structures, i.e. give a picture of things inside the crystal. They can cast shadows of broken bones and they were almost immediately exploited by medicine. One of the most famous pictures of the world of all times is exhibited in Fig. 1.2. Thus what is known as X-ray crystallography was formed, which has evolved into one of the most revolutions in science. Its most important application was the determination of the structure of DNA, one of the most important gifts of applied physics to genetics.

It turns out that this wavelength is larger than the size of the atom, so still one could not see inside the atom to test the atomic theory. It provided useful information about the atom, however, since it could be used to ionize gases, by emitting electrons (the electron was discovered in 1897 by Thomson). Thus positively charged entities, the ions, were left

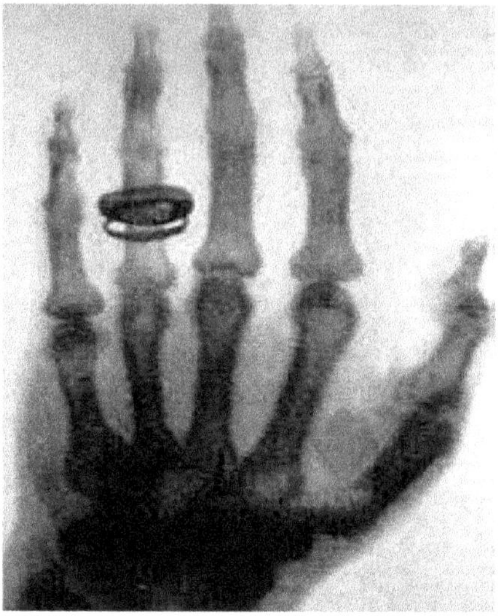

Fig. 1.2: An X-ray picture of the hand of Roentgen's wife with her wedding ring on it. It has become one of the most famous images in world photography. Within weeks, such a picture turned him into an international celebrity.

behind. So the atom was the carrier of the positive charge.

To summarize in order to see a small object of size d, we must probe it with a small wavelength

$$\lambda < d, \quad \lambda = \frac{h}{p} \rightarrow p > p_{\text{th}}, \quad p_{\text{th}} = \frac{h}{\lambda}$$

i.e. the momentum p of the probe must be larger than a threshold momentum p_{th}. The momentum depends on the (kinetic) energy

$$p_{\text{th}} = (1/c)[T_{\text{th}}(T_{\text{th}} + 2mc^2)]^{1/2}.$$

Thus:

$$d = \frac{2\pi\hbar c}{[T_{\text{th}}(T_{\text{th}} + 2mc2)]^{1/2}}.$$

The smaller the length scale the higher the energy! This is exhibited in Fig. 1.3.

Rutherford attempted to investigate what is inside the atom by scattering α particles on a dense target. This celebrated experiment will be discussed later. At this point we mention that, even though he did not

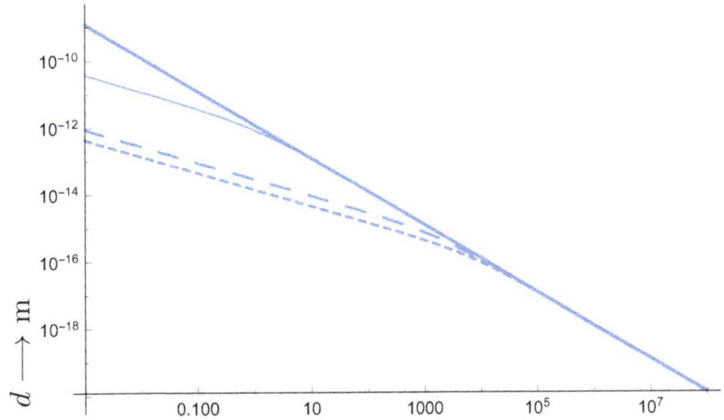

Fig. 1.3: The size of an object in meters that can be seen as a function of the minimum energy required in MeV. The thick solid, solid, dashed and dot-dashed curves correspond to a photon, electron, proton and α particle probe.

know it at the time, the α particle is the nucleus of a He atom, i.e. with a mass 4 times that of the proton. The de Broglie wavelength associated with an α particle of momentum p and energy of $T = 1$ MeV is:

$$\lambda = \frac{h}{p} = \frac{2\pi\hbar}{p} = \frac{2\pi\hbar c}{2\sqrt{\times 4 \times m_p c^2 T}} \approx \frac{200 \times 10^{-15}\ \text{m}}{\sqrt{10 \times 100}} \approx 1.4 \times 10^{-14}\ \text{m}$$

much smaller than the size of the atom. This is why the experiment worked. Subatomic scales (10^{-12} m) required energies at least 1 MeV. To see inside the nucleon (10^{-15} m) energies greater than 10 GeV are required. One day it may be possible to reach distances of 10^{-19} m!

The essential conclusion is that observation of very small scales requires very high energies. Initially energies that already existed in nature were exploited. Around the end of the 17th century chemistry became quantitative. In 1805 Dalton proposed the atomic theory, whereby all substances involved in chemistry and existed in nature could be formed by combining a certain number of elements called atoms, about 90, from the Greek word a-tomon $\alpha'\tau o\mu o\nu$, indivisible. In other words with the energies that were known to exist at the end of the nineteenth century the elementary constituents of matter were about 100, named atoms, organized in a convenient table by Medeleev (see Fig. 1.4).

Understanding the elementary constituents of nature, the **elementary particles**, is very important for an additional reason. It is the vehicle

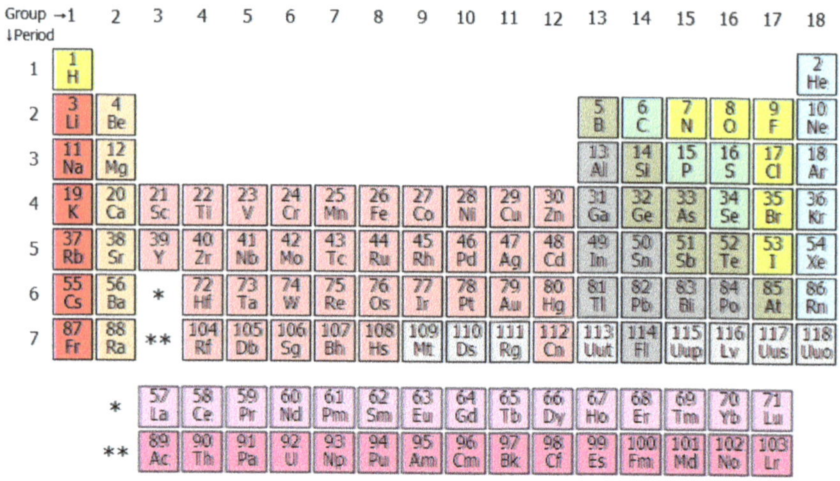

Fig. 1.4: The elementary particles at the end of the nineteenth century were the atoms organized in a table by Medleev shown here. Needless to say that not all of the elements shown here were known at that time.

to understand the elementary interactions that govern the structure and evolution of the world. These are expected to manifest themselves in a much simpler way at small distances. The elementary forces in the end of 19th century were very complicated, of the Van der Waals type. The achievement of higher energies, however, required the introduction of great ideas and depended on the technological progress.

1.3 The early years

At the end of the nineteenth century the following discoveries need be noticed:
1895: The discovery of X-rays as described above.
1896: The discovery of new radiation by Becquerrel.
1897: The discovery of the electron by Thomson.
1900: Rutherford showed that the radiation discovered by Becquerel, contained three components: α, β and γ. The latter was shown to be electromagnetic radiation, with much shorter wavelength compared to X-rays. The β which was identified with electrons via their bending in magnetic fields. The α radiation consisted of massive positive entities of unknown nature.

So the elementary particles of the time were the atoms and the electron. The electron provided the unit of negative charge. The unit of positive charge did not exist. In other words there existed a negative charge coming in discreet amounts, but a similar positive charge did not exist. The positive charge was assumed to be distributed uniformly inside the atom like a fluid, with the electrons capable of swimming in it. This was the Thomson model.

In the context of the classical Maxwell theory light was described as an electromagnetic wave. Planck in 1900 tried to explain the black body spectrum in the context of statistical mechanics. He faced the problem of the ultraviolet catastrophe, i.e. the power radiated would become infinite at very low wavelengths. The problem was solved assuming that the radiation was quantized, i.e. coming in amounts $\epsilon = h\nu$, with h a constant, adjusted to fit the data (to-day the value of this constant is $h = 626\times 10^{-27}$ erg s) and ν the frequency. Einstein, in order to understand the photoelectric effect, adopted a more radical view: what is quantized is the electromagnetic field itself and not bunches of energy having to do with the emission mechanism. When light hits a metallic surface it emits electrons with energy $E = w - h\nu$, with w the work function of the material. The maximum electron energy was found to be independent of the intensity of the electromagnetic radiation and depended on the frequency of light. Nobody was, however, prepared to accept at that time the corpuscular theory of light.

The issue was settled in 1923 with the Compton effect. Light scattered from an electron of mass m_e at rest is shifted in wavelength by:

$$\Delta\lambda = \lambda_c(1 - \cos\theta), \quad \lambda_c = \frac{h}{m_e c},$$

where θ is the scattering angle, i.e. the angle between the incident and the outgoing photon. This behavior results, if you treat light as a composed of particles of zero mass, called photons (if you cannot derive this formula, see Chapter 9, Section 9.4.1). λ_c is called Compton wavelength.

So the list of elementary particles was now expanded to include the photon.

1.4 The Rutherford experiment

This was the situation in 1911, the time when Rutherford undertook the task to check whether the Thomson model was the correct one, in a famous experiment that bears his name and is the prototype of high energy experiments even to day. A historical perspective in the history of elementary particles and a spicy sketch of the protagonists involved has been given

Veltman [Veltman (2003)], who participated in many of the activities and had a first hand knowledge of the personalities involved.

Rutherford's idea was to observe what happens when α particles were scattered off a target of Gold (see Fig. 1.5). He chose α particles, because among the available sources they had a very high mass and they could not be effected by the electrons, since they are so much lighter. Before undertaking this experiment Rutherford had a clear cut idea of how to interpret the experiment (see Fig. 1.6). The obtained results shot down the Thomson model. The α particles were scattered strongly; even backwards! From those that scattered backwards Rutherford estimated the minimum distance between the α particle and the positive charge constituents of the target. For this he used the fact that at the turning point all the energy

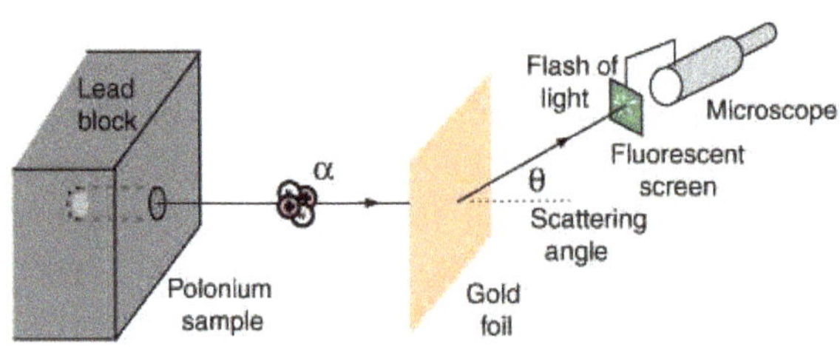

Fig. 1.5: Alpha particles are scattered off a Gold target. The resulting products are observed in a suitable detector

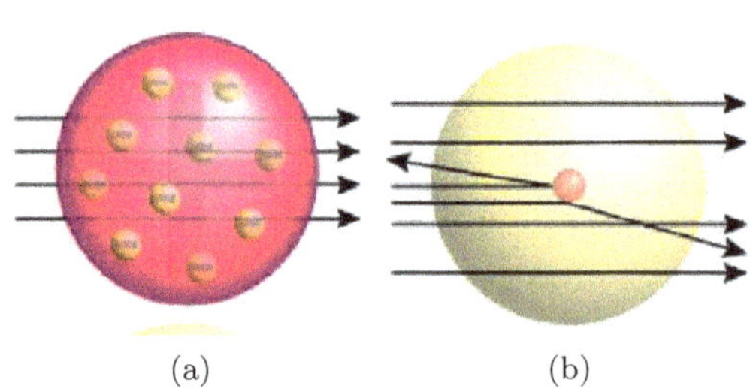

Fig. 1.6: (a) The scattering of alpha particles is not affected by the electrons, since they are light. It is affected by the positive charge. If the charge distribution is uniform like a fluid, the alpha particles are affected very little. (b) The scattering of alpha particles is affected strongly, if the target is composed of point particles. The α particle can scatter even backwards.

of the α particle, was potential energy. Thus applying Coulomb's law he measured distance between the α particle and the positive charged particle. This, by definition, was the size of the positively charged elementary particle, which was named nucleus. The result was astonishing: The nucleus occupies a tiny fraction of the atom's volume! The atom is essentially a void (empty inside), reminiscent of our solar system (see Fig. 1.7).

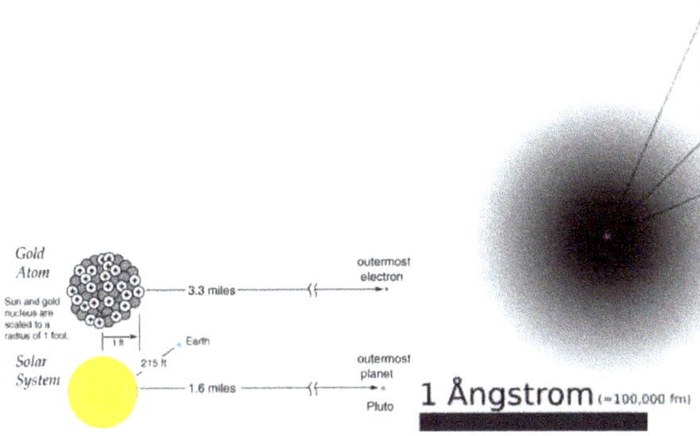

(a) (b)

Fig. 1.7: The size of the nucleus is a tiny fraction of the size of the atom. The atom is essentially a void, just like the solar system (a). The Helium atom (b).

Rutherford was able to explain the results of his experiment in terms of a differential cross section:

$$\frac{d\sigma}{d\Omega} = \frac{m^2 Z_1^2 Z^2 e^4}{p^4 \sin^4 \frac{\theta}{2}}, \tag{1.1}$$

which will be discussed elsewhere[1] [Vergados (2017)].

Some further developments followed:

- The discovery of cosmic rays (1932). This was going to play an important role, because it contained particles much more energetic than those produced by radioactivity.

[1]This will be referred to as the companion volume or SSB. There, from Fig. 7.1 of Section 7.3, it can be seen that, as anticipated by Rutherford, a substantial fraction of the scattering events take place at backward angles.

- The observation of Nuclear Reactions: α particles +nucleus$\rightarrow$ new nucleus + proton (1919). These demonstrated that the nucleus is made up of protons. Thus the nucleus is described by a mass number A, the number of protons, and charge Z, which arises by assuming that it contains in addition $A-Z$ electrons.

So for the first time the world looks quite simple. It is built up of only three elementary particles, namely the proton, the electron and the photon. From the electron and the proton the nuclei are built.[2] From the electron and the nuclei the atoms are made. From the atoms combined, the molecules are made etc.

This beautiful picture, was a bit spoiled in 1932 by the discovery of the neutron n by Chadwick, which is electrically neutral with spin 1/2, and a mass a bit larger than that of the proton. Irene Joliot Curie, daughter of Marie Curie and Frederic Joliot, had the first evidence for the neutron in an experiment bombarding the beryllium nucleus with α particles. She observed a neutral radiation, but she interpreted it as X-rays. She, perhaps inadvertently, discussed her experiment with Chadwick. He then repeated the experiment by adding a new essential ingredient to it: He placed paraffin wax far away from the target on the other side from the alpha beam and observed proton coming out of the parafin sheet after the neutral radiation hit it. X-rays could eject particles like electrons out of the parafin, but they were not able to eject heavy particles like protons. It had to be something as massive as the proton itself, but neutral. To-day we know that it is about 1.4 MeV/c^2 heavier than the proton.

Anyway the elementary particles now are 4, p, n, e^-, γ. In this scenario the nucleus does not anymore need the presence of electrons. It can be made up of Z protons and $N = A-Z$ neutrons. A new attractive force, however, stronger than the electromagnetic (EM), called **strong** interaction, acting between p and n is required to bind them. Unlike gravity and EM this force is of **short range** nature, and it does not manifest itself at distances greater than 1 fm $= 10^{-15}$ m. So it does not affect the macroscopic behavior of objects. We should note that a truly new elementary particle forced us to discover a new force. The story will repeat itself later.

In 1933 a new particle was discovered by Anderson. It looks like the electron but it has opposite charge. It was named positron and indicated as

[2] There were some problems, e.g. with the nitrogen nucleus, $^{14}_{7}$N. It was known to have spin 1, but 14 protons and 7 electrons = 21 particles should give rise to a half integral spin, but this had been overlooked.

e^+. This was indeed a great discovery, a triumph of Dirac's theory (1930–31), which predicted that, if a particle exists, its antiparticle, with opposite charge and equal mass, should also exist. So a better name for the positron would be anti-electron.

Now a natural question: Does the anti-proton exist? The anti-neutron? To answer these, one had to wait till energies high enough became available to produce such particles.

Suddenly, before the ashes settled, it was found that the neutron is mortal with a life time of about 898s. It was observed that:

$$n \to p + e^- + \text{(missing energy, momentum and angular momentum)}.$$

In the above decay in the energy conservation we also include the rest mass energy of the electron, $E = m_e c^2$, introduced by Einstein in the context of special theory of relativity. Was the conservation of these quantities violated? You must watch out before you can adopt such a radical view. In fact, in 1933 Pauli postulated the existence of an escaping particle, which was assumed to be massless, carry spin 1/2 and electrically neutral. It was named neutrino (tiny neutron). So the actual process was:[3]

$$n \to p + e^- + \tilde{\nu}_e. \tag{1.2}$$

For the accomplishment of neutron's death, a new force was required between the electron and the neutrino, much weaker than the Electromagnetic (EM), named **weak** interaction, but still much stronger than gravity. This did not have macroscopic consequences, since it was very short ranged. If the neutrino is indeed massless, weak interaction is the only force it feels.

At this point we should mention the neutrinos were directly detected for the first time by Reiness and Cowan in 1953 via the reaction:

$$\tilde{\nu}_e + p \to n + e^+.$$

It was also found that the neutrino produced in weak interactions is different from its anti-neutrino:

$$\nu_e +^{37}\text{Cl} \to^{37}\text{Ar} + e^-\text{(yes!)}, \quad \tilde{\nu}_e +^{37}\text{Cl} \to^{37}\text{Ar} + e^-\text{(No!)} \text{ (Davis, 1955)}.$$

Anyway we should mention now that the neutrino seemed to be useless. Bethe, however, in 1939 found a very important role for the neutrino, namely in the energy production in the stars (pp and CNO chain, which

[3]He actually used the symbol ν for neutrino. $\tilde{\nu}_e$ is our present notation, since we now know that it is the ant-neutrino associated with the the electron.

will be further discussed in Section 7.2.1). Effectively it involves fusion of protons into Helium:

$$4p \to \text{He} + 2\nu_e + 2e^+ + \text{light}.$$

Thus we have a symmetric picture. We have two heavy spin 1/2 particles, p and n, which behave similarly under the strong interaction, which we call **baryons** (from the Greek word $\beta\alpha\rho\upsilon'$ meaning heavy) or **nucleons**, since they are the building blocks of the nucleus. We also have two light ones named **leptons** (from the Greek word $\lambda\epsilon\pi\tau o'$ meaning thin, fine), which behave similarly under the weak interaction. Thus

$$\begin{pmatrix} p \\ n \end{pmatrix} \Leftrightarrow \begin{pmatrix} \nu_e \\ e^- \end{pmatrix}. \tag{1.3}$$

They are members of a doublet with the respective force acting vertically. They are normally written so that the member with the highest charge is on top. Clearly, of course, the two members of the doublet behave differently under the EM force.

In 1947 a new lepton μ^- was discovered in cosmic rays. A replica of the electron, but 208 times more massive. Is is also accompanied by its neutrino ν_μ, which was assumed to be different from ν_e, something that was confirmed at Brookhaven lab, USA, in 1962.

A short time after the discovery of muons in 1947, new particles, the $\pi^\pm$, π^0, were discovered. These are particles with zero spin and mass intermediate between leptons and baryons (nucleons) and, thus, they were named **mesons**. This was viewed as a triumph for the prediction of Yukawa (1936), since they were expected to be the carriers transmitting the strong interaction between the baryons. In field theory every interaction is mediated by such carriers with the range of the force being inversely proportional to its mass, $a = \hbar/(mc)$. The electromagnetic force is mediated by the photon, which is massless. Thus the interaction is of infinite range. The strong interaction has a range of about 1 fm $= 10^{-15}$ m, which implied that its carrier had a mass 300 times heavier the electron and about 1/6 of the proton mass, in agreement with the observation of the π mesons. The situation became a bit confused, however, since the discovery of other heavier mesons soon followed.[4]

The time had come to observe the anti-baryons. We witnessed the discovery of anti-proton at Berkeley in 1955 and of the anti-neutron in

[4]To-day we know that this meson exchange interaction is not fundamental. It is a Van der Waals type interaction obtained from QCD, see Section 9.15, Fig. 9.24. So other mesons can be involved.

1956. Even though the coming of these particles was not surprising, but expected as soon as the right energy was achieved, it lead to a Nobel prize to Emilio Segre and Owen Chamberlain in 1959. It can be viewed as another triumph for Dirac.

In the end we had only few elementary particles. In addition to p, n, e^-, ν_e and γ, we had the mesons π^+, π^- and π^0 needed to mediate the strong interaction and the, apparently, useless leptons μ^- and ν_μ ("who ordered them?"). Their number is almost doubled by the existence of their anti-particles (in the case the photon and the π^0, which have zero charge, the particle coincides with its antiparticle). This doubling, however, is not a source of concern, since their existence increased the symmetry in the formulation of the law's of nature.

1.5 The chaos is coming again — A new particle zoo

The order achieved in the world by its being built by only few elementary particles, did not last long. In the following decade about 150 particles were discovered all of them with equal claim to become candidates for elementary particles. A real zoo, without success in organizing them a la Medeleev. They seemed strange since:

- They were produced copiously (via strong interactions).
- They lived too long (they die via the weak interaction).
- They came in a great variety, either as mesons (K^+, K^-, K^0 etc.), with integral spin, or as baryons (Σ^+, Σ^-, Σ^0, Λ etc.), with half integral spin.

So they were named **strange.**

The puzzle was solved by Pais (1952) and, independently, Nishijima (1953) as follows:

- They carry a new additive quantum number (strangeness, s), which is conserved in strong and EM interactions.
- They are produced in pairs, with the members of each pair having opposite strangeness. Thus strangeness is conserved and the strong interaction remains operative.
- They die alone (like everyone of us!), via the strangeness violating weak interaction.

Examples of production:

$$\begin{array}{lllllll} & p & +p \to & p & +p & +K^+ & +K^- \\ \mathrm{s} & 0 & 0 & 0 & 0 & 1 & -1 \\ & p & +p \to & p & +p & +\Sigma^+ & +\Sigma^- \\ \mathrm{s} & 0 & 0 & 0 & 0 & -1 & 1 \\ & p & +\pi^- \to & p & & +K^- & +K^0 \\ \mathrm{s} & 0 & 0 & 0 & & -1 & 1 \\ & p & +\pi^+ \to & p & & +K^+ & +\bar{K}^0 \\ \mathrm{s} & 0 & 0 & 0 & & 1 & -1 \end{array}$$

The last reaction was actually predicted, before it was observed. The assignment of strangeness $+1$ to K^+ was arbitrary (not the most convenient choice, as it turned out). Then from reactions like those above the strangeness of all other particles is determined. Some decays of strange particles are as follows:

$$K^+ \to \mu^+ + \nu_\mu, \quad K^+ \to \pi^+ + \pi^0, \quad K^- \to \pi^- + \pi^0$$

$$\Sigma^+ \to p + \pi^0, \quad \Sigma^+ \to n + \pi^+, \quad \Sigma^- \to n + \pi^-, \quad \Sigma^0 \to \Lambda^0 + \gamma$$

etc.

Attempts to organize this zoo in terms of representations of continuous symmetry groups, mainly unitary groups $U(n)$, did not work as well as the Medeleef table did for atoms. It trained, however, particle physicists in group theory. Such topics will be discussed in more detail in Chapter 10. The observation that the fundamental representation of these symmetry groups did not materialize in nature, lead theoretical physicists to postulate the existence of new exotic particles named **quarks**.

1.6 The quark model

Recall that under the strong interaction we started with two baryons p and n and put them together $N = (p, n)$ as a douplet. With the spin analogy we can think of them as an isodoublet under a mathematical symmetry called isospin, i.e. think of them as characterized by $I = 1/2$ with projections $I_3 = \pm 1/2$. This, of course, is an internal symmetry and has nothing to do with spin. It has, however, analogous mathematical properties. Instead of this doublet they introduced a doublet $q = (u, d)$ together with another elementary entity s with strangeness quantum number $s = -1$. Thus

$$\begin{pmatrix} u \\ d \end{pmatrix} \Leftrightarrow \begin{pmatrix} Q = 2/3 \\ Q = -1/3 \end{pmatrix} \text{ and } s \text{ with } Q = -1/3. \tag{1.4}$$

Q is the electric charge, in units of the fundamental charge. The names stand as follows: u (for up), d (for down) and s (for strange). These particles were proposed by Sweig and Gell-Mann in 1964 [Sweig (1964)], [Gell-Mann (1964)]. The name **quark** was coined by Gell-Mann, evidently being inspired by "three quarks for Muster Mark!" found in the work of James Joyce, Finnegans Wake. Notice that the charge of the quarks was assumed fractional!

The following rules apply in the construction of hadrons (baryons and mesons):

- All Baryons are composed of three quarks, e.g. $p = uud$, $n = udd$, $\Sigma^+ = uus$ etc. This could only be achieved with the above quantization of the quark charge.
- All mesons are composed of a quark-antiquark pair, e.g. $\pi^+ = u\bar{d}$, $\pi^- = d\bar{u}$, $K^+ = u\bar{s}$ etc.
- The force between quarks is attractive with a strength, which is increasing with the distance. As a result the quarks are so tightly bound that they can never be found free.

With these rules all hadrons are guaranteed to have integral charges, consistent with observation.

This model worked quite well then, but new baryon resonances were soon found, e.g. the quartet with charges Δ^{++}, Δ^+, Δ^0 and Δ^-, $I = 3/2$ and a singlet, Ω^-, $I = 0$, all with spin 3/2. Then, e.g. $\Delta^{++} = uuu$, $Q = (2/3{+}2/3{+}2/3) = 2$, $\Omega^- = sss$, $Q = -1$. The problem is that the wave function describing each of them is symmetric, while the corresponding particle is a fermion!

The effective cure was the introduction of a new quantum number, called **color**, by Han and Nambu in 1965 [Han and Nambu (1965)]. So in describing a quark, three color indices r, g, b are needed, in addition to the flavor indices (u, d, s). This way the product $\Psi = \Psi_1(\text{flavor})\Psi_2(\text{color})$ can be antisymmetric, if the color part is made antisymmetric. More details will be given in Chapter 10.

Many people, including Bjorken (1964), essentially for aesthetic considerations, proposed the existence of one more quark flavor c, for charm, with charge Q = (2/3). Subsequently Glashow-Maiani-Iliopoulos (1970) [Glashow *et al.* (1970)] showed that such a quark was in fact necessary to provide a mechanism, named GIM mechanism after their initials, to explain the great suppression (10^6) of processes like:

$$\bar{K}^0 \to \mu^+ + \mu^-,\ K^+ \to \pi^+ + \nu + \bar{\nu}.$$

So with great excitement in 1974 the physics community learned that a narrow resonance $J/\Psi = c\bar{c}$ was found at Brookhaven and SLAC, Stanford. This resonance was unusually narrow barking for a new quantum number, which was c. This resonance had net charm zero, but a plethora of such resonances followed, some of them charmed. Also charmed baryons were detected in 1976 in SLAC. Thus it was established that:

$$\begin{pmatrix} u \\ d \end{pmatrix}, \begin{pmatrix} c \\ s \end{pmatrix} \Leftrightarrow \begin{pmatrix} Q = 2/3 \\ Q = -1/3 \end{pmatrix}. \tag{1.5}$$

Everything seemed to fit well, but in 1977 in FERMILAB, USA, a new quark, named b for beauty, with Q $= -1/3$ was unexpectedly born. Another meson was discovered of the type $Y = b\bar{b}$, with mass about 10 times that of the proton (but, still a meson!). Many such resonances with bound state spectra reminiscent of atomic and nuclear spectra (with a different energy scale, of course) soon followed. The short lived symmetry between quarks with charge $2/3$ and $-1/3$ was gone. Not for too long, however. In 1994 a third quark t with charge $2/3$ is discovered by CDF at FERMILAB, much heavier than expected (about 180 times heavier than the proton). Before then another lepton, named τ^- had been discovered (yes! called lepton even though it is almost two times heavier than the proton). Lepton now means that it does not exhibit strong interaction. Its neutrino, ν_τ was directly detected 20 years after its charged partner at FERMILAB.

This completes the story of elementary fermions. All non exotic (luminous) matter in the universe is made of these fermions (matter fields). The hadrons (quarks) as well as the leptons come in three generations. In addition each quark comes in three colors. The color is responsible for the strong interaction, sometimes called color charge.

On the theoretical side the Standard Model (SM) was developed by Glashow-Salam-Weinberg (GSW) 1967–71, [Glashow (1961)], [Weinberg (1967)], [Salam (1968)]. This model is central to particle physics and is discussed in many textbooks, e.g. in the companion book [Vergados (2017)]. Its essential features are the following:

(1) A symmetry. The adopted symmetry is not simple, but a product of three symmetries, $SU_I(2) \times U_Y(1) \times SU_c(3)$. It is thus characterized by three couplings g_2, g_1, g_3 and $3 + 1 + 8 = 12$ generators.
(2) This symmetry is a gauge symmetry. It thus implies the existence of $3 + 1 + 8 = 12$ gauge bosons, initially massless.

(3) It contains a Higgs particle, i.e. a particle with spin zero. They chose to put in a complex isodoublet representation of $SU_I(2)$. This corresponds to 4 real components.
(4) The $SU_I(2) \times U_Y(1)$ symmetry is broken spontaneously by this scalar, which acquires a non vanishing vacuum expectation value. Thus one component of the Higgs acquires a mass and survives as a physical particle. Its mass is not specified by the model.
(5) As a result of the symmetry breaking, three of the 4 $SU_I(2) \times U_Y(1)$ gauge bosons, namely the W^+, W^-, Z^0 acquire a mass about 100 times the mass of the proton. They mediate weak interactions.
(6) The fourth boson, A_μ, remains massless and it is identified with the photon. It mediates EM interactions.
(7) Eight gauge bosons g, called gluons, carrying color remain massless and mediate the strong interactions. Even though they are massless the do not lead to a long range force, because their interaction, unlike the EM force, is covered by a non Abelian symmetry.
(8) The fermion content, as given above, is put in by hand. The number of generations (families) is not specified by the symmetry. The left handed fermions are put in isodoublets and the right handed in isosinglets. The right handed neutrino does not exist in the SM. The fermion masses are not specified, except for the fact that the neutrinos are massless.

All the particles expected in view of the Standard Model have been discovered, for a more detailed exposition, see, e.g., the companion book [Vergados (2017)]. The neutral gauge boson Z was observed first in the decay channel $e^- e^+$ in the experiments UA1 and UA2 of CERN with a mass $M_Z = (91.187 \pm 0.007)$ GeV/c^2 in 1983. The charged gauge bosons were a bit harder to analyze, due to the fact that they decay to an escaping neutrino (anti-neutrino) and only one charged lepton, but they were detected by the same groups at CERN. The obtained mass was $m_W = 80.41 \pm 0.10$ GeV/c^2. The Higgs particle was discovered by the groups CMS and ATLAS of LHC at CERN [Aad *et al.* (2012)], [Chatrchyan *et al.* (2012)]. Its mass was later found to be 125.09 ± 0.21(stat.) ± 0.11(syst.) GeV/c^2 and its spin 0. This was considered a great discovery, in spite of the fact that it has been expected to occur, and lead to a Nobel Prize for the theoreticians, who proposed it, namely Peter Higgs and Francois Englert in 2013, [Englert and Brout (1964)], [Higgs (1964)].

The gluons, the mediators of strong interaction, have not been detected, since they are colored particles and they cannot be found free. Their mass is consistent with zero.

In summary, the SM is consistent with all the observed data, except two. It does not explain the observed matter anti-matter asymmetry in the cosmos, i.e. the fact that the visible universe is composed of matter. The only antimatter that exists is produced in the laboratory or reactions that are currently taking place. Also against the SM predictions, the neutrinos are not massless, as it is evidenced from neutrino oscillations to be discussed in Chapter 12. Furthermore it does not explain the bulk of the energy content of the universe, 95% of it is of unknown nature (dark energy and dark matter). For these reasons as well as theoretical arguments (the symmetry is not simple, it contains too many parameters not specified by it etc.), it is not viewed as the ultimate theory of nature.

1.7 Problems

For the charged particles, except for the proton, the charge is indicated by a superscript.

(1) Examine the processes:

$$e^+ + e^- \rightarrow \mu^+ + \mu^-, \quad n \rightarrow p + e^- + \tilde{\nu}_e, \quad \mu^\pm \rightarrow e^\pm + \tilde{\nu}_\mu + \nu_e$$

$$\mu^\pm \rightarrow e^\pm + \tilde{\nu}_e + \nu_\mu, \quad \nu_e + e- \rightarrow \nu_e + e-, \quad \nu_\mu + e^- \rightarrow \nu_\mu + e^-.$$

Considering the conservation of the additive quantum numbers charge and lepton flavor, which ones are allowed? For those allowed which of the $W^\pm$, Z and γ mediators[5] can, in principle, mediate them?

(2) Considering now, in addition, baryon and fermion number conservation (the latter essentially follows angular momentum conservation) indicate which of the processes

$$p + p \rightarrow W^- + W^+, \, p + p \rightarrow K^+ + p$$

are allowed.

(3) It is suggested that the processes

$$\Sigma^- \rightarrow \Lambda^0 + e^- + \tilde{\nu}_e, \, \Sigma^+ \rightarrow \Lambda^0 + e^+ + \nu_e, \, K^- \rightarrow \pi^- + \pi^+ + \pi^- \quad (1.6)$$

can occur. Do you agree? If you do, what interaction is responsible for the decays involved?

[5]Sometimes people write Z^0.

(4) Are the processes

$$\Sigma^- \to e^- + \gamma,\ K^0 \to 2\gamma$$

allowed? If not, why not?

(5) The decays of the type of Eq. (1.6) are not kinematically favored. One finds experimentally:

$$\Sigma^+ \to p + \pi^0\,(51.6\%),\ \Sigma^+ \to n + \pi^+(48.4\%)$$

$$\Sigma^0 \to \Lambda^0 + \gamma\,(100\%),\ \Sigma^- \to n + \pi^-(100\%).$$

Which interaction is responsible for each decay? Which decay is expected to have the shortest life time? (explain).

(6) Consider the processes:

$$\begin{aligned} p\bar{p} \to \pi^+\pi^-\pi^0\pi^+\pi^-,&\ \ \nu_e p \to e^+\Lambda^0 K^0 \\ D^+ \to K^-\pi^+\pi^+,&\ \ \nu_e p \to e^-\Sigma^+ K^+ \\ pK^- \to \Lambda^0\Sigma^0,&\ \ B^+ \to \bar{B}^0\nu_e e^+ \\ \nu_\mu p \to \mu^+, n&\ \ \Omega^- \to K^-\bar{K}^0 \\ \nu_\mu p \to e^+, n&\ \ \Sigma^0 \to \Lambda^0\gamma \\ D^0 \to K^-\pi^+\pi^-\pi^+,&\ \ e^-e^+ \to D^0\bar{D}^0 \\ \Lambda^0 \to p\pi^-,&\ \ \pi^- p \to \Lambda^0 K^0 \end{aligned}$$

(the D^+ and D^0 contain a charmed quark c). Which ones are allowed? If some of them are not allowed, what is the reason? If they take place, which interaction is responsible?

(7) What is the quark content of K^+, D^0, Λ^0 and Σ^+.

(8) Suppose that only three quarks exist:

$$u \leftrightarrow Q = 2/3,\ s = 0, I = 1/2, I_3 = 1/2,$$

$$d \leftrightarrow Q = -1/3,\ s = 0, I = 1/2, I_3 = -1/2,$$

$$s \leftrightarrow Q = -1/3,\ s = 1/2, I = 0$$

where s indicates the strangeness and I the iospin, together with their antiparticles. Sometimes the notation q is used for any of the quarks.

The allowed isospin combinations follow the same rules with the spin combinations.

- Construct all possible mesons, i.e. $q\bar{q}$ with their quantum numbers Q, s, I.
- Construct all possible baryons, q^3 with their quantum numbers Q, s, I.
- Have all the above been found? If, yes provide the common name used. If no, what is the reason?

  If the obtained results do not appear in this chapter, see the Review of Particle Physics (2018) and 2019 update, e.g. [Tanabashi *et al.* (2018a)].

Caution: For the hadrons in some tables one finds listed the quantum number Y, instead of the strangeness, where $Q = I_3 + Y/2$, $Y = B - s$.

(9) Suppose that in addition there exists the charmed quark c. Repeat the above procedure, restricting the search to baryons with only one charmed quark.

Chapter 2

Nuclear physics

In Chapter 1 we gave a panoramic view of elementary particles and their interactions. We have seen the role that these particles played in understanding the laws of nature. This was the result of many crucial factors. The high energies achieved, which allowed us to have a look at short distances. The advances in technology which allowed the construction of appropriate detection devices. The imagination of theoretical physicists in developing appropriate models for predicting new phenomena and last, not least, the ingenuity of the experimentalists in analyzing very complex and incredibly large amounts of data. It is clear that the progress made during the last 50 years has been remarkable.

Most of the matter in the universe, however, is composed of atoms, which are made of nuclei and electrons. The nuclei are mainly composed of nucleons, protons and neutrons. The nucleons are composed of u and d quarks. The quarks have not been seen directly as free and, as we have seen, they may never will, but everyone believes that they truly exist. So from the pantheon of the many particles encountered in Chapter 1, one would like focus on these exhibited in Fig. 2.1. So in the subatomic world we may begin with nuclei.

Nuclear physics has been a very active field of research during the last one hundred years. We are going to discuss the most important aspects of the developments of this field from an elementary point of view, with emphasis on those that serve as a bridge with particle physics and astrophysics. The reader interested in more technical discussions involving nuclear physics is referred to many existing books, to mention, e.g., [Konopinski (1966)], [Roy and Nigam (1967)], [deShalit and Feshbach (1974)], [Cottingham and Greenwood (1986)], [Brown (2020)] among others.

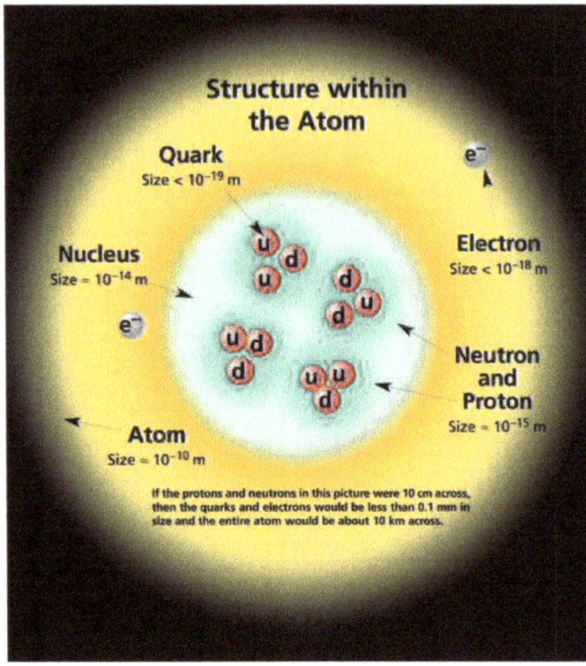

Fig. 2.1: The particles that most of the visible universe is made of (courtesy of Alex Brown [Brown (2020)]).

2.1 Some gross properties of nuclei

We have seen in the previous chapter that matter is composed by the formation of bound states of its constituents. These are held together by some kind of interaction. They are bound, because their energy is less than the sum of their rest energies. The energy required to separate them is called **binding energy**. Thus the binding energy characterizing the compound system is negative. Bound states of composite systems can easily be studied, when their binding energy is much smaller than the rest mass of the constituents. The simplest bound system is the hydrogen atom composed of a proton and an electron held together by the electromagnetic interaction. The binding energy of the hydrogen atom is -13.6 eV, whose absolute value is much smaller than the rest energy of the proton and the electron, in fact less than 2.7×10^{-5} of the electron rest energy. So the excited states of the atoms are very well studied.

The nucleus is an object composed of protons and neutrons. The binding is due to a strong spring like interaction. So it is quite a bit larger in absolute value than that encountered in atomic physics. The simplest nucleus is that of the He atom, composed of two protons and two neutrons with rest energies 938.272 and 939.563 MeV respectively to a total of 3755.67 MeV. It turns out that binding energy is about -28 MeV. So the binding energy is 0.7% of the total energy. It is not as small as in the case of the atom, but still quite small. We will see below that the binding energy changes from nucleus to nucleus. So it is more appropriate to study the binding energy per nucleon. For the He nucleus is $-28/7 = -7$ MeV. We will see below that this number is characteristic of stable nuclei, larger than 1.4 MeV, the difference of the rest mass of the proton and neutron. So the bound neutron cannot decay and such a nucleus is absolutely stable.

A nucleus is designated by the same name as the element with the same number of protons as it appears in Medeleev's table. Since the binding energy is small it is fairly easy to find the number of protons and neutrons in a nucleus. All one has to do is to measure its mass. There may exist many nuclei associated with the same element characterized by the number of neutrons they contain. Different nuclei with the same Z are called **isotopes**. Thus the nucleus is indicated by an upper index which corresponds to A, the total number of protons Z and neutrons N (A = A + Z), and a lower index indicating the number of protons. Thus $^{238}_{92}$U and $^{235}_{92}$U are two isotopes of the uranium element, with 92 protons and N = 146 and 143 neutrons respectively. There exist thee isotopes of the hydrogen nucleus p = $^{1}_{1}$H (proton), d $\equiv$ $^{2}_{1}$H (deutron) and t $\equiv$ $^{3}_{1}$H (triton).

2.1.1 *The stability line*

The nucleons, protons and neutrons, are held together under two forces. The repulsive electromagnetic force between the protons and the strong interaction acting between all nucleons. The latter in certain states is attractive. Thus the stable nuclei are expected to have about the same number of protons and neutrons. The repulsive electromagnetic force is long ranged, felt at very large (by nuclear standards) distances, while the strong interaction is of short range, i.e. it is felt by the nearby nucleons. Thus the number of protons cannot increase for ever and, usually, it remains smaller than the number of neutrons, (see Fig. 2.2). Even if nuclei with $N < Z$ are produced, they are unstable. The decay of any unstable nuclei is a phenomenon known as **radioactivity**. In recent years the nuclear landscape

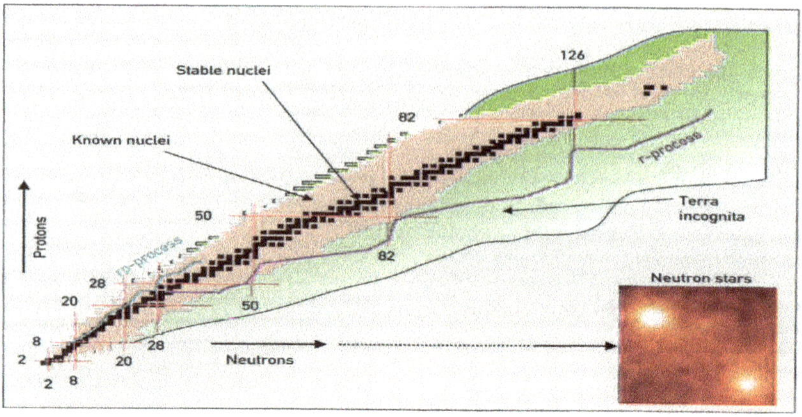

Fig. 2.2: The nuclear landscape that defines the territory for nuclear physics research as a function of N and Z. The black squares represent stable plus very long-lived nuclei that define the valley of stability. The pink region designates the range of unstable nuclei that has been explored by reactions employing stable nuclei by addition or removal of neutrons or protons. The area indicated as "terra incognito" corresponds to thousands of radioactive nuclei far from the valley of stability and remain to be explored. In particular, the explorable nuclear landscape is bounded by the proton drip line to upper the left and neutron drip line to the bottom right, both shown by thin black lines. The anticipated paths of astrophysical processes leading to nucleosynthesis are the r-process (purple line) and rp-process (turquoise line). The red lines show the magic numbers for the valley of stability. Finally neutron stars, which are compact objects consist mainly of neutrons ($\sim 90\%$) and a small fraction of protons ($\sim 10\%$) are also indicated.

has been expanded towards $N > 160$, see Fig. 2.3.

The volume of the nucleus is $V = AV_1$, where V_1 is the effective volume occupied by a single nucleon. Thus we expect the nuclear radius to scale with with the mass number. In fact

$$R = r_0 A^{1/3}, \text{ with } r_0 \approx 1.25 \text{ fm}, \ 1 \text{ fm (Fermi)} = 10^{-15} \text{ m}.$$

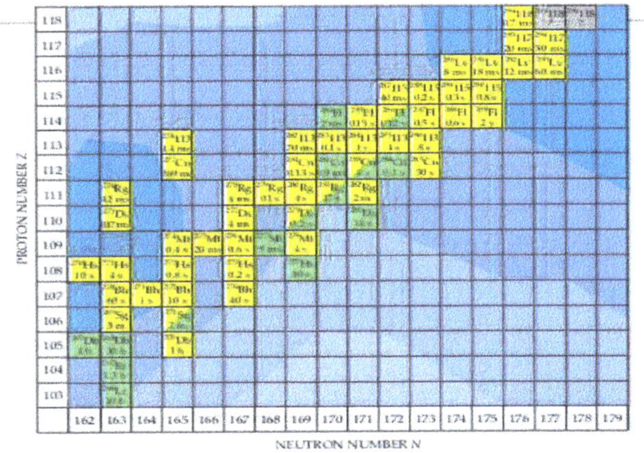

Fig. 2.3: The heaviest members of the nuclear landscape. The mode of decay (yellow for α emission and green for spontaneous fission) as well as the half life are indicated. The gray squares indicate data not established, but currently analyzed by US and Russian teams.

2.1.2 *Nuclear densities*

The nuclear density is usually assumed to be spherically symmetric described by a function $\rho(r)$, where r is the distance from the the nuclear center. If it represents the matter density, it is normalized to the number of nucleons, i.e. it is defined so that:

$$\int \rho(r)d\tau = 4\pi \int r^2 \rho(r)dr = A$$

where $d\tau$ is the volume element. It falls off to zero at the surface of the nucleus, but the way it approaches zero is characterized by a parameter called surface thickness. The reader is encouraged to study some simple densities, e.g., those included in the problem section at the end of this chapter, before proceeding with the realistic densities discussed here.

A reasonable nuclear density, which explains all data, has been derived phenomenologically and it is given by:

$$\rho(r) = -\frac{1}{\mathrm{Li}_3\left(-e^{\frac{R}{a}}\right)} \frac{A}{8\pi a^3 \left(e^{\frac{r-R}{a}} + 1\right)}, \tag{2.1}$$

where R is the nuclear radius given above and a is a length parameter, surface thickness, which is approximately 0.6 fm. $\mathrm{Li}_3\left(-e^{\frac{R}{a}}\right)$ is just a normalization constant. It is connected with the polylogarithm function.[1] This density profile, known as Fermi distribution or Woods-Saxon distribution, is exhibited in Fig. 2.4.

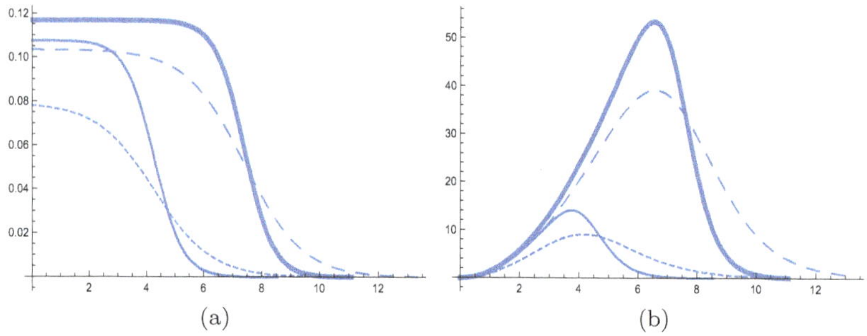

Fig. 2.4: (a) The nuclear particle density profile $\rho(r)$, in units of fm^3, for the cases: i) A = 40, a = 0.5 fm (solid line), ii) A = 40, a = 1.0 fm (dotted line), iii) A = 208, a = 0.5 fm (thick solid line) and iv) A = 208, a = 1.0 fm (dashed line). (b) The probability density $4\pi r^2\rho(r)$ normalized to A. The curves have been drawn with the same notation as in (a). In all cases the nuclear radius has been chosen to be $R = r_0 A^{1/3}$, $r_0 = 1.25$ fm.

The probability density,

$$P(r)dr = 4\pi r^2 \rho(r)dr$$

represents the probability to find A nucleons **within a spherical ring with length** dr at a given radius r from the center of the nucleus.

How can one determine the nuclear density? In an experiment a la Rutherford, see Section 1.4, using one of the probes indicated above. It turns out the cleanest approach is to use the electron probe, in which case one can get information the charge distribution inside the nucleus $\rho_{ch}(r)$. Now

$$\int \rho(r)_{ch} d\tau = 4\pi \int \rho(r)_{ch} r^2 dr = Z, \tag{2.3}$$

[1]This is defined as

$$\mathrm{Li}_n(z) = \sum_{k=1}^{\infty} \frac{z^k}{k^n}. \tag{2.2}$$

Mathematica indicates it as PolyLog$[n, z]$. For $n = 3$ the quantity PolyLog$[3, -E^{(R/a)}]$ used above can be computed only numerically (see Mathematica for its definition and use).

where Z is the number of protons in the nucleus. The charge probability density,

$$P(r)_{ch}dr = 4\pi r^2 \rho(r)_{ch}dr$$

represents the probability to find Z protons **within a spherical ring with length** dr at a given radius r from the center of the nucleus. To get this information regarding its ground state, one will need elastic electron-nucleus scattering.

2.1.3 *Cross section*

We want to know the number of particles $d\mathcal{N}$ scattered in a given direction per unit time. This is proportional to the number of scattering centers N and the flux Φ of the oncoming particles, i.e. the number of particles per unit area per unit time, in a given direction:

$$d\mathcal{N} = N\Phi d\sigma = N\sigma(\theta)d\Omega, \quad \sigma_{tot} = \int \sigma\theta d\Omega \tag{2.4}$$

see Fig. 2.5.

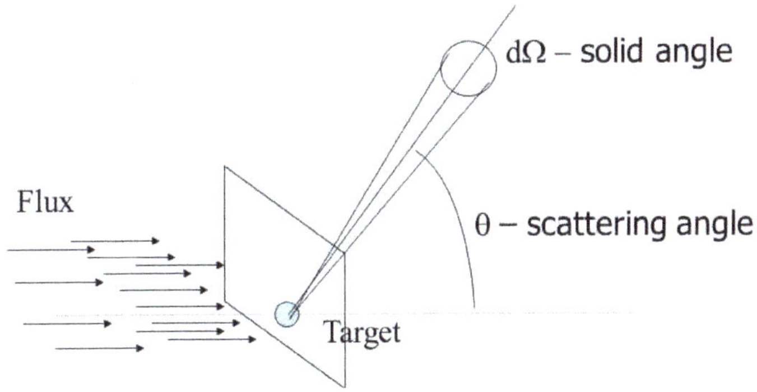

Fig. 2.5: The incident beam moving along the horizontal axis is scattered by the target. It is observed by a counter at an angle θ with respect to the incident beam, subtending a solid angle $d\Omega$.

The flux of incident mono-energetic particles is defined as the number of particles crossing an area perpendicular to the beam per unit time per unit area. It is usually written as $\Phi = n_i v_i$, where n_i is the particle density, i.e. the number of particles per unit volume, and v_i is the particle's velocity.

The quantity $d\sigma$ is called **differential cross section**, with $d\sigma = \sigma(\theta)d\Omega$, σ the **cross section** and σ_{tot} the total cross section. The cross section is one of the most important quantities in physics and has dimensions of length^2. In most cases the particle density is taken to be one per unit volume, $1/V$. In other words the cross section per scatterer is given by:

$$d\sigma = \frac{1}{\Phi} d\mathcal{N} = \frac{1}{v_i} V d\mathcal{N}. \tag{2.5}$$

We are not going to evaluate the cross section here. The interested reader is referred to Sections 7.1 and 7.2, 7.7.2 of the companion volume. The Standard Model and Beyond [Vergados (2017)], and Ref. [Frauenfelder and Henley (1974)], Section 6.2. Anyway

$$\mathcal{N} = N\Phi\sigma_{tot}. \tag{2.6}$$

Of particular interest is the case of the target being composed of elementary particles, like the Rutherford cross section.

$$\left.\frac{d\sigma}{d\Omega}\right|_{Ru} = \frac{m^2 Z_1^2 Z^2 e^4}{p^4 \sin^4 \frac{\theta}{2}}. \tag{2.7}$$

2.1.4 *Nuclear form factor*

For an extended nucleus the differential cross section is given by

$$\frac{d\sigma}{d\Omega} = \left.\frac{d\sigma}{d\Omega}\right|_{Ru} F^2(q). \tag{2.8}$$

Where $\left.\frac{d\sigma}{d\Omega}\right|_{Ru}$ is the differential cross section, if both target and projectile are assumed point like, and $F^2(q)$ is the form factor. In the case of charged particles it is the charge form factor:

$$F(q) = \frac{1}{Z}\int e^{(i/\hbar)\mathbf{q}.\mathbf{r}} \rho(r)_{ch} d^3\mathbf{r} = \frac{1}{Z} 4\pi \int \frac{\sin qr/\hbar}{qr/\hbar} \rho(r)_{ch} r^2 dr, \tag{2.9}$$

with the density normalized to the number of charged particles Z in the target. In other words the form factor is the Fourier transform of the density normalized so that it takes the value unity at zero momentum transfer. No factor of $1/Z$ appears, if the density is normalized to unity.

By definition for a point particle, $\rho(\mathbf{r})_{ch} = Z\delta(\mathbf{r})$, the form factor is unity. Any deviation the measured cross section in proton nucleus scattering from that obtained theoretically for point like particles, i.e. $F(q) \neq 1$, is a manifestation that the target has internal structure, the nucleus being composite of protons and neutrons. Similarly in deep inelastic electron proton scattering the appearance of a form factor implied that the proton is a composite particle made up of elementary particles, namely the quarks. One finds the the proton form factor is well given by the dipole shape expression:

$$F(q^2) = \frac{1}{\left(1 + \frac{q^2}{M_A^2 c^2}\right)^2} \tag{2.10}$$

where M_A is a parameter taking the value of about 0.84 GeV. This corresponds to a Yukawa-like density, see problem (2) at the end of this chapter, with range $a = \frac{\hbar c}{M_A c^2} = 0.23$ fm. Thus we find $\sqrt{\langle r^2 \rangle} \approx 0.8$ fm (see Eq. (2.122)). The dipole shape form factor is exhibited in Fig. 2.6.

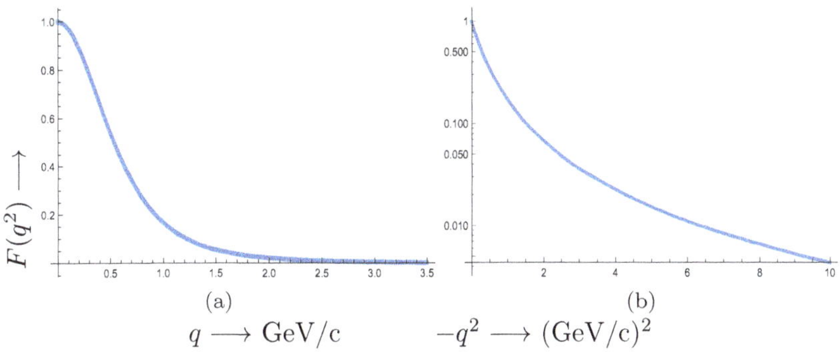

Fig. 2.6: (a) The dipole shape form factor associate with the proton. (b) The proton form factor as preferred by the experimentalists, i.e. in a logarithmic scale and using square of the 4-momentum transfer, $q^2 = -\mathbf{q}^2$.

In actual situations the form factors may exhibit oscillations at high momenta. This can be seen in Fig. 2.30 obtained with the constant density given by Eq. (2.116). This density, however, is not realistic. The density cannot abruptly go to zero at the nuclear surface.

Such form factor oscillations can occur in realistic densities, like the Fermi density given above. The obtained form factor $F(q^2)$ is exhibited in Fig. 2.7. We have chosen to use the dimensionless variable $u = (1/2)((qR)/\hbar)^2$ instead of q. The relation between them is $q = \frac{\hbar}{R}\sqrt{2u}$.

The above form factor has been obtained numerically. In some applications one would prefer an approximate analytic expression. In fact such an expression known as Helm's form factor has been obtained from elastic and inelastic electron scattering on nuclei [Helm (1956)] and has been found

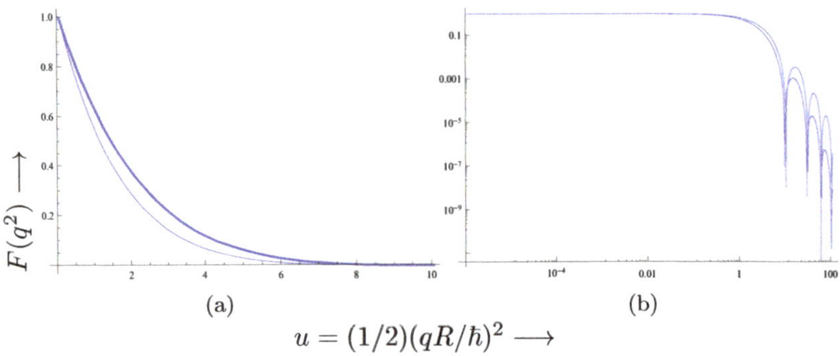

Fig. 2.7: The form factor for a W-S density as a function of $u = (1/2)((qR)/\hbar)^2$ for the A = 40 nucleus (solid line) and A = 208 (thick solid line) for small momentum transfer (a) and large momentum transfer (b). The value of $a = 0.6$ was employed.

useful in dark matter searches employing heavy nuclear targets, as e.g. in [Orrigo *et al.* (2016)]. It is given by:

$$F(q^2) = \frac{3e^{-\frac{1}{2}(q/\hbar)^2 a^2} j_1((q/\hbar)R)}{(q/\hbar)R}, \tag{2.11}$$

where R is the nuclear radius and a the surface thickness. The obtained form factors are exhibited in Fig. 2.8 The agreement between the the exact form factor and the approximate Helm form factor is reasonable.

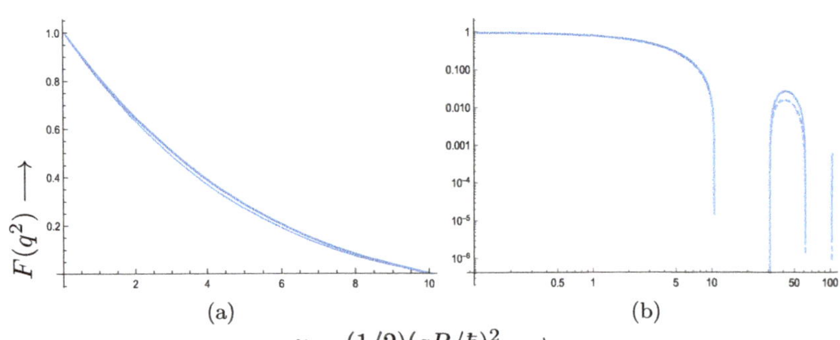

Fig. 2.8: The same as in Fig. 2.7, but employing the Helm form factor (for the definition see text).

The form factors which have been measured for $^{16}_{8}$O, $^{40}_{20}$Ca, $^{92}_{42}$Mo and $^{208}_{82}$Pb are shown in Fig. 2.9. The charge densities which have been extracted from these data are shown in Fig. 2.10. One observes various oscillations in the charge density attributed to the shell structure of nuclei. The surface gradually extends out for heavier nuclei but the diffuseness a is rather constant as assumed in the Fermi distribution model.

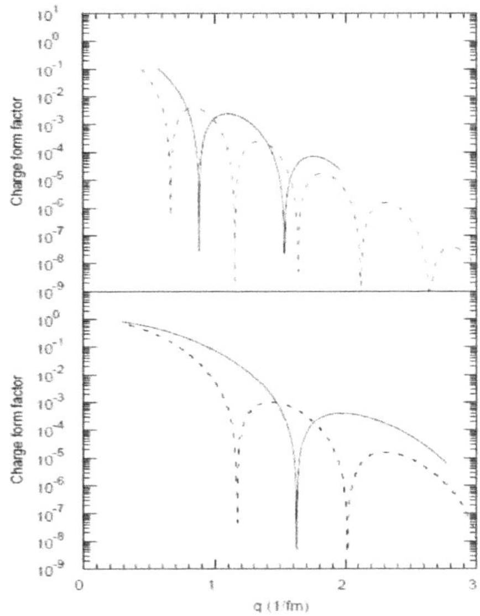

Fig. 2.9: The charged form factor extracted from the elastic electron-nucleus cross section. Experimental form factors for $^{16}_{8}$O (solid line in the lower panel), $^{40}_{20}$Ca (dashed line in the lower panel), $^{92}_{42}$Mo (solid line in the upper panel) and $^{208}_{82}$Pb (dashed line in the upper panel). The lines are plotted over the range of q values measured (courtesy of Alex Brown [Brown (2020)]).

2.1.5 *Binding energies of nuclei*

Nuclear binding energy is defined as the energy required to break up a given nucleus into its constituent parts of N neutrons and Z protons.

$$B(N, Z) = ZM_p c^2 + NM_n c^2 - M_{\text{nuclear}}(N, Z)c^2,$$

where $M_{\text{nuclear}}(N, Z)$ is the mass of the nucleus. It has, however, become customary to quote atomic rather than nuclear masses. The atomic mass

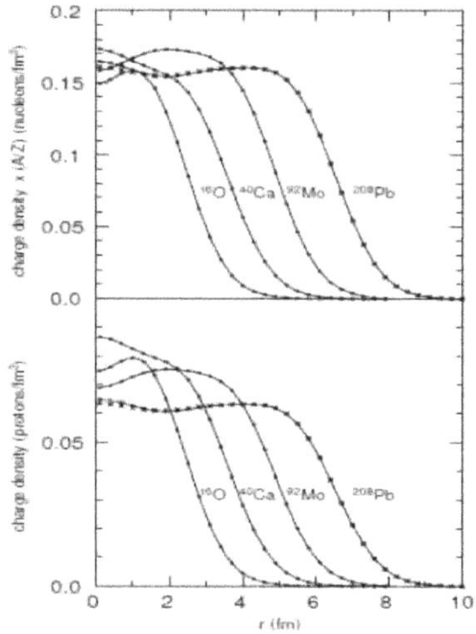

Fig. 2.10: Bottom: The experimental density distributions extracted from a fit to the elastic electron-nucleus scattering, i.e. the form factor of Fig. 2.9 for $^{16}_{8}$O, ^{40}20Ca, $^{92}_{42}$Mo and ^{208}82Pb. Top: experimental density distribution multiplied by (A/Z) (courtesy of Alex Brown [Brown (2020)]).

includes also the mass of the electrons. Furthermore the unit of the atomic mass unit (u) is defined to be 1/12 of the mass of the atom ^{12}C:

$$1u = 931.481 \text{ MeV} = 1.66043 \times 10^{-27} \text{ Kg}.$$

In terms of atomic masses $M(N,Z)$, the binding energy (BE) B(N, Z) can be written as

$$B(N,Z) = ZM_Hc^2 + NM_nc^2 - M(N,Z)c^2, \tag{2.12}$$

where M_H is the mass of the hydrogen atom and M_n is the mass of the neutron.[2] One also defines the quantity Δ

$$\Delta = M(N,Z)c^2 - Au \text{ (mass excess or mass deficiency)} \tag{2.13}$$

$B(N,Z)$, Eq. (2.12) and $-\Delta$, Eq. (2.13), essentially measure the same quantity. In tables the quantity Δ is given, because it follows directly from spectroscopic measurements. In terms of the mass excess the binding energy is

[2]This binding energy also includes contribution from the Coulomb interaction between electrons which is approximately given by $-1.43 \times 10^{-5}Z^{2.39}$ MeV. On the scale of nuclear binding this can usually be ignored. It is most important for heavy nuclei where, for example, for $Z = 120$ the electronic contribution is -1.34 MeV.

given by:

$$B(N,Z) = Z\Delta_H c^2 + N\Delta_n c^2 - \Delta(N,Z)c^2,$$

where $\Delta_H c^2 = 7.2890$ MeV and $\Delta_n c^2 = 8.0713$ MeV. The most interesting quantity is the BE/A, the binding energy per nucleon. This is exhibited in Fig. 2.11. We notice that the most stable nucleus is Fe. This has great significance in astrophysics regarding the formation of the elements in core if a collapsing star.

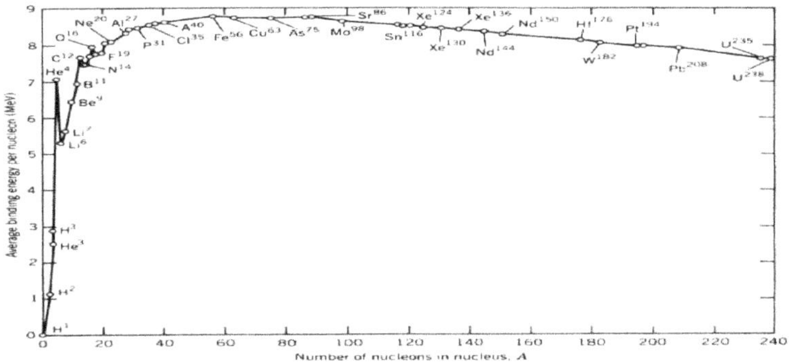

Fig. 2.11: The experimentally determined BE per nucleon. Note that the maximum occurs for the element Fe.

2.1.6 *The Bethe-Weizsäcker formula*

The fact that the BE per nucleon is essentially constant for $A \geq 50$ implies:

(1) There has to be a volume term E_V:

$$E_V = a_V A,\ a_V \text{ constant},\ a_V > 0.$$

(2) There has to be a negative surface term E_S:

$$E_S = -a_S A^{2/3},\ a_S \text{ constant},\ a_S > 0.$$

This comes about since the nucleons at the surface have fewer bonds and the finite size of the nucleus leads to a negative surface energy.

(3) The Coulomb energy between protons must be taken into account. For a uniform charge distribution the Coulomb interaction is

$$E_c = \frac{3}{5}\frac{1}{4\pi\epsilon_0}\frac{(Ze)^2}{R}.$$

Since $R = 1.25A^{1/3}$ we get

$$E_C = -a_C Z^2 A^{-1/3},\ a_C \text{ constant}, a_C > 0.$$

(4) The fact that only a narrow band of nuclei around $N \geq Z$ are stable requires the presence of a **symmetry energy**, which must be proportional to $(N - Z)^2$. The width of this band, however, does not increase as A increases. Thus:

$$E_{\text{sym}} = -a_{\text{sym}} \frac{(N-Z)^2}{A},\ a_{\text{sym}} \text{ constant}, a_{\text{sym}} > 0.$$

Thus

$$B(N, Z) = a_V A - a_S A^{2/3} - a_C Z^2 A^{-1/3} - a_{\text{sym}}(N - Z)^2/A$$

or

$$\frac{B(N, Z)}{A} = a_V - a_S A^{-1/3} - a_C Z^2 A^{-4/3} - a_{\text{sym}} \frac{(N-Z)^2}{A^2}. \quad (2.14)$$

The constants in the last equation have been determined experimentally by a fit to the data as follows:

$$a_V = 15.6\,\text{MeV}, a_S = 17.2\,\text{MeV}, a_C = 0.70\,\text{MeV}, a_{\text{sym}} = 22.5\,\text{MeV}. \quad (2.15)$$

One, of course, would like to predict these coefficients from a nuclear model and some progress has been made in this direction. In particular in the case of the first term, which is the most important. Indeed this term has extensively been studied from first principles in the context of **nuclear matter**. Nuclear matter is a state of matter of a very large nucleus. In fact it has successfully been applied in the study of neutron stars, nuclear explosions and astrophysics.

2.1.7 *Q-values*

In this section we consider the implications of energy conservation for nuclear transformations that include, for example, the fussion of two nuclei a and b into the combined system c

$$[N, Z]_a + [N, Z]_b \rightarrow [N, Z]_c$$

or the decay of nucleus a into two other nuclei b and c

$$[N, Z]_a \rightarrow [N, Z]_b + [N, Z]_c.$$

In general we consider the combinations

$$\sum_i [N, Z]_i \rightarrow \sum_f [N, Z]_f$$

where N and Z are conserved, i.e. $\sum_i N_i = \sum_f N_f$ and $\sum_i Z_i = \sum_f Z_f$. Such a process is characterized by the Q value given by:

$$Q = \sum_i M(N_i, Z_i)c^2 - \sum_f M(N_f, Z_f)c^2 = \sum_f B(N_f, Z_f) - \sum_i B(N_i, Z_i).$$

This is a very important parameter since a reaction involving a single initial nuclear state can take place spontaneously, i.e. it is allowed, if $Q > 0$. In the case of decay, energy is released in the form of the kinetic energy of the final products. Reactions involving two initial nuclei are endothermic (a net loss of energy), if $Q < 0$ and exothermic (a net release of energy), if $Q > 0$.

We can also consider the Q values associated with the removal of one or two nucleons from a nucleus. These are conventionally defined in terms of the one-nucleon and two-nucleon separation energies, S:

$$S_n = -Q_n = B(N,Z) - B(N-1,Z), S_p = -Q_p = B(N,Z) - B(N,Z-1),$$

$$S_{2n} = -Q_{2n} = B(N,Z) - B(N-2,Z), S_{2p} = -Q_{2p} = B(N,Z) - B(N,Z-2).$$

These values are very useful in astrophysics. A negative value for one of these quantities means that the nucleus can spontaneously decay by the emission of the corresponding neutrons or protons. The boundary between positive and negative values of S is called the drip line. Nuclei inside the drip lines are stable to the spontaneous emission of nucleons, whereas those outside the drip line can spontaneously decay by emission of one and/or two nucleons. The neutron drip line represents a sharp boundary between those nuclei just inside drip line that beta decay with a lifetime of the order of ms, and those outside the drip line that have lifetimes on the order of 10^{-20} s associated with the strong interaction decay widths of MeV. For heavy nuclei, the Coulomb barrier can greatly hinder the proton decay, and the lifetimes for the one or two proton decay of nuclei just outside the proton-drip line can be comparable to or even longer than the lifetimes for beta decay.

Figure 2.12(a) shows the nuclear chart for nuclei whose masses have been measured together with those that are predicted to lie within the neutron and proton drip lines from a microscopic nuclear model, see, e.g., refs. [Tondeur *et al.* (2000)], [Goriely *et al.* (2001)], called HF1. Measured one-neutron separation energies are also shown in Fig. 2.12(b).

The extension of the drip lines has been actively pursued in recent years, since it is expected to shed light in nuclear stability relevant in astrophysical processes.[3]

[3]While this book has been in preparation. A new result has been obtained [Ahn *et al.*

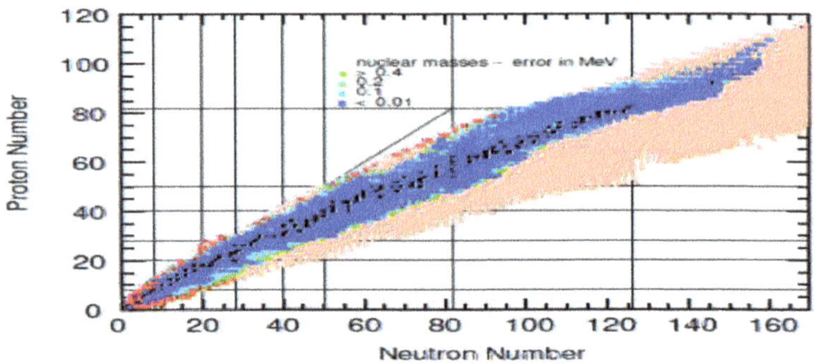

(a)

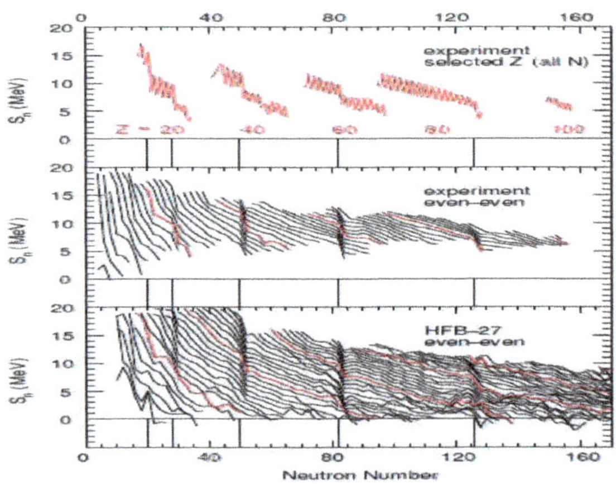

(b)

Fig. 2.12: The experimentally determined drip lines from the Audi-Wapstra-Thibault mass compilation together with the HF1 prediction for nuclei out to the neutron and proton drip lines. (a) Observed nuclei outside the drip lines are indicated. (b) HF1 results for the one-neutron separation energies (courtesy of Alex Brown [Brown (2020)]).

(2019)] in a search for the heaviest isotopes of fluorine, neon, and sodium by fragmentation of an intense ^{48}Ca beam at 345 MeV/nucleon at the RIKEN Radioactive Isotope Beam Factory: The ^{31}F and ^{34}Ne are the heaviest bound isotopes of fluorine and neon,

2.2 Probing the nucleus

The study of nuclei can be performed a la Rutherford using appropriate nuclei as targets and various accelerated particles as probes. In addition to projectiles found in nature, like α particles resulting as a result of radioactivity, charged particles produced in accelerators have been employed. Examples are:

- Electron accelerators
  Electrons produced in a device are accelerated via a combination of electric and magnetic fields in a long tube to high energies. As we have seen in the previous chapter, electrons of energy about 100 MeV can be used as probes to investigate the nucleus.[4] This is a clean probe, since the electromagnetic interaction is well understood.
- Proton and ion accelerators
  In 1932 Cockroft and Walton accelerators were used to yield protons of about 0.5 MeV, which were smashed on nuclei disintegrating them. Again, however, only protons of about 20 MeV kinetic energy can be used to probe the interior of the nucleus. This can be achieved by n cyclotrons, a circular accelerator devised by E. Lawrence at Berkeley accelerating protons to 1 MeV energy, which can easily accelerate protons up to 25 MeV. Similarly other fully ionized atoms of deuterium, tritium etc of about the same kinetic energy have been employed accelerated via Tandem Van der Graaf accelerators, using Van der Graaf generators, which can reach terminal voltage of about 10 or even 25 MeV i.e. able to accelerate protons of about 25 MeV, adequate for nuclear physics.[5]

---

respectively. Thus the neutron drip line has been experimentally confirmed up to neon for the first time since ^{24}O was confirmed to be the drip line nucleus nearly 20 years ago. These data provide new keys to understanding the nuclear stability at extremely neutron-rich conditions.

[4]In modern accelerators, see Chapter 11, high energies can be achieved. Thus energies up to 30 MeV per meter trasversed can be achieved in modern electron accelerators, which are linear to minimize energy losses. In such an accelerator the SLAC (Stanford Linear Accelerator) at Stamford with a length of about 3 km, an electron-positron colliding beam accelerator, energies up to 50 GeV have been achieved. Even higher energies about 100 GeV have been achieved by LEP (Large Electron accelerator) at CERN, a circular accelerator involving colliding electron- positron beams. Such devices allow the possibility to explore even the interior of the nucleons.

[5]The cycloctron idea has evolved to high energy synchrotrons and other modern colliding beam circular accelerators of proton-antiproton with a circumference of many

For more information regarding the experimental facilities and devices employed in nuclear physics experiments see Chapter 11.

2.3 Nuclear forces

In order to be able to ascertain the bulk properties of nuclei, including the nuclear structure and the binding energies, we must first understand the nature of nuclear forces between the nucleons. Knowledge of nuclear forces, however, cannot be simply obtained from first principles, like, e.g., the electromagnetic interaction. It mainly relies on experimental evidence as well as some physics insight. Thus one can phenomenologically obtain their character in terms of the nuclear parameters. A complete discussion of our phenomenological approach is beyond the scope of this book and we refer the reader to the book by [Roy and Nigam (1967)] and references there in.

2.3.1 *Nuclear potentials*

As a first approximation we consider that the two-nucleon potential is central and attractive and according to this consideration the following potential have been employed.

1) **Square-Well Potential:**

$$V_{SW}(r) = \begin{cases} -V_0 & r \leq r_0 \\ \\ 0 & r > r_0 \end{cases} \tag{2.16}$$

where in this case r and V_0 are the separation distance between the nucleons and the the depth of the potential respectively.

2) **Yukawa Potential:**

$$V(r) = -V_0 \left(\frac{r_0}{r}\right) \exp\left[-\frac{r}{r_0}\right]. \tag{2.17}$$

3) **Gaussian Potential**

$$V(r) = -V_0 \exp\left[-\frac{r^2}{r_0^2}\right]. \tag{2.18}$$

kilometers, like the superconducting Gatchina Synchrocyclotron at Petersburg Nuclear Physics yielding 1000 MeV protons. The highest energy accelerator is the proton-proton colliding beam set up with a circumference of 27 km at LHC at CERN, which has already reached energies of 2×7000 GeV = 14 TeV.

4) **Exponential Potential**

$$V(r) = -V_0 \exp\left[-\frac{r}{r_0}\right]. \tag{2.19}$$

5) **The Woods-Saxon potential**

The form of this potential resembles the nuclear Woods-Saxon density discussed in Section 2.1.2. The potential strength V_0 is usually taken to be 50 MeV.

$$V(r) = -V_0 \frac{1}{e^{\frac{r-R}{a}} + 1}. \tag{2.20}$$

The relative potential $V(r)/V_0$ is exhibited in Fig. 2.13.

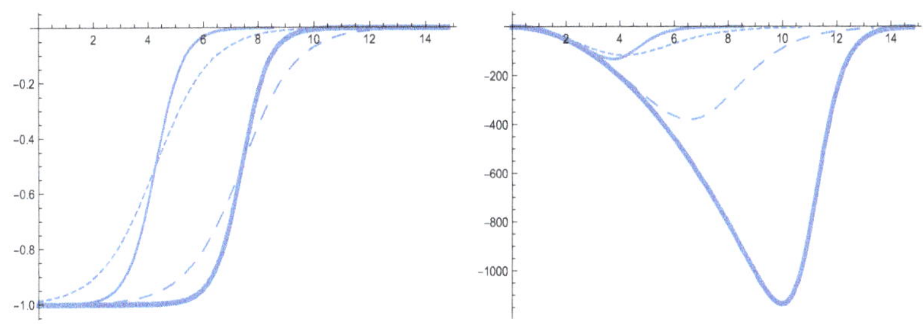

(a) (b)

$r \longrightarrow$ fm

Fig. 2.13: (a) The nuclear Woods-Saxon potential $V(r)/V_0$, for the cases: i) A = 40, a = 0.5 fm (solid line), ii) A = 40, a = 1.0 fm (dotted line), iii) A = 208, a = 0.5 fm (thick solid line) and iv) A = 208, a = 1.0 fm (dashed line). (b) The potential profile $4\pi r^2 V(r)/V_0$. The curves have been drawn with the same notation as in (a). In all cases the nuclear radius has been chosen to be $R = r_0 A^{1/3}$, $r_0 = 1.25$ fm.

Among the above potentials, the Square-Well potential is the simplest to use and it has been extensively employed, since it is a good approximation in a variety of problems. Furthermore it has the advantage that one can obtain the needed solutions exactly.

The next task it is to compare the theoretical predictions obtained by employing the above potentials with the experimental data. To this end from the most powerful tool, that is the scattering experiments, we have found that the nuclear forces exhibit short range behavior and they are of an exchange character. This simply means that they depend not only on the separation distance r, but on the relative direction of the vector $\mathbf{r}$ and the nucleon spin $\mathbf{s}$ as well. In other words one needs to extend the central potentials by incorporating the spin degrees of freedom. Thus the following interactions are commonly used:

i) The Majorana potential

This allows the exchange of the space coordinates, given by

$$V_{MAJ}(r) = -V_m(r)P_{ex} \tag{2.21}$$

where the space-exchange operator P_{ex} is defined as

$$P_{ex}\Psi(\mathbf{r}_1, \mathbf{r}_2) = \Psi(\mathbf{r}_2, \mathbf{r}_1) \tag{2.22}$$

where $\Psi(\mathbf{r}_1, \mathbf{r}_2)$ is the spatial part of the two nucleon wave function.

ii) The Bartlett potential

This allows for the possibility of exchange of the spin coordinates, defined as

$$V_B(r) = -V_b(r)P_\sigma \tag{2.23}$$

where the spin-exchange operator P_σ is defined as

$$P_\sigma = \frac{1}{2}(1 + \sigma_1 \cdot \sigma_2) \tag{2.24}$$

and has the property

$$P_\sigma \chi(\sigma_1, \sigma_2) = \chi(\sigma_2, \sigma_1) \tag{2.25}$$

where $\chi(\sigma_1, \sigma_2)$ is the spin part of the total two nucleons wave function and σ_1, σ_2 the Pauli spin matrices.

iii) The Heisenberg potential

This extends the previous by allowing the exchange of both the space and spin coordinates, defined as

$$V_H(r) = -V_h(r)P_{e,\sigma} \tag{2.26}$$

where the operator $P_{e,\sigma}$ acts both on the spatial and spin part of the two nucleon wave function and consequently is defined as

$$P_H = P_{ex}P_\sigma. \tag{2.27}$$

One can see that the antisymmetry of the wave function implies

$$P_{ex}P_\sigma = -P_\tau = -\frac{1}{2}(1 + \tau_1 \cdot \tau_2) \tag{2.28}$$

where τ_1, τ_2 the Pauli matrices in isospin space. Now, we have

$$P_H\Psi(\mathbf{r}_1, \mathbf{r}_2, \sigma_1, \sigma_2) = \Psi(\mathbf{r}_2, \mathbf{r}_1, \sigma_2, \sigma_1). \tag{2.29}$$

In total, the two nucleon potential takes the form

$$V(r) = -V_{SW}(r) - V_m(r)P_{ex} - V_b(r)P_\sigma + V_h(r)P_{e,\sigma}. \tag{2.30}$$

The exchange character of the nuclear forces is supported by experimental evidence, including a number of scattering experiments as well as measurements of the binding energy of the nucleon in the nucleus. Moreover, the saturation of the binding energy per nucleon, being independent of the mass number A, has also provided support for the exchange nature of the nuclear forces.

It is well known that of central potentials in Eq. (2.30) can explain the nucleon-nucleon interaction both in bound and unbound states. The use of central potentials, however, fails to predict, for example the quadrupole moment of the deuteron. The use of non-central tensor forces are necessary to explain the experimental data regarding the quadrupole moment of the deuteron.

In the following section we will briefly study the bound state of deuteron with the use of central forces.

2.3.2 *An example — The bound state of the deuteron*

The three possible states of the two nucleon system are the di-neutron (nn), di-proton (pp) and deuteron (np). However, only the deuteron is known to be a bound state with binding energy

$$\mathcal{E} = -2.226 \pm 0.0003 \text{ MeV}. \tag{2.31}$$

It is characterized by spin is $S = 1$ (in units of h) and a magnetic moment with a value determined experimentally to be

$$\mu_d = (0.85735 \pm 0.000003) \text{ nm}, \qquad \text{nm} = e\hbar/2m_p c.$$

The corresponding electric quadrupole moment is

$$Q = 2.82 \cdot 10^{-27} \text{ cm}^2.$$

Now, we proceed with the construction of the ground state of deuteron, assuming that the nuclear forces are central and attractive. We know, of course, that this approximation disagrees with some experimental results, but, at this point, our study is only instructive and we expect useful insights may be gained from it.

Let us assume that the deuteron is a spin triplet and orbital space singlet with total angular momentum $J = 1$, usually indicated as $^3\text{S}_1$. The Schrödinger equation in the center of mass system is given by

$$-\frac{\hbar^2}{2\mu}\nabla^2\psi(\mathbf{r}) + V(r)\psi(\mathbf{r}) = E\psi(\mathbf{r}) \tag{2.32}$$

where $\mathbf{r}$ is the relative distance while the reduced mass μ is defined as

$$\mu = \frac{m_p m_n}{m_p + m_n} \simeq \frac{M}{2} \tag{2.33}$$

and m_p, m_n and M are the mass of proton, neutron and their average mass respectively. After some algebra Eq. (2.32) can be written

$$-\frac{\hbar^2}{M}\frac{d^2u(r)}{dr^2} + V(r)u(r) = Eu(r) \tag{2.34}$$

where we replaced $\psi(r)$ by $=u(r)/r$ and assumed that the energy E is the binding energy of deuteron $\mathcal{E}$. Employing the square well potential for $V(r)$ Eq. (2.34) is divided in two differential equations that is

$$\frac{d^2u(r)}{dr^2} + \lambda^2 u(r) = 0, \qquad r < r_0 \tag{2.35}$$

and

$$\frac{d^2u(r)}{dr^2} - \gamma^2 u(r) = 0, \qquad r > r_0 \tag{2.36}$$

where the parameters λ and γ are given by

$$\lambda = \sqrt{\frac{M}{\hbar^2}(\mathcal{E} + V_0)} \tag{2.37}$$

and

$$\gamma = \sqrt{\frac{-\mathcal{E}M}{\hbar^2}}. \tag{2.38}$$

The general solution of Eq. (2.35) in each region is of the form:

$$u(r) = C_1 \sin(\lambda r) + C_2 \cos(\lambda r) \tag{2.39}$$

and using the condition $u(r = 0) = 0$ (in order $\psi(r)$ to be finite at the origin) we get

$$u(r) = C_1 \sin(\lambda r), \qquad r < r_0. \tag{2.40}$$

Moreover, the general solution of Eq. (2.36) is

$$u(r) = C_3 e^{-\gamma r} + C_4 e^{\gamma r} \tag{2.41}$$

and due to the condition $u(r) \to 0$ when $r \to \infty$ we get

$$u(r) = C_3 e^{-\gamma r}, \qquad r > r_0. \tag{2.42}$$

Now, since the wave function $u(r)$ as well as its first derivative must be continuous at $r = r_0$, we get the two following equations

$$C_1 \sin(\lambda r_0) = C_3 e^{-\gamma r_0} \tag{2.43}$$

and

$$C_1\lambda\cos(\lambda r_0) = -\gamma C_3 e^{-\gamma r_0}. \tag{2.44}$$

Dividing Eqs. (2.43) and (2.44) we obtain the equation

$$\lambda\cot\lambda r_0 = -\gamma. \tag{2.45}$$

The values of λ and γ depend on the values of $\mathcal{E}$ and the depth V_0. Equation (2.45) is a transcendental equation and can be solved only numerically. Now, for a given value of the binding energy and adopting the approximate value $r_0 = 2\cdot 10^{-13}$ cm, the resulting value of the potential depth is $V_0 \approx 36$ MeV. It is worth pointing out that the deuteron is a loosely bound system, since its binding energy is only 1.113 MeV per nucleon compared to the average value of 8 MeV per nucleon in nuclei.

Now considering also that $V_0 \gg -\mathcal{E}$ then Eq. (2.45) takes the simple form

$$\cot(-\lambda r_0) \simeq -\sqrt{\frac{-\mathcal{E}}{V_0}}. \tag{2.46}$$

For $-\mathcal{E} = 0$ $\lambda r_0 = \pi/2, 3\pi/2, \cdots$ and

$$V_0 r_0^2 \simeq \frac{\hbar^2\pi^2}{4M}, \frac{9\hbar^2\pi^2}{4M}, \cdots \tag{2.47}$$

the function $u(r)$ can not exhibit a node inside the well and, consequently, $\psi(r)$ is not the ground state unless only the first term is retained and

$$V_0 r_0^2 \simeq \frac{\hbar^2\pi^2}{4M}.$$

The ground-state wave function, the range and depth of the potential as well as the binding energy are shown in Fig. 2.14.

2.4 Simple nuclear models

The nucleus is a collection of strongly interacting protons and neutrons. One would have thought that one could apply the standard techniques in dealing with the many body aspects involved. This, however, is not so easy. The main reason is that the interaction involved is not so well understood. So one cannot supply a convenient set of single particle wave functions and energy levels. Furthermore the interaction between nucleons cannot be treated in a perturbative way. So the nucleus can only be described in terms of simplifying models.

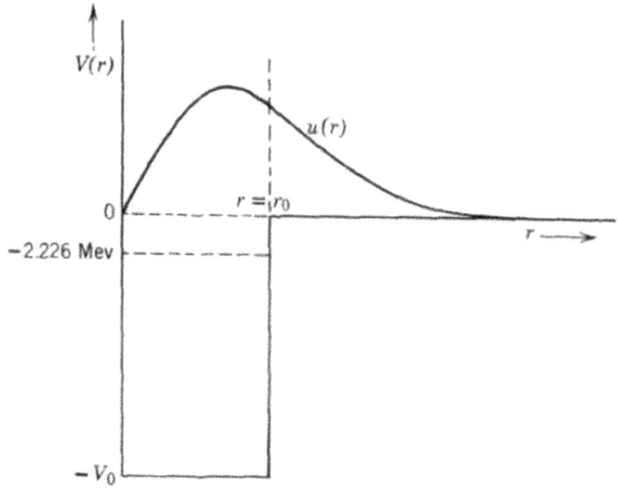

Fig. 2.14: The ground-state wave function of the deuteron, the depth of the potential as well as the binding energy.

2.4.1 *The Liquid drop model*

One of the striking properties of the nuclear medium is the constancy of its density (see Fig. 2.10) as A increases. As a result the volume of the nucleus is proportional to the number A of its constituents or, more specifically, proportional to the cube of the rms radius, which, in turn, is proportional to $A^{1/3}$ with a proportionality constant of about 1.25 fm. The same is true for liquids. In fact very early on Bohr and Weizsäcker developed a nuclear model patterned the liquid drop model [Bohr (1936)], [von Weizsäcker (1935)].

Another striking effect is the constancy of the binding energy per nucleon (see Fig. 2.11), with the exception of nuclei with very small A. It attains a maximum around $A = 60$ and from then on it drops slowly as A increases and drops rather sharply for small A. As a result in the case of a nucleus around, say, $A = 240$, one finds that, if it splits into two parts with $A \approx 120$, the binding energy of the two parts will be larger than that of the original nuclide and, thus, energy will be released. This mechanism, which is responsible for energy production in such a break up, is known as **fission**. Fission will be discussed in more detail in Section 6.5.

The history of fission is very intriguing. Neutrons, being electrically neutral, were used to bombard nuclei to obtain other nuclei and perform interesting investigations. To prevent neutrons from hitting the nucleus too hard. Enrico Fermi slowed them down by passing them through parafin. Thus he started with natural F producing artificial isotopes of that element and continued with over 40 heavier elements. In 1934 he reached the heavier known element U, uranium. He came close to a great discovery, but he missed it. He noticed that after hitting the uranium nucleus with neutrons, the neutron did not simply stick to it. He thought he produced a transuranic element, one place above uranium in Medeleev's table, which did not exist on Earth. He was taken up by such a glorious possibility and he missed the real explanation.

Otto Hahn and Lisa Meitner repeated the experiment in Germany, but Meitner, being Jewish, fled from Germany to Sweden, while Otto Hahn continued this research with Fritz Strassmann. They found that far from discovering a new element, the products in the experiment were ordinary stuff like barium with $Z = 56$. This was surprising. Would slow neutrons so disturb uranium to split it up?

Hahn communicated all this to Meitner, now in in Sweden, and, after she discussed it with Otto Frisch, they came up with the real explanation. The liquid drop is held together by surface tension, which cannot operate after a given droplet size. The nucleus is held together by a proper balance of the attractive strong force between neighboring protons and neutrons, only neighboring ones because the interaction is short ranged. On the other hand the electromagnetic repulsive force, being of long range, is exerted between all protons. As the number of protons exceeds a certain Z in the transuranic region, the nucleus is not stable. So Uranium itself is so delicately balanced that the absorption of a slow neutron makes the uranium nucleus wobble like a liquid drop and break up:

$$n +^{235}_{92} \mathrm{U} \rightarrow^{144}_{56} \mathrm{Ba} +^{89}_{36} \mathrm{Kr} + 3n. \tag{2.48}$$

As we have mentioned earlier this is an exothermic reaction. Note that 3 slow neutrons are produced, which can subsequently break up another uranium nucleus. We thus have a **chain reaction**, which can release enormous amounts of energy. This is illustrated in Fig. 2.15. The chain reaction can be maintained only if the mass of uranium is above a **critical mass**.

It is interesting from a historical point of view to observe that main ingredients for producing the atomic bombs were close to being understood in Italy and Germany almost 5 years before the beginning of world war II.

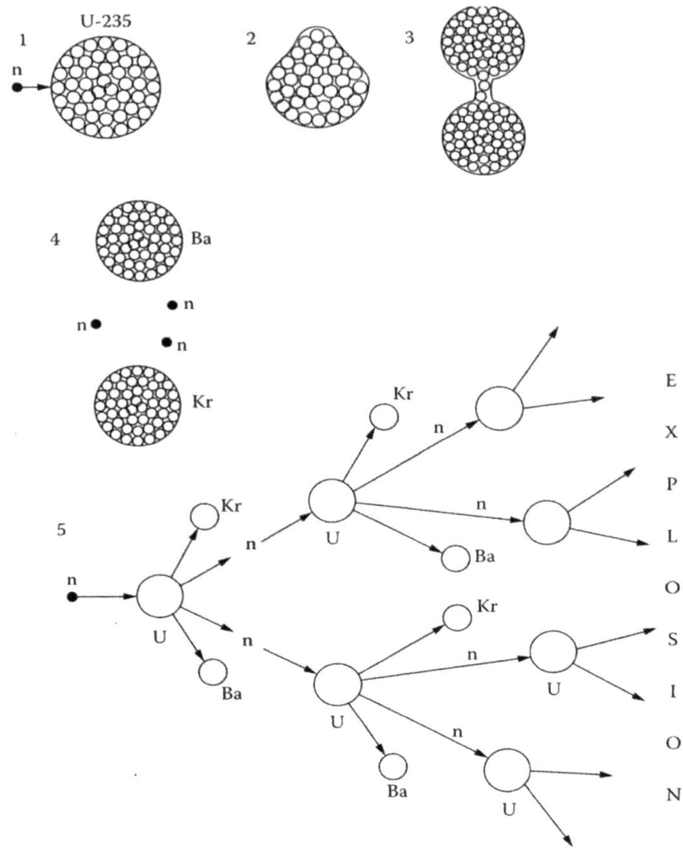

Fig. 2.15: Top: The process of fission. The initial neutron is absorbed 1. After absorption he nucleus gets deformed and wobbles 2. The two parts are loosely connected 3. The Ba fragment 4. The the Kr fragment 5. The three neutrons between 3 and 4. One of these neutrons hits another Uranium 5. The chain reaction follows.

However, a) Fermi missed the observation of fission and b) Meitner had fled from Germany. Had these not happened, Hitler might have been able produce the atomic bomb first and he might have won the war.

The most abundant element $^{238}_{92}$U does not fission. This handicap can be overcome in breeder reactors. A breeder reactor is a nuclear reactor capable of generating more fissile material than it consumes. In other words, it is a device able to achieve this, because its economy in producing neutrons is high enough to breed more fissile fuel than they use from fertile material such as uranium-$^{238}_{92}$U or thorium $^{232}_{90}$Th. So far, such a process [Donev *et al.* (2017)] is exhibited in Fig. 2.16. So far France has made the largest implementation of breeder reactors with their Super-Phenix fast breeder reactor.

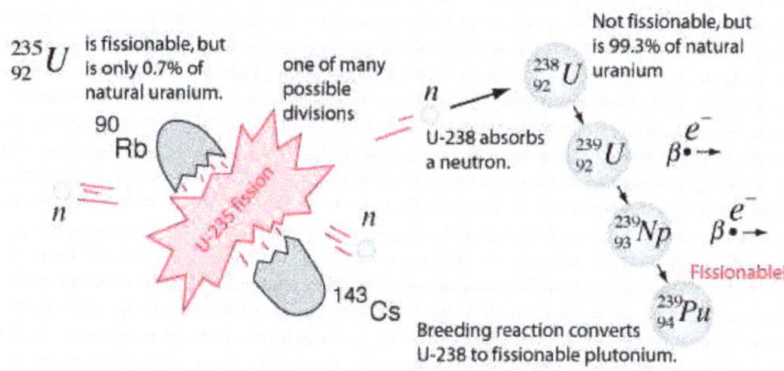

Fig. 2.16: The breeding of ^{239}Pu in a breeder reactor. The neutrons are supplied by the decay of ^{235}U, which transmute ^{238}U to plutonium (Courtesy of Rod Nave, HyperPhysics, GSU).

Breeders were at first found attractive because their fuel economy was better than light water reactors, but this interest declined after the 1960s as more uranium reserves were found and new methods of uranium enrichment reduced fuel costs. Adherents claim that, with seawater uranium extraction, there would be enough fuel for breeder reactors to satisfy our energy needs for 5 billion years at 1983's total energy consumption rate, thus making nuclear energy effectively a renewable energy. The real problem, however, is the radioactive nuclear fuel waste and the possibility of accidents, see Section 8.3. So let us hope that the fusion energy production will become economically viable.

2.4.2 *The Fermi gas model*

The liquid drop model described above is an oversimplification. The nucleus has many properties, which cannot be accounted for this way, as the result of a very strong interaction. Based on the experience with atomic physics we expect an independent particle model to be at work here as well. A given nucleon can be thought of as moving in some sort of average potential created by the rest of them. They occupy the available states, starting with those of the lowest energy subject to the Pauli exclusion principle, in such a way that no two of them can occupy the same level. The simplest such version is the Fermi model in which the nucleons are otherwise moving freely. The obtained results are familiar from the statistical mechanics for spin 1/2 particles, known as Fermi-Dirac Statistics. Of course, since the number of nucleons in the nucleus is small, we do not expect this approach to work better than 10%. In this context we know that the average number of particles with energy ϵ_i is:

$$f(\epsilon_i) \equiv \langle n \rangle_i = \frac{1}{e^{\beta(\epsilon_i-\mu)}+1}, \quad \beta = 1/kT, \tag{2.49}$$

where T is the absolute temperature, μ the chemical potential and k Boltzmann's constant. From Eq. (2.49) we see that we can define the chemical potential as the energy ϵ_i at which the occupation probability becomes 1/2. Of special interest is the case near absolute zero. Then

$$f(\epsilon)|\underset{T\to 0}{\rightarrow} \begin{cases} 1, & \epsilon < \mu(0), \\ 0, & \epsilon > \mu(0). \end{cases} \tag{2.50}$$

In other words the Fermi $f(\epsilon)$ at absolute zero takes the form of Fig. 2.17.

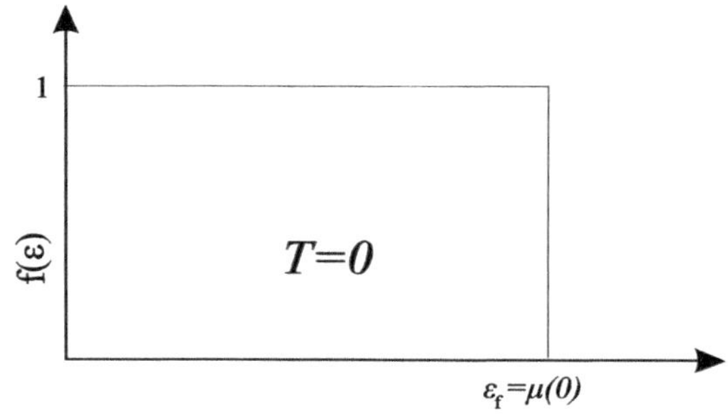

Fig. 2.17: The function $f(\epsilon)$ is exhibited as a function of the energy at absolute zero. The occupation probability of a state with energy $\epsilon < \epsilon_F$ is unity (certainty), while the occupation probability of one with $\epsilon > \epsilon_F$ is zero.

The value of the chemical potential at absolute zero is called **Fermi energy** and is designated as ϵ_F. This is the energy of the highest energy level occupied at absolute zero. In other words at $T = 0$ the energy levels with energy $\epsilon > \epsilon_F$ have zero occupation probability. We say that the Fermi gas is **completely degenerate**. Needles to say that at $T > 0$ some of the fermions near the Fermi energy can be excited to levels with energy $\epsilon > \epsilon_F$, i.e. some of the levels above the Fermi energy have a non zero probability to be occupied.

If we assume that the fermion spectrum is continuous, i.e. if the system is confined to be in a large box of volume V instead of $\langle n_i \rangle$ one talks about the particle distribution $\frac{dn}{d^3\mathbf{p}}$. It can be shown that:

$$\frac{dn}{d^3\mathbf{p}} = g_s V \frac{V}{h^3} f(\epsilon), \quad f(\epsilon) = \frac{1}{e^{\beta(\epsilon(\mathbf{p})-\mu)} + 1}, \tag{2.51}$$

where $g_s = 2s + 1 = 2$ is the degeneracy of the spin 1/2 system. Furthermore, since there is no preferred direction, we find:

$$\frac{dn}{dp} = 8\pi \frac{V}{h^3} p^2 f(\epsilon) = \frac{V}{\pi^2 \hbar^3} p^2 f(\epsilon), \quad f(\epsilon) = \frac{1}{e^{\beta(\epsilon(p)-\mu)} + 1}. \tag{2.52}$$

Returning to the nucleus we will assume that its constituents form a degenerate Fermi gas, which is non relativistic, i.e.

$$\epsilon(p) = \frac{p^2}{2m_N} \text{ or } p_F = \sqrt{2m_N \epsilon_F}.$$

We will further assume that protons and neutrons away from the walls are assumed to be moving freely in a potential well of constant depth, which is adjusted to fit the binding energy of the nucleus, see Fig. 2.18. The depth of the proton well is assumed to be different than that for neutrons due to the repulsive Coulomb potential. In addition to the shift upwards in the energy of protons, the proton well is characterized by a **Coulomb barrier**. Protons approaching the well from the outside can either tunnel in according to the laws of quantum mechanics or pass through essentially unaffected. The protons and neutrons are put in the energy levels in accordance with the two conditions: i) the total energy should be minimized and ii) the Pauli exclusion principle must be obeyed. Here, since the spin of the nucleons is 1/2, i.e. there exists a degeneracy of two associated with $s_z = \pm 1/2$, at most two identical nucleons can be put in the same level. The highest energy level occupied is the Fermi energy.

Let us now consider the neutron gas. The total number of neutrons is obtained is obtained by integrating Eq. (2.52) from zero to the maximum neutron momentum p_N. The result is:

$$N = \frac{1}{3\pi^2} \frac{V}{\hbar^3} p_N^3. \tag{2.53}$$

Taking now as V the volume of a sphere of radius $R = r_0 A^{1/3}$ and solving for p_N we get

$$p_N = \frac{\hbar}{r_0} \left(\frac{9\pi N}{4A} \right)^{1/3}. \tag{2.54}$$

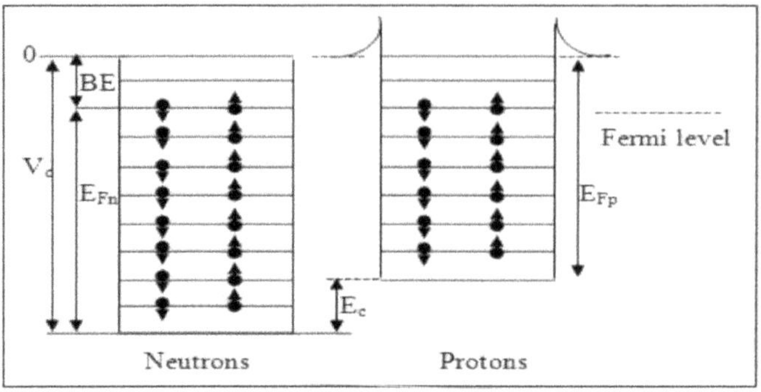

Fig. 2.18: The wells for neutrons and protons. The well parameters are determined by a fit to the binding energy. The proton well is different from that of the neutrons due to the coulomb interaction that shifts the proton energy by an amount E_c and creates a "coulomb barrier". Since the nucleons have spin 1/2 only at most two of them can be in the same state. The maximum energy level occupied depends on the number of nucleons. This maximum energy is called Fermi energy.

Similarly the maximum proton momentum is:

$$p_Z = \frac{\hbar}{r_0}\left(\frac{9\pi Z}{4A}\right)^{1/3}. \tag{2.55}$$

To estimate the Fermi energy we consider an appropriate system, e.g. the case $N = Z$. Then from Eq. (2.54) using $r_0 = 1.25$ fm one finds:

$$\epsilon_F = \frac{1}{2m_N}\left(\frac{\hbar}{r_0}\right)^2\left(\frac{9\pi}{8}\right)^{2/3} \approx 50 \text{ MeV}.$$

The average kinetic energy per nucleon is given by the well known formula:

$$\langle E\rangle = \frac{\int_0^{p_F}\epsilon(p)p^2dp}{\int_0^{p_F}p^2dp} = \frac{3}{5}\frac{p_F^2}{2m_N} = \frac{3}{5}\epsilon_F.$$

For $N \neq Z$ using Eqs. (2.54) and (2.55) one finds

$$\langle E(Z,N)\rangle = N\langle E_N\rangle + Z\langle E_Z\rangle = \frac{3}{10m_N}\left(N(p_F^2)_N + Z(p_F^2)_Z\right), \tag{2.56}$$

assuming the same proton and neutron mass. Furthermore taking the same radius for the neutron and proton wells, one finds:

$$\langle E(Z,N)\rangle = \frac{3}{10m_N}\left(\frac{\hbar}{r_0}\right)^2\left(\frac{9\pi}{4}\right)^{2/3}\frac{\left(N^{5/3}+Z^{5/3}\right)}{A^{2/3}}.$$

For a given $A = N + Z$ the minimum of $E(Z, N)$ occurs when $N = Z$. To see the general behavior we write

$$N - Z = \epsilon A \Rightarrow N = \frac{A}{2}(1+\epsilon),\ Z = \frac{A}{2}(1-\epsilon) \Rightarrow$$

$$\langle E(Z,N)\rangle = \frac{3}{10m_N}\left(\frac{\hbar}{r^0}\right)^2\left(\frac{9\pi}{4}\right)^{2/3}\left(\frac{1}{2}\right)^{5/3}\left((1+\epsilon)^{5/3} + (1-\epsilon)^{5/3}\right).$$

Using now the expansion

$$(1+x)^n \approx 1 + nx + \frac{1}{2}n(n-1)x^2.$$

We obtain

$$\langle E(Z,N)\rangle = \frac{3}{10m_N}\left(\frac{\hbar}{r_0}\right)^2\left(\frac{9\pi}{8}\right)^{2/3}\left(A + \frac{5}{9}\frac{(N-Z)^2}{A}\right).$$

The first term is the volume term leading to

$$a_V = \frac{3}{10m_N}\left(\frac{\hbar}{r_0}\right)^2\left(\frac{9\pi}{8}\right)^{2/3} \approx 30 \text{ MeV}.$$

The second term provides the symmetry energy with

$$a_{\text{sym}} = \frac{3}{10m_N}\left(\frac{\hbar}{r_0}\right)^2\left(\frac{9\pi}{8}\right)^{2/3}\frac{5}{9} = \frac{5}{9}a_V \approx 16.5 \text{ MeV}.$$

The first is twice as large and the second is only half of the values determined experimentally (see Eq. (2.15)). It is usually hard to obtain the absolute scale of the energies, here a_V, with simple nuclear models. The deviation in the a_{sym} is perhaps due to the approximation of assuming that the potential depth as given in Fig. 2.18 does not depend on $N - Z$. This approximation is not very good, because, due essentially to Pauli principle, the interaction between like particles is weaker than that between unlike particles. In fact the change in the depth of the potential has been estimated to be

$$\Delta V_0 \approx (30 \pm 10) \text{ MeV}\frac{Z-N}{A}$$

with this modification the a_{sym} can be brought into order, see [Frauenfelder and Henley (1974)] Section 14.2.

2.5 The nuclear shell model

In the previous section we discussed the early nuclear models. As the data on nuclear properties increased rapidly, however, these models could not provide a good description of all the accumulated data. As a result attempts were made to describe the nucleus in terms of closed shells a la atomic physics.

2.5.1 *The story of the magic numbers*

In fact Bartlett and Elsasser observed in 1934 that nuclei exhibit special behavior if Z and/or N coincide with the magic numbers

$$2, 8, 20, 28, 50, 82, \text{ and } 126.$$

The main evidence at that time being α particle emission energies and elemental abundance. He tried to explain this behavior in terms of a single particle potential, but he was unable to explain the stability of N or Z at 50 and 82 and N at 126. Thus his work was ignored.

Maria G. Mayer by 1948 was able to summarize the experimental data which indicated that nuclei with 20, 50, 82 or 126 neutrons or protons could be described in terms of closed shells obtained in a central potential, assumed to be harmonic. The occupation of the resulting levels was constrained by the Pauli Principle. Furthermore in 1949 the spin-orbit splitting was recognized and an appropriate one-body potential was proposed by Mayer [Mayer (1948)], independently, by Haxel, Jensen and Suess, see [Haxel *et al.* (1949)].

This one-body potential model has been the starting point for the nuclear Shell Model (SM). A short history of Mayer's contributions has also recently been given [Johnson (1986)]. The result is that these numbers are not anymore "magic", since they can be explained in terms of the Shell Model. At that time the model could adequately describe the then available data, such as the relative abundance and the separation energies described in Section 2.1.7 as shown in Fig. 2.12 (an update is given in Fig. 2.19).

We will see below that this model has been easily extended, see Section 2.5.2, and, coupled with developments in realistic effective nucleon-nucleon interactions, large computers and efficient computational methods, see Section 2.5.3, has become the most effective nuclear structure tool almost throughout the periodic table.

We know that the nucleons are held together by a potential that resembles the density. Such a potential is exhibited in Fig. 2.20(a). Ignoring surface effects, it is not different from that of a harmonic oscillator. The energy levels and wave functions obtained in a central potential is discussed in Chapter 14, Appendix A. In particular in the case of the harmonic oscillator potential $U(r) = (1/2)m\omega^2r^2$ are found to be

$$E_N = \left(N + \frac{3}{2}\right)\hbar\omega,\ N = 0, 1, 2\cdots,\ \hbar\omega \approx 41A^{-1/3}\ \text{MeV}.$$

These levels are exhibited in Fig. 2.20(b). They are degenerate, see Section 14.2, i.e. they do not depend on the angular momentum ℓ. The fact

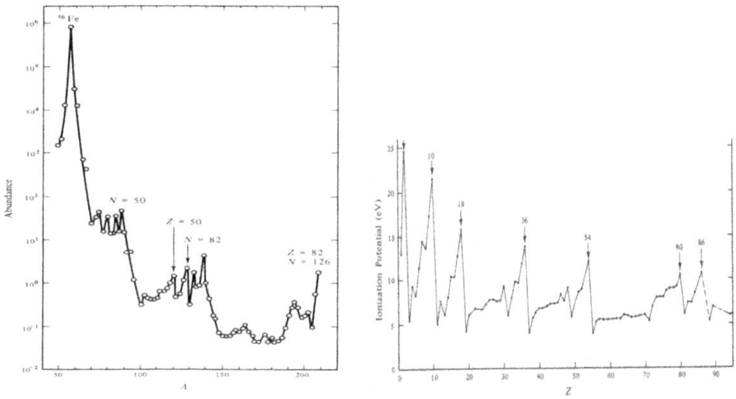

(a) (b)

Fig. 2.19: (a) The drip lines determined from the Audi-Wapstra-Thibault mass compilation, see [Audi *et al.* (2003); Wapstra *et al.* (2003)] for detailed tables. (b) The HF1 prediction for for the one-neutron separation energies.

that they do not depend on the magnetic quantum number m is a feature of all central potentials, due to rotational symmetry. The radial wave function $\phi_{n_r,\ell}(r)$, however, depends on two variables, n_r, giving the nodes of the wave function, i.e. the number of times it crosses the r-axis, and the angular momentum ℓ. The absence of dependence on ℓ is an accidental higher symmetry possessed by the harmonic oscillator potential[6] known as $SU(3)$ symmetry. As a result the energy depends only on the major harmonic oscillator quantum number N:

$$N = 2n_r + \ell,\ \ell = N, N-2, N-4, \cdots 0 \text{ or } 1.$$

Thus the number of degenerate states is:

$$g(N) = \frac{1}{2}(N+1)(N+2).$$

Obviously this degeneracy does not exist, if the potential is any other central potential. In addition there exists the spin degeneracy $g_s = 2s+1 = 2$ due the spin.

The magic numbers arise by putting particles in the lowest possible levels allowed by the Pauli principle. Thus

$$g_{\text{protons}} = 2g(N) = 2, 6, 12, 20, \ldots, \text{ for } = N = 0, 1, 2, 3 \ldots.$$

[6]The solutions of the Coulomb potential of atomic physics also possess an analogous degeneracy, due to a higher than simple rotational symmetry of this potential.

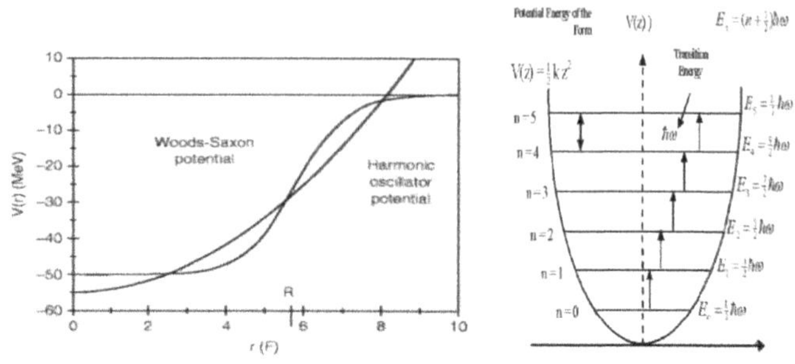

(a) (b)

Fig. 2.20: (a) A sketch of the nuclear potential (continuous line) together with the harmonic oscillator potential (dashed line). (b) The lowest energy levels in the 3-dim HO potential (in units of $\hbar\omega$).

$$Z = 2, 8, 20, 40, \ldots \text{ up to } N = 0, 1, 2, 3 \ldots$$

and similarly for neutrons.

The interaction, however, favors $N \approx Z$. Thus closed shells really observed are as follows:

$$A = 4, 16, 40, 80, \ldots \text{ up to } N = 0, 1, 2, 3 \ldots.$$

To put a nucleon in the next level costs extra energy $\hbar\omega$. So the stable nuclei are close to these "magic numbers". This was the first naive nuclear model.

2.5.2 *The spin-orbit interaction; The shell model*

The real breakthrough came, as we mentioned above, when the spin degrees of freedom entered the picture of the nucleus, initially via the spin orbit interaction. As we show in the appendix, see Section 14.3 and, e.g., [Brown (2020)], good angular momentum wave function are better in the description of the nucleus by combining the orbital angular momentum $\boldsymbol{\ell}$ and the spin $s = 1/2$ of the nucleon:

$$|n_r, \ell, j, m\rangle \equiv \psi_{n_r,\ell}(r) \left[Y^{\ell} \otimes \chi^{s}\right]^{j}_{m}, \; j = \ell + \frac{1}{2}, \left|\ell - \frac{1}{2}\right|. \tag{2.57}$$

This wave function is is a simultaneous eigenfunction of the commuting operators: $\mathbf{j}^2$, $\boldsymbol{\ell}^2$, $\mathbf{s}^2$ and $\mathbf{j}_z$, that is:

$$\mathbf{j}^2|n_r, \ell, j, m\rangle = j(j+1)|n_r, \ell, j, m\rangle, \; \boldsymbol{\ell}^2|n_r, \ell, j, m\rangle = \ell(\ell+1)|n_r, \ell, j, m\rangle,$$
$$\mathbf{s}^2|n_r, \ell, j, m\rangle = \frac{1}{2}\frac{3}{2}|n_r, \ell, j, m\rangle, \; \mathbf{j}_z|n_r, \ell, j, m\rangle = m|n_r, \ell, j, m\rangle. \tag{2.58}$$

Assuming a spin orbit interaction with constant strength $V_{\ell s}$, i.e. a potential of the form:

$$U_1 = -V_{\ell s} 2\boldsymbol{\ell}.\mathbf{s} = -V_{\ell s}\left[\mathbf{j}^2 - \boldsymbol{\ell}^2 - \mathbf{s}^2\right], V_{\ell s} > 0$$

we find that is diagonal in our basis and

$$\langle n_r, \ell, j, m | U_1 | n_r, \ell, j, m \rangle = V_{\ell s} \times \begin{cases} -\ell, \; j = \ell + 1/2 \\ \ell + 1, \; j = \ell - 1/2 \end{cases}. \tag{2.59}$$

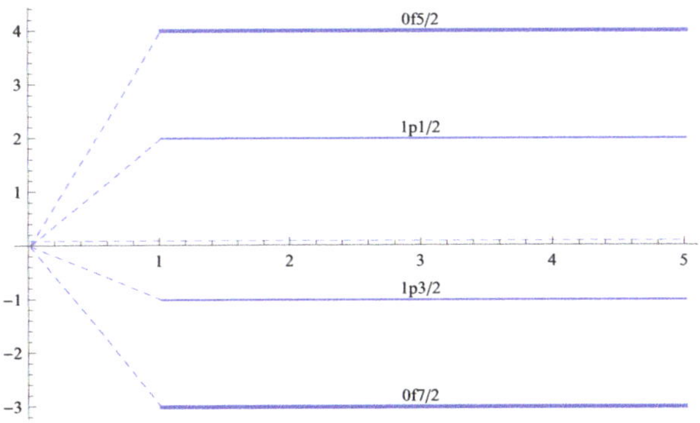

Fig. 2.21: The breaking of the degeneracy of the $N = 3$ levels due to the spin orbit interaction. The energy levels are measured from the energy at zero spin orbit interaction (dotted horizontal line) in units of $V_{\ell s}$.

The parameter $V_{\ell s}$ depends on A and can be different for protons and neutrons. The spin orbit splitting is exhibited in Fig. 2.21 for $N = 3$. Proceeding in an analogous way with realistic spin orbit interactions we obtain the picture of Fig. 2.22. We note that the lowest level of a given shell can come below some levels of the shell below. Also the highest level can move in the shell above.

Conclusion: Neutrons and protons are characterized by "magic numbers":

$$2, 8, 20, 28, 50, 82, 126.$$

Nuclei which have both neutron number and proton number equal to one of the magic numbers can be called "doubly magic", and are found to be particularly stable. The best examples are:

$${}^{4}_{2}\mathrm{He},\ {}^{16}_{8}\mathrm{O},\ {}^{40}_{20}\mathrm{Ca},\ {}^{48}_{20}\mathrm{Ca},\ {}^{56}_{28}\mathrm{Ni},\ {}^{208}_{82}\mathrm{Pb}.$$

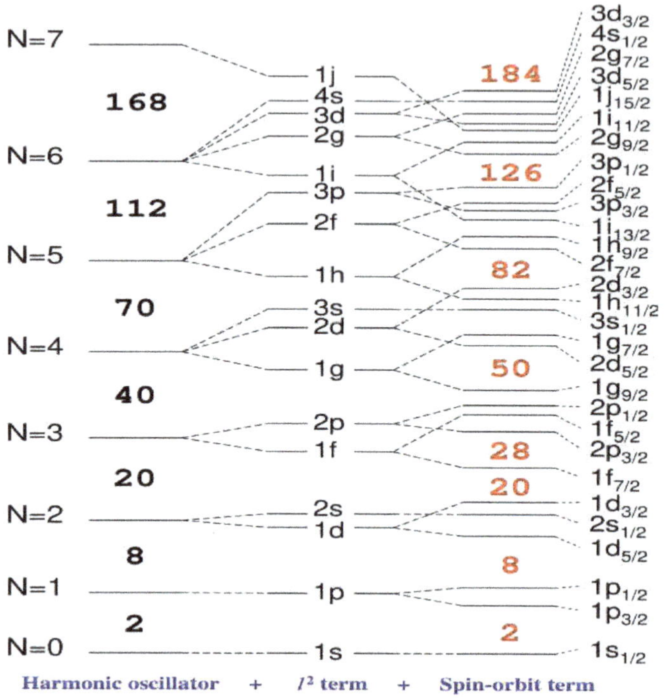

Fig. 2.22: Approximate single particle level scheme for nucleons. The harmonic oscillator levels are shown on the left and a splitting of the different ℓ values by a different central potential, with a term $D\ell^2$, is shown in the middle, while the splitting of the j-values due to the $\ell.\mathbf{s}$ is shown on the right. Protons and neutrons occupy the same levels up to $A = 50$. The magic numbers, indicating shell closure, are also indicated. Note that the radial quantum number n_r employed here is larger by 1 compared to the one employed in the text.

2.5.3 *The large basis shell model (LBSM)*

The above obtained shell model single particle wave functions can serve as a complete basis to describe the A-particle nuclear wave function as the hydrogenic wave functions do in atomic physics. We cannot elaborate here for this technique, but we will briefly outline the main steps:

(1) Determine the Hamiltonian describing the system. This takes the form of $H = H_{1b} + H_{2b}$. The first part, 1-body term, can be ad-

equately described in terms the shell model. The second part describes the two body interaction, involving the strong interaction between nucleons. In principle this can be obtained from the interaction between quarks mediated by gluons, see, e.g., Chapter 9, Section 9.15 and Fig. 9.24. For more details Chapter 5 on Quantum Chromodynamics, of the companion book [Vergados (2017)]. At the low energies involved in nuclear physics, however, this is really a formidable task. So one has to live with Van der Waals like interactions derived via various meson exchanges a la Yukawa. Quite a bit of success has been made in this direction and various phenomenologically successful interactions exist in the market. For comparison see Fig. 2.23.

(2) The A-particle product wave function symbolically written $\phi_\alpha(r_1)$ $\phi_\beta(r_2)....\phi_\alpha(r_A)$ must be properly antisymmetrized in the form of a Slatter determinant:

$$\Phi_n = \begin{vmatrix} \phi_\alpha(r_1) & \phi_\alpha(r_2) & \cdots & \phi_\alpha(r_A) \\ \phi_\beta(r_1) & \phi_\beta(r_2) & \cdots & \phi_\beta(r_A) \\ \cdots & \cdots & \cdots & \cdots \\ \phi_\omega(r_1) & \phi_\omega(r_2)\ , & \cdots & \phi_\omega(r_A) \end{vmatrix}, \tag{2.60}$$

where α, $\beta, \cdots \omega$ being all the single particle quantum numbers collectively as provided by the shell model.

The above simple expression holds in the case of the m-scheme, i.e. $\alpha = n_r, \ell, j, m$. This does not properly take into account the total angular momentum of the many particle wave function. This will automatically occur, however, with the obtained eigenfunctions, but one has to pay a price: one has to use a very large basis set. To reduce the space one must ensure that the basis is characterized by good angular momentum. In other words the basis must contain configurations of the type $|(j_i)^{n_i} J_i\rangle$ where n_i is the number of nucleons in the configuration i, j_i the angular momentum of the single particle of the configuration and J_i the angular momentum of the n_i nucleons. One must ensure, of course, that the allowed J_i are consistent with the Pauli principle. This is not a trivial problem but it can be dealt with elementary methods see Section 15.1, group theoretical techniques, see Section 15.5 or better as described in advanced textbooks [deShalit and Talmi (1963)]. After that one couples the angular momenta of the various configurations with the known rules of quantum mechanics to obtain the total angular momentum J of the state.

(3) One constructs the Hamiltonian matrix $H_{n'n} = \langle \Phi_{n'} | H | \Phi_n \rangle$.
(4) One must diagonalize the Hamiltonian to obtain the wave functions Ψ_k and the energies E_k.

$$\Psi_k = \sum_n C_k^n \Phi_n, \quad , C_k^n \text{ the mixing coefficients.}$$

These coefficients are obtained numerically from the diagonalization of the Hamiltonian matrix.
(5) The last step is easier said than done. The above basis is huge for even a few particles, especially if the basis of Eq. (2.60) is employed, instead of the j-shell configurations mentioned above. Thus some drastic truncation in the model space is necessary. Usually some approximations are necessary, e.g.:

- The inner nucleons, inner core, are ignored. They are made of closed shells with the nucleons put there according to the Pauli principle. Their possible effect is presumed to be taken into account by the one body shell model potential.
- The outer nucleons form the valence space. This is divided by a level corresponding to the Fermi surface. Above this all levels are empty. Due to the interaction n nucleons can be excited to higher levels above the Fermi (np) leaving behind them empty spaces called holes (nh). In this case we talk about np-nh excitations, see Fig. 2.24. Only few such excitations can be taken into account.

2.5.4 *A simple shell model calculation — The A = 14 system*

As an example let consider the case of the $A = 14$ system, i.e. ${}^{14}_{6}$C, ${}^{14}_{7}$N and ${}^{14}_{8}$O, which is a very interesting nuclear triad with very simple shell model description. The first involves two identical proton holes in the ${}^{16}_{8}$O closed shell and the last involves two neutron holes in the closed shell. Thus they are characterized by $I = 1$. ${}^{14}_{7}$N can be viewed as a proton and neutron hole in the closed shell and can have states with either $I = 0$ or $I = 1$. The active shell model space is spanned by the $N = 1, \ell = 1, j = 1/2 \leftrightarrow 0p_{1/2}$ and $N = 1, \ell = 1, j = 3/2 \leftrightarrow 0p_{3/2}$.

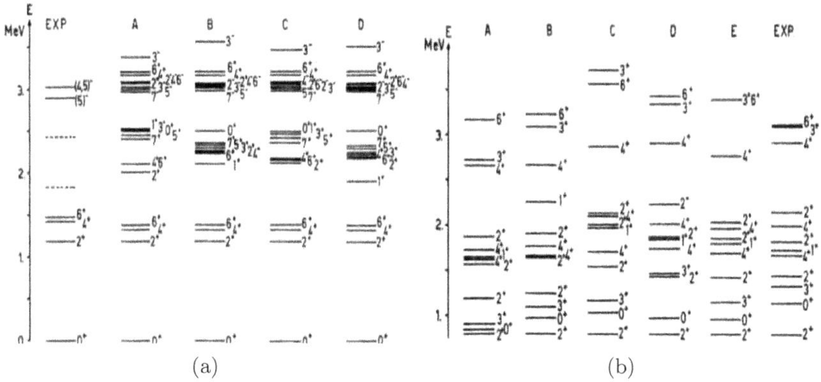

Fig. 2.23: The two neutron spectrum of ${}^{210}_{82}$Pb (a) and ${}^{206}_{82}$Pb (b). The first is considered as two particle (2p) and the second as two holes (2h) out of the closed shell nucleus ${}^{208}_{82}$Pb. The agreement between theory, for various values of an adjusted parameter, and experiment is not bad. Data taken from ref. [Hadermann and Alder (1969)].

2.5.4.1 *Antisymmetrization of two particle states*

Utilizing the results of Section 15.1 we can have the following two hole configurations for he $I = 1$ states

$$(0p_{1/2})^{-2}J = 0, (0p_{3/2})^{-2}J = 0, 2 \text{ and } \left((0p_{1/2})^{-1} \otimes (0p_{3/2})^{-1}\right) J = 1, 2.$$

Thus the $J = 0$ Hamiltonian matrix is 2-dimensional, the $J = 2$ is also 2-dimensional and the $J = 1$ is 1-dimensional.

In the $L - S$ coupling scheme we can have the following possibilities:
i) The $I = 1$ states are of the form:

$$|L = 0, S = 0, J = 0\rangle, |L = 1, S = 1, J = 0\rangle, |L = 2, S = 0, J = 2\rangle,$$

$$|L = 1, S = 1, J = 2\rangle, |L = 1, S = 1, J = 1\rangle.$$

The Hamiltonian matrix has, of course, the same dimension in this scheme as in the $j - j$ scheme.

ii) $I = 0$ states
In the $j - j$ scheme we have the possibilities

$$(0p_{1/2})^{-2}J = 1, (0p_{3/2})^{-2}J = 1, 3 \text{ and } \left((0p_{1/2})^{-1} \otimes (0p_{13/2})^{-1}\right) J = 1, 2.$$

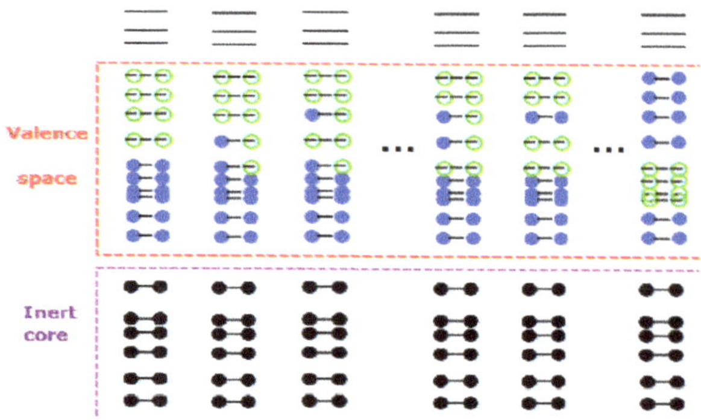

Fig. 2.24: The particle orbits of a shell model: i) An inert core. Nucleons in this region essentially play no role in the structure of the nucleus ii) the valence nucleons. Nucleons in the valence space occupy orbits according to the Pauli principle up to a given level forming the ground state of the nucleus. In the SM approach n nucleons can be excited to higher levels (np) leaving behind them empty spaces called holes (nh). In this case we talk about np-nh excitations. The first left column represents the gs of the system. The next two left columns correspond to 1p-1h excitations, while the right three columns are associated with 2p-2h.

In the $L-S$ coupling scheme we have the states:

$$|L=0,S=1,J=1\rangle,\ |L=1,S=0,J=1\rangle,\ |L=2,S=1,J=1\rangle,$$

$$|L=2,S=1,J=2\rangle,\ |L=2,S=1,J=3\rangle.$$

We note the absence of a $J=0$ state. The $J=1$ matrix is 3-dimensional, while the $J=2$ and $J=3$ are 1-dimensional.

2.5.4.2 *Expression of $j-j$ states in terms $L-S$ states*

It can be shown[7] that:

[7] The straightforward way to do this is to expand the two particle $j-j$ wave function, with the aid of Clebsch-Gordan (CG) coefficients, in terms of the relevant components in the $L-S$ coupling scheme, and perform the needed m-sums. More compactly this can be achieved by using the unitary UNine-J symbols (coefficients), which are expressed in terms of the usual Nine-J symbols, [Hecht (2000)], [deShalit and Talmi (1963)], [deShalit and Feshbach (1974)]. Both the C-G and the Nine-J symbols can be obtained using the package Mathematica.

i) $J = 0$ states:

$$\begin{pmatrix} (0p_{3/2})^{-2} \\ (0p_{1/2})^{-2} \end{pmatrix} \begin{pmatrix} \sqrt{\frac{2}{3}} & -\frac{1}{\sqrt{3}} \\ \frac{1}{\sqrt{3}} & \sqrt{\frac{2}{3}} \end{pmatrix} \begin{pmatrix} |L=0, S=0\rangle \\ |L=1, S=1\rangle \end{pmatrix}. \tag{2.61}$$

ii) $J = 1$ states:

$$\begin{pmatrix} (0p_{3/2})^{-2} \\ (0p_{1/2})^{-2} \\ (0p_{3/2})^{-1}(0p_{1/2})^{-1} \end{pmatrix} \begin{pmatrix} \frac{\sqrt{5}}{3} & \frac{\sqrt{10}}{3\sqrt{3}} & -\frac{\sqrt{2}}{3\sqrt{3}} & 0 \\ \frac{\sqrt{2}}{3} & -\frac{1}{3\sqrt{3}} & \frac{2\sqrt{5}}{3\sqrt{3}} & 0 \\ \frac{1}{3} & -\frac{2\sqrt{2}}{3\sqrt{3}} & -\frac{\sqrt{5}}{3\sqrt{6}} & \frac{1}{\sqrt{2}} \end{pmatrix} \begin{pmatrix} |L=1, S=0\rangle \\ |L=0, S=1\rangle \\ |L=2, S=1\rangle \\ |L=1, S=1\rangle \end{pmatrix}. \tag{2.62}$$

iii) $J = 2$ states:

$$\begin{pmatrix} (0p_{3/2})^{-2} \\ (0p_{3/2})^{-1}(0p_{1/2})^{-1} \end{pmatrix} \begin{pmatrix} \frac{1}{\sqrt{3}} & \frac{\sqrt{2}}{\sqrt{3}} & 0 \\ \frac{1}{\sqrt{3}} & -\frac{1}{\sqrt{6}} & \frac{1}{\sqrt{2}} \end{pmatrix} \begin{pmatrix} |L=2, S=0\rangle \\ |L=1, S=1\rangle \\ |L=2, S=1\rangle \end{pmatrix}. \tag{2.63}$$

iv) $J = 3$. This state is unique.

$$(0p_{3/2})^{-2} = |L=2, S=1, J=3\rangle. \tag{2.64}$$

We are not surprised that in the case of $(0p_{3/2})^{-1}(0p_{1/2})^{-1}$ there appears an extra $|L, S\rangle$ component, which does not satisfy the condition of antisymmetry. This is expected since the two orbits are different.

2.5.4.3 *Expression of $L - S$ states in terms $j - j$ states*

Proceeding as above one finds:
i) $J = 0$ states:

$$\begin{pmatrix} |L=0, S=0\rangle \\ |L=1, S=1\rangle \end{pmatrix} \begin{pmatrix} \sqrt{\frac{2}{3}} & \frac{1}{\sqrt{3}} \\ -\frac{1}{\sqrt{3}} & \sqrt{\frac{2}{3}} \end{pmatrix} \begin{pmatrix} |(0p_{3/2})^{-2} \\ (0p_{1/2})^{-2} \end{pmatrix}.$$

ii) $J = 1$ states:

$$\begin{pmatrix} |L=1, S=0\rangle \\ |L=0, S=1\rangle \\ |L=2, S=1\rangle \end{pmatrix} \begin{pmatrix} \frac{\sqrt{2}}{3} & \frac{\sqrt{5}}{3} & \frac{\sqrt{2}}{3} \\ -\frac{1}{3\sqrt{3}} & \frac{\sqrt{10}}{3\sqrt{3}} & -\frac{4}{3\sqrt{3}} \\ \frac{2\sqrt{5}}{3\sqrt{3}} & -\frac{\sqrt{2}}{3\sqrt{3}} & -\frac{\sqrt{5}}{3\sqrt{3}} \end{pmatrix} \begin{pmatrix} (0p_{3/2})^{-2} \\ (0p_{1/2})^{-2} \\ (0p_{3/2})^{-1}(0p_{1/2})^{-1} \end{pmatrix}$$

$$|L=1,S=1,J=1\rangle = (0p_{3/2})^{-1}(0p_{1/2})^{-1}J=1.$$

iii) $J = 2$ states:

$$\begin{pmatrix} |L=2,S=0\rangle \\ |L=1,S=1\rangle \end{pmatrix} \begin{pmatrix} \frac{1}{\sqrt{3}} & \sqrt{\frac{2}{3}} \\ \sqrt{\frac{2}{3}} & -\frac{1}{\sqrt{3}} \end{pmatrix} \begin{pmatrix} (0p_{3/2})^{-2} \\ (0p_{3/2})^{-1}(0p_{1/2})^{-1} \end{pmatrix}$$

$$|L=2,S=1,J=2\rangle = (0p_{3/2})^{-1}(0p_{1/2})^{-1}J=2.$$

iv) $J = 3$. This state is unique.

$$|L=2,S=1,J=3\rangle = (0p_{3/2})^{-2}J=3.$$

2.5.4.4 *Evaluation of the matrix elements*

We are going to evaluate the matrix elements of a central interaction in the case of the $A = 14$ system. We find it convenient to begin working with $L - S$ basis. Then we can express the matrix elements in the $j - j$ scheme in terms of those involving the $L - S$ scheme.

We should mention at this point that, for the evaluation of matrix elements of a central interaction, we found it convenient to express the two particle orbital wave function in terms of the relative and center of mass coordinate, something not too hard in the case of the harmonic oscillator wave functions. For p-shell nuclei this accomplished via Eqs. (14.29), (14.30), (14.32) of Chapter 14.

We will begin with the $I = 1$ states.

i) For the $J = 0$ we have two states $L = 0, S = 0$ and $L = 1, S = 1$. Since the interaction ic central

$$ME_a = \langle L=0,S=0,J=0|V|L=0,S=0,J=0\rangle$$

$$= \int\int d^3\mathbf{r}_1 d^3\mathbf{r}_2 \left([0p(\mathbf{r}_1)0p(\mathbf{r}_2)]^{L=0}|\right)^* V(|\mathbf{r}_1-\mathbf{r}_2|)\,[0p(\mathbf{r}_1)0p(\mathbf{r}_2)]^{L=0}.$$

Then using Eq. (14.29) we find that

$$[0p(\mathbf{r}_1)0p(\mathbf{r}_2)]^{L=0} = \frac{1}{\sqrt{2}}\left(0s(\mathbf{r})1s(\mathbf{R}) - 0s(\mathbf{R})1s(\mathbf{r})\right),$$

$$\mathbf{r} = \frac{1}{\sqrt{2}}(\mathbf{r}_1-\mathbf{r}_2),\ \mathbf{R} = \frac{1}{\sqrt{2}}(\mathbf{r}_1+\mathbf{r}_2).$$

Thus we get

$$ME_a = \int d^3\mathbf{r} d^3\mathbf{R}\frac{1}{2}\left(0s(\mathbf{r})1s(\mathbf{R}) - 0s(\mathbf{R})1s(\mathbf{r})\right)V(r\sqrt{2})(0s(\mathbf{r})1s(\mathbf{R}) - 0s(\mathbf{R})1s(\mathbf{r})).$$

Since the interaction is independent of $\mathbf{R}$, the integration over $\mathbf{R}$ gives unity in the case of the same wave functions of R and zero otherwise. Thus

$$ME_a = \int d^3\mathbf{r}\frac{1}{2}\left(0s(\mathbf{r})V(r\sqrt{2})0s(\mathbf{r}) + 1s(\mathbf{r})V(r\sqrt{2})1s(\mathbf{r})\right)$$
$$= \frac{1}{2}(ME_{0s} + ME_{1s})$$

where

$$ME_{0s} = \frac{4}{\sqrt{\pi}}\int_0^\infty x^2 e^{-x^2} V(bx\sqrt{2})dx,$$

$$ME_{1s} = \frac{8}{3\sqrt{\pi}}\int_0^\infty x^2 e^{-x^2}(3/2 - x^2)^2 V(bx\sqrt{2})dx.$$

ii) $J = 1$. Using Eq. (14.32) we find

$$[0p(\mathbf{r}_1)0p(\mathbf{r}_2)]^{L=1} = [0p(\mathbf{r})0p(\mathbf{R})]^{L=1}.$$

Thus

$$ME_b = ME_{0p},\ ME_{0p} = \frac{8}{3\sqrt{\pi}}\int_0^\infty x^4 e^{-x^2} V(bx\sqrt{2})dx.$$

iii) In the case of $J = 2$ we can have the $L = 2, S = 0$ and $L = 1, S = 1$. The matrix element involving the latter is the same with the one obtained above. Proceeding as above using Eq. (14.30) we find

$$(0p(\mathbf{r}_1 \otimes 0p(\mathbf{r}_2)^{L=2} = \frac{1}{\sqrt{2}}(0s(\mathbf{r})0d(\mathbf{R}) - 0s(\mathbf{R})0d(\mathbf{r})) \Rightarrow$$

$$ME_c = \int d^3\mathbf{r}\frac{1}{2}\left(0s(\mathbf{r})V(r\sqrt{2})0s(\mathbf{r}) + 0d(\mathbf{r})V(r\sqrt{2})1s(\mathbf{r})\right)$$
$$= \frac{1}{2}(ME_{0s} + ME_{0d})$$

with

$$ME_{0d} = \frac{16}{15\sqrt{\pi}}\int_0^\infty x^6 e^{-x^2} V(bx\sqrt{2})dx.$$

iv) $J = 3$ in this case $L = 2, S = 1$. The ME is

$$ME_d = \frac{1}{2}(ME_{0s} + ME_{0d}).$$

Using the HO oscillator wave functions given by Eq. (2.154) and the Yukava interaction, see Eq. (2.17) we obtain the results shown in Fig. 2.25. Using above expressions given by Eqs. (2.61)–(2.63) and the known ME in the LS coupling scheme we find the following two body Hamiltonian matrices

$$H_{J=0} = \begin{pmatrix} \frac{2}{3}\left(a + \frac{b}{2}\right) & \frac{1}{3}\sqrt{2}(a-b) \\ \frac{1}{3}\sqrt{2}(a-b) & \frac{1}{3}(a+2b) \end{pmatrix}, H_{J=2} = \begin{pmatrix} \frac{2b}{3} + \frac{c}{3} & \frac{c}{3} - \frac{b}{3} \\ \frac{c}{3} - \frac{b}{3} & \frac{b}{6} + \frac{5c}{6} \end{pmatrix}, H_{J=3} = (c)$$

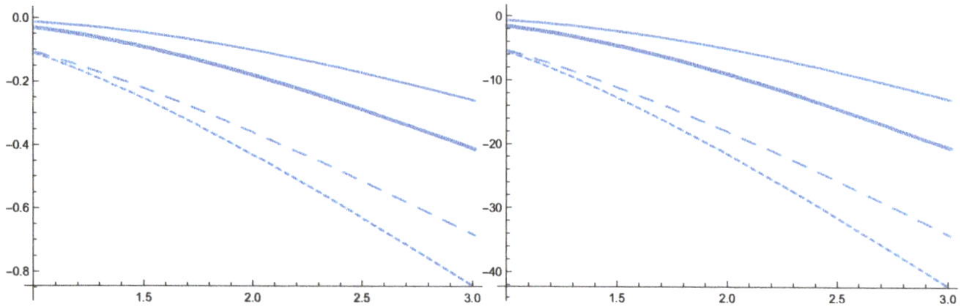

(a) (b)

Fig. 2.25: Sown are the integrals ME_{0s} (dotted line), ME_{1s} (dashed line). ME_{0p} (solid line) and ME_{0d} (thick solid line) obtained with Yukava exchange potential as a function r_0. (a) for $V_0 = 1$ MeV and (b) for 50 MeV.

$$H_{J=1} = \begin{pmatrix} \frac{10a}{27} + \frac{5b}{9} + \frac{2c}{27} & -\frac{\sqrt{10}a}{27} + \frac{\sqrt{10}b}{9} - \frac{2\sqrt{10}c}{27} & -\frac{4\sqrt{5}a}{27} + \frac{\sqrt{5}b}{9} + \frac{\sqrt{5}c}{27} \\ -\frac{\sqrt{10}a}{27} + \frac{\sqrt{10}b}{9} - \frac{2\sqrt{10}c}{27} & \frac{a}{27} + \frac{2b}{9} + \frac{20c}{27} & \frac{2\sqrt{2}a}{27} + \frac{\sqrt{2}b}{9} - \frac{5\sqrt{2}c}{27} \\ -\frac{4\sqrt{5}a}{27} + \frac{\sqrt{5}b}{9} + \frac{\sqrt{5}c}{27} & \frac{2\sqrt{2}a}{27} + \frac{\sqrt{2}b}{9} - \frac{5\sqrt{2}c}{27} & \frac{a}{27} + \frac{2b}{9} + \frac{20c}{27} \end{pmatrix}$$

with

$$a = \frac{1}{2}\left(ME_{0s} + (ME_{1s}\right), \; b = ME_{0p}, \; c = \frac{1}{2}\left(ME_{0s} + (ME_{0d}\right).$$

To this we must add the diagonal single particle energies. These are determined phenomenologically. For illustration purposes we will take them to be

$$\epsilon(p_{1/2})^{-1} = 0, \epsilon(p_{3/2})^{-1} = \epsilon.$$

Note that it takes less energy to shift the $p_{1/2}$ particle above the Fermi surface than the $p_{3/2}$. Thus taking the Yukava interaction with $V_0 = 50$ MeV, and $r_0 = 1.5$ fm, i.e. $a = -12$, $b = 4.6$, $c = -7.5$ MeV, and $\epsilon = 1$ MeV, we obtain

$$H_{J=0} = \begin{pmatrix} -7.4905 & -3.48318 \\ -3.48318 & -7.02752 \end{pmatrix}, H_{J=2} = \begin{pmatrix} -3.54974 & -0.985191 \\ -0.985191 & -7.02752 \end{pmatrix} \quad (2.65)$$

$$H_{J=1} = \begin{pmatrix} -5.52012 & 1.55772 & 2.20295 \\ 1.55772 & -7.02752 & 0 \\ 2.20295 & 0 & -6.02752 \end{pmatrix}, H_{J=3} = \begin{pmatrix} -5.52012 \end{pmatrix}. \quad (2.66)$$

One can now diagonalize the above matrices and obtain the eigenvectors and egenvalues, see problem at the end of this chapter, Eqs. (2.163)–(2.165).

The absolute value of the energy obtained this way is not reliable. So we are going to determine the excitation energy for each nucleus relative to the ground state. We thus find in the case of ${}^{14}_{6}\text{C}$ and ${}^{14}_{8}\text{O}$:

$$I = 1: \; Ex(2_1^+) = 3.48, \; Ex(0_1^+) = 6.98, \; Ex(2_2^+) = 7.45 \text{ MeV}.$$

In the case of ${}^{14}_{7}\text{N}$ we find

$$I = 0: \; Ex(1_1^+) = 2.07, \; Ex(3_1^+) = 3.21, \; Ex(1_2^+) = 5.55 \text{ MeV}.$$

In the last nucleus we also have four $I = 1$, $I_3 = 0$ excited states at energies

$$Ex(0_1^+) = \Delta, Ex(2_1^+) = \Delta + 3.48, Ex(0_1^+) = \Delta + 6.98, Ex(2_2^+) = \Delta + 7.45 \text{ MeV}.$$

Where Δ can be evaluated by a proper treatment of the Coulomb interaction.

The reader is encouraged to compare the results obtained here with those obtained with more realistic models discussed in Section 13.9 and, in particular, the energy spectrum of Fig. 13.6.

2.6 Relativistic mean field theory

2.6.1 *General concepts of RMF*

The Relativistic Mean Field Theory (RMF) is a phenomenological description of the nuclear many problem including both finite nuclei as well as infinite nuclear matter. This theory is based in the following three assumptions: 1) the nucleons are treated as point-like particles, 2) the theory is Lorentz invariant and 3) it obeys the causality condition (the speed of light is the highest in nature). In this framework the nucleons are treated as Dirac spinors ψ and they can interact between themselves only via the exchange of effective point-like particles, which are called mesons, ϕ_m. The mesons are characterized and described by a set of quantum numbers (including the spin J, parity P and isospin[8] T), mass m_m and the coupling constant g_m. In particular the masses and the coupling constants are determined by demanding to reproduce the experimental data. In addition RMF theory is essentially a mean field approximation in which the symmetries involved play an important role. Thus, some of the meson fields are more or less forbidden due to such symmetries.

[8]In nuclear physics usually the isospin is denoted by T with upper component $T_3 = 1/2$ associated with the neutron. In particle physics it is indicated by I with $|I = 1/2 I_3 = 1/2\rangle$ identified with the particle of the highest charge. We will try to use T in nuclear physics topics, whenever possible.

In what follows we provide the basic formalism of the RMF theory, starting from a number of fields symbolized as $q_j(x)$. The dynamical description of the fields is based on the Lagrangian density $\mathcal{L}(q, \partial_\mu q, t)$, after applying the variational principle in the form, [Ring (1996)],

$$\delta \int d^4x \mathcal{L}(q, \partial_\mu q, t) = 0. \tag{2.67}$$

This leads to the well known Euler-Lagrange equations of motion

$$\partial_\mu \left(\frac{\partial \mathcal{L}}{\partial(\partial_\mu q_j)} \right) - \frac{\partial \mathcal{L}}{\partial q_j} = 0 \tag{2.68}$$

for all the fields $q_j(x)$. Moreover, the energy-momentum tensor is defined as

$$T^{\mu\nu} = -g^{\mu\nu}\mathcal{L} + \frac{\partial L}{\partial(\partial_\mu g_j)}\partial^\nu g_j. \tag{2.69}$$

The energy-momentum tensor $T^{\mu\nu}$ obeys the continuity equation

$$\partial_\mu T^{\mu\nu} = 0 \tag{2.70}$$

and the four-momentum defined as

$$P^\nu = \int d^3\mathbf{r} T^{0\nu} \tag{2.71}$$

is conserved. The energy corresponds to the zeroth component of P^ν and is given by

$$E \equiv P^0 = \int d^3\mathbf{r}\mathcal{H}(\mathbf{r}) \tag{2.72}$$

where the Hamiltonian density $\mathcal{H}(\mathbf{r})$ is given by

$$\mathcal{H}(\mathbf{r}) = T^{00} = \frac{\partial \mathcal{L}}{\partial \dot{q}_j}\dot{q}_j - \mathcal{L}. \tag{2.73}$$

In the RMF theory, which describes both finite nuclei and infinite nuclear matter, the fields $q_j(x)$ are given by the wave functions of the nucleons, meson fields and the electromagnetic potential. In particular the nucleons are described by the Dirac spinors $\psi_i(x, s, T)$, see Chapter 16, with four components ($x = (t, \mathbf{r})$ is the space-time coordinate), s labels the four Dirac components (1,2 for the large components and 3,4 or the small ones) and $T = (p, n)$ indicates the isospin. Moreover, the index i runs over all the nucleons in the system. The nucleons interact via mesonic fields. The mesons are characterized by the quantum numbers (J, T, P), which correspond to its angular momentum, isospin and parity.

In particular, the exchange of the phenomenological scalar σ-mesons, with quantum numbers ($J = 0$, $T = 0$, $P = +1$), is associated with attractive forces between the nucleons. The corresponding field is a scalar one, $\sigma(x)$. The repulsive part of the interaction is the result of the exchange of vector mesons. Among them the most important is the ω-meson, with quantum numbers ($J = 1$, $T = 0$, $P = -1$). The corresponding field is a vector field $\omega^\mu(x)$. In addition the electromagnetic field is introduced via the photon which carries spin $J = 1$. It is described by the vector potential $A^\mu(x)$, whose time-like component represents the Coulomb interaction. The isospin character of the nuclear forces is introduced via the the exchange of ρ-mesons, with quantum numbers ($J = 1$, $T = 1$, $P = -1$). In principle there are many more mesons, which may play role on the description of the properties of nuclear forces. However, in order to simplify the study one treats only the most important fields that is $\sigma(x)$, $\omega^\mu(x)$, $\rho^\mu(x)$ and $A^\mu(x)$ and neglects all the others.

2.6.2 *The Lagrangian density*

In order to construct the basic equation which for the description of the structure of finite nuclei, as a starting point, we consider the following Lagrangian density

$$\mathcal{L} = \mathcal{L}_N + \mathcal{L}_M + \mathcal{L}_{\text{int}}. \tag{2.74}$$

The first term corresponds to the Lagrangian of the free nucleons and is defined, see Chapter 16, as

$$\mathcal{L}_N = \bar{\psi}(i\gamma^\mu \partial_\mu - m)\psi \tag{2.75}$$

where m is the bare mass of the nucleon and also ψ symbolizes the Dirac spinor. Moreover, the Lagrangian density for the free mesons $\mathcal{L}_M$ contains the following terms

$$\mathcal{L}_\sigma = \frac{1}{2}\left(\partial_\mu \sigma \partial^\mu \sigma - m_\sigma^2 \sigma^2\right) \tag{2.76}$$

$$\mathcal{L}_\pi = \frac{1}{2}\left(\partial_\mu \vec{\pi} \partial^\mu \vec{\pi} - m_\pi^2 \vec{\pi}^2\right) \tag{2.77}$$

$$\mathcal{L}_\omega = -\frac{1}{2}\left(\Omega_{\mu\nu}\Omega^{\mu\nu} - \frac{1}{2} m_\omega^2 \omega_\mu \omega^\mu\right) \tag{2.78}$$

$$\mathcal{L}_\rho = -\frac{1}{2}\left(\vec{R}_{\mu\nu}\vec{R}^{\mu\nu} - \frac{1}{2} m_\rho^2 \vec{\rho}_\mu \vec{\rho}^\mu\right) \tag{2.79}$$

$$\mathcal{L}_\omega = -\frac{1}{2} F_{\mu\nu} F^{\mu\nu} \tag{2.80}$$

where m_σ, m_ω, m_π and m_ρ are the rest masses of the mesons and also

$$\Omega^{\mu\nu} = \partial^\mu \omega^\nu - \partial^\nu \omega^\nu \tag{2.81}$$

$$\vec{R}_{\mu\nu} = \partial^\mu \vec{\rho}^\nu - \partial^\nu \vec{\rho}^\mu \tag{2.82}$$

$$F^{\mu\nu} = \partial^\mu A^\nu - \partial^\nu A^\mu \tag{2.83}$$

are the field tensors of the ω, the ρ mesons and the photon. The arrows indicate vectors in isospin space. The interaction lagrangian $\mathcal{L}_{\rm int}$ between the nucleons and mesons is given by the following ansatz

$$\mathcal{L}_{\rm int} = -g_\sigma \bar{\psi} \sigma \psi - i g_\pi \bar{\psi} \gamma_5 \vec{\tau} \vec{\pi} \psi - g_\omega \bar{\psi} \gamma_\mu \omega^\mu \psi - g_\rho \bar{\psi} \gamma_\mu \vec{\tau} \vec{\rho}^\mu \psi - e \bar{\psi} \gamma_\mu A^\mu \psi \tag{2.84}$$

where g_σ, g_π, g_ω, and g_ρ are the coupling constants and $\vec{\tau}$ the Pauli isospin matrices. In the following consideration we are not going to include the pion contribution. The above Lagrangian contains just linear coupling terms. In particular, if we neglect the pion and ρ-meson, we recover the original Walecka model, which describes, at least qualitatively, the bulk properties of finite nuclei [Serot and Walecka (1986)]. This model, however, fails to reproduce the experimental values of the incompressibility and the surface properties of finite nuclei. Later on, Boguta and Bodmer [Boguta and Bodmer (1977)] extended the model in order to include a non-linear self-coupling amongst the σ-mesons by replacing the term $1/2 m_\sigma^2 \sigma^2$ by the σ-potential,

$$U(\sigma) = \frac{1}{2} m_\sigma^2 \sigma^2 + \frac{1}{3} g_2 \sigma^3 + \frac{1}{4} g_3 \sigma^4 \tag{2.85}$$

where the parameters g_2 and g_3 are adjusted to reproduce the surface properties of nuclei. Since, however, the parameter g_3 gets negative values, the field theory treatment, which is based in this ansatz, is unstable. In particular the problems appear for high baryon densities.

2.6.3 *The relativistic mean field equations*

In order to derive the equation of motion, we introduce for each nucleon a classical spinor field ψ_i $(i = 1, \ldots, A)$. In this case the classical equation of motion for the variables $(q_j = \psi_1 \cdots \psi_A, \sigma, \vec{\pi}, \omega, \vec{\rho}, \cdots)$ and the corresponding Dirac equation for each spinor field ψ_i is written as

$$\left[\gamma_\mu (i\partial^\mu + V^\mu(x)) + m + S(x) + i\gamma_5 P(x)\right] \psi_i = 0, \tag{2.86}$$

where the relativistic fields $S(x)$, $P(x)$ and $V^\mu(x)$ are given by

$$S(x) = g_\sigma \sigma(x) \tag{2.87}$$

$$P(x) = g_\pi \times . \vec{\pi}(x) \tag{2.88}$$

$$V^\mu(x) = g_\omega \omega^\mu(x) + g_\rho \vec{\rho}^\mu(x) + eA^\mu(x). \tag{2.89}$$

Using the Lorentz gauge for the vector mesons we find finally the following Klein-Gordon equations

$$[\Box + m_\sigma]\,\sigma = -g_\sigma \rho_\sigma - g_2\sigma^2 - g_3\sigma^3 \tag{2.90}$$

$$[\Box + m_\pi]\,\vec{\pi} = -g_\pi \vec{\rho}_{ps} \tag{2.91}$$

$$[\Box + m_\omega]\,\omega^\mu = g_\omega j^\mu \tag{2.92}$$

$$[\Box + m_\rho]\,\vec{\rho}^\mu = g_\rho \vec{j}^\mu \tag{2.93}$$

$$\Box A^\mu = e j_c^\mu \tag{2.94}$$

where $\Box$ is the **d'Alembert operator** $\Box = \frac{1}{c^2}\frac{\partial^2}{\partial t^2} - \Delta$ and Δ the **Laplace operator** $\Delta f = \nabla^2 f$. Moreover, $\rho_s(x)$ is the scalar density

$$\rho_s(x) = \sum_{i=1}^{A} \bar{\psi}_i(x)\psi_i(x) \tag{2.95}$$

$\vec{\rho}_{ps}$ is the pseudo-scalar density

$$\vec{\rho}_{ps} = \sum_{i=1}^{A} \bar{\psi}_i(x)\gamma_5\vec{\tau}\psi_i(x) \tag{2.96}$$

$j^\mu(x)$ is the (isoscalar) baryon current

$$j^\mu(x) = \sum_{i=1}^{A} \bar{\psi}_i(x)\gamma^\mu\psi_i(x), \tag{2.97}$$

while $\vec{j}^\mu$ is the isovector current

$$\vec{j}^\mu = \sum_{i=1}^{A} \bar{\psi}_i(x)\gamma^\mu\vec{\tau}\psi_i(x) \tag{2.98}$$

and $j_c^\mu(x)$ is the electromagnetic current

$$j_c^\mu(x) \sum_{i=1}^{A} \bar{\psi}_i(x)\frac{1}{2}(1+\tau_3)\gamma^\mu\psi_i(x). \tag{2.99}$$

The four current $j^\mu = (\rho_v, \mathbf{j})$ contains the normal three-current density of the nucleons as well as the normal density of nucleons symbolized as ρ_v, where the index v indicates that it is the time-like component of a Lorentz vector (both isoscalar and isovector).

In order to describe the basic properties of a finite nuclei we have to specify a reference frame. Of course for any solution of the equations in one system, there is a corresponding solution in a new system. We consider now the special case of the static approximation where we assume time-independence for the mesons fields and also a time-dependent phase $e^{i\epsilon_i t}$ for each spinor ψ_i. Moreover, we consider the case with time-reversal invariance and with good parity (this is the case, for example, of the ground state of even-even nuclei). In this case the space-like components of all currents j, $\vec{j}$, j_c as well as the pion field vanish. Then we get the stationary relativistic mean field equations

$$[-i\alpha\nabla + \beta(m+S) + V]\,\psi_i = \epsilon_i\psi_i, \tag{2.100}$$

$$[-\triangle + m_\sigma]\,\sigma = -g_\sigma\rho_\sigma - g_2\sigma^2 - g_3\sigma^3 \tag{2.101}$$

$$[-\triangle + m_\omega]\,\omega^0 = g_\omega\rho_v \tag{2.102}$$

$$[-\triangle + m_\rho]\,\rho_3^0 = g_\rho\rho_3 \tag{2.103}$$

$$-\triangle\rho_c^0 = e\rho_c \tag{2.104}$$

where now $\rho_s(x)$ is the scalar density

$$\rho_s(x) = \sum_{i=1}^{A} \bar{\psi_i}(x)\psi_i(x) \tag{2.105}$$

$\rho_v(x)$ is the usual baryon density

$$\rho_v(x) = \sum_{i=1}^{A} \psi_i^\dagger(x)\psi_i(x) \tag{2.106}$$

$\rho_3(x)$ is the isovector density

$$\rho_3(x) = \sum_{i=1}^{A} \psi_i^\dagger(x)\tau_3\psi_i(x) \tag{2.107}$$

and $\rho_c(x)$ is the charge density

$$\rho_c(x) = \sum_{i=1}^{A} \psi_i^\dagger(x)\frac{1}{2}(1+\tau_3)\psi_i(x). \tag{2.108}$$

In Eq. (2.100) two potentials are included: i) The so-called vector potential $V(\mathbf{r})$ given by

$$V(\mathbf{r}) = g_\omega \omega^0(\mathbf{r}) + g_\rho \tau_3 \rho_3^0(\mathbf{r}) + eA^0(\mathbf{r}), \tag{2.109}$$

which corresponds to the Coulomb field. ii) The scalar potential $S(\mathbf{r})$ given by

$$S(\mathbf{r}) = g_\sigma \sigma(\mathbf{r}), \tag{2.110}$$

which contributes to the effective Dirac mass according to

$$m^*(\mathbf{r}) = m + S(\mathbf{r}) \tag{2.111}$$

Eqs. (2.100)–(2.104) are a set of coupled differential equations for the unknown fields ψ_i $(i = 1, \ldots, A)$, σ, ω^0, ρ_3^0 and the Coulomb field A^0. They can be solved numerically by iteration, according to the following recipe: We start with an initial guess for the values of the potentials V and S and solve the Dirac equations for the spinors ψ. The obtained results are used to calculate the densities ρ_s, ρ_v, ρ_3 and ρ_c which, in turn, from the sources in Eqs. (2.101)–(2.104), yield a calculation of the meson fields and a new set of potentials for $V(\mathbf{r})$ and $S(\mathbf{r})$. The iteration is repeated until convergence is achieved.

In particular, in the case of nuclear matter the above technique leads to an analytical solution of the problem, involving just one transcendental equation. For spherical nuclei one has to deal with usual differential equations, which are treated by standard techniques. In the case of axially symmetric or triaxial nuclei one has to solve a set of partial differential equations. Finally in the case of bound states one usually employs the technique of expanding the spinors (large and small components separately) as well as expressing the meson fields in terms of the complete set of eigenstates of a three dimensional oscillator in cylindrical coordinates, using coupled channel techniques.

After finding the solutions of the system of the RMF equations, one is able to calculate observable properties, including mainly the energy density, which is defined by

$$E[\psi_i, \bar{\psi}_i, \sigma, \omega^0, \rho_3^0, A^0] = \int d^3r \mathcal{H}(\vec{r}). \tag{2.112}$$

The solution can be given by the integral

$$
\begin{aligned}
E[\psi_i, \bar{\psi}_i, \sigma, \omega^0, \rho_3^0, A^0] = &\sum_{i=1}^{a} \int d^3r \psi_i^\dagger \left(-i\alpha\nabla + \beta m\right)\psi_i \\
&+ \int d^3r \frac{1}{2}\left((\nabla\sigma)^2 + U(\sigma)\right) \\
&- \int d^3r \frac{1}{2}\left((\nabla\omega^0)^2 + m_\omega^2 \omega_3^{02}\right) \\
&- \frac{1}{2}\int d^3r \left((\nabla\rho^0)^2 + m_\rho^2 \rho_3^{02}\right) \\
&- \frac{1}{2}\int d^3r (\nabla A^0)^2 \\
&+ \int d^3r (g_\sigma \rho_s \sigma + g_\omega \rho_v \omega^0 + g_\rho \rho_3 \rho_3^0 + e\rho_c A^0).
\end{aligned}
\tag{2.113}
$$

For the self-consistent solution we exploit the Klein-Gordon Eqs. (2.101)–(2.104) and the total energy takes the form

$$
E = \sum_{i=1}^{A} \epsilon_i - \frac{1}{2}\int d^3r \left(g_\sigma \rho_s \sigma + \frac{1}{3} g_2 \sigma^3 + \frac{1}{2} g_3 \sigma^4 + g_\omega \rho_v \omega^0 + g_\rho \rho_3 \rho_3^0 + e\rho_c A^0\right).
\tag{2.114}
$$

Actually we can derive the equations of motion also from the variation of the energy density functional (2.114) with respect to the field ψ_i, $\bar{\psi}_i$, σ, ω^0, ρ_3^0 and A^0, where the condition of normalization gives the single particle energies ϵ_i as Lagrangian multipliers.

2.6.4 *Applications*

In what follows, we present and discuss some applications of the RMF theory. In the case of finite nuclei, the input parameters required to carry out the calculations include: a) the baryon and the meson masses, b) the coupling constants of the meson fields to the nucleons and c) the parameters g_2 and g_3 describing the non-linear coupling of the σ-mesons among themselves. With these ingredients one can treat only doubly closed shell nuclei, where pairing correlations are negligible. In the case of open shell nuclei with deformation, the pairing effect must be included. This is done by the conventional method of the constant gap approximation, which is standard in density-dependent Hartree-Fock calculations of Skyrme type.

Various set of parameters of the Lagrangian have been found. In particular, the masses of the mesons and the coupling constants have been obtained by a fitting of experimental data of nuclear matter and a few spherical finite nuclei. In Table 2.1 we present various standard parameter sets and some results for symmetric nuclear matter, including the saturation density ρ_0 of symmetric nuclear matter, the banding energy per particle E/A, the incompressibility K, the symmetry energy at saturation density J and the effective mass ratio m^*/m.

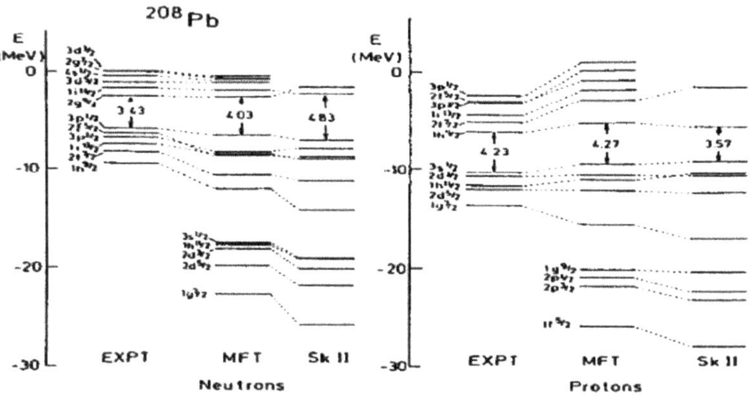

Fig. 2.26: The neutron and proton single particle spectrum for the nucleus ^{208}Pb calculated with the relativistic parameter set NL1, in comparison with the experimental data as well as those taken from the similar calculation based on the use of Skyrme forces [Gambhir *et al.* (1990)].

In Fig. 2.26 we show the neutron and proton single particle spectrum in the vicinity of the Fermi surface for the nucleus ^{208}Pb. In actuality, it is obvious that the experimental spectra are more compressed compared to the theoretical ones. In any case, at least qualitatively, the theoretical predictions are correct.

Finally in Fig. 2.27 we present the calculated charge density distribution for the nucleus ^{208}Pb in comparison with the experimental data. Obviously, in this case the theoretical prediction is in good agreement with the experimental data.

2.7 Problems

(1) Consider a uniform nuclear density in 3 dimensions.
In other words it extends uniformly up to $r = a$ and it falls

Table 2.1: A specific set of meson parameters and the resulting bulk properties of symmetric nuclear matter. The masses are given in MeV, while the parameter g_2 in fm^{-1}. The other coupling constants are dimensionless. The saturation density ρ_{sut} is given in fm^{-3}, the binding energy per nucleon E/A, the incompressibility K_{inc} and the symmetry energy at saturation density J are given in MeV. The effective mass m_b^* of the baryons is in units of the rest mass m_b.

Parameters	HS	NL1	NL2	NLSH
m_N	939	938	938	939
m_σ	520	492.25	504.89	526.059
m_ω	783	795.359	780	783
m_ρ	770	763	763	763
g_σ	10.47	10.138	9.111	10.444
g_ω	13.8	13.285	11.493	12.945
g_ρ	4.035	4.976	5.507	4.383
g_2	0.0	-12.172	-2.304	-6.9099
g_3	0.0	-36.265	13.783	-15.8337
Bulk properties				
ρ_{sut}	0.148	0.151	0.146	0.146
E/A	-15.731	-16.426	-17.018	-16.346
K_{inc}	546.3	211.1	399.2	355.4
J	34.9	43.5	45.1	36.1
m_b^*/m_b	0.541	0.573	0.67	0.597

suddenly to zero. For convenience we take it to be normalized to unity.

- Show that:

$$\rho(r) = \begin{cases} \frac{3}{R^3}, \ 0 \le r \le R \\ 0, \ \ \text{otherwise} \end{cases} \tag{2.115}$$

where R is the nuclear density. The angular factor of $1/(4\pi)$ is understood, when we perform the angular integration.

- Show that the form factor associated with this density is

$$F(k^2) = \frac{3}{R^3}\int_0^R e^{i\mathbf{k}.\mathbf{r}} r^2 dr \frac{d\Omega}{4\pi}, \ \mathbf{k} = \frac{\mathbf{q}}{\hbar} \Rightarrow$$

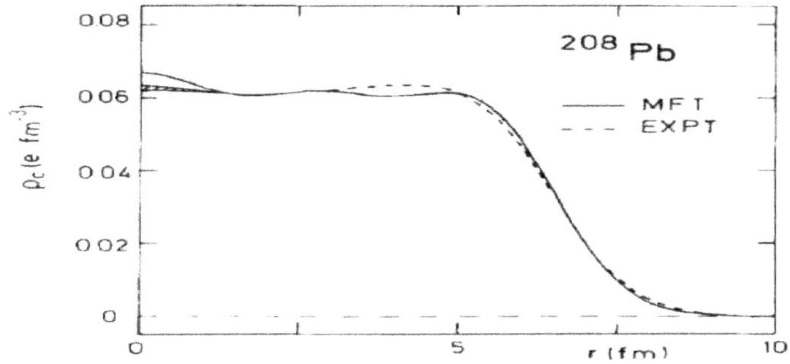

Fig. 2.27: The charge density distribution for the nucleus ^{208}Pb derived in the framework of the RMF theory. The calculated results are compared with the experimental data [Gambhir *et al.* (1990)].

$$F(k^2) = \frac{3}{R^3}\frac{1}{k}\int_0^R r\sin krdr = \frac{3}{R^3}\frac{1}{k}(-)\frac{d}{dk}\int_0^R \cos krdr$$

$$F(k^2) = \frac{3(\sin(Rk) - Rk\cos(Rk))}{R^3k^3} \Rightarrow$$

$$F(q^2) = \frac{3(\sin(R(q/\hbar)) - R(q/\hbar)\cos(R(q/\hbar)))}{R^3(q/\hbar)^3}. \qquad (2.116)$$

Taking now $R \approx 1.25A^{\frac{1}{3}}$ fm we obtain the form factors shown in Fig. 2.28.

(2) Proceeding as in the previous problem show that the following pairs, density and form factor hold:

- Yukawa like

$$\rho(r) = \frac{e^{-\frac{r}{a}}}{2a^3},\ F(q^2) = \frac{1}{\left(a^2(q/\hbar)^2 + 1\right)^2}. \qquad (2.117)$$

- Gaussian like

$$\rho(r) = \frac{4e^{-\frac{r^2}{a^2}}}{\sqrt{\pi}a^3},\ F(q^2) = e^{-\frac{1}{4}a^2(q/\hbar)^2}. \qquad (2.118)$$

The above form factors for[9] $a = A^{1/6}$ are illustrated in Fig. 2.29 for various nuclei. Can you identify which curves corresponds to which nucleus and density model?

[9] Strictly speaking this is true in the case of the harmonic oscillator $a = \frac{\hbar c}{\sqrt{\hbar\omega m_N c^2}}$, $\hbar\omega \approx 41$ MeV $A^{-1/3}$. But we use it for the Yukava density as well.

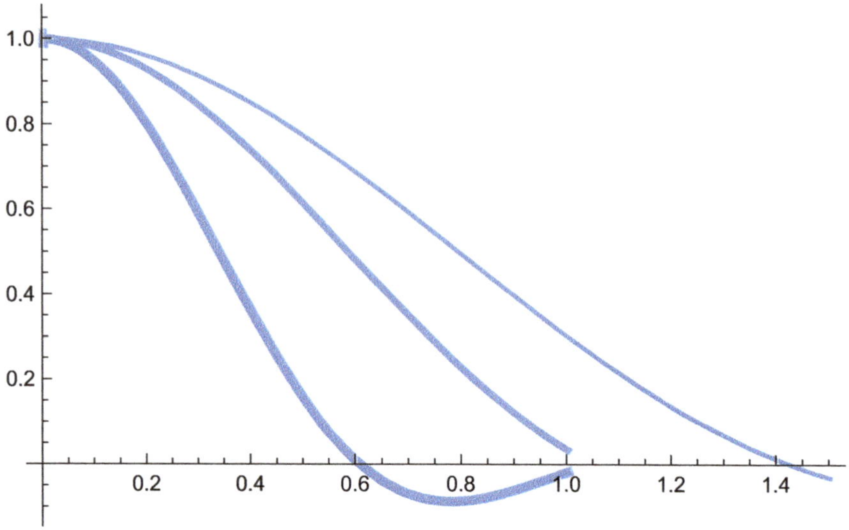

Fig. 2.28: The form factors for $A = 16$ (fine line), $A = 40$ (thin line) and $A = 208$ (thick line) in the case of constant density for low momentum transfers.

(3) Even if you did not find the form factors of the previous problem, can you identify the graphs of Fig. 2.29 by making use of the fact that, if the density is compressed in space, the form factor is extended in momentum space and vice versa. For example to a delta function density, point particle, the form factor is unity (it extends all over the momenta)

(4) Repeat the procedure outlined above in the density of (2.118) for the densities obtained from the $1s$ and $1p$ wave functions of Eq. (2.154).

(5) We have seen that the form factor can be determined from Eq. (2.8), if the cross section for a two particle collision is measured experimentally and the corresponding one for target and projectile both point-like can be calculated.

One interesting quantity is the root mean square distance $\langle r^2 \rangle$. There exist two ways of obtaining it:

i) directly, of course, from the nuclear density, if it is known.

ii) From the form factor, if it is known.

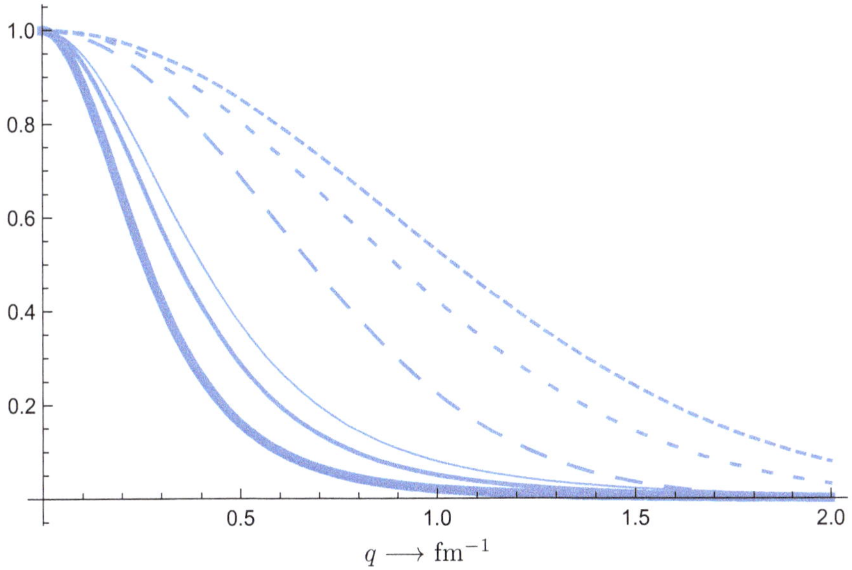

Fig. 2.29: Shown are the form factors associated with the Yukawa and Gaussian densities for various nuclei. Those with one density are continuous and those with the other appear dashed. In each case there appear three graphs corresponding to the nuclei with $A = 16, 40, 208$.

In fact

$$\begin{aligned} F(q^2) &= \int_0^\infty \frac{\sin(q/\hbar)r}{(q/\hbar)r} r^2 \rho(r) dr \\ &= \sum_{\ell=0}^{\infty} (-1)^\ell \frac{1}{(2\ell+1)!} (q/\hbar)^{2\ell} \int_0^\infty r^{2\ell} \rho(r) dr \\ &= \sum_{\ell=0}^{\infty} (-1)^\ell \frac{1}{(2\ell+1)!} (q/\hbar)^{2\ell} \langle r^{2\ell} \rangle = 1 - \frac{1}{6}(q/\hbar)^2 \langle r^2 \rangle + \cdots \end{aligned} \tag{2.119}$$

On the other hand a Taylor expansion yields:

$$F(q^2) = 1 + \frac{1}{2} \left. \frac{d^2 F}{dq^2} \right|_{q=0} q^2 + \cdots \tag{2.120}$$

Combining the two equations we get

$$\langle r^2 \rangle = -3\hbar^2 \left. \frac{d^2 F}{dq^2} \right|_{q=0} \tag{2.121}$$

which proves our assertion. Similar relations hold for the higher moments in the series.

For this to be useful experimentally, the form factor must be accurately measured at low momentum transfers. Thus:
i) Obtain and compare the results by both methods for the densities given in the above problems.
ii) Confirm that $\sqrt{\langle r^2 \rangle}$ in the case of the dipole form factor for the proton, Eq. (2.10), is given by

$$\sqrt{\langle r^2 \rangle} = 2\sqrt{3}a = 2\sqrt{3}\frac{\hbar c}{M_A c^2} \approx 0.8 \text{ fm}. \tag{2.122}$$

(6) Oscillating form factors.
In Fig. 2.28 we exhibited the low momentum form factor associated with the constant density given by Eq. (2.116). It is clear that at high momenta this form factor can exhibit oscillations. The correct picture is given in Fig. 2.30.

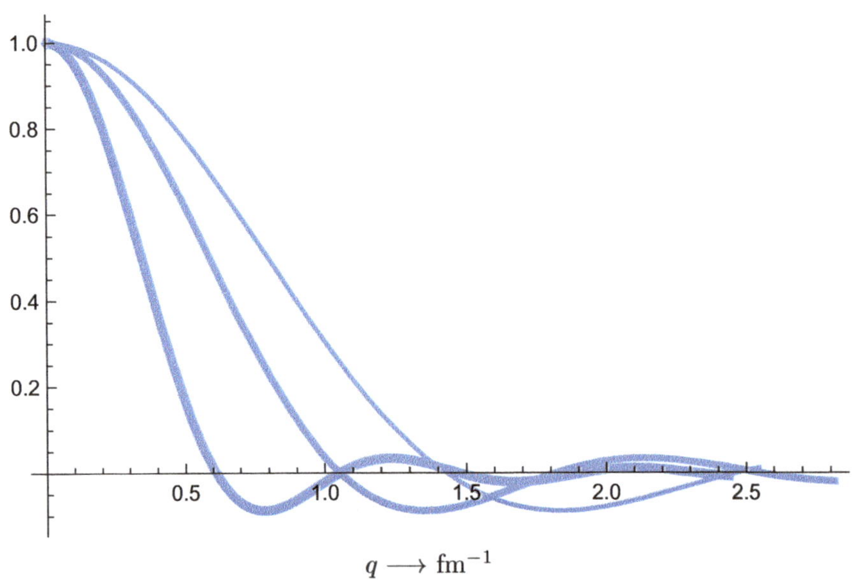

Fig. 2.30: The form factors for $A = 16$ (fine line), $A = 40$ (thin line) and $A = 208$ (thick line) in the case of constant density. One can see the oscillations at high momenta.

(7) The size of the nucleus can be determined mainly by the following three ways: a) electron scattering experiments, b) the energy levels

of muonic atoms, and c) the ground state energies of the isotropic spin multiplet. Discuss each case and show the dependence in each case to the radius of the nucleus R.

Solution:

a) The nucleus form factor is given by the formula

$$F(q^2) = \frac{(d\sigma)_{\rm exp}}{(d\sigma)_{\rm p}} \tag{2.123}$$

where q is the momentum transferred to the nucleus, $(d\sigma)_{\rm exp}$ is the experimental value of the differential cross section and $(d\sigma)_{\rm p}$ is the theoretical differential cross section where the nucleus is considered as a point particle. Now, considering that $\rho(\mathbf{r})$ is the charge density distribution of the nucleus, then in the framework of the Born approximations we get

$$\begin{aligned}
F(q^2) &= \int \rho(\mathbf{r}) e^{i\mathbf{q}\cdot\mathbf{r}} d^3\mathbf{r} \approx \int \rho(r) \left[1 + \frac{1}{2}(i\mathbf{q}\cdot\mathbf{r})^2\right] d^3\mathbf{r} \\
&= 1 - \frac{1}{2}\int \rho(r)(\mathbf{q}\cdot\mathbf{r})^2 d^3\mathbf{r} \\
&= 1 - \frac{q^2}{2}\int_{r=0}^{\infty} \rho(r) r^4 dr \cdot \int_{\phi=0}^{\pi} d\phi \cdot \int_{\theta=0}^{\pi} \cos^2\theta \sin\theta d\theta \\
&= 1 - \frac{2\pi q^2}{3}\int_{r=0}^{\infty} \rho(r) r^4 dr = 1 - \frac{1}{6} q^2 \langle r^2 \rangle, \quad \langle r^2 \rangle \int \rho(r) r^2 d^3\mathbf{r}
\end{aligned} \tag{2.124}$$

where we considered that the charge density exhibits a spherical distribution ($\rho(\mathbf{r}) = \rho(r)$) and moreover $\mathbf{q}\cdot\mathbf{r} \ll 1$ and consequently in the series expansion of $e^{i\mathbf{q}\cdot\mathbf{r}}$ we keep just the first two terms. In addition, without loss of generality we considered that the vector $\mathbf{q}$ is parallel to the z-axis. Experimental measurements of $(d\sigma)_{\rm exp}$ leads to the estimation of $F(q^2)$ and consequently according to Eq. (2.124) on the radius of the nucleus $R = \langle r^2 \rangle$.

b) This method based on the measure of the differences between the excited states and the ground state of a muonic atoms. For example, since the muonic mass is $m_\mu \approx 210 m_e$ the radius of the muonic atom is approximately $R_\mu \approx (1/210)a_0$ (where a_0 is the Bohr radius). The Hamiltonian of the s-state where the nucleus can be considered as a point charge then

$$\hat{H} = -\frac{1}{2m_\mu}\nabla^2 + V_0(r), \quad V_0(r) = -\frac{e^2}{r}. \tag{2.125}$$

However, if we considered the more realistic case where the nuclear charge is uniformly distributed in a sphere of radius R, then we get for the potential $V(r)$

$$V(r) = \begin{cases} -\dfrac{e^2}{2R^3}(3R^2 - r^2), & 0 < r < R \\ \\ -\dfrac{e^2}{r}, & r > R. \end{cases} \tag{2.126}$$

The perturbation on the initial hamiltonian is given by

$$\hat{H}' = V(r) - V_0(r) = \begin{cases} -\dfrac{e^2}{2R^3}(3R^2 - r^2) + \dfrac{e^2}{r}, & 0 < r < R \\ \\ 0, & r > R, \end{cases} \tag{2.127}$$

and the contribution of the perturbation on the energy reads

$$\Delta E = \langle \Psi_0 | \hat{H}' | \Psi_0 \rangle = 4\pi \int_{r=0}^{R} |\Psi_0(r)|^2 \hat{H}' r^2 dr. \tag{2.128}$$

Now, considering that the unperturbed wave function Ψ_0 of the ground state is

$$\Psi_0(r) = \sqrt{\frac{1}{\pi a_\mu^3}} e^{-r/a_\mu}$$

and to a very good approximation that $R/a_\mu \ll 1$, that is

$$e^{-2r/a_\mu} \approx 1 - \frac{2r}{a_\mu},$$

then easily we get

$$\Delta E = \frac{2}{5}\left(\frac{e^2}{2a_\mu}\right)\left(\frac{R}{a_\mu}\right)^2.$$

Now, since we are able to measure the energy E_X of the X-rays emitted in the transition form the first excited state to the ground state

$$E_X = (E_1 - E_0) - \frac{2}{5}\left(\frac{e^2}{2a_\mu}\right)\left(\frac{R}{a_\mu}\right)^2,$$

the values of the radius R can be inferred.

c) We consider that the nuclear structure of the same isotopic spin multiplet are the same so that the mass difference in the multiplet arises from electromagnetic interactions and the proton-neutron mass difference. Thus we have

$$\begin{aligned}\Delta E &\equiv [M(Z,A) - M(Z-1,A]\,c^2 = \Delta E_e - (m_n - m_p)c^2 \\ &= \frac{3c^2}{5R}\left[Z^2 - (Z-1)^2\right] - (m_n - m_p)c^2 \qquad (2.129)\end{aligned}$$

and consequently for known values of Z we get

$$R = \frac{1}{\Delta E}\left(\frac{3c^2}{5}\left[Z^2 - (Z-1)^2\right] - (m_n - m_p)c^2\right). \qquad (2.130)$$

(8) Calculate the electrostatic energy of a charge q distributed uniformly throughout a sphere of radius R. As an application, since the ground state for the *mirror* nuclei $^{27}_{14}$Si and $^{27}_{13}$Al are the same, estimate their radius, if their mass differencc is 6 MeV.

Solution:

The electric field intensity at distance r form the center of uniformly charged sphere is given by

$$E(r) = \begin{cases} \dfrac{qr}{2R^3}, & r < R \\ \\ \dfrac{q}{r^2}, & r > R. \end{cases} \qquad (2.131)$$

The electrostatic energy $\mathcal{W}$ which corresponds to the above electric field intensity is determined by employing Eq. (2.131) as following

$$\begin{aligned}\mathcal{W} &= \int_V \frac{1}{8\pi}E^2 \mathrm{d}V = \frac{q^2}{8\pi}\left[\frac{4\pi}{R^6}\int_{r=0}^{R} r^4 dr + 4\pi\int_{r=R}^{\infty}\frac{1}{r^2}dr\right] \\ &= \frac{q^2}{2}\left[\frac{1}{R^6}\int_{r=0}^{R} r^4 dr + \int_{r=R}^{\infty}\frac{1}{r^2}dr\right] = \frac{q^2}{2}\left(\frac{1}{5R} + \frac{1}{R}\right) = \frac{3q^2}{5R}.\end{aligned} \qquad (2.132)$$

Considering that the mass difference between the two nuclei is due just to the difference in electrostatic energy we get

$$\Delta\mathcal{W} = \frac{3e^2}{5R}\left(Z_1^2 - Z_2^2\right). \qquad (2.133)$$

Now, since $Z_1 = 14$, $Z_2 = 13$, $\Delta\mathcal{W} = 6$ MeV we found

$$\begin{aligned}R &= \frac{3e^2}{5\Delta\mathcal{W}}(14^2 - 13^2) = \frac{3\hbar c}{5\Delta\mathcal{W}}\left(\frac{e^2}{\hbar c}\right)(14^2 - 13^2) \\ &= \frac{3 \times 197.327 \text{ MeV fm}}{5 \times 6 \text{ MeV}} \times \left(\frac{1}{327}\right) \times (14^2 - 13^2) = 3.88 \text{ fm}.\end{aligned} \qquad (2.134)$$

(9) Using the Bethe-Weizsäcker formula for the binding energy to a) derive a relationship between A and Z for naturally occurring isobaric nuclei b) use the Fermi gas model to estimate the value of the asymmetry energy coefficient $a_4 = a_{\text{asym}}$, assuming $A \neq 2Z$ and also that the radius of the nucleus is given by the simple formulae.

Solution:

a) The Bethe-Weizsäcker formula for the binding energy of a nucleus reads

$$\mathcal{BE}(A,Z) = a_1 A - a_2 A^{2/3} - a_3 Z^2 A^{-1/3} - a_4 (A-2Z) A^{-1} \quad (2.135)$$

where a_1, a_2, a_3, a_4 are the volume, surface, Coulomb and symmetry energy coefficients respectively. In the case of isobaric nuclei for a fixed A and different Z, the stable nuclides should satisfy the relation

$$\frac{\partial(\mathcal{BE}(A,Z))}{\partial Z} = -2A^{-1/3} a_3 Z + 4a_4 A^{-1}(A-2Z) = 0 \quad (2.136)$$

which leads to

$$Z = \frac{A}{2 + \frac{a_3}{2a_4} A^{2/3}}$$

or considering that $a_3 = 0.714$ and $a_4 = 23.2$ then

$$Z = \frac{A}{2 + 0.0154 A^{2/3}}.$$

b) The total energy of a Fermi gas (protons or neutrons gas) of volume V at zero temperature is given by

$$E = \frac{2V}{h^3} \frac{4\pi}{3} \frac{p_0^3}{2m} \quad (2.137)$$

while the particle number N (for example in the case of neutron gas) is given by

$$N = \frac{4\pi}{3} \frac{2V}{h^3} p_0^3 \quad (2.138)$$

and consequently the corresponding momentum

$$p_0 = h \left(\frac{3}{8\pi} \frac{N}{V} \right)^{1/3}. \quad (2.139)$$

The energy is written also using Eqs. (2.138) and (2.139)

$$E = \frac{3}{40} \left(\frac{3}{\pi} \right)^{2/3} \frac{h^2}{m} V^{-2/3} N^{5/3}. \quad (2.140)$$

Considering that in the nucleus volume V neutrons and protons gases coexist independently then the total energy is given by

$$E = \frac{3}{40}\left(\frac{3}{\pi}\right)^{2/3}\frac{h^2}{m}\left(\frac{N^{5/3}}{V^{2/3}} + \frac{Z^{5/3}}{V^{2/3}}\right). \tag{2.141}$$

Considering in addition that

$$V = \frac{4\pi}{3}R_0^3 A, \; R_0 = 1.2 \text{ fm}$$

we found

$$E = \mathcal{C}\left(\frac{N^{5/3}}{V^{2/3}} + \frac{Z^{5/3}}{V^{2/3}}\right), \quad \mathcal{C} \approx 31.7 \text{ MeV}. \tag{2.142}$$

Moreover, assuming that for stable nuclei $N \approx Z$ and setting

$$N = \frac{1}{2}A(1+\delta/A), \qquad Z = \frac{1}{2}A(1-\delta/A), \quad \delta = N - Z, \quad \delta/A \ll 1$$

a Taylor expansions leads to the

$$\left(1 + \frac{\delta}{A}\right)^{5/3} = 1 + \frac{5\delta}{3A} + \frac{5\delta^2}{9A^2} - \dots \tag{2.143}$$

$$\left(1 - \frac{\delta}{A}\right)^{5/3} = 1 - \frac{5\delta}{3A} + \frac{5\delta^2}{9A^2} + \dots \tag{2.144}$$

and also to

$$N^{5/3} + Z^{5/3} \approx 2\left(\frac{A}{2}\right)^{5/3}\left(1 + \frac{5\delta^2}{9A^2}\right).$$

The total energy reads now

$$E \approx \frac{\mathcal{C}}{2^{2/3}}A + \frac{5}{9}\frac{C}{2^{-2/3}}\frac{(N-Z)^2}{A}. \tag{2.145}$$

The symmetry energy coefficients is

$$a_4 = \frac{5}{9}\frac{C}{2^{-2/3}} \approx 11 \text{ MeV}$$

which is by a factor 2 smaller that the realistic value.

(10) The numbers of protons an neutrons are toughly equal for stable lighter nuclei while the umber of neutrons is substantially greater than the number of protons for stable heavy nuclei. Moreover, in light nuclei the energy required to remove a proton or neutron from the nucleus is roughly the same while in heavy nuclei required more energy to remove a proton than a neutron. Can you explain this behavior?

Solution:

The energy which is required to remove a nucleon (proton or neutron) for a stable nucleus (Z, A) is given by

$$\mathcal{E}_{\mathrm{p}} = \mathcal{BE}(Z, A) - \mathcal{BE}(Z-1, A-1) \tag{2.146}$$

and

$$\mathcal{E}_{\mathrm{n}} = \mathcal{BE}(Z, A) - \mathcal{BE}(Z-1, A-1) \tag{2.147}$$

respectively. Using the Bethe-Weizsäcker formula for the binding energy $\mathcal{BE}$ of a nucleus, that is

$$\mathcal{BE}(Z, A) = a_v A - a_s A^{2/3} - a_c Z^2 A^{-1/3} - a_a \left(\frac{A}{2} - Z\right)^2 A^{-1} + a_p \delta A^{-1/2} \tag{2.148}$$

where the last term related with the pairing forces then we get

$$\mathcal{E}_{\mathrm{p}} - \mathcal{E}_{\mathrm{n}} = -a_c(2Z-1)(A-1)^{-1/3} + a_a(A-2Z)(A-1)^{-1} \tag{2.149}$$

where $a_c = 0.714$ MeV and $a_a = 92.8$ MeV. We have found also that in the case of stable nuclei we have (see problem 3)

$$Z = \frac{A}{2 + \frac{2a_c}{a_a}A^{2/3}} \approx \frac{A}{2}\left(1 - \frac{2a_c}{a_a}A^{2/3}\right). \tag{2.150}$$

Now, finally we get

$$\mathcal{E}_{\mathrm{p}} - \mathcal{E}_{\mathrm{n}} \approx \frac{a_c}{A-1}\left[A^{5/3} - (A-1)^{5/3} + \frac{a_c}{a_a}A^{5/3}(A-1)^{2/3}\right]. \tag{2.151}$$

Obviously, for heavy nuclei where $A \gg 1$ we get

$$\mathcal{E}_{\mathrm{p}} - \mathcal{E}_{\mathrm{n}} \approx 5.5 \times 10^{-3} A^{4/3},$$

that is in heavy nucleus more energy is needed to dissociate protons than neutrons.

(11) The most stable nucleus with $A = 101$ is $^{101}_{44}Ru$.
Using this knowledge, the semi-empirical mass formula and the Coulomb coefficient in this formula, calculate the asymmetry parameter a_4 ($a_3 = 0.697$ MeV).

Solution:
The stability condition is given by the equation $\dfrac{\partial \mathcal{BE}(Z,A)}{\partial Z} = 0.$
Considering that according to the semi-empirical formula

$$\mathcal{BE}(A,Z) = a_1 A - a_2 A^{2/3} - a_3 Z^2 A^{-1/3} - a_4 \left(\frac{A}{2} - Z\right)^2 A^{-1} \tag{2.152}$$

where a_1, a_2, a_3, a_4 are the volume, surface, Coulomb and symmetry energy coefficients respectively. The stability condition leads to the equation

$$-\frac{2a_3}{A^{1/3}} - \frac{a_4(A-2Z)}{A} = 0 \Rightarrow a_4 = \frac{2a_3 Z A^{2/3}}{(A-2Z)} \approx 6\text{MeV}. \tag{2.153}$$

(12) Using the results of Appendix A, Chapter 14, Eq. (14.15), show that the lowest normalized radial Harmonic oscillator wave functions are as follows:

$$\psi_{0s}(x) = \frac{2e^{-\frac{x^2}{2}}}{\sqrt[4]{\pi}},\ \psi_{1s}(x) = \frac{2\sqrt{\frac{2}{3}}e^{-\frac{x^2}{2}}\left(\frac{3}{2} - x^2\right)}{\sqrt[4]{\pi}},$$
$$\psi_{2s}(x) = \frac{2\sqrt{\frac{2}{3}}e^{-\frac{x^2}{2}}\left(\frac{3}{2} - x^2\right)}{\sqrt[4]{\pi}},$$
$$\psi_{0p}(x) = \frac{2\sqrt{\frac{2}{3}}e^{-\frac{x^2}{2}}x}{\sqrt[4]{\pi}},\ \psi_{1p}(x) = \frac{4e^{-\frac{x^2}{2}}x\left(\frac{5}{2} - x^2\right)}{\sqrt{15}\sqrt[4]{\pi}},$$
$$\psi_{0d}(x) = \frac{4e^{-\frac{x^2}{2}}x^2}{\sqrt{15}\sqrt[4]{\pi}},\ \psi_{1d}(x) = \frac{4\sqrt{\frac{2}{105}}e^{-\frac{x^2}{2}}x^2\left(\frac{7}{2} - x^2\right)}{\sqrt[4]{\pi}}$$
$$\psi_{0f}(x) = \frac{4\sqrt{\frac{2}{105}}e^{-\frac{x^2}{2}}x^3}{\sqrt[4]{\pi}},\ \psi_{1f}(x) = \frac{8e^{-\frac{x^2}{2}}x^3\left(\frac{9}{2} - x^2\right)}{3\sqrt{105}\sqrt[4]{\pi}} \tag{2.154}$$

in all the above equations $x = r/b$.

Plot the density associated with each of the above wave functions. Make sure it is normalized to unity.

(13) Write the total wave function associated with each of the above functions, by including the angular part associated with angular momentum projection $m = 0$.

(14) As it is shown in the Appendix A, Chapter 14, Eq. (14.19), that in the presence of spin orbit potential $a_\ell U_1(r)$ in addition to the spin independent harmonic potential $U(r)$ one must solve only one differential equation. This Equation for $U_1(r) \neq U(r)$ can only be solved numerically. A simple solution can be found, if one assumes that $U_1(r) = -CU(r)$, C = constant, $C > 0$, $U(r)$ = harmonic. Show that, for sufficiently small C, the solution is that of the harmonic oscillator found in the Appendix A with the following modifications:

$$\hbar\omega \to \hbar\omega(1 + \delta(C, \ell, j)),\ b \to \frac{b}{\sqrt{1 + \delta(C, \ell, j)}},$$

$$\delta(C, \ell, j) = \begin{cases} -C\ell, & j = \ell + 1/2 \\ C(\ell + 1), & j = \ell - 1/2 \end{cases}.$$

Table 2.2: The energy shifts $\delta\epsilon$ in the spin orbit partners for $0p_j$, $1p_j$, $0d_j$ and $1d_j$ from their unperturbed location ϵ_0. All energies in units of $\hbar\omega$.

nlj	ϵ_0	$\delta\epsilon$	nlj	ϵ_0	$\delta\epsilon$
$0p_{3/2}$	5/2	−1/4	$0d_{5/2}$	7/2	−7/10
$0p_{1/2}$	5/2	1/2	$0d_{3/2}$	7/2	21/20
$1p_{3/2}$	9/2	−9/20	$1d_{5/2}$	11/2	−11/10
$1p_{1/2}$	9/2	9/10	$1d_{3/2}$	11/2	33/20

With these modifications and $C = 1/10$ one can proceed to obtain the energy shifts $\delta\epsilon(n_r, \ell, j)$, see Table 2.2, and the wave functions by otherwise retaining their form as give by Eq. (2.154), except possibly for normalization:

i) Obtain the $0p_j$ and $1p_j$, $j = 1/2, 3/2$, and plot them.

ii) Repeat for $0d_j$ and $1d_j$, $j = 3/2, 5/2$.

iii) Confirm that the above wave functions are represented as in Fig. 2.31.

iv) Compare the obtained results with the pertinent ones of Figs. 14.1 and 14.2 of the standard HO wave functions.

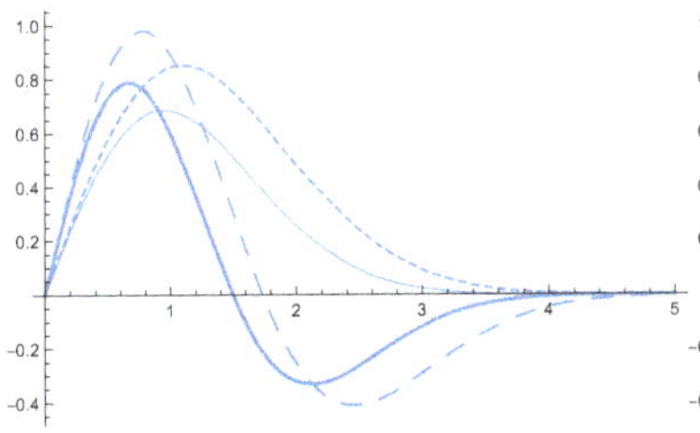

(a)

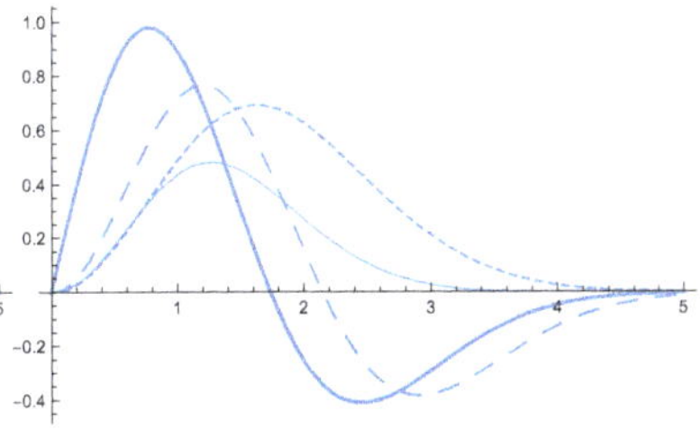

(b)

$q \longrightarrow \text{fm}^{-1}$

Fig. 2.31: (a) The harmonic oscillator wave function for $0p_j$, $j = 3/2$ and $j = 1/2$ indicated by solid and dotted lines respectively. Also $1p_j$ for $j = 3/2$ and $j = 1/2$ for thick solid and dashed lines respectively. (b) The same for $0d_j$ $j = 5/2$, $j = 3/2$ and $1d_j$ $j = 5/2$, $j = 3/2$.

(15) In a good approximation, a medium weight nucleus can be regarded as a cubical box of length equal to the nuclear diameter. Consider a nucleus of iron-56 which has 28 protons and 28 neutrons. Estimate the kinetic energy of the highest energy nucleon. Assume a nuclear diameter of 10^{-12} cm.

Solution:

The potential where a nucleon is moving in the case of a cubic box is given by

$$V(x,y,z) = \begin{cases} \infty, & |x|, |y|, |z| > \dfrac{L}{2} \\ \\ 0, & |x|, |y|, |z| < \dfrac{L}{2} \end{cases} \tag{2.155}$$

where L is the nuclear diameter. The Schrödinger equation reads

$$-\frac{\hbar^2}{2m}\nabla^2\Psi(x,y,z) + V(x,y,z)\Psi(x,y,z) = E\Psi(x,y,z) \tag{2.156}$$

and can be easily solved considering the method of separated in the variables by letting

$$\Psi(x,y,z) = \Psi(x)\Psi(y)\Psi(z).$$

The separation leads to the one-dimensional Schrödinger equation

$$-\frac{\hbar^2}{2m}\frac{d^2}{dx_i^2}\Psi(x_i) + V(x_i)\Psi(x_i) = E_i\Psi(x_i), \quad x_1 = x, x_2 = y, x_3 = z \tag{2.157}$$

where

$$V(x_i) = \begin{cases} \infty, & |x_i| > \frac{L}{2} \\ 0, & |x_i| < \frac{L}{2} \end{cases} \tag{2.158}$$

and $E = E_1 + E_2 + E_3$. Solving the equations we find

$$\Psi(x_i) = A_i \sin(k_i x_i) + B_i \cos(k_i x_i), \quad k_i = \frac{\sqrt{2mE_i}}{\hbar}. \tag{2.159}$$

Moreover, the boundary conditions $\Psi(x_i)|_{x_i=\pm\frac{L}{2}} = 0$ leads to

$$\Psi(x_i) = \begin{cases} A_i \sin\left(\frac{n\pi}{L} x_i\right), & \text{for even } n \\ B_i \cos\left(\frac{n\pi}{L} x_i\right), & \text{for odd } n \end{cases} \tag{2.160}$$

and also

$$E_{x_i} = \frac{\pi^2 n_{x_i}^2 \hbar^2}{2mL^2}, \quad n_{x_i} = 1, 2, 3, \ldots \tag{2.161}$$

and in total

$$E = E_0\left(n_x^2 + n_y^2 + n_z^2\right), \quad E_0 = \frac{\pi^2\hbar^2}{2mL^2} \approx 2.04 \text{ MeV}. \tag{2.162}$$

In the case of the nucleus ^{56}Fe, and with respect to the Pauli principle (each state can accommodate one pair of neutrons and protons and one of protons). According to the Table 2.3 we find that the energy of the highest energy nucleon is $E_{\max} = 14E_0 = 2.04 \times 14 = 28.6$ MeV.

(16) Obtain the eigenvalues and eigenvectors of the matrices given by Eqs. (2.65). Construct the diagonalizing matrix S.
Hint:

$$0^+ : \lambda_1 = -10.75,\, \lambda_2 = -3.77,\, S = \begin{pmatrix} -0.730176 & 0.68326 \\ -0.68326 & -0.730176 \end{pmatrix} \tag{2.163}$$

$$2^+ : \lambda_1 = -7.28,\, \lambda_2 = -3.29,\, S = \begin{pmatrix} 0.254891 & -0.96697 \\ 0.96697 & 0.254891 \end{pmatrix}. \tag{2.164}$$

Table 2.3: Energies of $^{56}_{28}Fe$ in the case of a cubical box of length equal to the nuclear diameter 10^{-12} cm.

(n_x, n_y, n_z)	Numbers of nucleons	Energy
$(1,1,1)$	4	$3E_0$
$(2,1,1)$	4	$6E_0$
$(1,2,1)$	4	$6E_0$
$(1,1,2)$	4	$6E_0$
$(2,2,1)$	4	$9E_0$
$(1,2,2)$	4	$9E_0$
$(2,1,2)$	4	$9E_0$
$(3,1,1)$	4	$11E_0$
$(1,3,1)$	4	$11E_0$
$(1,1,3)$	4	$11E_0$
$(2,2,2)$	4	$12E_0$
$(1,2,3)$	4	$14E_0$
$(1,3,2)$	4	$14E_0$
$(2,3,1)$	4	$14E_0$
$(2,1,3)$	4	$14E_0$
$(3,1,2)$	4	$14E_0$
$(3,2,1)$	4	$14E_0$

(17) Do the same for the matrix given by Eq. (2.66).
Hint:

$$1^+ : \lambda_1 = -8.75,\ \lambda_2 = -6.66,\ \lambda_3 = -3.18,$$

$$S = \begin{pmatrix} -0.633076 & -0.179704 & -0.752942 \\ 0.577733 & -0.757071 & -0.305071 \\ 0.515208 & 0.628133 & -0.583104 \end{pmatrix}. \qquad (2.165)$$

(18) Express the eigenvectors obtained above in $L-S$ coupling scheme utilizing the transformations found in Section 2.5.4.2. Compare the results you found with those of the more realistic calculations given in Section 13.9.

(19) Find the eigenvalues and eigenvectors of the $A = 14$ system discussed in the text, using a Gaussian type interaction.

(20) Repeat the previous problem not by expanding the HO wave functions, but by expending the interaction in terms of products of functions of $\mathbf{r}_1$ and $\mathbf{r}_2$.

Chapter 3

Introduction to special theory of relativity

In this chapter we are going to present some features of special theory of relativity, which are of interest to nuclear and particle physics. Features of relativistic quantum mechanics, like the Dirac theory, will be given in the Appendix C, see Chapter 16. Here emphasis will be given on Lorentz transformations and Lorentz invariant quantities, appearing in particle dynamics.

3.1 Lorentz transformations

We will consider an inertial frame[1] Σ and another one Σ' moving with respect to Σ with constant velocity V (see Fig. 3.1). The Lorentz transformation connects the coordinates ($x^0 = ct$, x^1, x^2 and x^3) as measured in Σ, with the corresponding ones ($x'^0 = ct'$, x'^1, x'^2 and x'^3) as measured in Σ', demanding the invariance of the quantity:

$$(x^0)^2 - (x^1)^2 + (x^2)^2 + (x^3)^2) = (x'^0)^2 - [(x'^1)^2 + (x'^2)^2 + (x'^3)^2]. \quad (3.1)$$

The relevant transformation is:

$$\begin{aligned} x'^0 &= \gamma(x^0 - \beta x^3), \quad \text{or } t' = \gamma\left(t - \frac{V}{c^2}x^3\right) \\ x'^3 &= \gamma(-\beta x^0 + x^3) \\ x'^1 &= x^1 \\ x'^2 &= x^2 \end{aligned} \quad (3.2)$$

[1] A system is inertial, if a particle viewed in this system is moving with constant velocity provided that no force is exerted on it. Furthermore, if we know that a system is inertial, any other system moving with respect to it with constant velocity is also an inertial frame.

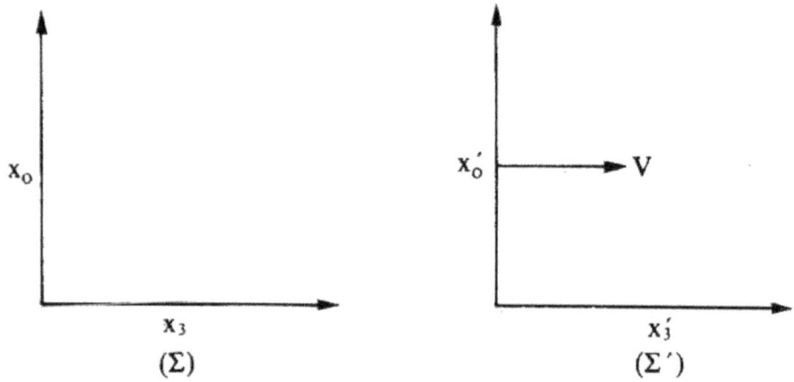

Fig. 3.1: Two frames moving with respect to each other with constant velocity V.

where

$$\beta = \frac{V}{c} \quad \text{and} \quad \gamma = \frac{1}{\sqrt{1-\beta^2}}. \tag{3.3}$$

If we now define

$$x^\mu \equiv (x^0, \mathbf{x}) \equiv (x^0, x^1, x^2, x^3) \equiv x,$$

the transformation (3.2) take the form

$$x'^\mu = \sum_\nu \Lambda^\mu_\nu x^\nu \quad (\text{or } x' = \Lambda x) \tag{3.4}$$

where Λ^μ_ν is a matrix with μ labeling the rows and ν the columns, i.e.

$$\Lambda = \begin{pmatrix} \gamma & 0 & 0 & -\beta\gamma \\ 0 & 1 & 0 & 0 \\ 0 & 0 & 1 & 0 \\ -\beta\gamma & 0 & 0 & \gamma \end{pmatrix}. \tag{3.5}$$

Suppose now that Σ' is moving with constant velocity $\mathbf{V}$ in arbitrary direction. Then we decompose $\mathbf{x}$ into one component parallel to the velocity and one perpendicular to it, namely

$$\mathbf{x}_{||} = (\mathbf{V}\cdot\mathbf{x})\frac{\mathbf{V}}{c^2} \qquad \mathbf{x}_\perp = \mathbf{x} - \mathbf{x}_{||},$$

the Lorenz transformation becomes

$$\mathbf{x}'_{||} = \gamma(\mathbf{x}_{||} - \boldsymbol{\beta} x^0),\ \boldsymbol{\beta} = \frac{\mathbf{V}}{c} \tag{3.6}$$

$$\mathbf{x}'_\perp = \mathbf{x}_\perp \tag{3.7}$$

$$x'^0 = \gamma(x^0 - \boldsymbol{\beta}\cdot\mathbf{x}). \tag{3.8}$$

3.1.1 *Lorentz invariant quantities*

The vector x^μ transforming as in Eq. (3.2) or (3.4) is called contravariant vector. Equation (3.1) can also be written as

$$x^\mu g_{\mu\nu} x^\nu = x'^\mu g_{\mu\nu} x'^\nu,$$

with the Einstein convention understood, i.e. repeated indices imply a summation over them. $g_{\mu\nu}$ is the metric tensor given by:

$$(g_{\mu\nu}) = \begin{pmatrix} 1 & 0 & 0 & 0 \\ 0 & -1 & 0 & 0 \\ 0 & 0 & -1 & 0 \\ 0 & 0 & 0 & -1 \end{pmatrix} \qquad \mu, \nu = 0, 1, 2, 3 \tag{3.9}$$

$g_{\mu\nu}$ is often called metric. Another metric is given by $\tilde{g}_{\mu\nu} = -g_{\mu\nu}$. We note that unlike the metric involved in Euclidian spaces, it is not possible to transform it into a multiple of the identity through a similarity transformation.

With the aid of the metric tensor we define the x_μ:

$$x_\mu = g_{\mu\nu} x^\nu.$$

It is easy to see that

$$x_\mu = (x^0, -\mathbf{x}).$$

Furthermore the invariant quantity $x^\mu g_{\mu\nu} x^\nu$ takes the form

$$x^\mu g_{\mu\nu} x^\nu = x_\mu x^\mu = x_0^2 - \mathbf{x}^2. \tag{3.10}$$

This coincides with the basic invariant quantity of Eq. (3.1), i.e. Eq. (3.10) is manifestly Lorentz invariant. These equations can be written in infinitesimal form:

$$ds^2 = c^2(dt)^2 - (dx^1)^2 - (dx^2)^2 - (dx^3)^2 = (cdt)^2 - (d\mathbf{x})^2, \tag{3.11}$$

which in practice is more useful.

The elementary proper time $d\tau$ is defined by

$$d\tau = \frac{\sqrt{(cdt)^2 - (d\mathbf{x})^2}}{c}. \tag{3.12}$$

It has the meaning of the time difference between two events, if they occur at the same point of space, $d\mathbf{x} = 0$.

We can now define the velocity 4-vector by:

$$u^\mu = \frac{dx^\mu}{d\tau}.$$

The quantity u^μ indeed transforms like x^μ, since $d\tau$ is invariant. Similarly one can define u_μ, which, under Lorentz transformations, has the same transformation properties with x_μ. As a result the $u^\mu u_\mu$ is invariant.

Let us now consider a particle moving with velocity $\boldsymbol{v}$ with respect to Σ, e.g. in the direction of x^3. In the system Σ the event is described by $x^\mu = (x^0, \mathbf{x})$ and we have

$$d\tau = \frac{1}{c}\sqrt{(cdt)^2 - (d\mathbf{x})^2} = dt\sqrt{1 - \left(\frac{d\mathbf{x}}{dt}\right)^2} = dt\sqrt{1-\beta^2} = \frac{dt}{\gamma}$$

that is

$$u^\mu = \gamma\frac{dx^\mu}{dt} = \gamma(c, \boldsymbol{v})$$

$$u^\mu = \gamma(c, -\boldsymbol{v}) \tag{3.13}$$

$$u^\mu u_\mu = \gamma^2(c^2 - v^2) = c^2\gamma^2(1-\beta^2) = c^2.$$

We are now ready to define the momentum 4-vector as follows:

$$p^\mu = mu^\mu = (\gamma mc, \gamma m\boldsymbol{v}), \quad m = \text{mass} \tag{3.14}$$

$$p_\mu = mu_\mu = (\gamma mc, -\gamma m\boldsymbol{v}),$$

yielding

$$p^2 = p^\mu p_\mu = (mc)^2. \tag{3.15}$$

Furthermore, if we define the energy: $E = cp^0 = \gamma mc^2$, we see that it depends an the velocity $\boldsymbol{v}$. For $\boldsymbol{v} = 0$ we get $E = mc^2$. This is called **rest energy**. It depends, of course, on the mass of the particle. The momentum of the particle is

$$\mathbf{p} = \gamma m\boldsymbol{v}.$$

Hence

$$p^2 = \left(\frac{E}{c}\right)^2 - \mathbf{p}^2$$

that is

$$\left(\frac{E}{c}\right)^2 - \mathbf{p}^2 = (mc)^2 \quad \text{(invariant)}. \tag{3.16}$$

This can be written as

$$E = \sqrt{(\mathbf{p}c)^2 + (mc^2)^2}. \tag{3.17}$$

The kinetic energy is defined by

$$K = E - mc^2 = \sqrt{(\mathbf{p}c)^2 + (mc^2)^2} - mc^2. \tag{3.18}$$

In the classical limit ($v << c$) this becomes $K = \mathbf{p}^2/2m$ as expected. The 4-vector p^μ transforms under Lorentz transformations like x^μ, Eq. (3.2), with the correspondence

$$\frac{E}{c^2} \leftrightarrow t, \quad \frac{E'}{c^2} \leftrightarrow t', \quad \mathbf{p} \leftrightarrow \mathbf{x}, \quad \mathbf{p}' \leftrightarrow \mathbf{x}'.$$

3.1.2 *Covariant and contravariant derivatives*

In general an invariant quantity constructed out of two 4-vectors A^μ and B^μ can be written on the form

$$AB = A_\mu B^\mu = A^\mu g_{\mu\nu} B^\nu = A_0 B_0 - \mathbf{A}.\mathbf{B}. \tag{3.19}$$

One such vector can be the derivative with respect to the coordinates:

$$\partial_\mu = \frac{\partial}{\partial x^\mu} = \left(\frac{\partial}{\partial x^0}, \frac{\partial}{\partial x^1}, \frac{\partial}{\partial x^2}, \frac{\partial}{\partial x^3}\right) = \left(\frac{\partial}{\partial x_0}, -\frac{\partial}{\partial x_1}, -\frac{\partial}{\partial x_2}, -\frac{\partial}{\partial x_3}\right). \tag{3.20}$$

Note the upper and lower indices. To remember this rule think as follows: A term ∂_μ (lower index) is generated by the derivative acting on an invariant $x^\mu x_\mu$. So clearly in the process x^μ must be eliminated. This is what the derivative $\partial/\partial x^\mu$ actually does. Since, by the same token, one can define

$$\partial^\mu = \left(\frac{\partial}{\partial x_0}, \frac{\partial}{\partial x_1}, \frac{\partial}{\partial x_2}, \frac{\partial}{\partial x_3}\right) \tag{3.21}$$

we find

$$\partial^\mu \partial_\mu = \frac{\partial^2}{\partial x_0^2} - \nabla^2 = -\Box^2, \quad \Box^2 = \text{D'Alembertian operator}. \tag{3.22}$$

Chapter 4

Resonance-decay widths

In nuclear and particle physics certain systems are not stable, but they can decay into various other products. Often the decay proceeds via a resonance. One parameter characterizing such systems is their decay time, the decay probability per unit time and the decay width. In this chapter we are going to see how these variables are related.

4.1 The notion of a resonance

The resonance in subatomic physics is a quantum mechanical extension of the classical enhancement in the observation of an event, if the energy of the system becomes appropriate. We will consider the decay through a resonance. For a resonance in elastic scattering see Section 9.9.

It is known from elementary quantum theory that the number of unstable particles decaying in time dt is given by:

$$dN(t) = -\lambda N(t)dt. \tag{4.1}$$

Integrating this we obtain

$$N = N_0 \cdot e^{-\lambda t}. \tag{4.2}$$

The quantity $\tau \equiv 1/\lambda$ is known as the average life time since:

$$\tau = \langle t \rangle = \frac{N_0 \int_0^\infty t e^{-\lambda t} dt}{N_0 \int_0^\infty e^{-\lambda t} dt} = \frac{\frac{1}{\lambda^2}}{\frac{1}{\lambda}} = \frac{1}{\lambda}. \tag{4.3}$$

The half life $T_{1/2}$ is defined so that $N(T_{1/2}) = \frac{N_0}{2}$

$$\frac{N_0}{2} = N_0 e^{-\lambda T_{1/2}},$$

that is

$$T_{1/2} = \frac{1}{\lambda} \ln 2 = \tau \ln 2. \tag{4.4}$$

How are the above related to the wave function describing the particle? In its rest frame ($\mathbf{p} = 0$) it takes the form:

$$\Psi(t) = \Psi(0)e^{-i\frac{E}{\hbar}t}.$$

Thus if the energy is real the probability of finding the particle is a constant independent of time:

$$|\Psi(t)|^2 = |\Psi(0)|^2 \text{ independent of } t.$$

In other words such a particle does not decay, it lives forever.

In the above formalism a decay occurs only if the energy has an imaginary part, that is,

$$E = E_0 - i\frac{\Gamma}{2}, \quad E_0, \Gamma = \text{real},$$

(the factor of $\frac{1}{2}$ was introduced for later convenience). In this case we get

$$|\Psi(t)|^2 = |\Psi(0)|^2 e^{-\frac{\Gamma}{\hbar}t}. \tag{4.5}$$

In other words it is in agreement with Eq. (4.2), if

$$\frac{\Gamma}{\hbar} = \lambda \text{ or } \Gamma\tau = \hbar. \tag{4.6}$$

The wave function $\Psi(t)$ takes the form

$$\Psi(t) = \begin{bmatrix} \Psi(0)e^{-i\frac{E_0}{\hbar}t - \frac{\Gamma}{\hbar}t}, & t > 0 \\ 0, & t < 0 \end{bmatrix}. \tag{4.7}$$

The appearance of a complex energy is at first sight disturbing, since we learned in non relativistic quantum mechanics that the eigenvalues of a hermitian operator are real. Hermiticity, however, is linked with the boundary conditions. In quantum mechanics we assumed that the wave function vanishes at infinity. This, however, does not include the decay, since then there is an outgoing spherical way to infinity. Complex boundary conditions yield complex eigenvalues. An analogous situation occur in relativistic field theory [Majore (2015)]. Now the Feynman propagators can have complex energies.

The above discussion aside, what is the meaning of the parameter Γ? Before we give an answer to this, let us ask ourselves what is the probability for the particle to have energy between E and $E + dE$? To this end we must find the wave function in terms of the frequency ω, with $E = \hbar\omega$. In other words we must find the Fourier transform of Eq. (4.7). This way we find:

$$\begin{aligned}\tilde{\Psi}(\omega) &= \frac{1}{\sqrt{2\pi}}\int_{-\infty}^{+\infty}\Psi(t)e^{i\omega t}dt\\ &= \frac{\Psi(0)}{\sqrt{2\pi}}\int_{0}^{\infty}e^{-i\frac{E_0}{\hbar}t-\frac{\Gamma}{2\hbar}t}e^{i\omega t}dt\\ &= \frac{\Psi(0)}{\sqrt{2\pi}}\left.\frac{e^{i\left(\omega-\frac{E_0}{\hbar}\right)t-\frac{\Gamma}{2\hbar}t}}{\left(\omega-\frac{E_0}{\hbar}\right)i-\frac{\Gamma}{2\hbar}}\right|_0^{\infty}\end{aligned}$$

or

$$\tilde{\Psi}(\omega) = \frac{\Psi(0)}{\sqrt{2\pi}}\frac{i\hbar}{(\hbar\omega - E_0)+i\frac{\Gamma}{2}}. \tag{4.8}$$

Recalling that $E = \hbar\omega$, we get

$$\tilde{\Psi}(E) = \frac{\Psi(0)}{\sqrt{2\pi}}\frac{i\hbar}{(E - E_0)+i\frac{\Gamma}{2}}$$

or

$$|\tilde{\Psi}(E)|^2 = \frac{\hbar^2}{2\pi}\frac{|\Psi(0)|^2}{(E-E_0)^2+\frac{\Gamma^2}{4}}.$$

The condition

$$\int_{-\infty}^{+\infty}|\tilde{\Psi}(E)|^2 dE = 1$$

yields

$$|\tilde{\Psi}(0)|^2 = \frac{\Gamma}{\hbar^2} \tag{4.9}$$

(Eq. (4.9) can be obtained from the normalization condition of Eq. (4.7)). Thus we finally get:

$$\rho(E) = |\tilde{\Psi}(E)|^2 = \frac{\Gamma}{2\pi}\frac{1}{(E-E_0)^2+\left(\frac{\Gamma}{2}\right)^2} \tag{4.10}$$

as the probability of finding the particle with energy between E and $E+dE$. In other words the parameter Γ widens the state of the particle, so that it no longer has definite energy and this leads to its destruction. The width of a state that arises from the decay of the particle is called natural width. The shape of Fig. 4.1β is called Lorentzian or Breit-Wigner. Γ is the width at half the maximum value of $\rho(E)$ (see Fig. 4.1β).

From the above we see that

$$\tau = \frac{\hbar}{\Gamma}. \tag{4.11}$$

4.2 Applications — The determination of very short lifetimes

The notion of resonance is very important in subatomic physics and we are going to encounter it in many places of this book, in particular in Chapter 9. We will see that, from the observation of a resonance, one can infer the mechanism responsible for the process taking place, see Section 9.10.1 and, thus, use it as test of the existing theoretical models. Some particles are even named resonances as in Section 10.1.2. If in a process a resonance is formed, the process proceeds extremely fast. Thus from knowledge of the width Γ of a resonance one can determine the lifetime of particles, which are too short lived for their lifetime to be measured directly.

Thus the fanciest application of the notion of resonances is the determination of very short life times. In fact, as we will see in Section 9.10.2 and Fig. 9.9, even from the first resonance experiment, which yielded $\Gamma = 125$ MeV, a life time measurement of

$$\tau = 6 \times 10^{-24} \text{ s}$$

was obtained for the ρ meson, which is very close to the present mean value of

$$4.5 \times 10^{-24} \text{ s.}$$

Yes, via observation of resonances, physicists can measure such short life times! Great process requires a new idea, not just refining of existing ones!

An avalanche of resonances followed, which, as we will see in Section 10.5 and in Section 10.6, radically changed our view of the world with the introduction of charm and beauty quantum numbers and, consequently, the necessity to introduce new symmetries of nature.

Before concluding this section we should mention that from the relation $\Delta E \cdot \Delta t \approx \hbar$ we see that, in order to measure the energy with accuracy $\Delta E = \Gamma$, the uncertainty in time must be $\Delta t = \tau$.

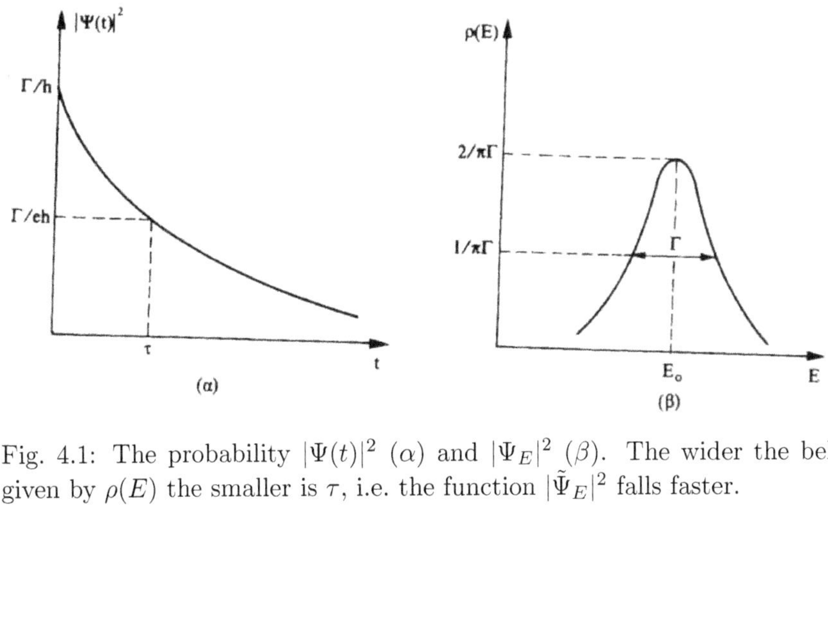

Fig. 4.1: The probability $|\Psi(t)|^2$ (α) and $|\Psi_E|^2$ (β). The wider the bell given by $\rho(E)$ the smaller is τ, i.e. the function $|\tilde{\Psi}_E|^2$ falls faster.

Chapter 5

Elementary scattering theory

A brief presentation of elementary scattering theory will be presented in this chapter. For more details the reader is referred to the literature, e.g. quantum mechanics [Hecht (2000)] and subatomic physics [Frauenfelder and Henley (1974)].

5.1 Elements of scattering theory

We will consider a beam of particles represented by a plane wave traveling in the z direction colliding with a target particle or scattering center. For simplicity we will assume that the target is spinless. The beam can be described as:

$$\psi_j = e^{kz}$$

where

$$k = \frac{1}{\lambda\!\!\!^{-}}, \; \lambda\!\!\!^{-} = \frac{\lambda}{2\pi}, \lambda = \text{the Compton wavelength of the beam particle.}$$

The familiar term $e^{-i\omega t}$ is understood, but has been omitted for brevity. The plane wave will be represented as a superposition of spherical waves:

$$\psi_i = e^{kz} = \sum_{\ell} (2\ell + 1) j_\ell(kr) P_\ell(\cos\theta)$$

(for the chosen coordinate system the angle ϕ does not enter). $j_\ell(kr)$ is the spherical Bessel function of order ℓ and P_ℓ are the familiar Legendre polynomials. For large values of r we get:

$$j_\ell(kr) \approx \frac{1}{kr} \cos kr - (\ell + 1)\frac{\pi}{2}, \; kr >> 1.$$

This condition is easily satisfied since the distance from the scattering center is typically a few cm and $k \geq 10^{13}\text{cm}^{-1}$. Thus

$$\psi_i = \frac{i}{2kr} \sum_{\ell} (2\ell + 1) \left\{(-1)^{\ell} e^{-kr} - e^{kr}\right\} P_\ell(\cos\theta), \tag{5.1}$$

i.e. we have a superposition of two waves. As it can be seen, taking into account the time dependence, the first term in the curly bracket represents an oncoming wave, while the second an outgoing wave. Clearly the oncoming wave is not affected by the scattering potential, but the outgoing wave is, see Fig. 5.1. The most important effect is a phase change, **phase shift**, denoted by $2\delta_\ell$, but the amplitude may be affected by a positive factor η, $0 \le \eta \le 1$. In other words:

$$\psi_f = \frac{i}{2kr} \sum_\ell (2\ell+1) \left\{ (-1)^\ell e^{-kr} - \eta_\ell e^{2\delta_\ell} e^{kr} \right\} P_\ell(\cos\theta). \tag{5.2}$$

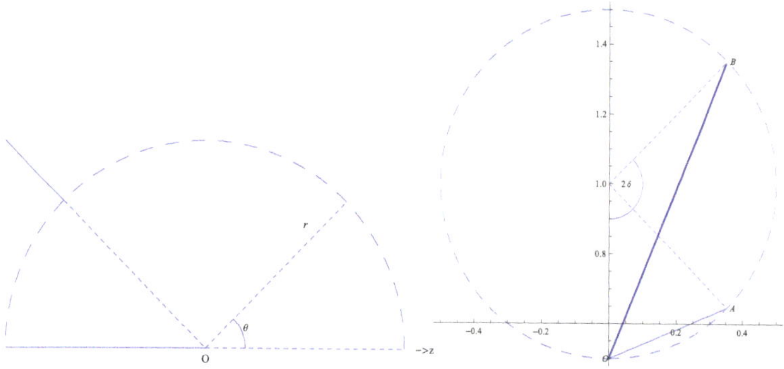

(a) (b)

Fig. 5.1: (a) The oncoming wave is coming from the left traveling along the z axis and is scattered at an angle θ. For very large r both waves are spherical. (b) The scattering amplitude f(E) in the case of no absorption. It is plotted as a vector in the complex plane (Ref, Imf) for two values of δ, $\vec{OA}$ for $\delta = \pi/8$ and $\vec{OB}$ for $\delta = 3\pi/8$. Obviously as the energy changes the vector is rotating, since the phase shift is a function of the energy. The maximum is attained when $\delta = \pi/2$.

Thus the part due to scattering becomes:

$$\psi_{\text{scatt}} = \psi_f - \psi_i = e^{kr} \frac{1}{kr} \sum_\ell (2\ell+1) \frac{1}{2i} \left(\eta_\ell e^{2\delta_\ell} - 1 \right) P_\ell(\cos\theta) = \frac{1}{r} e^{kr} F(\theta) \tag{5.3}$$

with

$$F(\theta) = \frac{1}{k} \sum_\ell (2\ell+1) \frac{1}{2i} \left(\eta_\ell e^{2\delta_\ell} - 1 \right) P_\ell(\cos\theta). \tag{5.4}$$

We note that $F(\theta)$ depends only on the momentum of the initial particle. This can be true only in the center of momentum system or if the target does not absorb energy. The outgoing flux in a solid angle $d\Omega$ through a sphere of radius r, see Fig. 5.1, is:

$$\Phi = v_0 \psi^*_{\text{scatt}} \psi_{\text{scatt}} r^2 d\Omega = v_0 \left| F(\theta) \right|^2 .$$

But, by definition, this is the same as the incident flux times the cross section, $v_i \psi_i \psi_i^* d\sigma$. Thus since $v_i = v_0$ we recover the celebrated formula:

$$\frac{d\sigma}{d\Omega} = \left| F(\theta) \right|^2 . \tag{5.5}$$

The integration over the angles is trivial, since the Legendre polynomials are orthogonal and normalized as follows:

$$\int P_\ell(\cos\theta) P_{\ell'}(\cos\theta) \sin\theta d\theta = \int P_\ell(\xi) P_{\ell'}(\xi) d\xi = \delta_{\ell,\ell'} \frac{2}{2\ell+1}.$$

We thus get

$$\sigma = 4\pi \lambda\!\!\!^{-2} \sum_\ell (2\ell+1) \left| \frac{\eta_\ell e^{2\delta_\ell} - 1}{2i} \right|^2 . \tag{5.6}$$

Of particular interest is the case of no absorption, $\eta_\ell = 1$. Then

$$F_{el}(\theta) = \frac{1}{k} \sum_\ell f^\ell_{el}(\theta),\ f^\ell_{el}(\theta) = (2\ell+1) \frac{1}{2i} \left(e^{2\delta_\ell} - 1 \right) P_\ell(\cos\theta) \tag{5.7}$$

$$\sigma_{el} = 4\pi \lambda\!\!\!^{-2} \sum_\ell (2\ell+1) \left| \frac{e^{2\delta_\ell} - 1}{2i} \right|^2 = 4\pi \lambda\!\!\!^{-2} \sum_\ell (2\ell+1) \sin^2 \delta_\ell . \tag{5.8}$$

This is referred to as *elastic scattering*. For a given channel ℓ it attains the maximum value which is $4\pi \lambda\!\!\!^{-2} (2\ell+1)$.

Of particular interest for the study of a resonance in a given channel ℓ (see previous subsection) is the function:

$$f(E) = \frac{e^{2\delta_\ell(E)} - 1}{2i} = \frac{1}{2} \sin\left(2\delta_\ell(E)\right) + i \left(1 - \frac{1}{2} \cos\left(2\delta_\ell(E)\right) \right) . \tag{5.9}$$

The behavior of $f(E)$ is exhibited in Fig. 5.1, known as Argand plot. The maximum is attained at $\delta_\ell(E) = \pi/2$.

For completeness we will briefly review some more facts about potential scattering. The reaction cross section is:

$$\sigma_r = \int d\Omega \left\{ \left| (\psi_f) \right|^2_{\delta_\ell=0} - \left| (\psi_f) \right|^2_{\delta_\ell \neq 0} \right\} = 2\pi \lambda\!\!\!^{-2} \sum_\ell (2\ell+1)(1-\eta_\ell) \cos 2\delta_\ell ,$$

in an obvious notation, see Eq. (5.2). It attains for a given channel a maximum value $2\pi\lambda\!\!\!^{-2}(2\ell+1)$, when $\eta_\ell = 0$, $\delta_\ell = 0$.

Combining the above results we get:

$$\sigma_T = \sigma_{el} + \sigma_r = 2\pi\lambda\!\!\!^{-2}\sum_\ell (2\ell+1)(1-\eta_\ell\cos 2\delta_\ell).$$

Noting now that $P_\ell(1) = 1$ for all ℓ we get:

$$ImF(0) = \frac{1}{2k}\sum_\ell (2\ell+1)(1-\eta_\ell\cos 2\delta_\ell),$$

see Eq. (5.4), which leads to relating the total cross section to the forward scattering amplitude, that is:

$$\sigma_T = \frac{4\pi}{k} ImF(0), \quad \text{(the optical theorem)}. \tag{5.10}$$

We note that, for a given channel ℓ, the total cross section attains a maximum value of $4\pi\lambda\!\!\!^{-2}(2\ell+1)$, whenever the elastic one does.

5.2 Hard sphere scattering

The key quantities to calculate the cross sections are the phase shifts $\delta_l = \delta_l(E)$. Firstly, considering a scattering by a central potential $V(r)$, one has to solve the differential equation (for a review solving Schrödinger equation see Section 14.1)

$$y_l'' + \left(k^2 - U(r) - \frac{l(l+1)}{r^2}\right) y_l = 0, \quad U(r) = 2mV(r)/\hbar^2. \tag{5.11}$$

In particular, the phase shift are given by

$$\delta_l = \phi_l - \phi_l^{(0)} = \phi_l + \frac{l\pi}{2} \tag{5.12}$$

where the phase ϕ_l is defined by the asymptotic behavior of the solution y_l in the infinity, that is

$$y_l \sim \lim_{r\to\infty} \sin(kr + \phi_l). \tag{5.13}$$

In the special case where $l = 0$, the differential equation (5.11) takes the very simple form

$$y_0'' + \left(k^2 - U(r)\right) y_0 = 0. \tag{5.14}$$

The differential equation (5.14) has the advantage that can be solved analytically for a various case of the potential $V(r)$. Let as consider the simple case of a scattering by a hard sphere potential. In this case the potential is defined as

$$V(r) = \begin{cases} \infty, & r < a \\ \\ 0, & r > a \end{cases} \tag{5.15}$$

and consequently the particle is moving out of the sphere of radius a. Then ones has to solve the equation

$$y_0'' + k^2 y_0 = 0 \tag{5.16}$$

for $r > a$ with the boundary condition $y(a) = 0$. Now, since the general solution of Eq. (5.16) is on the form

$$y_0 \sim \sin(kr + \phi_0)$$

we get

$$y_0(a) \sim \sin(ka + \phi_0) = 0 \to \phi_0 = -ka + n\pi, \qquad n = 0, \pm 1, \pm 2, \cdots .$$

Moreover, according to Eq. (5.12) the phase shifts are given by

$$\delta_0 = \phi_0 = -ka + n\pi.$$

Since the contribution σ_l is given by

$$\sigma_l = \frac{4\pi}{k^2}(2l+1)\sin^2\delta_l \tag{5.17}$$

then in the specific case of scattering by a hard sphere and for the $l = 0$ component we get

$$\sigma_0 = \frac{4\pi}{k^2}\sin^2\delta_0 = \frac{4\pi}{k^2}\sin^2(-ka + n\pi) = \frac{4\pi}{k^2}\sin^2(ka). \tag{5.18}$$

In Fig. 5.2 we display the dependence of σ_0 on the wave number k. It is interesting also that in the limit $k \to 0$, we get

$$\lim_{k\to 0}\sigma_0(k) = 4\pi a^2.$$

Actually, since for the rest components σ_l holds that $\lim_{k\to 0}\sigma_l(k) = 0$, the contribution on the total cross section is due only on the $l = 0$ component.

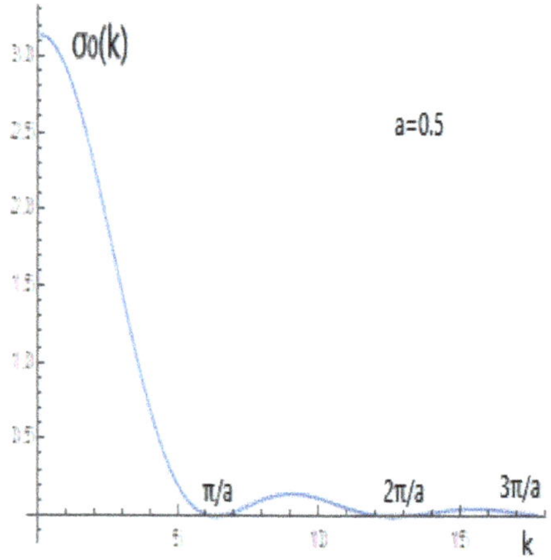

Fig. 5.2: The cross section $\sigma_0(k)$ (in units fm^2) as a function of the wave number k (in units fm^{-1}). The values $a = 0.5$ fm for the hard sphere radius is implemented. For the special case $k = 0$ we get the value $\sigma_0(0) = \pi\ \text{fm}^2$.

Chapter 6

Nuclear instability

We will consider some processes which cause nuclear instability: alpha, gamma and beta decays, which are due to the main known forces in nature, namely strong, electromagnetic and weak interaction respectively.

6.1 Alpha decay

It is known that the binding energy per nucleon has a maximum value close to the iron nucleus, but in heavier elements it is reduced considerably. This means that heavy nuclei may be unstable against spontaneous disintegration. Indeed in most cases the break-up of a heavy nucleus leads to the emission of the $^4_2$He nucleus. It is worth mentioning here that at the beginning the emitted light nucleus was called α-particle. Rutherford was the fist to identify the α-particle as the $^4_2$He nucleus. The binding energy of $^4_2$He is approximately 28.3 MeV and this value plays an important role in the alpha decay process.

The starting point of the alpha decay studies is the fundamental equality

$$E_{\alpha-\text{kin}}(A, Z) = E_B(A-4, Z-2) + 28.3 \text{ MeV} - E_B(A, Z). \qquad (6.1)$$

This is a semi-empirical mass formula obtained after neglecting the pairing energy term. The resulting $E_{\alpha-\text{kin}}(A, Z)$ is presented in Fig. 6.1 as a function of the atomic number Z.

In this case the negative values of $E_{\alpha-\text{kin}}(A, Z)$ indicate stability against α-decay. Actually, the predictions of the semi-empirical mass formula are very close to the experimental values (with a very small deviation). It is worth mentioning that the main deviations are for nuclei around the double-closed shell nucleus $^{208}_{82}$Pb. In general from Eq. (6.1) it is concluded that nuclei with $Z \geq 66$ are, in principle, unstable against α-decay.

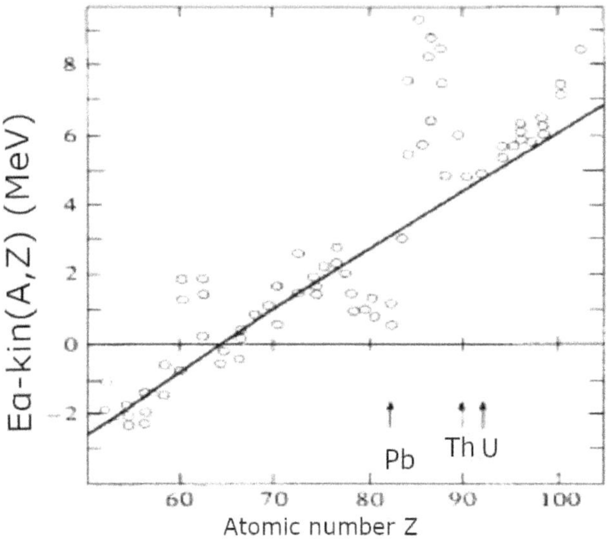

Fig. 6.1: The quantity $Q = E_{\alpha-\text{kin}}(A, Z)$ is plotted as a function of Z. One can see that that nuclei with $Z \geq 66$ are unstable against α-decay [Cottingham and Greenwood (1986)].

6.1.1 *The formulation of α-decay*

The staring point of the formulation of α-decay is the realization that the presence of Coulomb force is the reason for the limitation of the α-decay (even when it is energetically favorable). As an example we consider the case of the ${}^{238}_{92}$U decay to ${}^{234}_{90}$Th according to the reactions [Cottingham and Greenwood (1986)]

$$ {}^{238}_{92}\text{U} \longrightarrow {}^{234}_{90}\text{Th} + {}^{4}_{2}\text{He} + 4.47 \text{ MeV}. \tag{6.2} $$

In Fig. 6.2 we plot the electrostatic potential energy of the α-particle moving inside the nucleus ${}^{238}_{92}$U, where r is the distance of the daughter nucleus ${}^{234}_{90}$Th.

In Fig. 6.2 r_c is the distance where the Coulomb barrier appears and r_s is the distance where the a-particle escapes form the mother nucleus. This figure helps us understand, at least qualitatively, the escape of the a-particle of the mother nuclei, since the Coulomb barrier is dramatically higher compared to the kinetic energy of ${}^{4}_{2}$He. The explanation is given by Quantum Mechanics through the phenomenon of tunneling.

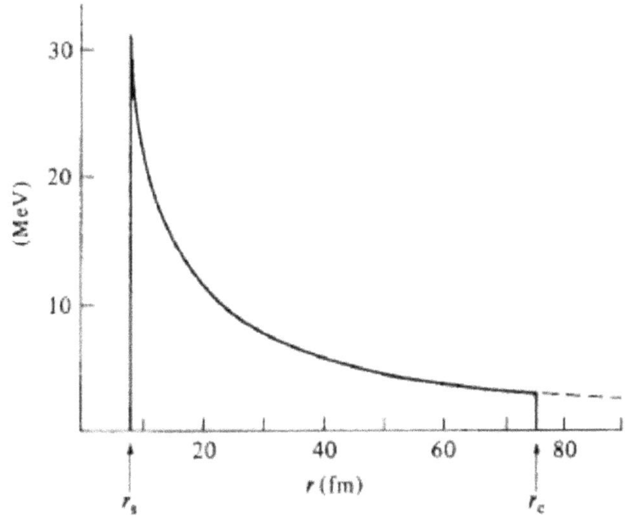

Fig. 6.2: We exhibit the electrostatic potential energy of the α-particle moving inside the nucleus ${}^{238}_{92}$U, where r is the distance of the daughter nucleus ${}^{234}_{90}$Th.

6.1.2 *The tunneling effect*

We will provide a simple explanation of the tunneling viewed as one dimensional problem. We consider an orthogonal barrier potential (see Fig. 6.3) where the barrier potential is defined as [Zettili (2009)]

$$V(x) = \begin{cases} 0 & x < 0 \quad \text{region I} \\ V_0 & 0 \leq x \leq a \quad \text{region II} \\ 0 & x > a \quad \text{region III} \end{cases} . \tag{6.3}$$

We consider that a particle with mass m is moving with energy $E < V_0$ from the left to the right side. In the regions II and III the Schrödinger equation takes the simple form

$$\frac{d^2\psi}{dx^2} + \frac{2mE}{\hbar^2}\psi = 0 \tag{6.4}$$

and the solutions are given by

$$\Psi_{\rm I}(x) = e^{ikx} + Re^{-ikx}, \; k = \frac{\sqrt{2mE}}{\hbar} \tag{6.5}$$

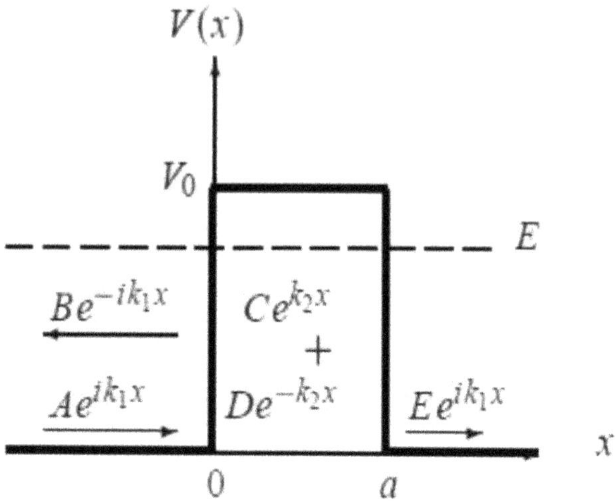

Fig. 6.3: We exhibit in the case of the one-dimensional square well potential, with a non zero value in the interval $0 < x < a$, the forms of the various waves expected in the various regions, assuming that a plane wave initially started in the left moving rightwards. For the notation see text.

$$\Psi_{\text{III}}(x) = Te^{ikx} + Me^{-ikx} \tag{6.6}$$

where

$$k = \frac{\sqrt{2mE}}{\hbar}$$

and R, T, and M are the amplitudes of the corresponding waves, which are related to the scattering problem. In particular, R and T are the amplitudes of the reflected and transmitted wave respectively. The factor M is the amplitude of the returning wave. Since, there is no reason for the wave to return in region III we consider from the beginning that $M = 0$.

Now, concerning the region II the Schrödinger equation reads

$$\frac{d^2\psi}{dx^2} - \frac{2m}{\hbar^2}(V_0 - E)\Psi = 0 \tag{6.7}$$

with the solution

$$\Psi_{\text{II}}(x) = Ae^{ik'x} + Be^{-ik'x} \tag{6.8}$$

where

$$k' = \frac{\sqrt{2m(V_0 - E)}}{\hbar}$$

A and B are constants. What we need is to calculate the factors T and R. We can proceed by demanding that the wave function and its derivative are continuous in all regions. We concentrate mainly on the points 0 and a by demanding

$$\begin{aligned}
\Psi_{\rm I}(0) &= \Psi_{\rm II}(0) \\
\Psi'_{\rm I}(0) &= \Psi'_{\rm II}(0) \\
\Psi_{\rm II}(a) &= \Psi_{\rm III}(a) \\
\Psi'_{\rm II}(a) &= \Psi_{\rm II}(a)
\end{aligned} \tag{6.9}$$

where Ψ' denotes the first derivative of Ψ. Equations (2.43) lead to

$$\begin{aligned}
&1 + R = A + B \\
&ik(1 - R) = k'(A - B) \\
&Ae^{k'a} + Be^{-k'a} = Te^{ika} &(6.10)\\
&k'(Ae^{k'a} - Be^{-k'a}) = ikTe^{ika}. &(6.11)
\end{aligned}$$

Now, solving the above system of equations, we find that the transmitted wave amplitude is

$$T = e^{-ik'a} \frac{2kk'}{2kk' \cosh(k'a) - i(k^2 - k'^2)\sinh(k'a)} \tag{6.12}$$

and this implies that the probability of transition is

$$|T|^2 = \left[1 + \frac{1}{4}\left(\frac{k^2 + k'^2}{kk'}\right)^2 \sinh^2(k'a)\right]^{-1}. \tag{6.13}$$

Now, since

$$\left(\frac{k^2 + k'^2}{kk'}\right)^2 = \frac{V_0^2}{E(V_0 - E)}$$

the transmission probability read

$$|T|^2 = \left[1 + \frac{1}{4\epsilon(1 - \epsilon)} \sinh^2\left(\lambda\sqrt{1 - \epsilon}\right)\right]^{-1} \tag{6.14}$$

where

$$\lambda = a\sqrt{2mV_0/\hbar^2}, \quad \text{and} \quad \epsilon = E/V_0.$$

The meaning of the approximation (6.14) is clear: since $|T|^2$ is finite, even though the energy E of the particle is less than the potential barrier V_0 ($E \ll V_0$), the probability for the transmission of the particle into the region $x \geq a$, is not zero (in classical physics this possibility is zero). This is a purely quantum effect, known as a **tunneling effect.** This barrier penetration effect has many and important applications in modern physics, including especially particle and nuclear physics as well as atomic physics.

There two special cases:
1) $E \ll V_0$. Hence $\epsilon \ll 1$ (or $\lambda\sqrt{1-\epsilon} > 1$). Then to a very good approximation one finds

$$\sinh\left(\lambda\sqrt{1-\epsilon}\right) \simeq \frac{1}{2}e^{\lambda\sqrt{1-\epsilon}}$$

and the transition factor becomes asymptotically equal to

$$|T|^2 \simeq 16\epsilon(1-\epsilon)e^{-2\lambda\sqrt{1-\epsilon}} = \frac{16E}{V_0}\left(1-\frac{E}{V_0}\right)e^{-(2a/\hbar)\sqrt{2m(V_0-E)}}. \quad (6.15)$$

According to Eq. (6.15) the transition factor is finite (but not zero as in the classical case) and consequently there is a tunneling beyond the barrier.
2) In the classical limit $\hbar \to 0$ the transition factor approaches its classical value, that is $|T|^2 \to 0$.

We will study now the general case of the tunneling effect which consists of a particle going through a region where the potential energy has a spatial dependence of the form $V(x)$, while its energy is smaller than the potential energy (that is $E < V(x)$). Obviously, in the simplest case of the square barrier, we managed to find an analytic solution for the transmission factor. However, when the barrier potential exhibits arbitrary spatial dependence it is almost impossible to obtain analytic expressions (see Fig. 6.4). In this general case we employ an approximation which is known as Wentzel-Kramers-Brillouin (WKB) method, see any quantum mechanics book, e.g. [Hecht (2000)]. In this case the transmission factor for the barrier potential $V(x)$ is given by

$$|T|^2 \sim \exp\left[-\frac{2}{\hbar}\int_{x_1}^{x_2}\sqrt{2m(V(x)-E)}dx\right]. \quad (6.16)$$

The above relation can be obtained by dividing the classically forbidden region $x_1 < x < x_2$ (see Fig. 6.4) into a series of small intervals Δx_i. Then it is a good approximation to consider that the potential in this small region is a square potential barrier with the corresponding transition probability

$$T_i \sim \exp\left[-\frac{2\Delta x_i}{\hbar}\sqrt{2m(V(x_i)-E)}\right]. \quad (6.17)$$

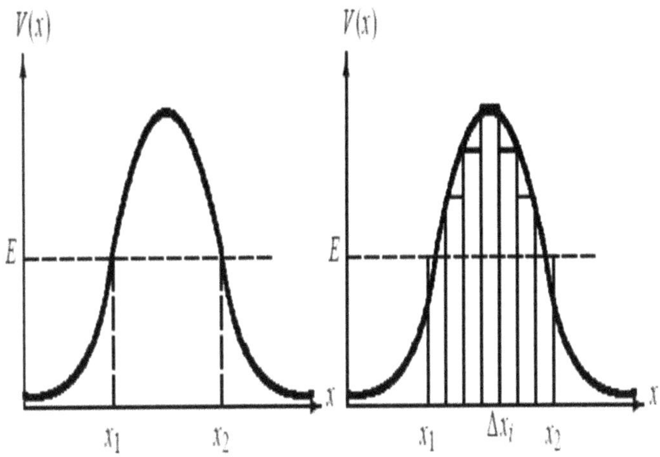

Fig. 6.4: We exhibit how, in the context of the WKB approximation, the classically forbidden region $x_1 < x < x_2$ is split into a series of small intervals Δx_i. Then in each interval Δx_i one obtains the transmission factor T_i assuming that the potential remains essentially constant.

Now, the transition probability for the whole potential of Fig. 6.4 is given by:

$$\begin{aligned} T &\sim \lim_{N\to\infty} \prod_{i=1}^{N} \exp\left[-\frac{2\Delta x_i}{\hbar}\sqrt{2m(V(x_i)-E)}\right] \\ &= \exp\left[-\frac{2}{\hbar}\lim_{\Delta x_i\to 0}\sum_i \Delta x_i\sqrt{2m(V(x_i)-E)}\right] \\ &\to \exp\left[-\frac{2}{\hbar}\int_{x_1}^{x_2}\sqrt{2m(V(x)-E)}dx\right]. \end{aligned} \tag{6.18}$$

It is worth realizing that Eq. (6.18) holds only in the case where the potential $V(x)$ is a smooth and slowly varying function of x.

6.2 Electromagnetic transitions — Gamma decay

The Gamma decay,[1] (Γ-decay), mainly deals with the emission of energetic photons from a nucleus. It occurs, in particular, when a nucleus de-excites from some high-lying excited state to the ground state or low excited states. In contrast to the case of alpha and beta decay the mother and daughter nucleus are the same. Actually, what takes place is re-ordering of the nucleons within the nucleus which leads to lowering of the mass of the excited state M_{ex} to, e.g., the lowest value M_{gr}. The conservation of energy then reads (see Fig. 6.5)

$$M_{ex}c^2 = M_{gs}c^2 + E_\gamma + T_N \tag{6.19}$$

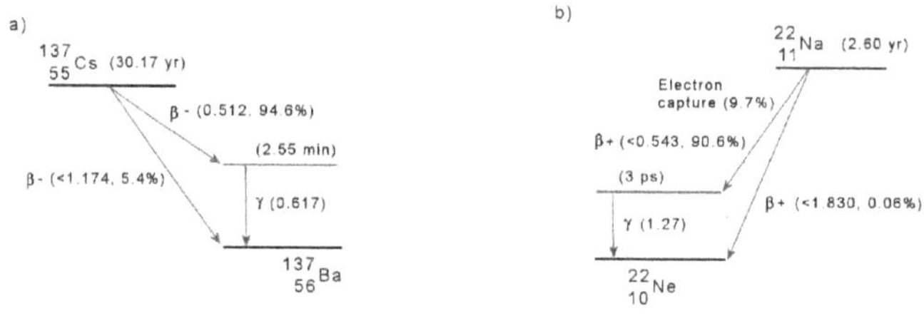

Fig. 6.5: (a) The nucleus ${}^{127}_{55}$Cs undergoes β (negaton, β^-) decay to the ${}^{127}_{55}$Ba excited state. (b) The nucleus ${}^{22}_{55}$Na undergoes β^+ (positon) decay to the ${}^{22}_{10}$Ne nucleus. In both cases the populated excited state $M_0^* c^2({}^A_Z X_N)$ undergoes gamma decay (where E_γ is the energy of photons) to the ground state $M_0 c^2$.

where E_γ is the energy of the emitted photon and T_N the kinetic energy of the recoiling nucleus. The corresponding conservation of momentum reads

$$\mathbf{P}_\gamma + \mathbf{P}_N = 0 \tag{6.20}$$

where $\mathbf{P}_\gamma$ and $\mathbf{P}_N$ are the momentum of the photon and nucleus respectively. The recoil energy (considering non-relativistic kinematics) is given by

$$T_N = \frac{\mathbf{P_N^2}}{2M_{gs}} = \frac{\mathbf{P_\gamma^2}}{2M_{gs}} \tag{6.21}$$

[1]The present section is based mainly on the book [Heyde (1999)].

and considering that $P_\gamma = E_\gamma/c$ we find that

$$T_N = \frac{E_\gamma^2}{2M_{gs}c^2}. \tag{6.22}$$

In general the recoiling energy is small (a few eV), but it is enough to shift the photon out of resonance condition. The other case, photon emission without recoil, is possible considering the nucleus in a lattice. Then, the recoil energy is taken from the whole lattice and not by a isolated nucleus. Now, considering that the excitation energy (of the whole system) to the first excited state, $\hbar\omega_{lat}$, is much larger than the single-nucleon recoil energy, i.e.

$$\hbar\omega_{lat} \gg T_N, \tag{6.23}$$

the emitted photon takes away the total energy difference given by Eq. (6.19). This effect, known as **Mössbauer effect**, see Section 6.4, has many applications in various domains of physics and other sciences.

6.2.1 *The dipole approximation*

For a detailed study of gamma decay one must make use of quantum mechanics. The simplest case is to consider the one photon emission process. In this case, the quantum system considered, found in the initial state $|i\rangle$, through the electromagnetic interaction operator, undergoes transition to the final state $|f\rangle$. The application of time perturbation theory leads to the following ansatz for the transition probability P_{if}

$$P_{if} = \frac{2\pi}{\hbar}\left|\langle f|\hat{H}_{int}|i\rangle\right|^2 \frac{dN}{dE} \tag{6.24}$$

where dN/dE is the density of the final states (including the radiation of the photon with momentum P_γ and energy E_γ). In particular, for the emission of the photon in solid angle $d\Omega$ this density is given by

$$\frac{dN}{dE} = \frac{VE_\gamma^2}{(2\pi\hbar c)^3}d\Omega \tag{6.25}$$

where V is the total volume. Moreover, the operator which corresponds to the Hamiltonian of electromagnetic interaction $\hat{H}_{int}$ is of the form

$$\hat{H}_{int} = \frac{e}{m}\hat{p}\cdot\mathbf{A} \tag{6.26}$$

where the momentum operator $\hat{p}$ is defined as $\hat{p} = -i\hbar\boldsymbol{\nabla}$. The treatment of the vector potential $\mathbf{A}$, strictly speaking, demands the use of relativistic quantum mechanics. However, we can employ an approximation by considering that the vector potential $\mathbf{A}$ describes the wave function of the photon and has the form

$$\mathbf{A} = a_0 \boldsymbol{\epsilon} \cos(\mathbf{k}\cdot\mathbf{r} - \omega t)\,. \tag{6.27}$$

In Eq. (6.27) $\boldsymbol{\epsilon}$ and a_0 are the polarization vector and the amplitude of one photon respectively. The energy of the plane wave, normalized to volume V, is given by

$$E = \epsilon_0 |\mathbf{E}|^2_{Aver} V = \epsilon_0 \frac{a_0^2\omega^2}{2} V \tag{6.28}$$

and equating it with the energy of a photon $E_\gamma = \hbar\omega$ we find

$$a_0 = \sqrt{\frac{2\hbar^2}{\epsilon_0 E_\gamma V}}. \tag{6.29}$$

In order to proceed further it is convenient to rewrite the vector potential $\mathbf{A}$ as

$$\mathbf{A} = \sqrt{\frac{2\hbar^2}{\epsilon_0 E_\gamma V}} \boldsymbol{\epsilon} \left(e^{i\mathbf{p}_\gamma\cdot\mathbf{r} - E_\gamma t)/\hbar} + e^{-i(\mathbf{p}_\gamma\cdot\mathbf{r} - E_\gamma t)/\hbar} \right). \tag{6.30}$$

Now, considering that the initial $|i\rangle$ and final $|f\rangle$ states correspond to the wave functions Ψ_i and Ψ_f respectively and using the operator $\hat{H}_{int}$ one finds

$$\hat{H}_{int} = \frac{e}{m}(-i\hbar\boldsymbol{\nabla})\cdot\mathbf{A}. \tag{6.31}$$

Then the matrix element corresponding to this interaction can be written as

$$\langle f|\hat{H}_{int}|i\rangle = \int \Psi_f^* \hat{H}_{int} \Psi_i dV = -\frac{ie\hbar}{m}\int \Psi_f^* \boldsymbol{\nabla}(\Psi_i)\cdot\mathbf{A}dV. \tag{6.32}$$

We can now proceed further

i) Employing the dipole approximation, that is expanding the plane waves

$$\exp\left(\pm i\frac{\mathbf{p}_\gamma\cdot\mathbf{r}}{\hbar}\right) = 1 \pm i\frac{\mathbf{p}_\gamma\cdot\mathbf{r}}{\hbar} \pm \frac{1}{2}\left(i\frac{\mathbf{p}_\gamma\cdot\mathbf{r}}{\hbar}\right)^2 + \ldots \tag{6.33}$$

and keeping only the constant first term in Eq. (6.30).

ii) Making the additional assumption that the particles involved are spinless with initial and final energy E_i and E_f respectively.

iii) Making use of the conservation of energy

$$E_i = E_f + E_\gamma \tag{6.34}$$

and
iv) Employing the condition

$$t \gg 2\pi\hbar/E_\gamma.$$

Then, after some algebra, we finally obtain an approximate expression for matrix element

$$\langle f|\hat{H}_{int}|i\rangle = -i\frac{\hbar^2 e}{m}\sqrt{\frac{1}{2\epsilon_0 E_\gamma V}}\left(\int \Psi_f^* \boldsymbol{\nabla}\Psi_i dV\right)\cdot \boldsymbol{\epsilon} \tag{6.35}$$

and the transition probability takes the form

$$P_{if}(\theta,\phi) = \frac{e^2 E_\gamma}{8\pi^2 m^2 c^3}\left|\left(\int \Psi_f^* \boldsymbol{\nabla}\Psi_i dV\right)\cdot \boldsymbol{\epsilon}\right|^2 d\Omega. \tag{6.36}$$

In order to proceed further with the calculation P_{if} it is necessary to have a knowledge of the particle wave functions. These wave function are eigenfunctions of the unperturbed Hamiltonian $\hat{H}_0$ defined as

$$\hat{H}_0 = \frac{\hat{p}^2}{2m} + V(\mathbf{r}) \tag{6.37}$$

where $V(\mathbf{r})$ is a mean field potential felt by the particle. The wave functions Ψ_i and Ψ_f are eigenfunctions of the Hamiltonian $\hat{H}_0$, that is

$$\hat{H}_0\Psi_i = E_i\Psi_i, \qquad \hat{H}_0\Psi_f = E_i\Psi_f.$$

Employing now the well known commutation relation

$$\left[\mathbf{r},\hat{H}_0\right] = \frac{\hbar^2}{m}\boldsymbol{\nabla},$$

we get

$$\int \Psi_f^* \vec{\nabla}\Psi_i dV = \frac{m}{\hbar^2}E_\gamma \int \Psi_f^* \mathbf{r}\Psi_i dV = \frac{m}{\hbar^2}E_\gamma \langle f|\mathbf{r}|i\rangle. \tag{6.38}$$

The probability distribution can be written in the more elegant form

$$P_{if}(\theta,\phi) = \frac{e^2}{8\pi^2\hbar^4 c^3}\frac{E_\gamma^3}{\epsilon_0}\left|\boldsymbol{\epsilon}\cdot\langle f|\mathbf{r}|i\rangle\right|^2 d\Omega. \tag{6.39}$$

Without loss of generality we can take the direction of the photon momentum $\mathbf{p}_\gamma$ as pointing in the z-direction, while the other two vectors $\langle f|\mathbf{r}|i\rangle$ and $\boldsymbol{\epsilon}$ are as drawn in Fig. 6.6.

Summing over the two photon polarization states the probability P_{if} becomes

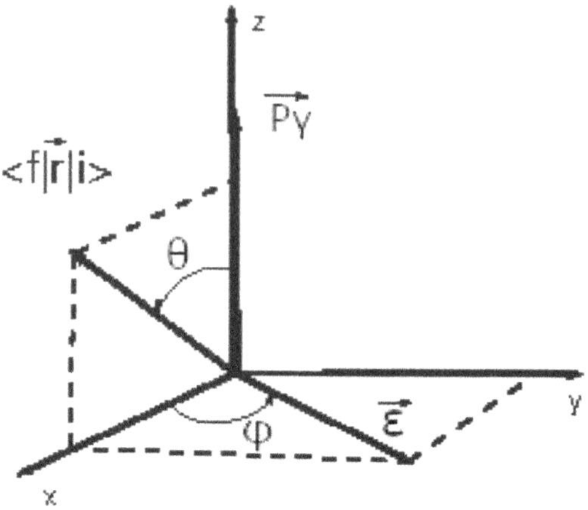

Fig. 6.6: Presentation of the various vector quantities related with the gamma decay process (for more details see text).

$$
\begin{aligned}
P_{if}(\theta, \phi) &= \frac{e^2}{8\pi^2\hbar^4 c^3}\frac{E_\gamma^3}{\epsilon_0} 2\left|\langle f|\mathbf{r}|i\rangle\right|^2 \sin^2\theta\cos^2\phi d\Omega \\
&= \alpha\frac{1}{\pi}\left(\frac{E_\gamma}{\hbar c}\right)^3 c\left|\langle f|\mathbf{r}|i\rangle\right|^2 \sin^2\theta\cos^2\phi d\Omega,\ \alpha = \frac{e^2}{\hbar c\, 4\pi\epsilon_0} \approx \frac{1}{137}.
\end{aligned}
\tag{6.40}
$$

The angular integration is easy:

$$
\int_{\theta=0}^{\pi}\int_{\phi=0}^{2\pi} \sin^2\theta\cos^2\phi\sin\theta d\theta d\phi = \int_1^1 d\xi(1-\xi^2)\int_0^{2\pi} d\phi\frac{1+\cos 2\phi)}{2} = \frac{4}{3}\pi.
\tag{6.41}
$$

Thus the total probability distribution becomes

$$
P_{if} = \frac{4}{3}\alpha\left(\frac{E_\gamma}{\hbar c}\right)^3 c\left|\langle f|\mathbf{r}|i\rangle\right|^2. \tag{6.42}
$$

What remains to be done is the calculation of the matrix element $|\langle f|\mathbf{r}|i\rangle|$. Since the transition operator is of rank unity and independent of the spin we have the following selection rules:

$$\Delta J = |J_i - J_f| \leq 1,\ \pi_i\pi_f = -1 \text{ (parity change)},\ \Delta L = \pm 1,\ |S_i - S_f| = 0.$$

These selection rules apply to the quantum numbers of the single particle states involved in the interaction. Furthermore, in this case, the HO oscillator major quantum number cannot change by more the one unit.

In any case we expect this ME to scale with the radius of the nucleus. We thus write

$$P_{if} = \frac{4}{3}\alpha\left(\frac{E_\gamma}{\hbar c}\right)^3 c\left(A^{1/3}r_0\right)^2 \frac{|\langle f|\mathbf{r}|i\rangle|^2}{\left(A^{1/3}r_0\right)^2}. \tag{6.43}$$

It is more convenient to express the photon energy in units of MeV and write

$$\begin{aligned} P_{if} &= \frac{4}{3}\alpha\left(\frac{E_\gamma}{1\text{ MeV}}\right)^3\left(\frac{1\text{ MeV}}{\hbar c}\right)^3 c\left(A^{1/3}r_0\right)^2 \frac{|\langle f|\mathbf{r}|i\rangle|^2}{\left(A^{1/3}r_0\right)^2} \\ &= 5.8\times 10^{14}\ \text{s}^{-1}\left(\frac{E_\gamma}{1\text{ MeV}}\right)^3 A^{2/3}\frac{|\langle f|\mathbf{r}|i\rangle|^2}{\left(A^{1/3}1.24\text{ fm}\right)^2}. \end{aligned} \tag{6.44}$$

This is a general formula giving the transition probability in terms of the photon energy in MeV, the nuclear mass number A and the matrix element in units of $A^{1/3}1.24$ fm.

Thus assuming that $|\langle f|\mathbf{r}|i\rangle| \simeq 1.24A^{1/3}$, taking a photon energy of $E_\gamma = 1$ MeV and using $A \simeq 100$ we find

$$P_{if} = 1.26\times 10^{16}\ \text{s}^{-1}. \tag{6.45}$$

The corresponding width is 0.25 eV. In nuclear physics the following modifications are needed:

i) Transition operator involves all protons

$$\mathbf{D} = \sum_{i=1}^{Z}\mathbf{r}_i = \sum_{i=1}^{A}\mathbf{r}_i\frac{1-\tau_3(i)}{2}.$$

ii) The scale involved is the harmonic oscillator size parameter:

$$b = \frac{\hbar c}{\sqrt{m_N c^2\hbar\omega}},\ \hbar\omega \approx 41A^{-1/3}\text{ MeV} \Rightarrow b \approx A^{1/6}\text{ fm}.$$

Thus

$$P_{if} = 3.8\times 10^{14}\ \text{s}^{-1}\left(\frac{E_\gamma}{1\text{ MeV}}\right)^3 A^{1/6}\left|\langle f|\mathbf{D}/b|i\rangle\right|^2. \tag{6.46}$$

iii) One must sum over the final angular momentum projections and average over the initial ones

$$\frac{1}{2J_i+1}\sum_{Mi,M_f}|\langle J_f,M_f|\mathbf{D}|J_i,M_i\rangle|^2$$

$$=\frac{1}{2J_i+1}\frac{1}{2J_f+1}\sum_{Mi,M_f,\rho}\langle (J_i,m_i),(1,\rho)|(J_f,m_f)\rangle^2\,|\langle J_f||\mathbf{D}||J_i\rangle|^2$$

$$=\frac{1}{2J_i+1}\,|\langle J_f||\mathbf{D}||J_i\rangle|^2\,.$$

The double bar ME is independent of the angular momentum projections and is known as reduced matrix element, see any book discussing angular momentum, e.g. [Hecht (2000)], or any advanced nuclear physics book, e.g. [deShalit and Feshbach (1974)]. It is defined for any tensor T_q^k of rank k via the equation

$$\langle J_f||T^k||J_i\rangle=\frac{\sqrt{2J_f+1}}{\langle (J_i,M_i),(k,q)|(J_f,M_f)\rangle}.$$

It can be computed by any convenient but allowed combination of M_f, q, M_i. Examples of its use can be found in Section 13.4.

In the last step the summation was easily performed using the orthonormality properties of the C-G coefficients. Thus

$$P_{if}=3.8\times 10^{14}\ \mathrm{s}^{-1}\left(\frac{E_\gamma}{1\ \mathrm{MeV}}\right)^3 A^{1/6}\frac{1}{2J_i+1}\,|\langle \alpha_f\, J_f||\mathbf{D}/b||\alpha_i\, J_i\rangle|^2 \quad (6.47)$$

where α_i and α_f any quantum numbers, other than angular momentum, needed to specify the initial and final states. In the case of a simple harmonic oscillator wave function the reduced matrix element can be evaluated using the techniques of Section 14.4. In the special case of $0d_{3/2}\to 0p_{3/2}$ we find that the U function and the L-reduced ME are unity, see Table 14.3. So

$$\langle 0p_{3/2}||\mathbf{D}/b||0d_{3/2}\rangle=-\langle 0p_{|}|r/b|0d\rangle=-\sqrt{\frac{5}{2}}. \quad (6.48)$$

Assuming now $A=16$ and estimating $E_\gamma\approx 10$ MeV we find:

$$P_{if}=3.8\times 10^{14}\ \mathrm{s}^{-1}\ 16^{(1/6)}\frac{1}{4}\frac{5}{2}\times 10^3=3.77\times 10^{17}\ \mathrm{s}^{-1}.$$

6.2.2 *Excitation of electrons in the dipole approximation*

In this section, we will discuss an additional process which can take place, i.e. the excitation energy of the nucleus transferred to a bound electron. As a result the electron can move into an unbound state with kinetic energy E_k^e

$$E_k^e = (E_i - E_f) - B_e \tag{6.49}$$

where B_e is the electron binding energy. The K electrons are characterized by a larger probability to participate in this process, since their wave function is concentrated close to the nucleus. As an example we consider the transition matrix element for the 1s electron converted into a free electron which described by a plane wave. In this case we have

$$T_{if} \sim \sum_i \int e^{-i\mathbf{k}_e \cdot \mathbf{r}_e} \Psi_f^*(\mathbf{r}_i) \frac{e^2}{4\pi\epsilon_0 |\mathbf{r}_e - \mathbf{r}_i|} b_0^{-3/2} e^{-\frac{r_e}{b_0}} \Psi_i(\mathbf{r}_i) dV_i dV_e \tag{6.50}$$

where sum goes over all protons. $\mathbf{k}_e$ is the momentum of the outgoing electron, Ψ_i, Ψ_f are the initial and final nuclear wave functions and b_0 is the Bohr radius. Now, considering that $r_e > r_i$, we employ an expansion of the Coulomb potential in terms of the Legendre polynomials $P_l(\cos\theta)$. These can be written [Hecht (2000)] as

$$P_l(\cos\theta) = \frac{4\pi}{2l+1} \sum_{m=-l}^{l} Y_l^{*m}(\Omega_l) Y_l^m(\Omega_l)$$

with Y_l^m the well known spherical harmonics. Thus we get

$$\sum_i \frac{1}{|\mathbf{r}_e - \mathbf{r}_i|} = \frac{1}{r_e} \sum_i \sum_l \left(\frac{r_i}{r_e}\right)^l P_l(\cos\theta) \tag{6.51}$$

where θ is the angle between the vectors $\mathbf{r}_i$ and $\mathbf{r}_e$. Finally, the transition matrix element T_{if} takes the form

$$T_{if} \sim \frac{4\pi e^2}{b_0^{3/2}} \sum_{m=-l}^{l} \sum_l \frac{k_e^{l-2}}{2l+1} \int e^{-i\mathbf{k}_e \cdot \mathbf{r}_e} \left(\frac{1}{k_e r_e}\right)^{l+1} Y_l^{*m}(\Omega_e) e^{-\frac{r_e}{b_0}} T_{if}(l,m) dV_e \tag{6.52}$$

where

$$T_{if}(l,m) = \sum_i \int \Psi_f^*(\mathbf{r}_i) Y_l^m(\Omega_i) \Psi_i^*(\mathbf{r}_i) dV_i. \tag{6.53}$$

In general the calculation of the matrix element T_{if}, given by Eq. (6.52) is a hard problem. If we consider the simplest case where both the photon and the electron (from 1s state) can be emitted, we can express the ratio of the transition to the conversion probability in terms of the coefficient $\mathcal{C}_k$ defined as

$$\mathcal{C}_k = \frac{T_{if}(e)}{P_{if}(\gamma)}. \tag{6.54}$$

One can show that, in the context of the above mentioned simple case, the coefficient $\mathcal{C}_k$ takes the form

$$\mathcal{C}_k = \frac{Z^3}{2}\left(\frac{e^2}{\hbar c 4\pi e_0}\right)^4 \left(\frac{2m_0c^2}{E_\gamma}\right)^{7/2}. \tag{6.55}$$

6.3 Nuclear magnetic moments — Magnetic M_1 transitions

One can proceed as in the electric dipole case with the interaction now being due to the magnetic field

$$H_{int} = \boldsymbol{\mu}_N.\mathbf{B},\ \boldsymbol{\mu}_N = \mu_0\left(g_\ell \boldsymbol{\ell} + g_s\mathbf{s}\right),\ \mu_0 = \frac{e\hbar}{2m_N c} \tag{6.56}$$

with

$$g_\ell = \begin{cases} 1, \text{ for protons} \\ 0, \text{ for neutrons} \end{cases},\quad g_s = \begin{cases} 5.5858, & \text{for protons} \\ -3.8263, & \text{for neutrons} \end{cases}. \tag{6.57}$$

Then

$$\mathbf{B} = \boldsymbol{\nabla}\times\mathbf{A} \Rightarrow i\mathbf{k}\times\boldsymbol{\epsilon} \Rightarrow H_{int} = \mu_0\mathbf{G}.(i\mathbf{k}\times\boldsymbol{\epsilon}),\ \mathbf{G} = g_\ell\boldsymbol{\ell} + g_s\mathbf{s}$$

$$P_{if} = \mu_0^2\frac{1}{2k}\frac{1}{(2\pi)^2}k^2\left|\int\langle\Psi_f|\mathbf{G}.(i\mathbf{k}\times\boldsymbol{\epsilon})|\Psi_i\rangle\right|^2 d\Omega. \tag{6.58}$$

The vector $(\mathbf{k}\times\boldsymbol{\epsilon})$ now is orthogonal to both $\mathbf{k}$ and $\boldsymbol{\epsilon}$, but the angular integral is the same as in the electric dipole case. So summing over the helicity states we obtain

$$P_{if} = \mu_0^2\frac{1}{2k}\frac{1}{(2\pi)^2}k^2k^2 2\frac{4\pi}{3}\left|\langle\Psi_f|\mathbf{G}|\Psi_i\rangle\right|^2 = \frac{1}{3\pi}\mu_0^2E_\gamma^3\left|\langle\Psi_f|\mathbf{G}|\Psi_i\rangle\right|^2. \tag{6.59}$$

Proceeding as above in the case of shell model calculations we obtain in terms of the reduced ME the transition probability

$$P_{if} = \frac{1}{3\pi}\mu_0^2E_\gamma^3\frac{1}{2J_i+1}\left|\langle\alpha_f\, J_f||\mathbf{G}||\alpha_i\, J_i\rangle\right|^2,$$

$$\mathbf{G} = \sum_i^A\left(g_\ell\boldsymbol{\ell}_i\frac{1-\tau_3(i)}{2} + \frac{g_{sp}+g_{sn}}{2}\mathbf{s}_i + \frac{g_{sn}-g_{sp}}{2}\mathbf{s}_i\tau_3(i)\right) \tag{6.60}$$

where α_i and α_f any quantum numbers, other than angular momentum, needed to specify the initial and final states. We see that the isovector spin contribution is expected to be dominant. In natural units

$$\mu_0^2 = \frac{4\pi\alpha}{(2m)^2} = \frac{\pi\alpha}{m_p^2}$$

$$P_{if} = \frac{\alpha}{3}\frac{E_\gamma^3}{m_p^2}\frac{1}{2J_i+1}\left|\langle\alpha_f\, J_f||\mathbf{G}||\alpha_i\, J_i\rangle\right|^2$$
$$\approx 8.26\times 10^{14}\ \mathrm{s}^{-1}\left(\frac{E_\gamma}{1\ \mathrm{MeV}}\right)^3\frac{1}{2J_i+1}\left|\langle\alpha_f\, J_f||\mathbf{G}||\alpha_i\, J_i\rangle\right|^2. \quad (6.61)$$

The corresponding width is

$$\Gamma \approx 0.00275\ \mathrm{eV}\left(\frac{E_\gamma}{1\ \mathrm{MeV}}\right)^3\frac{1}{2J_i+1}\left|\langle\alpha_f\, J_f||\mathbf{G}||\alpha_i\, J_i\rangle\right|^2. \quad (6.62)$$

As an example let us consider the decay $0d_{3/2}\to 0d_{5/2}$. Then

$$\langle 0d_{5/2}||\mathbf{G}||0d_{3/2}\rangle = g_s\langle 0d_{5/2}||\mathbf{s}||0d_{3/2}\rangle + g_\ell\langle 0d_{5/2}||\boldsymbol{\ell}||0d_{3/2}\rangle.$$

From the data of Table 14.2 for $\ell = 2$ we obtain

$$\langle 0d_{5/2}||\mathbf{G}||0d_{3/2}\rangle = g_s\frac{2\sqrt{3}}{\sqrt{5}} + g_\ell\frac{\sqrt{6}}{\sqrt{5}} = \begin{cases} 9.75, & \text{for proton} \\ -5.93, & \text{for neutron}\end{cases}.$$

Taging $E_\gamma = 3$ MeV for $J_i = 3/2$ we obtain

$$\Gamma = \begin{cases} 1.8\ \mathrm{eV}, & \text{for proton} \\ 0.65\ \mathrm{eV}, & \text{for neutron}\end{cases}. \quad (6.63)$$

At this point we should mention that the magnetic moment of a single particle $|n\ell j\rangle$ is defined as follows:

$$\mu_j = \frac{1}{j+1}$$
$$\left(g_\ell\frac{j(j+1)+\ell(\ell+1)-s(s+1)}{2} + gs\frac{j(j+1)-\ell(\ell+1)+s(s+1)}{2}\right). \quad (6.64)$$

Thus one finds:

$$\mu_{\ell+1/2} = g_\ell\ell + g_s\frac{1}{2},\ \mu_{\ell-1/2} = \frac{1}{\ell+1/2}\left(g_\ell(\ell+1) - g_s\frac{1}{2}\right),$$

being different for protons and neutrons. For shell model states it is defined analogously.

6.4 Mössbauer effect

We begin by noting that, when a system decays with a finite lifetime, the decay proceeds via a resonance, see Section 4.1. It was found that the probability of finding the particle with energy between E and $E + dE$ is given by:

$$\rho(E) = |\tilde{\Psi}(E)|^2 = \frac{\Gamma}{2\pi} \frac{1}{(E - E_0)^2 + \left(\frac{\Gamma}{2}\right)^2}. \tag{6.65}$$

In other words the parameter Γ widens the state of the particle, so that it no longer has definite energy and this leads to its destruction. The width of a state that arises from the decay of the particle is called natural width. The shape of Fig. 4.1β is called Lorentzian or Breit-Wigner. Γ is the width at half the maximum value of $\rho(E)$.

We know that in particle physics experiments very short life times have been determined from the measurements of the width of the corresponding resonances, see Section 9.10.1 for the ρ meson and in Section 10.1.2 for baryon resonances, Section 10.5 for the charmed resonances etc.

In the previous sections we have studied the emission and absorption of radiation by a nucleus. We have seen that the emitted photon has enough energy to cause a nuclear recoil. Since the nuclear momentum is opposite to that of the photon, the recoiling energy of the nucleus is $E_r = \frac{E_\gamma^2}{2Am_p^2}$. In other words, since this energy is taken by the nucleus, the photon will not have enough energy to be absorbed by another nucleus. This was a puzzle when the Mössbauer effect was observed, i.e. almost recoil-less decays, for which he won the 1961 Nobel Prize for his discovery.

We will investigate this phenomenon using the standard example of the 14.4 keV transition, see Fig. 6.7, between nuclear states of a rather uncommon, but stable, isotope of iron, i.e. $^{57}_{26}$Fe (57 total nucleons whereas the most abundant iron isotope is $^{56}_{26}$Fe).

This isotope is the product of the decay via electron capture of a radioactive cobalt isotope, ^{57}Co. By employing the Mössbauer effect you can use an absorber containing $^{57}_{26}$Fe as a probe to generate a very precise energy spectrum of this 14.4 keV transition, revealing a resonance with an extremely high quality factor, Q. The half-life of the first excited state $3/2^-$ is found to be 9.8×10^{-8} s which, see Eq. (4.11), leads to

$$\Gamma = \frac{\hbar}{\tau} = \frac{\hbar \ln 2}{t_{1/2}} = 4.67 \times 4.710^{-9} \text{ eV} \to \frac{\Gamma}{E} = 3 \times 10^{-13}.$$

The question now is: Can γ ray emission have such a narrow width as the above natural width implied by the uncertainty principle? We cannot

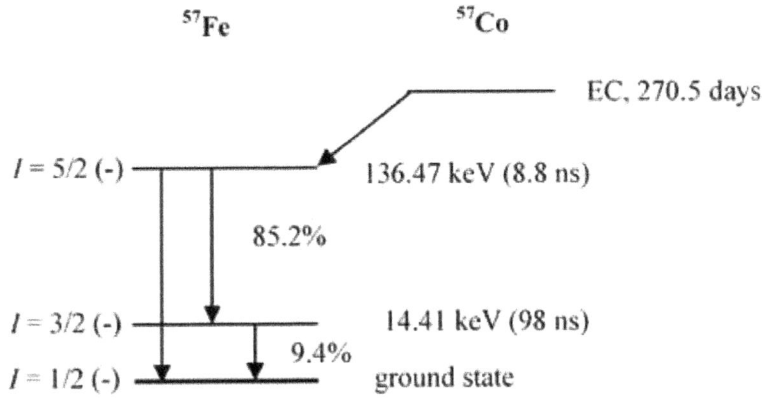

Fig. 6.7: The main features of the A = 57 low energy spectrum.

answer this question by the traditional ways of detecting γ rays. We can only do it by exploiting the idea of a recoil-less emission. This can happen since the nuclei are not free, but they are part of composite systems. But one cannot avoid imparting some energy to the recoiling nucleus. This is admittedly less than in the case of a free nucleus, but this is not adequate. Furthermore the thermal motion of the nuclei broadens the emitted γ ray. The solution to the problem case with Mössbauer proposal in 1957 [Mössbauer (1958)], namely to consider a nucleus as part of a solid. The solid is a quantum mechanical system and it is possible for the transition without the solid absorbing a significant energy. The probability for this happening [Frauenfelder and Henley (1974)] is

$$f = e^{\frac{3R}{2k\Theta_D}}, \quad R = \frac{E_\gamma^2}{2Am_pc^2}, \quad R = \text{the recoiling energy of a free nucleus,} \tag{6.66}$$

with Θ_D the Debye temperature of the solid and $k = 8.64 \times 10^{-3}$ eV/^{0}K is Boltzmann's constant. Thus for 14.4 keV transition $R \approx 1.9 \times 10^{-3}$ eV. The Debye temperature of most solids is around 200^0K leading to $k\Theta_D \approx 1.7 \times 10^2$ eV. Thus $f \approx 1$ and the γ ray will be emitted without a loss.

The process $3/2^- \to 1/2^-$ cannot occur via electric dipole transitions, discussed in Section 6.2.1, due to parity selection rules. It can occur via M_1 transitions discussed in Section 6.3 or via the next order electric transitions, known as quadrupole E_2 transitions. Detailed discussion of E_2 transitions is beyond the goals of this book and the interested reader is referred to more advanced books, e.g., [deShalit and Feshbach (1974)], [Hecht (2000)].

The main feature of the successful experimental apparatus is the Mössbauer spectrometer, which combines γ emission by a nucleus with the inverse process of γ absorption by another identical nuclear system. This, of course, is possible only, if the γ emission is towards the ground state. The basic elements of a Mössbauer spectrometer are: a source, the sample, the detector and a drive to move the source or absorber [Dyar *et al.* (2006)]. Most commonly, this is done by moving the source toward and away from the sample, while varying its velocity linearly with time. For example, for ^{57}Fe, moving the source at a velocity of 1 mm/sec toward the sample increases the energy of the emitted photons by about ten natural line-widths. For simplicity, "mm/sec" is the conventional "energy" unit in Mössbauer spectroscopy. This is exhibited in Fig. 6.8. One can, of course, move the source with velocity v towards or away from the absorber. In this the emitted γ ray will be Doppler shifted by an amount $\Delta E = \frac{v}{c}E_0$ and the corresponding intensity at $E_0 + \Delta E$ can be measured. The situation is sketched in Fig. 6.9. This provides an interesting signature.

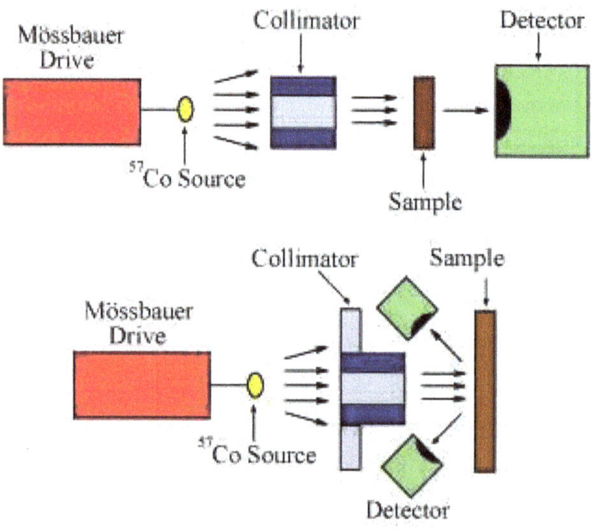

Fig. 6.8: A schematic presentation of the essential components of a Mössbauer spectrometer, courtesy of M. Darby Dyar [Dyar *et al.* (2006)].

One can, of course break the m-degeneracy of the levels by putting the sample in a magnetic field, the Zeeman effect, or in a quadrupole electric

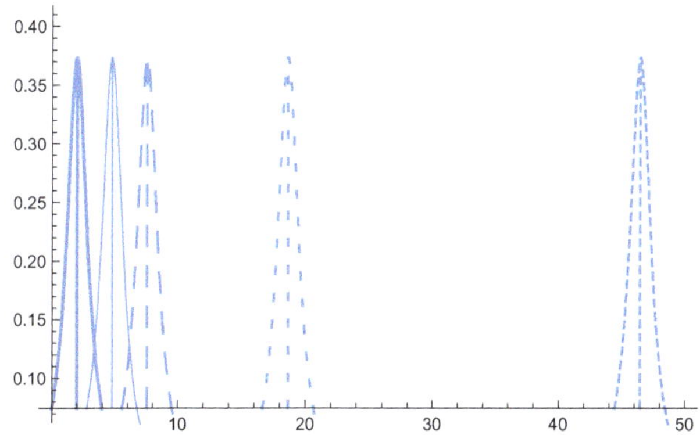

Fig. 6.9: The absorbed intensity for various values of the velocity, the thick solid line, the solid line, the long dashed, the long-short dash and the short dash correspond to velocities 0.0, 0.125, 0.250, 0.75 and 2.00 mm/s. Note the shift of the average values are in units of the width.

field. These signatures are important in the study of various materials as of chemical and biological systems, but we are not going to elaborate here (for such aspects, see [Frauenfelder and Henley (1974)]).

6.5 Nuclear fission

As we have mentioned in the introduction, see Section 2.4.1, after the discovery of the neutron in 1932, the researchers began experimental studies including mainly reactions induced by the bombardment of very heavy nuclei with neutrons. The experimental results, from bombardment of uranium with neutrons, explained few years later, in 1939 by Hahn and Strassmann. In particular the above reactions produced nuclei, such as barium, which have atomic number much smaller than that of uranium. Very soon Meitner and Frisch suggested that the uranium nucleus, after neutron capture, divided into two nuclei with almost equal size. This process was named fission. It is worth pointing out that the fission process exhibited similarities with the division of a small liquid drop into droplets. The suggested similarities played important role on the development of the theoretical explanation of the fission process. Bohr and Wheeler provided the theoretical understating of fission, a theory which remains the basic theoretical tool to study this process.

In the following we will provide a simplified model for the fission process based on the following reaction

$$n + {}^{235}_{92}\mathrm{U} \longrightarrow {}^{236}_{92}\mathrm{U} \longrightarrow {}^{141}_{56}\mathrm{Ba} + {}^{92}_{36}\mathrm{Kr} + 3n. \tag{6.67}$$

The fission process, as given in Eq. (6.67), is a two step process. Firstly, the ^{235}U nucleus captures a neutron forming the ^{236}U. Secondly, the compound nucleus ^{236}U is highly excited. The vibration amplitude becomes very large leading to the separation in two pieces. Then the repulsive Coulomb forces drive the two pieces apart with large energy. The rest energy is stored as internal energy of the two fragments. This energy leads to the evaporation of a few neutrons. In total, the fission process leads to the creation of two nuclei and a few neutrons.

The fission process described above is displayed in a schematic way in Fig. 6.10.

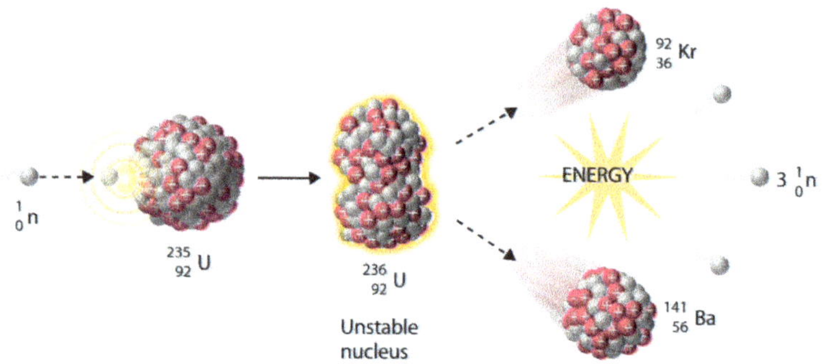

Fig. 6.10: Various stages of the fission process.

In following we will describe the most striking features of the fission process on the basis of the very successful liquid drop model, Section 2.4.1. The presentation, although semiquantitative, gives the main insights of the process. The starting point is the ascertainment that the medium-weight nuclei are more tightly bound than the heavier ones. For example, while the binding energy for $A = 125$ nuclei is 8 MeV per nucleon, for $A = 250$ nuclei it becomes 7 MeV per nucleon. The difference of 250 MeV is the energy released in the fission process. Now, employing the Bethe-Weizsäcker formalism, we find that for the process $(A, Z) \to 2(A/2, Z/2)$ the energy which released is given by

$$E_{\mathrm{rel}} = 2B\left(\frac{A}{2}, \frac{Z}{2}\right) - B(A, Z) = -4.5A^{2/3} + 0.26Z^2A^{-1/3} \quad (\mathrm{MeV}), \tag{6.68}$$

where for the specific case of ^{235}U becomes $\approx$ 180 MeV. In spite of the approximations involved in its derivation, the above equation gives at least quantitatively reasonable values for $E_{\rm rel}$. It fails, however, to explain what kind of fragments are produced and how much energy must be supplied. For example it so happens that a large amount of energy is deposited as internal energy sufficient to produce neutrons.

In the following we will consider a dynamical model to explain with more accuracy the fission process. In particular, we will assume that initially the nucleus is spherical. After the bombardment, its binding energy decreases as the nucleus gets deformed. This leads to the fission process. In the case of a small deformation, considering a prolate nucleus, the major and the minor semi-axes are given by

$$R_{\rm major} = R(1+\epsilon) \qquad \text{and} \qquad R_{\rm minor} = R\left(1-\frac{1}{2}\epsilon\right) \tag{6.69}$$

but the total volume of the nucleus remains constant. According to the Bethe-Weizsäcker formulae the surface energy $E_{\rm surf}$, to a good approximation, is given by

$$E_{\rm surf} = -a_s A^{2/3}\left(1+\frac{2}{5}\epsilon^2+\cdots\right) \tag{6.70}$$

where the parameter ϵ related to the deformation parameter δ via the expression

$$\delta = \epsilon\left(1+\frac{1}{2}\epsilon\right).$$

Moreover, the Coulomb energy of the deformed nuclei is given by

$$E_{\rm Coul} = -a_c Z^2 A^{-1/3}\left(1-\frac{1}{5}\epsilon^2+\cdots\right). \tag{6.71}$$

In the case of a small deformation ($\delta \ll 1$), the energy difference between the spherical and deformed nuclei is given by

$$\Delta E = \mathcal{C}\epsilon^2. \tag{6.72}$$

It turns out, by combining Eqs. (6.70) and (6.71), that the parameter $\mathcal{C}$ is given by

$$\mathcal{C} = \frac{1}{5}\left(a_c Z^2 A^{-1/3} - 2a_s A^{2/3}\right). \tag{6.73}$$

From Eq. (6.73) it is concluded that the coefficient $\mathcal{C}$ (and consequently the energy difference ΔE) is positive only when

$$\frac{Z^2}{A} > \frac{2a_s}{a_c} \simeq 49. \tag{6.74}$$

The meaning of Eqs. (6.72) and (6.73) is the following: While the Coulomb force tends to deform the spherical nucleus the surface tensions tend to act in a way maintaining the spherical shape of the nucleus. Obviously, the inequality (6.74) holds when the Coulomb forces dominate.

Now, we can approach the fission process in a different way, that is via the consideration of the potential energy $U(r)$ of the fissioning nucleus. In particular we made an effort to smoothly interpolate the region between the small deformation of the nucleus and after the situation called scission (after the scission point the process of actual break into two fragments begins).

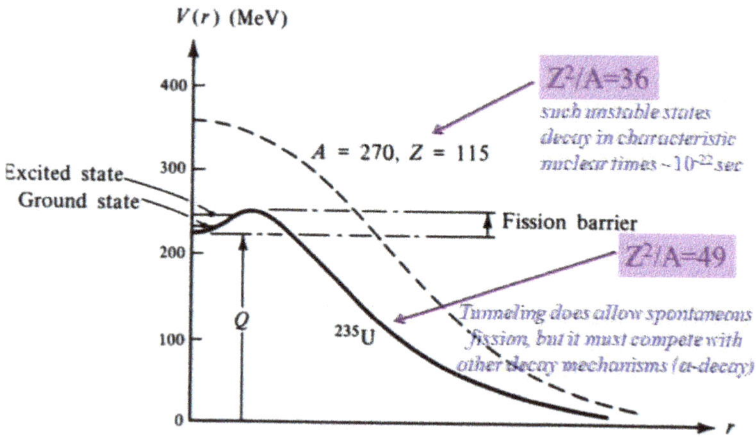

Fig. 6.11: The potential $U(r)$ as a function of the separation distance between the fragments for the indicated two special cases.

The potential energy $U(r)$, which is plotted in Fig. 6.11, is defined as

$$U(r) = C - E_{\rm B} \tag{6.75}$$

where r is the separation distance, $E_{\rm B}$ is the binding energy while the condition $U(\infty) = 0$ holds. Now, in the case of a small deformation the potential energy for $r \ll R$, where R is the nucleus radius, reads

$$U(r) = E_{\rm rel} - \Delta E = E_{\rm rel} - \frac{C}{4}\left(\frac{r}{R}\right)^2 . \tag{6.76}$$

It is worth pointing out that, as expected, after scission, when the two fragments are apart the potential energy is simply the Coulomb energy

$$U(r) = \frac{Z_1 Z_2 e^2}{r} . \tag{6.77}$$

In Fig. 6.11 the potential energy $U(r)$ is displayed for two cases: a) for $Z^2/A = 36$ and b) for $Z^2/A = 49$. For example in the first case a nucleus with $Z = 115$ and $A = 270$ is formed, which decays rapidly in the standard way. The second case corresponds to an unstable nucleus, which can decay by α emission as usual, while at the same time fission may occur, like the classical case of ^{235}U. Even for $Z^2/A < 49$, spontaneous fission can occur, due to quantum mechanical tunneling effects, leading to fragmentation of a nucleus. In general, although the approximation (6.74) holds only for small deformation, it is a very useful guide. For instance it can be shown that the lifetime of a nucleus (A, Z) against spontaneous fission depends on the ratio Z^2/A. In particular, the behavior of the lifetimes exhibits an approximately exponential dependence as a function of Z^2/A.

Finally, there are some very crucial properties of the fission for understanding of the co-called chain reaction in nuclear reactors. These include the following three properties: a) the number of neutrons emitted per fission, b) the energy distribution of these neutrons and c) the fission cross section as a function of neutron energy. Moreover, if scission occurs for small deformation, then the kinetic energy of the fragments will be very large. However, for large deformation the largest part of the energy is stored in the distorted fragments and, consequently, only a small part goes to kinetic energy.

As a conclusion the liquid drop model describes nicely the basic properties and characteristics of the complex phenomenon of the fission process. A more detailed study of the fission process requires the consideration of more elaborate models, which is beyond the goals of this book.

6.6 Beta decays

Beta decay has a long history. It first appeared in 1933, when it was found that the neutron was unstable and decayed into a proton, an electron and a neutrino. The observed life time was very long and Fermi understood that the interaction responsible for the decay must be very weak. He named it weak and the name stuck. It was also realized that the weak interaction can manifests itself only when a neutrino, the only particle which does not feel the strong and electromagnetic interaction, is involved.[2] The second remarkable fact was that the weak interaction was of extremely short range. So in 1934 Fermi proposed that it was a **current-current contact**

[2]The neutrino can be either a real particle or virtual state like in neutrinoless double beta decay.

interaction of the form (in a modern notation)

$$H_W = G_F\left((J^+)^\lambda(j^-)_\lambda + (J^-)^\lambda(j^+)_\lambda\right) \tag{6.78}$$

with

$$(J^+)^\lambda = \bar{p}\gamma^\lambda n \leftrightarrow (n \rightarrow p), (j^-)_\lambda = \bar{e}\gamma_\lambda\nu_e \leftrightarrow \nu_e \rightarrow e^- \text{ or } (e^-, \tilde{\nu}_e) \text{ production} \tag{6.79}$$

where γ_λ are the familiar gamma matrices of the Dirac theory, see Appendix C, Chapter 16. They are discussed in detail in Section 16.2.3. p, n, e and ν are the proton, neutron, electron and neutrino fields respectively, represented by 4-spinors, which are also discussed in Chapter 16. At that time, of course, no distinction was made between neutrino and antineutrino. In addition

$$(J^-)^\lambda = \bar{n}\gamma^\lambda p \leftrightarrow (p \rightarrow n), (j^+)_\lambda = \bar{\nu}\gamma_\lambda e \leftrightarrow e^- \rightarrow \nu_e \text{ or } (e^+, \nu_e) \text{ production.} \tag{6.80}$$

The process $p \rightarrow n + e + \tilde{\nu}_e$ cannot take place in free space.

In this context nuclear processes studied were the following

$$A(N,Z) \rightarrow A(N-1,Z+1)+e^-+\bar{\nu}_e \text{ or } A(N,Z) \rightarrow A(N+1,Z-1)+e^++\nu_e \tag{6.81}$$

the first is called negaton decay or simply nuclear decay and the second positon decay. Another process caused by the weak interaction is electron capture, in which a bound electron of the atom is absorbed by the nucleus

$$e_b^- + A(N,Z) \rightarrow A(N,Z-1) + \nu. \tag{6.82}$$

This is an interesting process and it will be discussed below.[3]

Processes involving neutrino-nucleus scattering

$$\nu_e + A(N,Z) \rightarrow A(N-1,Z+1) + e^-, \, \tilde{\nu}_e + A(N,Z) \rightarrow A(N+1,Z-1) + e^+ \tag{6.85}$$

were not considered at that time.

[3]We are not going to consider here other very interesting decays, second order weak processes like:

$$A(N,Z) \rightarrow A(N-2,Z+2) + 2e^- + 2\bar{\nu}_e, 2\nu\beta\beta \text{ decay} \tag{6.83}$$

or the lepton violating

$$A(N,Z) \rightarrow A(N-2,Z+2) + 2e^-, 0\nu\beta\beta \text{ decay}. \tag{6.84}$$

The last process is very interesting, since it provides the best way for the determination of the scale of neutrino mass (see Chapter 12). The interested reader is referred to the literature, e.g. [deShalit and Feshbach (1974)], [Vergados *et al.* (2012a)].

But then came the news that shocked the word: The fall of parity conservation. It was proposed by Lee and Yang [Lee and Yang (1956)] and experimentally found by Wu and co-workers [Wu *et al.* (1957)]. Furthermore the neutrino is very prejudiced: always left handed! This fact was found experimentally by Goldhaber and co-workers [Goldhaber *et al.* (1958)]. As a result, after efforts by a lot of people, including Lee, Yang, Feynman etc., the contact Fermi interaction was improved and became Vector-Axial Vector (V-A) theory.

$$H_W = \frac{G_F}{\sqrt{2}} \left((J^+)^\lambda (j^-)_\lambda + (J^-)^\lambda (j^+)_\lambda \right) \tag{6.86}$$

with

$$(J^+)^\lambda = \bar{p}\gamma^\lambda (g_V - g_A\gamma_5) n, \; (j^-)_\lambda = \bar{e}\gamma_\lambda (1-\gamma_5)\nu_e \tag{6.87}$$

where γ_5 is the familiar from the Dirac theory gamma-5 matrix, see Section 16.2.3.

This was fine except that it violates unitarity and leads to divergence contributions to the cross section at high energies. After many attempts a solution to this problem came with the invention of the Standard Model (SM) of particle physics or GSW [Glashow (1961)], [Weinberg (1967)], [Salam (1968)]. This model is discussed in detail in many books on particle theory, e.g. in the companion book [Vergados (2017)]. So the interaction for hadrons at the quark level becomes

$$\frac{g}{2\sqrt{2}} \left(\bar{u}\gamma^\lambda (1-\gamma_5) d \right) W^-_\lambda, \quad \frac{g}{2\sqrt{2}} \left(\bar{d}\gamma^\lambda (1-\gamma_5) u \right) W^+_\lambda \tag{6.88}$$

or

$$\frac{g}{2\sqrt{2}} \left(\bar{q}\gamma^\lambda (1-\gamma_5) \tau_\pm q \right) W^\mp_\lambda, \tag{6.89}$$

where τ_+, τ_- are the isospin raising and lowering operators i.e. $\tau_+|d\rangle = |u\rangle$, $\tau_-|u\rangle = |d\rangle$.

For leptons it is of the form

$$\frac{g}{2\sqrt{2}} \left(\bar{\nu}_e\gamma^\lambda (1-\gamma_5) e \right) W^-_\lambda, \quad \frac{g}{2\sqrt{2}} \left(\bar{e}\gamma^\lambda (1-\gamma_2) \nu_e \right) W^+_\lambda. \tag{6.90}$$

Where g is a coupling constant $W^\pm_\lambda$ are two charged heavy bosons with helicity λ.

It is easy to show that at the nucleon level the interaction takes the form

$$\frac{g}{2\sqrt{2}} \left(\bar{N}\gamma^\lambda (g_V - g_A\gamma_5) \tau_\pm N \right) W^\pm_\lambda, \tag{6.91}$$

where now we use the nuclear convention for isospin,[4] namely

$$\left|I = \frac{1}{2},\ I_3 = \frac{1}{2}\right\rangle = n,\ \left|I = \frac{1}{2},\ I_3 = -\frac{1}{2}\right\rangle = p,\ I_+|p\rangle = |n\rangle,\ I_-|n\rangle = |p\rangle. \tag{6.92}$$

Since the mediator between the currents is massive the propagator becomes

$$H_W = \left(\frac{g}{2\sqrt{2}}\right)^2 \frac{1}{\mathbf{q}^2 + m_W^2} \left((J^+)^\lambda (j^-)_\lambda + (J^-)^\lambda (j^+)_\lambda\right). \tag{6.93}$$

For very small momentum transfer this equation is reduced to Eq. (6.86) with

$$\frac{g^2}{m_W^2} = 4\sqrt{2} G_F. \tag{6.94}$$

We thus recover the old current-current contact interaction.

At this point we should mention that the standard model predicts the existence of a neutral gauge boson Z^0. This, of course, leads to a new interaction at the nucleon level of the form

$$H_W = \frac{G_F}{2\sqrt{2}} (J^Z)^\lambda (j^Z)_\lambda \tag{6.95}$$

where

$$(j^Z)_\lambda = \bar{\nu}\gamma_\lambda(1-\gamma_5)\nu,\ (J^Z)^\lambda = \bar{N}\gamma^\lambda(1-\gamma_5)\tau_3 N. \tag{6.96}$$

These currents became known as neutral currents. Their existence was confirmed experimentally in 1973, in a neutrino experiment in the Gargamelle bubble chamber at CERN [Hajert *et al.* (1973)]. This was a triumph for the SM. The $W^\pm$ and Z^0 were discovered in 1983 at CERN [Arnison *et al.* (1983a)], [Arnison *et al.* (1983b)], [Arnison *et al.* (1983c)]. The accepted values of their mass are

$$m_W = 80.425(38)\ \mathrm{GeV/c^2}, \quad m_{Z^0} = 91.1876(21)\ \mathrm{GeV/c^2}.$$

The hadronic current J^λ gives rise to the nuclear matrix element $M_{i\to f}$ as will be discussed below.

Nuclear beta decay is discussed in a number of books, to mention, e.g., [Konopinski (1966)], [Roy and Nigam (1967)], [deShalit and Feshbach (1974)], [Cottingham and Greenwood (1986)] among others.

---

[4] In nuclear physics the isospin indicated by T rather I, but we are not going to use this notation here.

6.6.1 *Kinematics of beta decay*

It can be shown, using standard techniques, as given, e.g., in the companion book [Vergados (2017)], and similar techniques as those of the nuclear physics books [Roy and Nigam (1967)], [deShalit and Feshbach (1974)], that for a non polarized nucleus the invariant amplitude takes the form:

$$|ME_{i\to f}|^2 = |ME_{i\to f}|_F^2 \frac{2(E_e E_\nu + \mathbf{p}_e.\mathbf{p}_\nu)}{m_e m_\nu} + |ME_{i\to f}|_{GT}^2 \frac{2(E_e E_\nu - \mathbf{p}_e.\mathbf{p}_\nu)}{m_e m_\nu} \tag{6.97}$$

E_e, E_ν are the total electron and neutrino energies respectively with $\mathbf{p}_e$ and $\mathbf{p}_\nu$ the corresponding momenta of the electron and the neutrino. The two nuclear matrix elements denoted by F (Fermi) and GT (Gamow-Teller) will be discussed in the next section.

The width, see Chapter 4, for beta decay can be cast in the form[5]

$$d\Gamma = |ME_{i\to f}|^2 G_F^2 \frac{1}{(2\pi)^5}(4\pi)^2\sqrt{E_e^2 - m_e^2}E_e dE_e(\Delta - E_e)^2. \tag{6.98}$$

For a derivation of this formula see Section 13.2. In the above expression Δ is the available energy.[6] This expression is normally written in standard units and with the energy in units of $m_e c^2$ as follows:

$$d\Gamma = |ME_{i\to f}|^2 \frac{G_F^2 m_e^4}{2\pi^3}\frac{c^2}{\hbar^6} m_e c^2 \sqrt{\epsilon^2 - 1}\epsilon(\epsilon_0 - \epsilon)^2 d\epsilon \tag{6.99}$$

with $\epsilon_0 = \Delta/m_e$ and $\epsilon = E_e/m_e$.

The total width is

$$\Gamma = |ME_{i\to f}|^2 \frac{G_F^2 m_e^4}{2\pi^3}\frac{c^2}{\hbar^6} m_e c^2 f(\epsilon_0),$$

$$f(\epsilon_0) = \int_1^{\epsilon_0} \rho(\epsilon_0)d\epsilon,\ \rho(\epsilon_0) = \sqrt{\epsilon^2 - 1}\epsilon(\epsilon_0 - \epsilon)^2. \tag{6.100}$$

$$G_F m_e^2 = N\left(G_F m_p^2\right)\frac{m_e^2}{m^2}\frac{\hbar^3}{c},$$

see Section 9.1, where $N\left(G_F m_p^2\right)$ is the numerical value of the quantity which is $\approx 1.03 \times 10^{-5}$. We thus obtain

$$\Gamma = K|ME_{i\to f}|^2 \frac{1}{2\pi^3} m_e c^2 f(\epsilon_0),\ K \approx 9.265 \times 10^{-24} \tag{6.101}$$

[5]We find it convenient to use here the natural system of units, see Section 9.1.

[6]The available energy Δ is normally given as the energy difference between the initial and final nuclear ground states. For transitions to excited states one should use $\Delta - E_x$ with E_x the excitation energy.

the life time, see Eq. (4.11), is

$$\tau = \frac{1}{K\left|ME_{i\to f}\right|^2 f(\epsilon_0)}(2\pi^3)\frac{\hbar}{m_e c^2}. \qquad (6.102)$$

The half-life is given by

$$\tau_{\frac{1}{2}} = \ln(2)\frac{1}{K\left|ME_{i\to f}\right|^2 f(\epsilon_0)}(2\pi^3)\frac{\hbar}{m_e c^2}. \qquad (6.103)$$

From this we get the celebrated expression ft value

$$ft \equiv f(\epsilon_0)t_{\frac{1}{2}} = \ln(2)\frac{1}{K\left|ME_{i\to f}\right|^2}(2\pi^3)\frac{\hbar}{m_e c^2},\ \frac{\hbar}{m_e c^2} \approx 1.285\times 10^{-21}\ \text{s}. \qquad (6.104)$$

This expression is very useful. It allows one to extract the nuclear matrix element from the measurement of the half life and the spectral function f. In doing so, one must be careful. In Ref. [deShalit and Feshbach (1974)], e.g., one finds

$$t_{\frac{1}{2}} = \ln(2)\frac{1}{\left|ME_{i\to f}\right|^2}\frac{2\pi^3}{\Gamma^2}\frac{\hbar}{m_e c^2}$$

with a different definition of Γ.

At this point we should mention that the above derivation assumes that the outgoing charged lepton, e^- or e^+, is described by a plane wave. This is not true, since the Coulomb field causes a distortion given by the Fermi function $F(Z,\epsilon,R)$. Thus

$$\rho(\epsilon_0,Z,R) = \sqrt{\epsilon^2-1}\,\epsilon(\epsilon_0-\epsilon)^2F(Z,\epsilon,R)d\epsilon,\ f(\epsilon_0,Z,R) = \int_1^{\epsilon_0}\rho(\epsilon,Z,R)d\epsilon. \qquad (6.105)$$

Furthermore we have neglected the small neutrino mass. If that is taken into account one obtains

$$\begin{aligned} \rho(\epsilon_0,Z,\nu) &= \sqrt{\epsilon^2-1}\sqrt{(\epsilon_0-\epsilon)^2-\nu^2}\,\epsilon(\epsilon_0-\epsilon)F(Z,\epsilon), \\ f(\epsilon_0,Z,\nu) &= \int_1^{\epsilon_0-\nu}\rho(\epsilon_0,Z,\nu)d\epsilon, \quad \nu = \frac{\sqrt{\langle m_\nu\rangle^2}}{m_e}. \end{aligned} \qquad (6.106)$$

The precise meaning of $\langle m_\nu\rangle^2$ may become clear in the discussion of triton decay in Section 13.4. If needed, both of these corrections are understood in the ft value given in the expression (6.104).

The Fermi function at the origin [deShalit and Feshbach (1974)]

$$F(Z,\epsilon,0) = \left|\frac{\eta}{1-e^{-2\pi\eta}}\right|, \quad \eta = \pm\frac{\alpha Z\epsilon}{\sqrt{\epsilon^2-1}}. \qquad (6.107)$$

The above simple approximation is not adequate, if relativistic corrections must be taken into account. In this case one adopts a Fermi function, which is evaluated at some appropriate distance from the origin. The appropriate expression now becomes [Landau and Lifshitz (1977)], [deShalit and Feshbach (1974)]:

$$F(Z,\epsilon,R) = (2kR)^{2(s-!)}\frac{2(1+s)}{s^2+\eta^2}e^{\pi\eta}\left|\frac{\Gamma(s+1+i\eta)}{\Gamma(2s+1)}\right|^2, \quad s=\sqrt{1-(\alpha Z)^2}.$$

It can be shown [Landau and Lifshitz (1977)] that the most appropriate distance to use is given by:

$$R \approx \frac{1}{2}\left(\eta^2 - s(s-1)\right)\frac{1}{m_e\alpha Z}.$$

Thus

$$k = \alpha Z m_e \Rightarrow 2kR = \eta^2 - s(s-1).$$

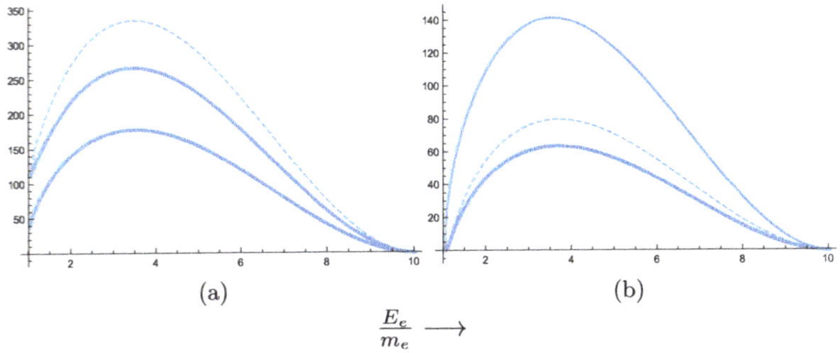

Fig. 6.12: The distribution ρ, spectrum of $e^{\pm}$ in the case of beta decay, as a function of the energy in units of the electron mass. (a) in the case of electron emission and (b) in the case of positron emission. The solid line corresponds to zero charge (no Fermi function correction), while the thick solid line corresponds to the Fermi function with $R = 0$ and Z = 30. The dotted line corresponds to $R \neq 0$, as discussed in the text, also with Z = 30.

In Fig. 6.12 we exhibit the spectrum ρ for three cases, no Coulomb correction, with Coulomb correction with $R = 0$ and with Coulomb correction with $R \neq 0$: (a) electron and (b) positron emission. A typical value of $Z = 30$ has been employed. In Fig. 6.13 we exhibit the same quantity using a Fermi function with $R \neq 0$ for various charges.

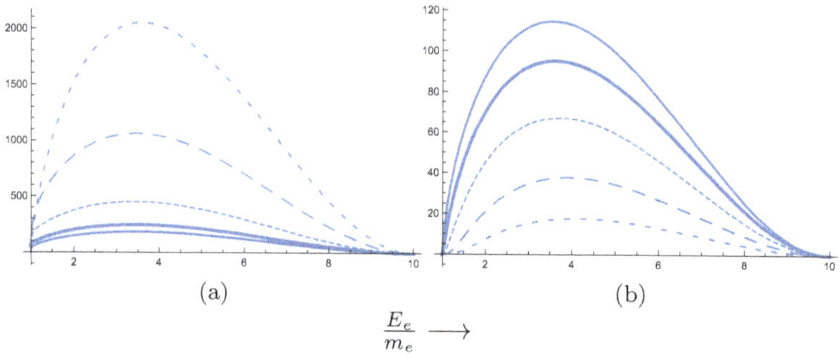

Fig. 6.13: The distribution ρ, spectrum of $e^{\pm}$ in the case of beta decay, as a function of the energy in units of the electron mass. (a) in the case of electron emission and (b) in the case of positron emission. In both cases we used a Fermi function with $R \neq 0$, as discussed in the text. The solid, thick solid, dotted, dashed and dot-dashed curves correspond to $Z = 20$, 40, 70 and 100.

It is interesting to consider the expected spectra of triton

$$^3_1\text{H} \rightarrow ^3_2\text{He} + e^- + \bar{\nu}_e \tag{6.108}$$

with an end point electron kinetic energy of 18.591 keV. Because of its small end point energy, this decay is considered to be the best place to extract the absolute scale of neutrino mass, see Section 12.9. This kinetic energy corresponds to $\epsilon_0 = 1.03638$. In this case the expected spectra are exhibited in Fig. 6.14(a) for massless neutrino and in Fig. 6.14(b), if the neutrino has a non zero mass, say of 1 eV. Both spectra are exhibited in Fig. 6.14(c) near the end point. One notices the suppression in the expected rate if the neutrino has a mass.

6.6.2 *The beta decay nuclear matrix element*

It can be shown, see Chapter 13 that the two matrix elements appearing in beta decays, known as Fermi and Gamow-Teller, involve the operators, which at at nucleon level are given as

$$g_V \Omega_F, \Omega_F = \sum_n \tau_{\pm}(n) = T_{\pm} \text{ Fermi operator}$$

$$g_A \Omega_{GT}, \Omega_{GT} = \sum_n \tau_{\pm}(n) \boldsymbol{\sigma}(n) \text{ Gamow-Teller operator.}$$

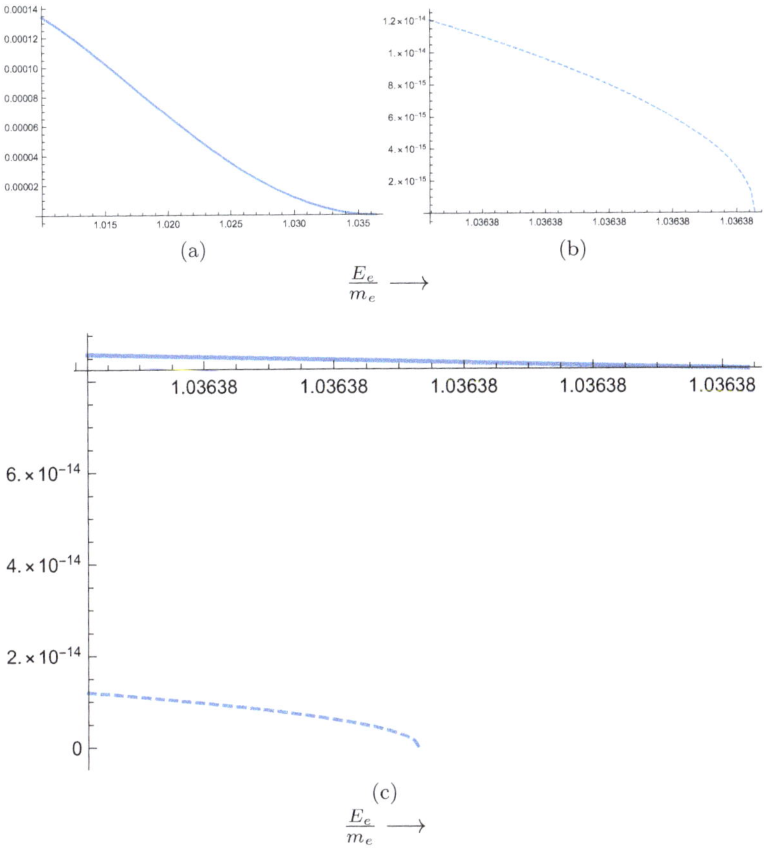

Fig. 6.14: (a) The expected distribution ρ, spectrum of e^- in the case of the triton beta decay, assuming zero neutrino mass. (b) The expected spectrum assuming a neutrino with average mass of 1 eV. One notices the dip of the spectrum near the end point. (c) The previous 2 spectra in the region near the endpoint. One sees the suppression of the spectrum in the case of a massive neutrino. All spectra are plotted as a function of the energy in units of the electron mass.

We will evaluate the nuclear matrix elements considering transitions between states characterized by good angular quantum numbers, i.e. from a state $|J_i M_i\rangle$ to a state $|J_f M_f\rangle$. The nuclear matrix element is defined by

$$|ME_{i\to f}|_F^2 = g_V^2 \frac{1}{2J_i+1} \sum_{M_i,M_f} |\langle J_f, M_f|\Omega_F|J_i M_i\rangle|^2 ,$$

$$|ME_{i\to f}|_{GT}^2 = g_A^2 \frac{1}{2J_i+1} \sum_{M_i,M_f} |\langle J_f, M_f|\Omega_{GT}|J_i M_i\rangle|^2 . \quad (6.109)$$

In the above expression we followed the standard procedure by averaging over the initial magnetic sub-states and summing over all the final ones.

Sometimes the notation

$$|M_F|^2 = \frac{|ME_{i\to f}|_F^2}{g_V^2},\ |M_{GT}|^2 = \frac{|ME_{i\to f}|_{GT}^2}{g_A^2} \quad (6.110)$$

is employed.

Let us begin with Fermi transition. The operator is spin independent and, as a result, the angular momentum cannot change, $J_f = J_i$, $M_f = M_i$ and the matrix element is independent of the of M_i and M_f. As a result the summation over the sub-states yields $(2J_i + 1)$. Thus

$$|ME_{i\to f}|_F^2 = g_v^2 \left| \langle J_f | \sum_n \tau_\pm(n) | J_i \rangle \right|^2 . \quad (6.111)$$

The nuclear states have a good isospin projection I_3 and, usually, good isospin I. In such a case we get:

$$|ME_{i\to f}|_F^2 = g_V^2 \left((I \pm I_3)(I \mp +I_3 + 1) \right), \text{ for } e^\mp \quad (6.112)$$

where I_3 characterizes the initial state. In the case of the neutron decay the isospin matrix element is unity. Thus

$$|ME_{n\to p}|_F^2 = g_V^2 .$$

In the case of Gamow-Teller transition

$$|ME_{i\to f}|_{GT}^2 = g_A^2 \frac{1}{2J_i+1}$$
$$\sum_{M_i,M_f} \langle J_f, M_f| \sum_n \tau_\pm(n)\boldsymbol{\sigma}(n)|J_i M_i\rangle . \langle J_i, M_i| \sum_n \tau_\pm(n)\boldsymbol{\sigma}(n)|J_f M_f\rangle^* \quad (6.113)$$

Eq. (6.113) now becomes:[7]

$$|ME_{i\to f}|^2_{GT} = g_A^2 \frac{1}{2J_i+1} \sum_{M_i,M_f,q} (-1)^q$$

$$\left\langle J_f, M_f \left| \sum_n \tau_\pm(n)\boldsymbol{\sigma}_q(n) \right| J_i M_i \right\rangle . \left\langle J_i, M_i \left| \sum_n \tau_\pm(n)\boldsymbol{\sigma}_{-q}(n) \right| J_f M_f \right\rangle^* \tag{6.115}$$

or

$$|ME_{i\to f}|^2_{GT} = g_A^2 \frac{1}{2J_i+1} \sum_{M_i,M_f,q}$$

$$\left\langle J_f, M_f \left| \sum_n \tau_\pm(n)\boldsymbol{\sigma}_q(n) \right| J_i M_i \right\rangle \left\langle J_f, M_f \left| \sum_n \tau_\pm(n)\boldsymbol{\sigma}_q(n) \right| J_f M_f \right\rangle \tag{6.116}$$

$$|ME_{i\to f}|^2_{GT} = g_A^2 \frac{1}{2J_i+1}$$
$$\frac{1}{2J_f+1} \langle J_f||T^1||J_i\rangle^2 \sum_{M_i,M_f,q} \langle J_i M_i 1q|J_f, M_f\rangle^2 \tag{6.117}$$

but from the unitarity of the C-G coefficients we get

$$\sum_{M_f} \left(\sum_{M_i,q} \langle J_i M_i 1q|J_f, M_f\rangle^2 \right) = \sum_{M_f} = 2j_f + 1.$$

Thus finally

$$|ME_{i\to f}|^2_{GT} = g_A^2 \frac{1}{2J_i+1} \langle J_f||T^1||J_i\rangle^2. \tag{6.118}$$

The evaluation of the many body reduced matrix element can be obtained in the context of the shell model. This general case is beyond the scope of this book. The following selection rules hold:

$$\Delta J = |J_f - J_i| \le 1,\ J_i = 0, J_f = 0 \text{ forbidden, no parity change}$$
$$\Delta I = |I_f - I_i| \le 1,\ I_i = 0, I_f = 0 \text{ forbidden },\Delta I_3 = \pm 1 \text{ for } e\mp. \tag{6.119}$$

[7]We now generalize this result for a general tensor operator T^k_q:

$$\langle J_f M_f|T^k_q|J_i M_i\rangle = \frac{1}{\sqrt{2J_f+1}} \langle J_i M_i, kq|J_f, M_f\rangle \langle J_f||T^k||J_i\rangle \tag{6.114}$$

where $\langle J_f||T^k||J_i\rangle$ is the reduced ME of the operator, see, e.g., [deShalit and Feshbach (1974)], [Hecht (2000)] and Section 14.4. It depends on its nature, but it is independent of the M's.

Table 6.1: The Gamow-Teller ME $|M_{GT}|^2 = \frac{|ME_{J_i \to J_f}|^2_{GT}}{g_A^2}$ for single particle states $n\ell J$. They are independent of n. For $\ell = 0$ only the first entry should be considered.

	$J_i = \ell + \frac{1}{2}$	$J_i = \ell - \frac{1}{2}$
$J_f = \ell + \frac{1}{2}$	$\frac{J_f+1}{J_f}$	$\frac{2J_f+1}{J_f}$
$J_f = \ell - \frac{1}{2}$	$\frac{2J_f+1}{J_f+1}$	$\frac{J_f}{J_f+1}$

For the Fermi transitions the isospin selection rules hold, but $J_i = J_f$.

In the special case of the neutron decay we find:

$$|ME_{n \to p}|^2_{GT} = g_A^2 \frac{1}{2} (\sqrt{6})^2 = 3g_A^2.$$

(This can also be obtained using Table 6.1.)

In the case the nucleus can be described by a single shell model states $n\ell J_i$, $J_i = \ell \pm 1/2$ and $n\ell J_f$, $J_f = \ell \pm 1/2$ the reduced matrix element can be evaluated, see, e.g., [deShalit and Feshbach (1974)], [Hecht (2000)] and Section 14.4. Using such reduced ME one finds the G-T matrix element given in Table 6.1. They are independent of n.

6.6.3 *Single particle or single hole nuclear systems*

We will begin by evaluating the nuclear matrix elements for the triton decay, ${}^3_1\text{H} \to {}^3_2\text{He} + e^- + \tilde{\nu}_e$. For the Fermi transition we clearly have $|M_F|^2 = 1$ and so we need evaluate $|M_{GT}|^2$.

In the simple shell model description the A = 3 system is composed of three $0s1/2$ nucleons. This can be considered as a single hole 0s1/2 hole on the closed os shell. So the matrix elements are the same as those of the p-n system. Thus

$$|M_F\left({}^3_1\text{H} \to {}^3_2\text{He} + e^- + \tilde{\nu}_e\right)|^2 = \frac{1}{2}\left(-\langle 1/2||1||1/2\rangle\right)^2 = 1, \tag{6.120}$$

$$|M_{GT}\left({}^3_1\text{H} \to {}^3_2\text{He} + e^- + \tilde{\nu}_e\right)|^2 = \frac{1}{2}\left(-\langle 1/2||\boldsymbol{\sigma}||1/2\rangle\right)^2 = 3, \tag{6.121}$$

$$|ME_{n \to p}|^2 = f_V^2 + 3f_A^2 \approx 5.613. \tag{6.122}$$

This can be used for other nuclei with 1 particle less than the the closed shell. As we mentioned the single particle approach can be used for one particle outside a closed shell, see Fig. 2.22. This way one can easily evaluate with the aid of Table 6.1 the G-T matrix elements for the following transitions:

- ${}^{11}_{6}\text{C} \to^{11}_{5}$ B, (one hole, ${}^{12}_{6}$C assumed as closed shell).
- ${}^{13}_{7}\text{N} \to^{13}_{6}$ C, (one particle, ${}^{12}_{6}$C assumed as closed shell).
- ${}^{15}_{8}\text{O} \to^{15}_{7}$ N, (one hole, ${}^{16}_{8}$O assumed as closed shell).
- ${}^{17}_{9}\text{F} \to^{17}_{8}$ O, (one particle, ${}^{16}_{8}$O assumed as closed shell) etc.

The evaluation of the Fermi and Gamow-Teller matrix elements is left as an exercise. The goodness of the approximation depends on the goodness of the closed shell description. At some point this approach dos not work, since the protons and neutrons occupy single particle levels, which do not have the same n and/or ℓ. Then the beta decays are not of the allowed Fermi and Gamow-Teller type discussed here, with the selection rules given above. Such decays are are of a different type and hey are called **forbidden beta decays**. The reader interested on such topics is referred to the literature, e.g. [Konopinski (1966)], [deShalit and Feshbach (1974)].

Anyway using the fact that:

$$|ME_{n\to p}|^2 = f_V^2 + 3f_A^2 \approx 5.613$$

from Eq. (6.104) we obtain the $ft = 1062$ s. One can check the value of the matrix element obtained above in the one nucleon approximation by comparing the corresponding ft values with those of Table 6.2. One must first check, however, whether the processes indicated in this table are beta decays or electron captures.

6.6.4 *Electron capture*

The relevant process is indicated by

$$e_b^- + A(N,Z) \to A(N,Z-1) + \nu, \tag{6.123}$$

where the absorbed electron is in a bound atomic orbit. Electron capture is the primary decay mode for isotopes with a relative superabundance of protons in the nucleus, but with insufficient energy difference between the isotope and its prospective daughter (the isobar with one less positive charge) for the nuclide to decay by emitting a positron. Electron capture is always an alternative decay mode for radioactive isotopes that do not have sufficient energy to decay by positron emission.

It is an old process considered soon after Fermi's theory of beta decay. The theory of electron capture was first discussed by Gian-Carlo Wick in 1934 and further developed by Hideki Yukawa and others. K-electron capture was first observed by Luis Alvarez in vanadium-48 [Alvarez (1937)].

Table 6.2: The ft values for various weak transitions in the case of mirror nuclei taken from Konopinski, [Konopinski (1966)].

parent	daughter	single particle transition	ft[s]
${}^{3}_{1}$H	${}^{3}_{2}$He	$0s_{1/2} \to 0s_{1/2}$	1092
${}^{7}_{4}$Be	${}^{7}_{3}$Li	$0p_{3/2} \to 0p_{3/2}$	1730
${}^{11}_{6}$C	${}^{11}_{5}$B	$0p_{3/2} \to 0p_{3/2}$	1730
${}^{13}_{7}$N	${}^{13}_{6}$C	$0p_{1/2} \to 0p_{1/2}$	3876
${}^{15}_{8}$O	${}^{15}_{7}$N	$0p_{1/2} \to 0p_{1/2}$	3876
${}^{17}_{9}$F	${}^{17}_{8}$0	$0d_{5/2} \to 0d_{5/2}$	1915
${}^{19}_{10}$Ne	${}^{19}_{9}$F	$1s_{1/2} \to 1s_{1/2}$	1092
${}^{25}_{13}$Al	${}^{25}_{12}$Mg	$0d_{5/2} \to 0d_{5/2}$	1915
${}^{27}_{14}$Si	${}^{27}_{13}$Al	$0d_{5/2} \to 0d_{5/2}$	1915
${}^{29}_{15}$P	${}^{29}_{14}$Si	$1s_{1/2} \to 1s_{1/2}$	1092
${}^{31}_{16}$S	${}^{31}_{15}$P	$1s_{3/2} \to 1s_{1/2}$	1092
${}^{33}_{17}$Cl	${}^{33}_{16}$S	$0d_{3/2} \to 0d_{3/2}$	3090
${}^{35}_{18}$Ar	${}^{35}_{17}$Cl	$0d_{3/2} \to 0d_{3/2}$	3090
${}^{39}_{20}$Ca	${}^{39}_{19}$K	$0d_{3/2} \to 0d_{3/2}$	3090
${}^{41}_{20}$Ca	${}^{41}_{19}$K	$0f_{7/2} \to 0f_{7/2}$	8476

Some light nuclei weak transitions, which can only proceed via electron capture, are:

$$e_b^- +^7_4 \text{Be} \to^7_3 \text{Li}$$

$$e_b^- +^{26}_{13} \text{Al} \to^{26}_{12} \text{Mg}$$

$$e_b^- +^{37}_{18} \text{Ar} \to^{37}_{17} \text{Cl}$$

$$e_b^- +^{40}_{19} \text{K} \to^{40}_{18} \text{Ar}$$

$$e_b^- +^{59}_{28} \text{Ni} \to^{59}_{27} \text{Co}.$$

It can be shown, see Chapter 13, that the width for this process is

$$\Gamma = \left|ME_{i\to f}\phi_b\right|^2 G_F^2 \frac{4\pi}{(2\pi)^2}(\Delta - \epsilon_b)^2. \tag{6.124}$$

To a good approximation the electron can be evaluated at the origin and thus

$$\left|ME_{i\to f}\phi_b\right|^2 = \left|ME_{i\to f}\right|^2 \phi_b^2(0). \tag{6.125}$$

Where the nuclear matrix element can be evaluated as in positon decay. We thus can write (in natural units)

$$\Gamma = \kappa(Z)\left|ME_{i\to f}\right|^2(G_F^2 m_e^4)\frac{1}{2\pi^3}m_e f(\epsilon_0),$$

$$f(\epsilon_0) = \left(\epsilon_0 - \frac{\epsilon_b}{m_e}\right)^2, \quad \kappa(Z) = 2\pi^2\frac{\phi_b^2(0)}{m_e^3}. \tag{6.126}$$

Thus all the above formulas hold except that

$$\left|ME_{i\to f}\right|^2 \to \kappa(Z)\left|ME_{i\to f}\right|^2.$$

Thus, e.g.,

$$f(\epsilon_0)t_{\frac{1}{2}} = \ln(2)\frac{1}{K\kappa(Z)\left|ME_{i\to f}\right|^2}(2\pi^3)\frac{\hbar}{m_e c^2}, \; \frac{\hbar}{m_e c^2} \approx 1.285\times 10^{-21}\text{ s}. \tag{6.127}$$

The $N = 1(n_r = 0, \ell = 0)$ hydrogenic wave function is given by:

$$\phi_b(r) = \frac{a^{3/2}}{\sqrt{2}}e^{-ar}, \; a = 2\alpha Z\frac{m_e c^2}{\hbar c} \tag{6.128}$$

$$\kappa(Z) = 2\pi^2\frac{1}{2}(2\alpha Z)^3 = 8\pi^2(\alpha Z)^3.$$

For $(\alpha Z) < 0.233$ one finds $\kappa < 1$.

$\kappa(Z)$ can also be evaluated for other electron orbits. If necessary, especially for high Z, distortion factors can be incorporated taking into account the finite nucleus effects but we will elaborate here on such matters, see, e.g., [Konopinski (1966)].

Let us now consider the decay of ${}^{41}_{20}$Ca. In this case the available energy is only 0.44 ± 0.02 MeV. So positron emission is energetically forbidden. For the allowed electron capture the nuclear transition is a single particle one, $0f_{7/2} \to 0f_{7/2}$, with $\kappa(20) = 0.246$. The nuclear matrix element can be obtained from Table 6.1, but the ft value in Table 6.2 must be obtained using this value of κ.

6.6.5 *Neutrino nucleus scattering*

In this case a flux of neutrinos, measured in particles per unit area per unit time, is falling on a nuclear target. The process in question is:

$$\nu_e + A(N,Z) \to A(N-1,Z+1) + e^- \text{ or } \tilde{\nu}_e + A(N,Z) \to A(N+1,Z-1) + e^+ \tag{6.129}$$

involving the charged current and

$$\nu + A(N,Z) \rightarrow A(N,Z) + \nu \text{ or } \tilde{\nu}A(N,Z) \rightarrow A(N,Z)\tilde{\nu}, \quad \text{neutral current} \tag{6.130}$$

(with ν any neutrino flavor). If the neutrino is of sufficient energy, the final nucleus can be in an excited state. The daughter nucleus can also break up, but we will not concern ourselves here with this last possibility. The neutrino nucleus scattering is a vary interesting process and it will be examined in Chapter 13

6.7 Problems

(1) The decay law for N radioactive nuclei as a function of time t is given by:

$$\frac{dN(t)}{dt} = -\lambda N(t), \; \lambda = \text{constant}.$$

Consider now a k chain of subsequent decays involving $n = k+1$ nuclei:

$$\frac{dN_1(t)}{dt} = -\lambda_1 N_1(t), \; \frac{dN_2(t)}{dt} = -\lambda_2 N_2(t) + \lambda_1 N_1(t)$$

$$\cdots$$

$$\cdots$$

$$\frac{dN_{n-1}(t)}{dt} = -\lambda_{n-1}N_{n-1}(t) + \lambda_{n-2}N_{n-2}(t), \; \frac{dN_n(t)}{dt} = \lambda_{n-1}N_{n-1}(t)$$

with initial conditions $N_1 = N_0$, $N_k = 0, k = 2, 3, \cdots n$, while the nth nucleus is stable (the chain stops there).

Consider now specifically $n = 3$. Show that:

$$N_1 = N_0 e^{-\lambda_1 t}, \; N_2 = N_0 \frac{\lambda_1}{\lambda_2 - \lambda_1}\left(e^{-\lambda_1 t} - e^{-\lambda_2 t}\right),$$

$$N_3 = N_0 \left(1 - \frac{\lambda_2}{\lambda_2 - \lambda_1} e^{-\lambda_1 t} + \frac{\lambda_1}{\lambda_2 - \lambda_1} e^{-\lambda_2 t}\right).$$

Find an approximation tho these expressions:

(i) For relatively small times and $\lambda_2 \approx \lambda_1$.

(ii) For $\lambda_1 t \gg 1$ $\lambda_2 \gg \lambda_1$.

(iii) For $\lambda_2 t \gg 1$ $\lambda_2 \gg \lambda_1$.

(2) Consider now a $k = 3$ chain with $n = 4$ nuclei. The population of N_1 and N_2 is as above. Show that:

$$N_3(t) = N_0\lambda_1\lambda_2$$

$$\left(\frac{e^{-\lambda_1 t}}{(\lambda_2-\lambda_1)(\lambda_3-\lambda_1)} + \frac{e^{-\lambda_2 t}}{(\lambda_3-\lambda_2)(\lambda_1-\lambda_2)} + \frac{e^{-\lambda_3 t}}{(\lambda_2-\lambda_3)(\lambda_1-\lambda_3)}\right)$$

$$N_4(t) = N_0\left(\frac{\lambda_2\lambda_3}{(\lambda_2-\lambda_1)(\lambda_3-\lambda_1)}(1-e^{-\lambda_1 t})\right.$$

$$\left.+\frac{\lambda_1\lambda_3}{(\lambda_1-\lambda_2)(\lambda_3-\lambda_2)}(1-e^{-\lambda_2 t} + \frac{\lambda_1\lambda_3}{(\lambda_2-\lambda_3)(\lambda_1-\lambda_3)}(1-e^{-\lambda_3 t}))\right).$$

(3) A well known $k = 3$ chain is the following:

$$\begin{aligned} {}^{79}_{38}\mathrm{Sr} &\rightarrow {}^{79}_{37}\,\mathrm{Rb} + e^+ + \nu_e \ (2.25\ \mathrm{min}) \\ {}^{79}_{37}\mathrm{Rb} &\rightarrow {}^{79}_{36}\,\mathrm{Kr} + e^+ + \nu_e \ (22.9\ \mathrm{min}) \\ {}^{79}_{36}\mathrm{Kr} &\rightarrow {}^{79}_{35}\,\mathrm{Br} + e^+ + \nu_e \ (35.04\ \mathrm{hr}). \end{aligned} \tag{6.131}$$

The final nucleus is stable.
(i) Plot he population of each nucleus as a function of time.
(ii) Since the life time of ${}^{79}_{36}$Kr is much longer than the other two, can one consider it as a $n = 3$ system?

(4) The heaviest naturally-occurring radioactive nuclei are basically unstable because of the Coulomb repulsion of their protons.

Discuss why is alpha decay favored over other modes of disintegration (like fission) in terms of Coulomb barrier to be penetrated.

Solution:

In Fig. 6.15 displayed the Coulomb potential energy of a nucleus of charge $Z_1 e$ and a fragment of charge $Z_2 e$. We consider that the nucleus is broken up into the two mentioned fragments. The probability P_f of penetrating the Coulomb barrier by a fragment of energy E_f is given by

$$P_f = \exp\left[-\frac{2}{\hbar}\int_R^{R_c}\left(2\mu\left[\frac{Z_1 Z_2 e^2}{r} - E_f\right]\right)^{1/2} dr\right] \tag{6.132}$$

where μ is the rest mass and R_c is defined as

$$R_c = \frac{Z_1 Z_2 e^2}{E_f}.$$

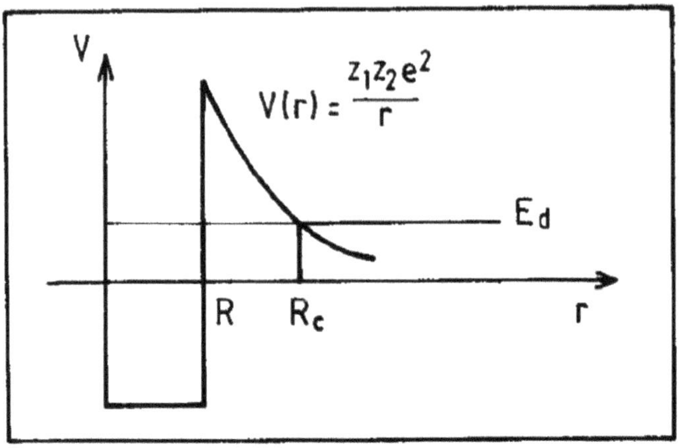

Fig. 6.15: The Coulomb potential energy of a nucleus of charge $Z_1 e$ and a fragment of charge $Z_2 e$.

The probability can be written also as

$$P_f = \exp[-\mathcal{F}]$$

where

$$\mathcal{F} = \frac{2\sqrt{2\mu E_f}}{\hbar} \int_R^{R_c} \left(\sqrt{\frac{R_c}{r}} - 1 \right) dr. \tag{6.133}$$

Defining $x = r/R_c$, the integral takes the form

$$I = \int_R^{R_c} \left(\sqrt{\frac{R_c}{r}} - 1 \right) dr = R_c \int_a^1 \sqrt{\frac{1}{x}} - 1 dx$$
$$= R_c \left(-\sqrt{a - a^2} + \pi/4 - \frac{1}{2} \tan^{-1} \left[\frac{2a-1}{2\sqrt{a - a^2}} \right] \right), \quad a = R/R_c. \tag{6.134}$$

Considering that $a = R/R_c \ll 1$ and expanding Eq. (6.134) we find that

$$I \approx R_c \left(\frac{\pi}{2} - \left(\frac{R}{R_c} \right)^{1/2} \right)$$

and consequently

$$\mathcal{F} \approx \frac{2\sqrt{2} Z_1 Z_2 e^2}{\hbar} \sqrt{\frac{\mu}{E_f}} \left(\frac{\pi}{2} - \left(\frac{R}{R_c} \right)^{1/2} \right). \tag{6.135}$$

In the case of fission the energy release is around 50 times larger than that α-decay, the reduced mass is 20 times larger and also the factor $Z_1 Z_2$ is 5 times larger then the value of $\mathcal{F}$ is almost 4 times larger and consequently the barrier penetrating probability is considerably lower compared to the case of α-decay.

(5) Instability (radioactivity) of atomic nuclei with respect to α-particle emission is a comparatively common phenomenon among the very heavy nuclei, while proton-radioactivity is virtually nonexistent. Give relevant quantitative arguments to explain this difference.

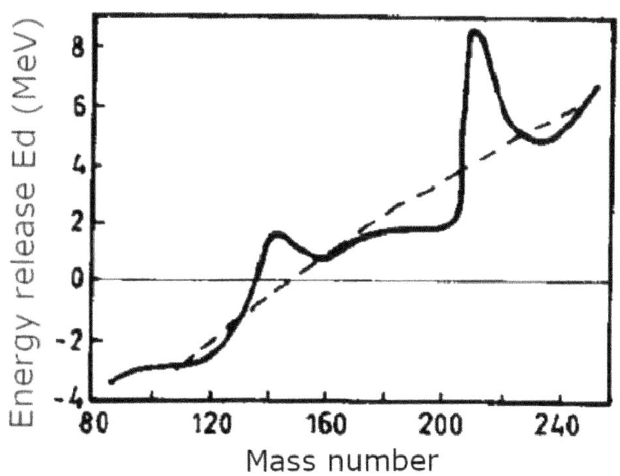

Fig. 6.16: The energy E_d as a function of the mass number A.

Solution:

In the case of α-decay of heavy nucleus (A, Z) the energy release is given by the semi-empirical formula

$$\mathcal{E}_d = -\mathcal{BE}(A,Z) + \mathcal{BE}(A-4, Z-2) + \mathcal{BE}(4,2)$$

$$= -a_s\left[A^{2/3} - (A-4)^{2/3}\right] - a_c\left[Z^2A^{-1/3} - (Z-2)^2(A-4)^{-1/3}\right]$$

$$-a_a\left[\left(\frac{A}{2} - Z\right)^2 A^{-1} - \left(\frac{A-4}{2} - Z + 2\right)^2 (A-4)^{-1}\right]$$

$$+\mathcal{BE}(4,2) - 4a_v. \qquad (6.136)$$

Since for heavy nuclei $Z \gg 2$ and $A \gg 4$ hold, the energy release $\mathcal{E}_d$ to a good approximation is given by

$$\mathcal{E}_d \approx \frac{8}{3}a_s A^{-1/3} + 4a_c Z A^{-1/3}\left(1 - \frac{Z}{3A}\right)$$

$$-a_a\left(1 - \frac{2Z}{A}\right)^2 + 28.3 - 4a_v$$

$$= 48.9A^{-1/3} + 2.86ZA^{-1/3}\left(1 - \frac{Z}{3A}\right)$$

$$-92.8\left(1 - \frac{2Z}{A}\right)^2 - 35.04 \quad (\text{MeV}). \qquad (6.137)$$

Now, considering that for stable nuclei

$$Z = \frac{A}{2 + 0.0154A^{2/3}},$$

we can plot $\mathcal{E}_d$ as a function of A using Eq. (6.137) (Fig. 6.16, dashed line). In the same figure we also plot the actual values of $\mathcal{E}_d$ for naturally occurring nuclei (solid line). We require $\mathcal{E}_d > 0$ in order α-decay to takes place. In particular, the α-radioactive isotopes ${}^{147}_{62}\text{Sm}$ and ${}^{144}_{60}\text{Nd}$ are observed.

In the case of proton-decay we find

$$\mathcal{E}_d \approx -\mathcal{BE}(A,Z) + \mathcal{BE}(A-1, Z-1) < 0 \qquad (6.138)$$

where $\mathcal{E}_d$ is about -7 MeV for heavy nuclei. Since the decay energy is negative, proton-decay cannot take place.

(6) Assume a uranium nucleus breaks up spontaneously into two roughly equal parts. Estimate the reduction in electrostatic energy of the nuclei and relate it with the total change in energy (assume uniform charge distribution and nuclear radius = $1.2 \times 10^{-13} A^{1/3}$ cm).

Solution:

We consider the uranium nucleus where $^{236}_{92}\text{U}$ ($Z_0 = 92$ and $A_0 = 236$) with radius $R_0 = 1.2 \times 10^{-13}\ A_0^{1/3}$ cm. When it breaks up into two roughly equal parts, each part has

$$Z = \frac{Z_0}{2}, \quad A = \frac{A_0}{2}, \quad R = 1.2 \times 10^{-13}\ A_0^{1/3}\ \text{cm}.$$

Considering that the electrostatic energy of a sphere of a uniformly distributed charge Q is $\mathcal{E}_{el} = \dfrac{3Q^2}{5R}$ where R is the radius. Thus, for uranium fission, the electrostatic energy reduction reads

$$\Delta\mathcal{E}_{el} = \frac{3}{5}\left(\frac{(Z_0 e)^2}{R_0} - 2\frac{(Ze)^2}{R}\right) = \frac{3(Z_0 e)^2}{R_0}\left(1 - \frac{1}{2^{2/3}}\right) \approx 364\,\text{MeV}. \tag{6.139}$$

The reduction energy is the source of the energy related in uranium fission. In any case, more elaborate calculations are needed in order to improve the accuracy of the reduction energy.

Chapter 7

Nuclear astrophysics

7.1 Cosmic rays

The cosmic rays, which are one of the main sources of energy originating from the galaxy, are defined as the charged particles which reach the Earth from the space. The cosmic rays consist mainly of a flux of nuclei, electrons, positrons, photons and neutrons [Frauenfelder and Henley (1974)]. In particular, for energies of a few GeV, the flux of hydrogen and helium dominates, including other stable nuclei. Moreover, one of the most important components of the cosmic rays is a flux of electrons. The detection of the cosmic rays is achieved by the study of the different interactions that cosmic rays undergo during their propagation from their source to Earth. These include: a) the electromagnetic interaction of charged particle, b) the inelastic hadronic interaction and c) the nuclear interaction, where heavy nuclei are split into lighter ones. These lead to the production and possible detection of the secondary particle fluxes.

It is worth pointing out that the origin of the cosmic rays is related to the dynamics of star formation, stellar evolution, supernova explosions as well as other astrophysical phenomena.

The cosmic rays were discovered by Victor Hess in 1912 by observing the ionization in an electrometer carried in a manned balloon. Through the years, the cosmic rays have been studied systematically. In particular, these studies have been focused on the composition, energy spectrum, spatial and temporal variation. As an example we consider the case when a cosmic ray proton, with very high energy, enters the top of the Earth's atmosphere. The proton interacts with an nitrogen or oxygen nucleus and a sequence of processes begins. A sketch of a process is presented in Fig. 7.1.

The interaction leads to the creation of of a large number of hadrons,

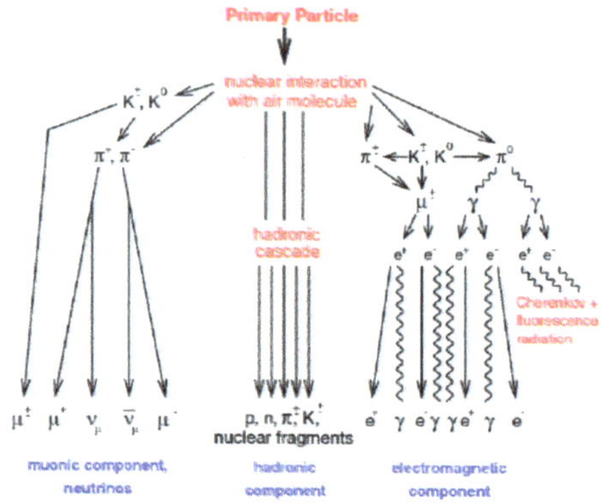

Fig. 7.1: An incident high-energy proton strikes the top of the atmosphere producing cascade showers.

e.g., pions, antinucleons, kaons and hyperons. The created hadrons interact also with the nitrogen or oxygen nucleus, while the unstable ones will also decay leading to the creation of electrons, muons, neutrinos and photons. In addition, the photons produce electron-positron pairs, while the muons decay producing charged leptons and neutrinos. In total, the cosmic ray protons lead to the production of a large number of photons and leptons. To have a picture of the cosmic ray effect, we should mention that cosmic rays can cover a region of many Km^2 at the Earth's surface.

It is of great interest to study the composition of the nuclear components in cosmic ray (see Fig. 7.2). In the same figure the universal distribution of this components is plotted. The most striking features are summarized as following: The elements Li, Be and B are about 10^5 times more abundant in cosmic rays compared to the universality, the ratio ^{3}He/^{4}He is 300 times larger in cosmic rays and finally in cosmic rays the presence of heavy nuclei is more prominent. This is exhibited in Fig. 7.3.

The energy spectrum, that is the number of primary particles as a function of their energy, has been measured with a good accuracy and in a large energy range. It was found that a good fit to the measurements (except for low energy) leads to the dependence

$$\mathcal{I}(E) \sim E^{-2.6} \tag{7.1}$$

where $\mathcal{I}(E)$ is the intensity of the nuclear component with energy E. A similar spectrum exhibited by the electrons, especially for energies above

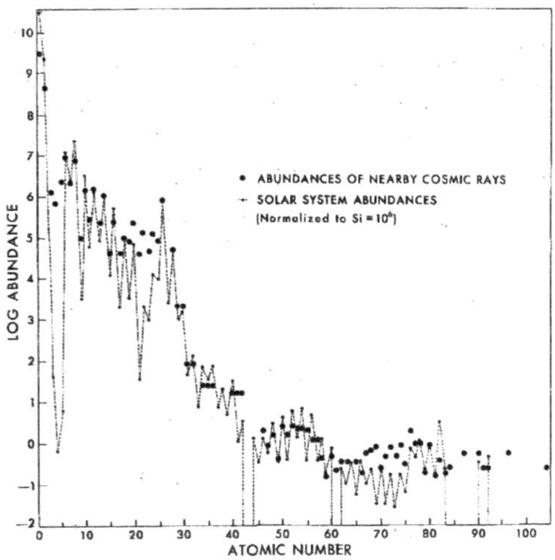

Fig. 7.2: The synthesis of the nuclear component of the primary cosmic rays is exhibited. The universal reference distribution has also been included for comparison.

1 GeV. It also worth mentioning two specific features of cosmic rays. The fist one is the isotropy of cosmic rays, i.e. measurements indicate that the flux of cosmic rays is isotropic, and the second is their constancy over a long period of time. Indeed, the intensity of cosmic rays is approximately constant for a period about 10^9 years.

Cosmic rays have already played a crucial role in strophysics, see the excellent books [Gaisser (1990); Berezinskii *et al.* (1990); Ginzburg and Syrovatskii (1964)]. The detection of ultra high energy particles is becoming of great interest, neutrino telescopes are expected to become the best probes of the sky.

7.2 Nucleosynthesis

The elements which exist in the universe are found on the sun's surface, on the surfaces of distance stars (including elements dredged up from the interior by mixing), and in the interstellar medium. One can divide the elements into roughly five groups:

(1) The most abundant elements in the universe are H about 75% and

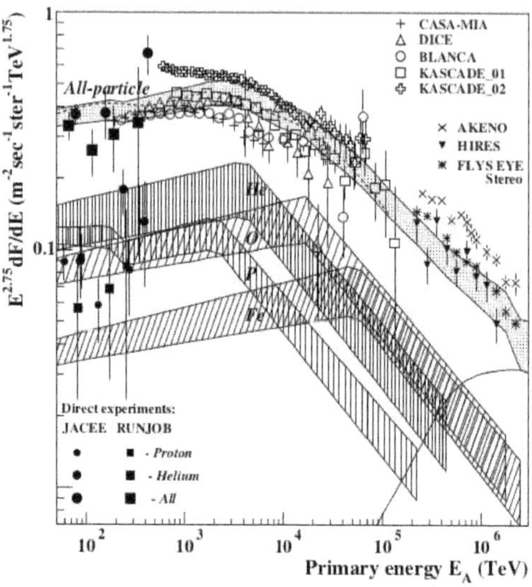

Fig. 7.3: Expected all-particle energy spectrum and energy spectra of different nuclei (shaded areas) according to a 3-component model of primary cosmic rays. The solid line is the expected (third) extragalactic component and the dashed lines correspond to the all-particle spectrum from [Antonyan and Biermann (2001)]. The data JACEE, RUNJOB, CASA, DICE, BLANCA, KASCADE01 and KASCADE02 have been taken from [Swordy and Fortson (2002)].

He 25%. The He elements are now produced in stars. But the stars are too young to have produced so much He. So most He has been produced primordially, i.e. after the big-bang when the conditions in the universe were similar to those in the interior of stars.

(2) Light elements.
In contrast to the abundance of H and He, the lighter "1p-shell" nuclei, i.e. those heavier than He but lighter than C are relatively rare. Relative to H their abundances are lower by 8–10 orders of magnitude:

$$^{6}_{3}\mathrm{Li} \approx 7.75 \times 10^{-10}$$

$$^{7}_{3}\mathrm{Li} \approx 1.13 \times 10^{-8}$$

$$^{9}_{4}\mathrm{Be} \approx 3.13 \times 10^{-10}$$

$$^{10}_{5}\mathrm{B} \approx 5.22 \times 10^{-10}$$

$$^{11}_{5}\mathrm{B} \approx 2.30 \times 10^{-9}.$$

These rare elements have a fascinating heritage. $^{7}_{3}$Li can be produced in the Big Bang, while $^{6}_{3}$Li, Be, and B can also be produced in the interstellar medium, when energetic cosmic-ray protons collide with elements like C, N, and O. This process, therefore, connects to two important sections, one on the origin of cosmic rays, the other on supernovae (which on exploding enrich the interstellar medium with the products of stellar burning, such as C, N, and O). Some stars have intense winds that similarly eject elements into the interstellar medium. But Li, Be, and B can also be synthesized by core-collapse supernovae, directly by the interactions of neutrinos in the carbon shells of such stars. Finally, elements like Li are notoriously fragile, easily burned in stars. We know, e.g., that our sun has only (1/100) of the Li we would anticipate. The result is a very interesting, but the history for these rare elements is complicated. By carefully measuring such abundances in our solar system, in pristine material, i.e. in its original condition, in the interstellar medium as well as on the surfaces of stars of varying ages, we can deduce their evolution history. Thus, the goal of nuclear astrophysics is to use that history towards understanding how the various processes mentioned above have operated since the Big Bang.

(3) The α stable elements.

The α-stable elements C, N, O, Ne, Mg, Si are relatively more abundant. Here α-stable is slang for a nucleus whose (N, Z) is a multiple of $^{4}_{2}$He. An important property of the nuclear force is pairing: nucleons are fermions with spin 1/2. They, thus, obey the Pauli principle and, as we have seen, they fill the lowest levels of the nuclear potential, they gain additional energy from the attractive interaction of two like nucleons (two protons or two neutrons) residing in the same level that is, having the same spatial wave function. Two like-nucleons can exist in the same spatial level because there are two spin choices. The α-stable nuclei are those that, naively, could be envisioned as fully paired. These nuclei tend to be more tightly bound than their neighbors with broken pairs. As a result, these nuclei, in the thermal furnaces inside the cores of stars, tend to be the thermodynamically-favored products of nuclear burning.

These are the final nuclear states that yield the most energy, when daughter nuclei fuse. The relative abundances

$$^{12}_{6}\mathrm{C} \approx 3.87 \times 10^{-3}$$

$$^{14}_{7}\mathrm{N} \approx 0.94 \times 10^{-3}$$

$$^{16}_{8}\mathrm{O} \approx 8.55 \times 10^{-3}$$

$$^{20}_{10}\mathrm{Ne} \approx 1.34 \times 10^{-3}$$

$$^{24}_{12}\mathrm{Mg} \approx 0.58 \times 10^{-3}$$

$$^{28}_{14}\mathrm{Si} \approx 0.75 \times 10^{-3}.$$

(4) There is an abundance peak near iron-group nuclei.
Elements near Fe and Ni have the largest binding energy/nucleon. Thus such elements are the last possible products of a sequence of fusion reactions that produce more tightly bound nuclei, thus liberating energy to maintain the gas pressure of a star. The synthesis of such elements requires hot, dense conditions, rather explosive conditions, because of the large Coulomb barriers that must be overcome for nuclear fusion. High temperatures, and thus high bombarding energies, are needed.

(5) The heavy elements, $A > 100$.
Many of these elements are very rare. Yet an examination of abundances as a function of A or N shows interesting patterns, several abundance peaks. In fact, on closer examination, each peak is actually a double peak, with the two components split by ≈ 10 mass units. The huge Coulomb barriers that inhibit the fusion of heavy nuclei argues that another mechanism must be at work. The only reasonable mechanism is (n, γ) reactions. We will see that heavy elements can be produced in ordinary stars with relatively weak neutron sources (thus the capture on a given nucleus occurs slowly), or under explosive conditions (such as a supernova or neutron star collision), when extremely high neutron densities can be produced for short times. The origin of the elements is shown in Fig. 7.4. The solar system abundance of elements is shown Fig. 7.5. How these elements were formed? Could $^{4}_{2}\mathrm{He}$, 25% by mass, be generated by stellar burning during the lifetime of our galaxy?

Eddington first proposed 4p $\rightarrow^4_2$He as an energy source for our sun

$$4p \rightarrow^4_2 \text{He} + 2e^+ + 2\nu_e. \quad (7.2)$$

The mass difference for this reaction is about 25.7 MeV, while the typical kinetic energy carried off by neutrinos about 0.4 MeV. Therefore about 25 MeV is released per four protons consumed. This is the energy that keeps the electron gas in our sun hot: energy is produced in the solar core at the required rate, just about balancing the energy that is carried off the sun by the photons emitted from the photosphere. Thus we can estimate the rate of fusion in the solar core from the measured solar constant. The solar constant, i.e. the light energy flux on earth, is $\Phi = 0.033$ cal/sec/cm^2. The sun-Earth distance is $r = 1.49 \times 10^{13}$ cm. Thus the power output is

$$4\pi r^2 \Phi \approx 0.92 \times 10^{26} \text{ cal/s} = 2.4 \times 10^{39} \text{ MeV/s}.$$

Since, however, 4 protons are consumed for every 25 MeV produced, the sun consumes about 4×10^{38} protons/s. Now the sun has a mass equal to that of about 1.19×10^{57} protons. Let us assume a solar age of 5 billion years (b.y.) (the sun is estimated to be about 4.6 b.y. in age) and it is burning at the current power level over that period. Then we can estimate the number of protons consumed over that lifetime 5×10^9 y $= 3.156 \times 10^7 \times 5 \times 10^9 = 1.57 \times 10^{17}$ s. Thus

$$1.57 \times 10^{17} \text{ s} \times 4 \times 10^{38} \text{ protons/s} \approx 0.6310^{56} \text{ protons}.$$

So this is 0.63 /11.9 = 5:3% of the sun's mass. Thus only about 5% of protons converted in 5 b.y. Furthermore, of course, this $^4_2$He is locked in the core of our sun, not in places like the interstellar medium where it could be counted by those interested in determining abundances. In addition many protons are not residing in stars.

Thus the tentative conclusion is that stellar burning contributes to, but cannot account for all, of the $^4_2$He observed in the universe. In fact, this conclusion can be put on much firmer ground by looking at the $^4_2$He abundance as a function of stellar metallicity, which is kind of a "clock" for the galaxy. Very metal poor stars presumably were formed very early (before supernovae and novae created many of the metals). The surfaces of such stars should not know about the $^4_2$He synthesis in the core, but rather be representative of the star at its birth. So, even if the surface shows a large $^4_2$He abundance, one would conclude that that $^4_2$He was primordial. We know to-day more about 4p$\rightarrow^4_2$He. It was first described by Bethe and Critchfield in 1939 and was further developed by Salpeter and by Burbidge, Fowler, and Hoyle in the 1950's.

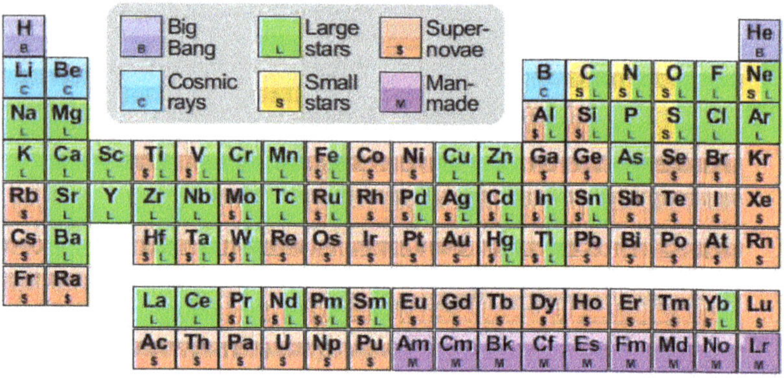

Fig. 7.4: The periodic table of the elements indicating their main source of origin.

We have seen that the Universe is dominated by H and $^4_2$He. The following quantities are employed:

$$X_4 = \frac{N_{^4_2\text{He}}}{N_{\text{Hy}}} = \frac{1}{2}\frac{n}{p}\left(1 - \frac{n}{p}\right) \text{ or } Y_4 = \frac{4X_4}{1+4X_4} = 2\frac{n}{p}\left(1 + \frac{n}{p}\right),$$

where n and p the initial number of protons and neutrons respectively, which depends on the neutron proton mass difference of about 0.8 MeV and the temperature of the universe $kT \approx 0.41$ MeV, i.e. $(n/p) = e^{-\Delta m/kT} = 1/7$. For $n/p = 1/7$ one finds $X_4 = 0.06$ and $Y_4 = 0.33$, i.e. one finds a ratio of proton to He 3 to 1. An equilibrium between protons and neutrons is established via the reactions:

$$n + p \leftrightarrow d + \gamma$$

$$p + e^- \leftrightarrow n + \nu_e$$

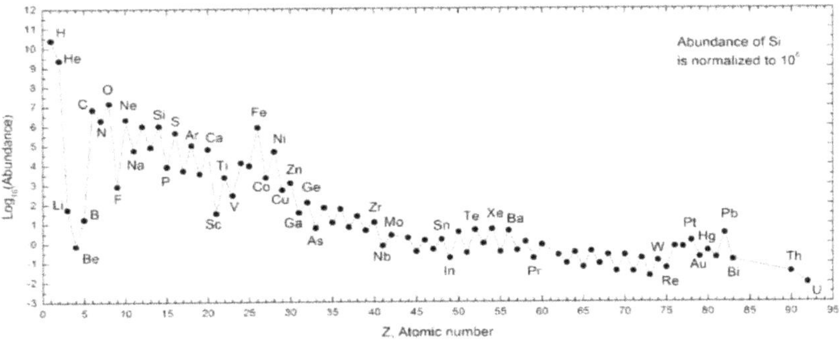

Fig. 7.5: Abundances of the chemical elements in the Solar System. Hydrogen and helium are most common, residuals within the paradigm of the Big Bang, see, e.g., [Weinberg (2008)], [Vergados (2017)]. The next three elements (Li, Be, B) are rare because they are poorly synthesized in the Big Bang and also in stars. The two general trends in the remaining stellar-produced elements are: (1) an alternation of abundance of elements according to whether they have even or odd atomic numbers, and (2) a general decrease in abundance, as elements become heavier. Within this trend is a peak at abundances of iron and nickel, which is especially visible on a logarithmic graph spanning fewer powers of ten, say between logA = 2 (A = 100) and logA = 6 (A = 1,000,000).

$$n + e^+ \leftrightarrow p + \bar{\nu}_e$$
$$n \rightarrow p + e^- + \bar{\nu}_e.$$

Thus at the end of the first minute we have 75% protons and 25% neutrons. At this point the neutron, with a mean lifetime of $\tau = 885. \pm 0.9$ s, is considered as stable. Basically, the hydrogen-helium abundance helps us to model the expansion rate of the early universe. If it had been faster, there would be more neutrons and more helium. If it had been slower, more of the free neutrons would have decayed before the deuterium stability point and there would be less helium.

Stable nuclei of ${}^4_2$He begin to form in the first three minutes of the age of the Universe.

7.2.1 *Fusion of hydrogen to helium in stars*

Before proceeding further with the **Big Bang Nucleosynthesis (BBN)** let us elaborate a bit on what we know to-day about the fusion of hydrogen to helium in stars. In fact we know that there exist two main processes:

- The pp cycle.

$$ {}^1_1\mathrm{H} + {}^1_1\mathrm{H} \to {}^2_2\mathrm{He} + \gamma $$

$$ {}^2_2\mathrm{He} \to {}^2_1\mathrm{H} + e^+\nu_e $$

$$ {}^2_1\mathrm{H} + {}^1_1\mathrm{H} \to {}^3_2\mathrm{He} + \gamma\,(5.49\ \mathrm{MeV}) $$

$$ {}^3_2\mathrm{He} + {}^3_2\mathrm{He} \to {}^4_2\mathrm{He} + 2\,{}^1_1\mathrm{H} + 12.86\ \mathrm{MeV}. $$

The above process takes place 86% of the time and is known as pp-I cycle. Instead of the last reaction one could have:

$$ {}^3_2\mathrm{He} + {}^4_2\mathrm{He} \to {}^7_4\mathrm{Be} + \gamma, $$

followed by

$$ {}^7_4\mathrm{Be} + e^- \to {}^7_3\mathrm{Li} + \nu_e, \quad {}^7_3\mathrm{Li} + p \to 2\,{}^4_2\mathrm{He}\ (\text{pp-II})\,99.89\% $$

or

$$ {}^7_4\mathrm{Be} + p \to {}^8_5\mathrm{B} + \gamma, \quad {}^8_5\mathrm{B} \to {}^8_4\mathrm{Be}^* + e^+ + \nu_e, $$

$$ {}^8_4\mathrm{Be}^* \to 2\,{}^4_2\mathrm{He}\ (\text{pp-III})\,0.11\%. $$

The above process is quite slow, since the second reaction of the above pp cycle is very slow.

- The CNO cycle

$$ {}^{12}_6\mathrm{C} + p \to {}^{13}_7\mathrm{N} + \gamma $$

$$ {}^{13}_7\mathrm{N} \to {}^{13}_6\mathrm{C} + e^+ + \nu_e $$

$$ {}^{13}_6\mathrm{C} + p \to {}^{14}_7\mathrm{N} + \gamma $$

$$ {}^{14}_4\mathrm{N} + p \to {}^{15}_8\mathrm{O} + \gamma $$

$$ {}^{15}_8\mathrm{O} \to {}^{15}_7\mathrm{N} + e^+ + \nu_e $$

$$ {}^{15}_7\mathrm{N} + p \to {}^{16}_8\mathrm{N}^* \to {}^{12}_6\mathrm{C} + {}^4_2\mathrm{He}. $$

So ${}^{12}_6\mathrm{C}$ acts as a catalyst. The net result is the same as in the pp cycle. Four protons are consumed to produce two positrons, two neutrinos and an α particle plus energy.

The CNO process was independently proposed by Carl von Weizsäcker and Hans Bethe in 1938 and 1939, respectively.

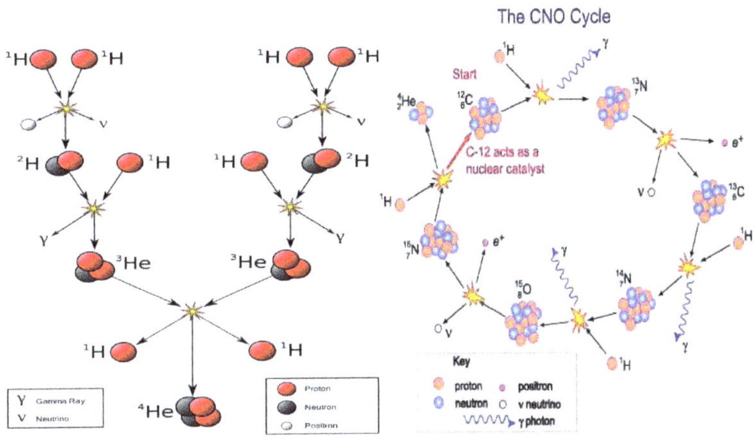

(a) (b)

Fig. 7.6: The fusion of hydrogen to He in stars. (a) The pp cycle and (b) the CNO cycle.

The above reactions are shown pictorially in Fig. 7.6. Additional energy is produced by the pair annihilation of the positrons:

$$e^+ + e^- \to 2\gamma \leftrightarrow (1.05 \text{ MeV}).$$

There exist some slight variations of the above cycles (e.g. CNO-I, CNO-II etc.).

Theoretical models suggest that the CNO cycle is the dominant source of energy in stars more massive than about 1.3 times the mass of the Sun. The proton-proton chain is more important in stars the mass of the Sun or less. This difference stems from temperature dependent differences between the two reactions; pp-chain reactions start occurring at temperatures around $4 \times 10^{6}\,{}^0$K or 4 MK (1 MK = 1 megakelvin), making it the dominant energy source in smaller stars, see, e.g., [Salaris and Cassisi (2005)]. A self-maintaining CNO chain starts occurring at approximately 15 MK, but its energy output rises much more rapidly with increasing temperatures. At approximately 17 MK, the CNO cycle starts becoming the dominant source of energy. The Sun has a core temperature of around 15.7 MK, and only 1.7% of ${}^4_2$He nuclei being produced in the Sun are born in the CNO cycle.

The above light elements have also been produced primordially. First deuterium, D or ${}^2_1$H, nuclei begin to form:

$$p + n \to D + \gamma, \tag{7.3}$$

with a binding energy of 2.2 MeV. Even though the deuterium breaks up easily,

$$\gamma + D \rightarrow p + n, \tag{7.4}$$

each time a fraction of it manages to fuse yielding ${}^4_2\mathrm{He}$

$$D + D \rightarrow {}^4_2\mathrm{He} + \gamma\,. \tag{7.5}$$

This is also achieved via the reactions

$$D{+}p \rightarrow {}^3_2\,\mathrm{He}{+}\gamma, \quad D{+}n \rightarrow {}^3_1\mathrm{H}{+}\gamma, \quad D{+}D \rightarrow {}^3_2\mathrm{He}{+}n, \quad D{+}D \rightarrow {}^3_1\mathrm{H}{+}p, \tag{7.6}$$

$${}^3_1\mathrm{H} + p \rightarrow {}^4_2\mathrm{He} + \gamma, \quad {}^3_1\mathrm{H} + n \rightarrow {}^4_2\mathrm{He} + \gamma, \tag{7.7}$$

$${}^3_1\mathrm{H} + D \rightarrow {}^4_2\mathrm{He} + n, \quad {}^3_2\mathrm{H} + D \rightarrow {}^4_2\mathrm{He} + p. \tag{7.8}$$

With a much slower rate heavier elements are also produced

$${}^4_2\mathrm{He} + D \rightarrow {}^6_3\mathrm{Li} + \gamma, \quad {}^4_2\mathrm{He} +{}^3_1\mathrm{H} \rightarrow {}^7_3\mathrm{Li} + \gamma, \quad {}^4_2\mathrm{He} +{}^3_2\mathrm{He} \rightarrow {}^7_4\mathrm{B} + \gamma. \tag{7.9}$$

The process ends up here since time is running out and there no stable nuclei with 8 nucleons. Indeed the most suitable is the process:

$${}^4_2\mathrm{He} +{}^4_2\mathrm{He} \rightarrow {}^8_4\mathrm{Be} + \gamma\,. \tag{7.10}$$

The produced nucleus, however, has a mean life time $\tau = 2 \times 10^{-16}$ s. As a result the BBN (Big Bang nucleosynthesis) proceeds as in Fig. 7.7.

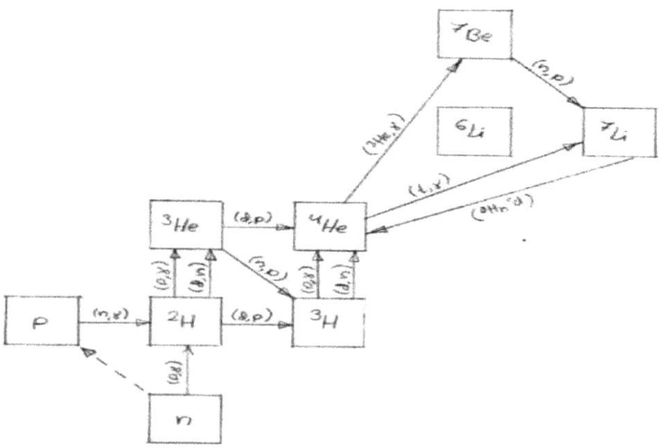

Fig. 7.7: A sketch of the nuclear reactions leading to BBN.

7.2.2 *Solar neutrinos*

We have seen above that in the various cycles electron neutrinos are produced. The main production processes are:

- The pp neutrinos. These are produced as

$$p + p \to d + \nu_e, \quad E_{\nu_e} \leq 0.42 \text{ MeV}, 86\%. \tag{7.11}$$

- The pep neutrinos produced as

$$p + p + e^- \to d + \nu_e, \quad E_{\nu_e} = 1.44 \text{ MeV}. \tag{7.12}$$

- The ^{7}Be neutrinos produced as

$$^7\text{Be} + e_b^- \to ^7\text{Li} + \nu_e, \quad E_{\nu_e} = 0.86 \text{ MeV}(90\%),\ 0.35 \text{ MeV}(10\%). \tag{7.13}$$

- The ^{8}B neutrinos produced as

$$^7\text{Be} + p \to ^8\text{B} + \gamma,\ ^8\text{B} \to ^7\text{Be}^* + e^+ + \nu_e, \quad E_{\nu_e} \leq 18.8 \text{ MeV}, 0.02\%. \tag{7.14}$$

- The hep neutrinos produced as

$$^3\text{He} + p + e^- \to ^4\text{He} + e^+ + \nu_e, E_{\nu_e} \leq 20 \text{ MeV}. \tag{7.15}$$

Unfortunately the most abundant pp and pep neutrinos are characterized by low energy, a fact which makes their detection extremely difficult, below the threshold energy of standard detectors. The same fate awaits the less abundant mono-energetic ^{7}Be neutrinos, as well as the neutrinos arising from the positon decay of ^{13}N, ^{15}O and ^{19}F, with energies less than 1.1 MeV. The neutrinos that can be employed in nuclear physics experiments are the boron neutrinos. (The hep neutrinos have similar energy spectrum but their flux is smaller.) So in general either the flux or the energy is low.

The solar neutrino fluxes on earth are not well understood experimentally and they have been derived theoretically based on the Standard Solar Model of the Sun (SSM), density and temperature profile of the sun, as well as information about the cross sections of the above reactions, which are based on nuclear physics experiments. The thus obtained fluxes, found e.g. in Ref. [Bahcall *et al.* (2005)] are exhibited in Fig. 7.8.

The actual experimental activities concerning the detection of solar neutrinos is based on the recent developments in neutrino oscillations and they will be reviewed in Chapter 12.

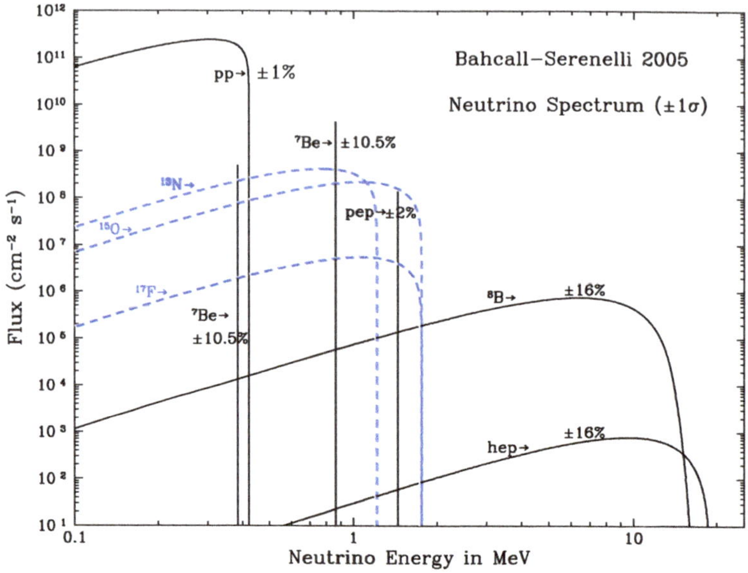

Fig. 7.8: The spectrum of solar neutrinos.

7.2.3 *Summary on nucleosynthesis*

Various elements/isotopes are made via different processes:

- $^{4}_{2}$He by H fusion.
- $^{2}_{1}$H, Li, Be, B non-thermal processes (spallation)

$$^{4}_{2}\mathrm{He} +^{4}_{2} \mathrm{He} \to^{8}_{4} \mathrm{Be}^{*}.$$

- $^{14}_{7}$N, $^{13}_{6}$C, $^{15}_{7}$N, $^{17}_{8}$O: CNO processing.
- $^{12}_{6}$C, $^{16}_{8}$O, $^{20}_{20}$Ne: Helium burning, e.g.

$$^{4}_{2}\mathrm{He} +^{8}_{8} \mathrm{Be} \to^{12}_{6} \mathrm{C}$$

$$^{4}_{2}\mathrm{He} +^{12}_{6} \mathrm{C} \to^{16}_{8} \mathrm{O} + \gamma.$$

- $^{18}_{8}$O, $^{22}_{10}$Ne α: captures on $^{14}_{7}$N (He burning).
- $^{20}_{10}$Ne, Na, Mg, Al, $^{28}_{14}$Si: Partly from carbon burning

$$^{12}_{6}\mathrm{C} +^{12}_{6} \mathrm{C} \to \begin{cases} ^{20}_{10}\mathrm{Ne} + \alpha \\ ^{23}_{11}\mathrm{Na} + p\,. \\ ^{24}_{12}\mathrm{Mg} \end{cases}$$

This requires temperatures temperatures $\geq 10^{9}\,{}^{0}$K. Such temperatures can occur in exploding stars.

- Mg, Al, Si, P, S: Partly from oxygen burning

$$^{16}_{8}\mathrm{O} + ^{16}_{8}\mathrm{O} \to \begin{cases} ^{28}_{14}\mathrm{Si} + \alpha \\ ^{31}_{15}\mathrm{P} + p \\ ^{31}_{16}\mathrm{Mg} + n \end{cases} .$$

  This populates $32 \leq A \leq 42$, but temperatures $\geq 3.6 \times 10^{9}\,{}^{0}\mathrm{K}$ are required. It also occurs in very massive exploding stars.
- Ar, Ca, Ti, Cr, Fe, Ni: Partly from silicon burning. This can populate elements up to $^{56}_{28}\mathrm{Ni}$. As the formation of elements reaches iron, no new element appears. We have seen that the binding energy per nucleon reaches a maximum at the iron group. Beyond this region the binding energy per nucleon deceases. So the iron group cannot serve as a fuel, i.e. burning ceases as soon as Fe is formed. This explains why the elements around iron are more abundant than others.
- Most other elements: neutron capture.

For many applications, the above can be divided into 1) CNO processed elements, 2) α-elements, 3) Fe-peak elements, and 4) neutron-capture elements.

Neutron-capture elements can further be divided into 2 (overlapping) groups: s-process elements (slow process) and r-process elements (rapid process).

To understand the production of heavy elements, consider the locations of stable elements in a plot of proton number versus neutron number, Fig. 7.9. s-process elements are created when neutrons are released slowly during nucleosynthesis, so that the timescale for neutron capture is long compared to that for β-decay (which is typically hours to days). In this case a nuclide (A, Z) is formed via (n, γ). Then a beta decay changes into a nucleus $(A, Z+1)$. What happens is a question of what is the rate for (n, γ) compare to beta decay to $(A, Z+1)$ (see Fig. 7.10)). This generally begins during helium-shell burning with the reactions

$$^{13}_{6}\mathrm{C} + ^{4}_{2}\mathrm{He} \to ^{16}_{8}\mathrm{O} + n \text{ and } ^{22}_{10}\mathrm{Ne} + ^{4}_{2}\mathrm{He} \to ^{25}_{12}\mathrm{Mg} + n.$$

The r-process occurs when neutrons are released rapidly (i.e., during a supernova). The elements acquire a maximum number of neutrons and decay later.

p-process elements (proton capture) are rarer, since such processes must overcome electrostatic repulsion.

Such processes are exhibited in Fig. 7.10.

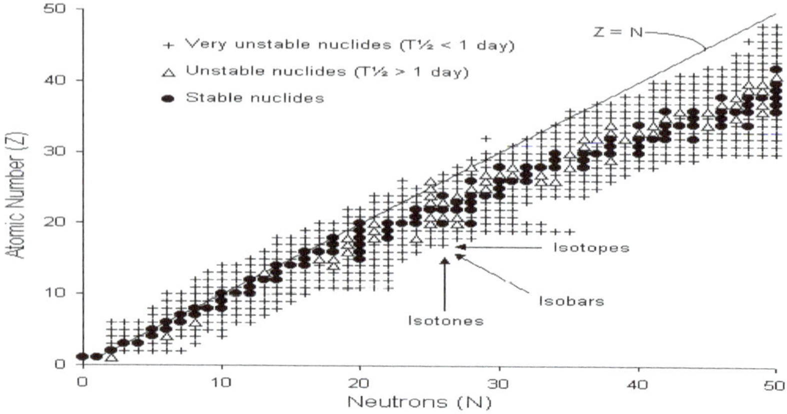

Fig. 7.9: A stability diagram in the (Z, N) plane. There exist a stability valley the nuclei are stable. At its edges they become radioactive.

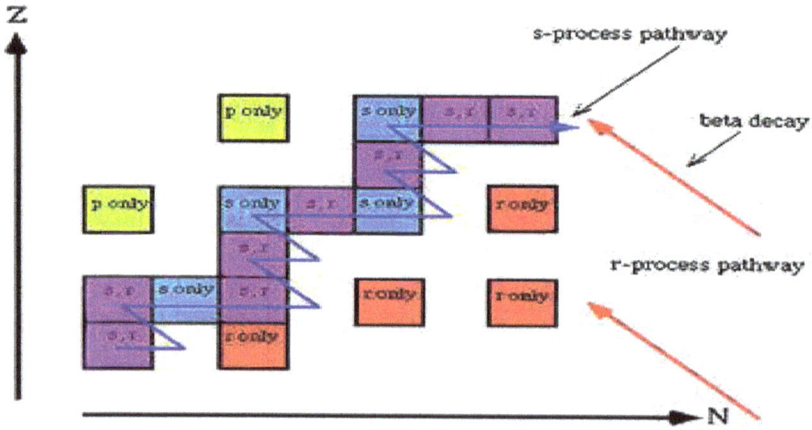

Fig. 7.10: The r, s processes and β decay competition. The blue line indicates the starting and end nuclei after a sequence of r, s and β processes.

7.3 Supernova neutrinos

Supernovae (SN) are considered as powerful explosions in the universe, as the final state of some stars [Henley and Garcia (2007)]. In this case a few solar masses of matter are ejected in the interstellar space with a very large kinetic energy. The ejected matter contains mainly heavy elements which are crucial for the evolution of galaxies and stars. All heavy elements found in our solar system originated from the condensation of the stable nuclei produced this way. In some cases the explosion leads to creation of neutron stars or black holes.

The most remarkable event detection was that of the supernovae SN1987A, which occurred on 23 February 1987 in the Large Magellanic Cloud. One of the most striking features of this observation was its detection through its neutrino burst. This was important not only for the study of the evolution of supernovae but for the useful information gained concerning the neutrino properties as well.

A very important problem, related with the detection of supernovae neutrinos, is the correct estimation of SN rates especially in our galaxy. Such an explosion could produce neutrino bursts with high statistics in neutrino telescopes. However, this estimate for our galaxy (Milky Way), a few per 100 years, is very uncertain and, consequently, additional studies are necessary.

7.3.1 *Core-collapse superonova and neutrino ejection*

Supernova, which is produced by the collapse of the core of very massive stars, is expected to be a source of a very large fluxes of neutrinos. These neutrinos may be detected by terrestrial experiments. We know that, when the core collapses to a proto-neutron star, a chock wave is created and, as a result, it ejects the stellar envelope.

From the core of the proto-neutron star neutrinos of all flavors are produced, according the following process:
a) electron-positron pair annihilation

$$e^- + e^+ \to \nu + \bar{\nu} \tag{7.16}$$

b) electron-nucleon bremsstrahlung

$$e^\pm + N \to e^\pm + N + \nu + \bar{\nu} \tag{7.17}$$

c) nucleon-nucleon bremsstrahlung

$$N + N \to N + N + \nu + \bar{\nu} \tag{7.18}$$

d) plasmon decay

$$\gamma \to \nu + \bar{\nu} \tag{7.19}$$

e) photoannihilation

$$\gamma + e^\pm \to e^\pm + \nu + \bar{\nu}. \tag{7.20}$$

It is worth it to point out that in spite of their weak interaction, the neutrinos are trapped in a SN core (due to their short mean free path inside the very dense nuclear matter). Only, neutrinos created far from the core (in densities lower than 10^{11} gr/cm^3) can escape from the star.

7.3.2 *The supernova neutrino flux*

Supernova neutrinos can be directly detected by the large neutrino telescopes. Such neutrinos originating in our galaxy can be detected by smaller devices like spherical TPC detectors, see Section 11.4.6.2. Large event rates are expected, if the detector can exploit neutral currents [Giomataris and Vergados (2006a)] using a neutron rich target, large N. We have seen in Section 13.8 that the coherent contribution leads to rates proportional to N^2. These detectors have no problem measuring the needed nuclear recoil. The neutral current detectors allow the contribution of all flavors and do not suffer from complications arising from neutrino oscillations. The possibility of detection depends, of course, on the expected neutrino flux on earth. It is normally assumed that the neutrino spectrum can be described be a Fermi-Dirac distribution with a given temperature T and a chemical potential $\mu = aT$. The constants T and a are normally treated as free parameters. Thus

$$f_{sp}(E_\nu, T, a) = \mathcal{N}\frac{1}{1+\exp\left(E_\nu/T - a\right)} \tag{7.21}$$

where $\mathcal{N}$ is a normalization constant. The temperature T is taken to be 3.5, 5 and 8 MeV for electron neutrinos (ν_e), electron antineutrinos ($\tilde{\nu}_e$) and all other flavors (ν_x) respectively. The parameter a will be taken to be $0 \leq a \leq 5$. The number of emitted neutrinos [Tomas *et al.* (2004)], [Janka *et al.* (2007)] can be obtained from the total emitted energy $U_\nu = 3 \times 10^{53}$ erg

$$N_\nu = \frac{U_\nu}{\langle E_\nu \rangle}. \tag{7.22}$$

The quantity $\langle E_\nu \rangle$ is estimated for each species from the distribution, Eq. (7.21)

7.3.3 *Neutrino mass*

The constraining of neutrino mass is a fundamental in particle physics. Actually from neutrino oscillation data, see Chapter 12, we know the differences of the squares of their masses quite well, Table 12.1. The absolute

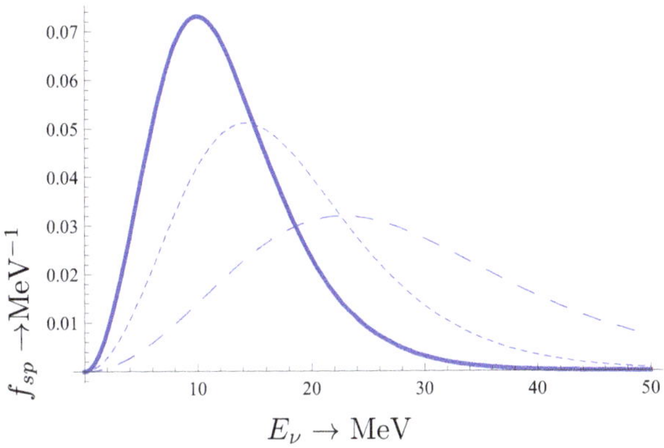

Fig. 7.11: The normalized to unity SN spectrum for $a = 3$. The continuous, dotted and dashed curves correspond to $T = 3.5$ (ν_e), 5 ($\tilde{\nu}_e$) and 8 (ν_x) respectively.

scale, however, of their mass is uncertain. Various laboratory experiments as well as astronomical observations are under way to settle this issue, see Section 12.9.

After the detection of gravitational waves, however, another intriguing possibility arises: the observation of the time of flight of SN neutrinos together with that of gravitational waves may help to settle this problem. Lets consider a relativistic neutrino ($m_\nu c^2 \ll E_\nu$) propagates with velocity

$$v = \frac{p_\nu}{E_\nu} = \sqrt{1 - \frac{m_\nu^2 c^4}{E_\nu^2}} \simeq 1 - \frac{m_\nu^2 c^4}{4E_\nu^2}. \tag{7.23}$$

Now, considering an emitted massless particle (photon or graviton) form the same source at a distance d then the time-of-flight of these relative to that of neutrinos is

$$\Delta t = \frac{d}{v} - \frac{d}{c} = 2.57 \left(\frac{m_\nu c^2}{\text{eV}}\right)^2 \left(\frac{E_\nu}{\text{MeV}}\right)^{-2} \frac{d}{50kpc} \text{ s}. \tag{7.24}$$

This can be written as

$$\Delta t \approx \frac{d}{2c} \left(\frac{m_\nu c^2}{E_\nu}\right)^2. \tag{7.25}$$

The time of flight difference is plotted for various values of $r = (m_\nu c^2)/E_\nu$, for a signal originating within our galaxy, as a function of the distance, in units of the radius of our galaxy, in Fig. 7.12. This, of course, assumes that

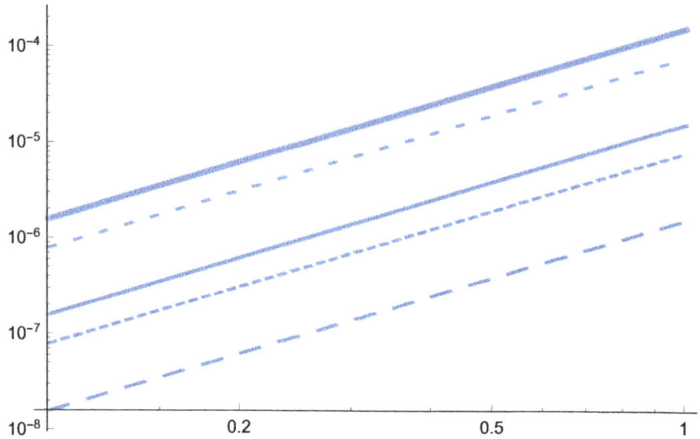

Fig. 7.12: The time of flight difference, in seconds, between the neutrino and the graviton originating from a distance d as a function of d in units of the radius of our galaxy. The long dashed, the short dashed, the solid, the long-short dashed and the thick solid lines correspond to $r/10^{-6} = 0.1$, 0.5, 1.0, 5.0 and 10 respectively.

the gravitational wave and the neutrinos originated essentially at the same point.

The above formula does not take into account the burst, which leads to the radiation time for neutrinos.

Thus, assuming that the burst, which leads to the radiation of neutrinos, has duration ΔT_b, the duration of the observation of neutrinos must be larger than this time. Furthermore, if the mean value and the width of the energy spectrum of neutrinos are E_ν and ΔE_ν respectively, then neutrinos which are produced at the same time with different energies reach the detector in the time interval

$$\Delta T \simeq \frac{m_\nu^2 c^4}{4E_\nu^2} D \frac{\Delta E_\nu}{E_\nu}. \tag{7.26}$$

Where D is the source detector distance. Thus, if ΔT_{obs} is the observed time interval, the following inequality must hold

$$\Delta T \leq \Delta T_b \leq \Delta T_{obs}. \tag{7.27}$$

The inequality (7.27) leads to the definition of the upper mass limit of the neutrino according to

$$m_\nu c^2 \nu \leq E_\nu \sqrt{\frac{E_\nu}{\Delta E_\nu} \frac{\Delta T_{obs}}{D}}. \tag{7.28}$$

Consequently, the upper mass limit of neutrino is written

$$m_\nu^{\max} = 14 \text{ eV} \left(\frac{E_\nu}{10 \text{ MeV}}\right) \left(\frac{E_\nu}{\Delta E_\nu} \frac{\Delta T_{obs}}{10 \text{ s}} \frac{50 Kpc}{D}\right)^{1/2} . \tag{7.29}$$

The energy of neutrinos from a supernovae explosion is of the order of 10 MeV and, at the moment, the existing detectors allow observation only of neutrinos originating from our galaxy, at most from its satellites including the Small and Large Magellanic Clouds. As an example let us consider the SN1987A, from the Large Magellanic Clouds at a distance around 50 kpc form our solar system. The measured neutrino burst had an (average) energy $E_\nu \simeq 15$ MeV, a corresponding width $\Delta E_\nu \simeq 15$ MeV and observation time $\Delta T_{obs} \leq 12$ s. In such a case Eq. (7.28) implies the following model independent bound for an electron neutrino

$$m_{\nu_e} c^2 \leq 30 \text{ eV}. \tag{7.30}$$

Some other more elaborate calculations (considering model-dependent assumptions) leads to bound in the range 5–30 eV.

7.4 Stellar evolution

A star is characterized by its mass and its temperature. Based on them one defines the absolute luminosity, which is the power output of the star. The mass and the luminosity are measured in units of the sun's mass and the sun's luminosity respectively. An interesting plot is the star's luminosity as a function of its temperature known as the Hertzsburg Russel (H-R) diagram. Most stars belong to the main sequence, see Fig. 7.13. A star which is in the main sequence has, depending on its mass, essentially three possibilities to evolve: i) to a white dwarf, which then explodes to a supernova, ii) to a neutron star, the central remnant of a supernova explosion, iii) a black hole, the remnant of a very massive star after it explodes to supernova (see Fig. 7.14). The essential picture of a star's evolution to the Giant Branch is as follows: As soon the stellar material is accumulated, it continues to contract due to gravity and its temperature rises. One has a high density core and an outside shell called mantle.

- Roughly speaking, a star's main sequence hydrogen-to-helium burning core consists of the inner $\approx 10\%$ (by mass) of the star.
- When all the core hydrogen is used up the core begins to collapse, since there is no longer an energy source to maintain its central pressure.

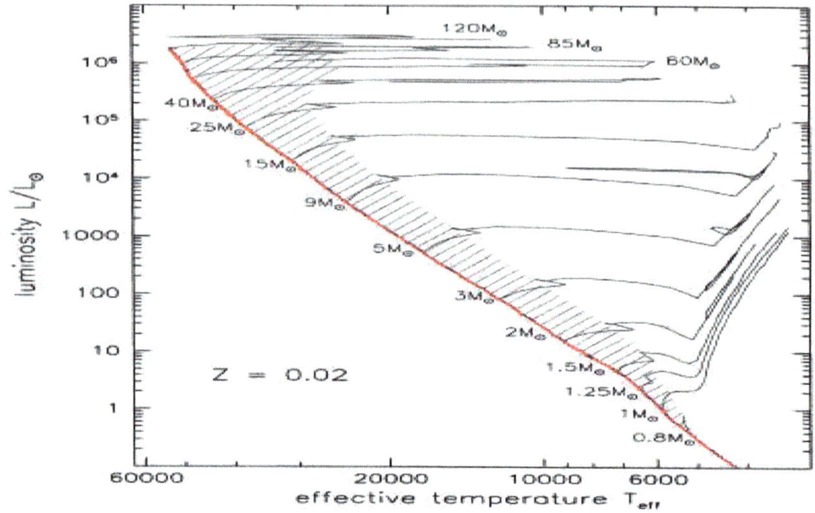

Fig. 7.13: The main sequence of stars in an H-R diagram. The luminosity is measured in units of the sun's luminosity.

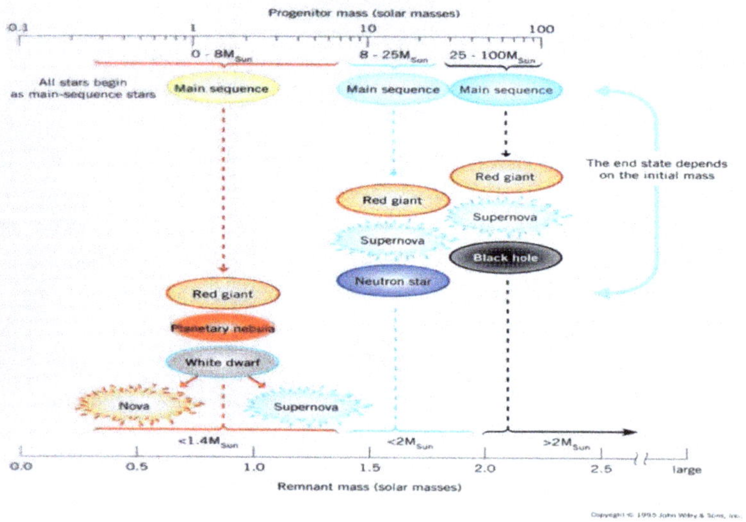

Fig. 7.14: A star belonging to the main sequence has, depending on its mass, essentially three possibilities to evolve: i) to a white dwarf, ii) to a neutron star, iii) to a black hole.

- The core's gravitational potential energy is converted into thermal energy.
- The core collapse also increases the pressure just outside the core, causing hydrogen fusion in a thick shell around the core (5% of the star). This shell will eventually get very thin ($\approx 0.5\%$ by mass).
- As conditions in the shell become more extreme, fusion proceeds via CNO burning, with the rate being entirely a function of the core mass, i.e., $L \propto M^\gamma$, with $\gamma \approx 8$.
- The radiation pressure associated with shell burning pushes matter away in all directions.

Quick rule of thumb: shells follow an expansion. If the core is contracting, the material outside a shell is expanding (see Fig. 7.15).

We should mention that H^- is the dominant source of opacity in the atmospheres of cool stars. The requisite free electrons come from metals, i.e. elements other than hydrogen and helium; consequently, the lower the star's metal abundance, the lower the opacity, the less energy is trapped in the star. Thus the less PdV work is done in expanding the atmosphere, and the hotter the star. The following cases are of special interest:

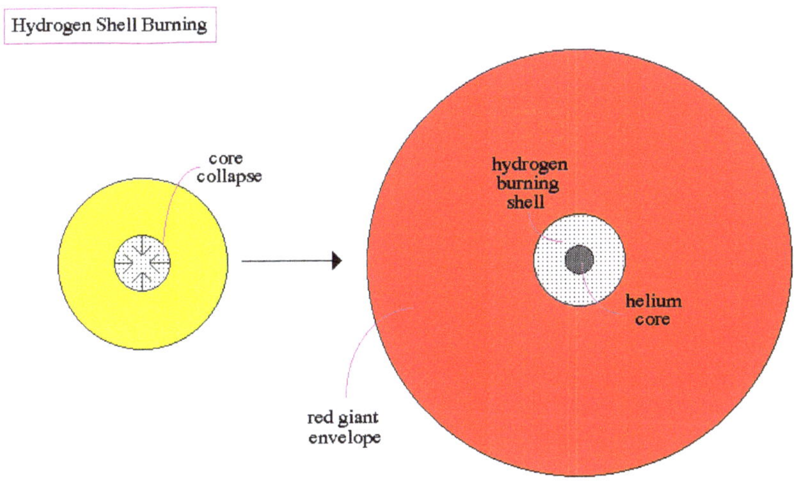

Fig. 7.15: A star in the post main sequence level having reached the red giant stage.

- Horizontal branch stars
  The star does not collapse under its own gravity due to the high opposing pressure resulting from electrons, which form a degenerate electron Fermi gas. Low mass stars ($M \leq 1M_{\odot}$) (where $M_{\odot}$ the mass of the sun), which ignite helium degenerately, settle onto the "Horizontal branch," with luminosity of $L \approx 0.75L_s$ (see Fig. 7.17(a)). The temperature of these stars depends on their envelope mass: the more matter on top of the hydrogen burning shell, the cooler the star. After helium ignition, helium fusion occurs in the center of the (previously inert) helium core. Hydrogen continues to fuse in a shell outside the core (see Fig. 7.16).

  Helium fusion is much less efficient than hydrogen fusion, so the horizontal phase is relatively short (< −1 Gyr). Hydrogen fusing lasts during most of life time of the star.

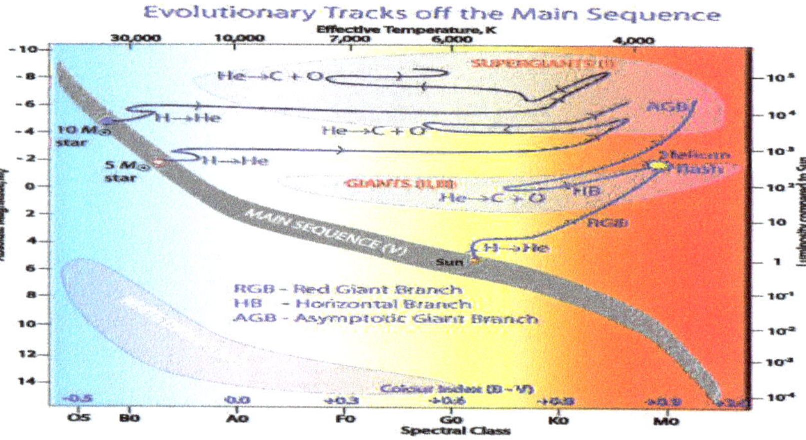

Fig. 7.16: The corresponding H-R diagram relevant for horizontal branch stars.

- The Asymptotic Giant Branch (AGB).
  When helium in the core runs out, the core again contracts, and a thick helium shell burning stage begins. The star starts its trip to the Asymptotic Giant Branch (AGB). The C-O burning begins, see Fig. 7.17(b).

What happens to massive stars?

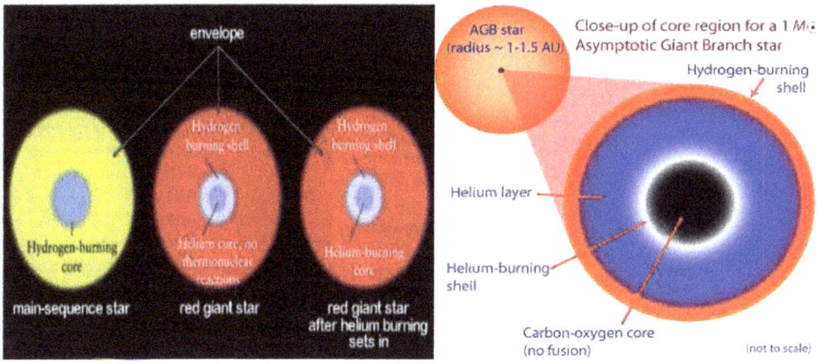

(a) (b)

Fig. 7.17: (a) The sequence from hydrogen to helium burning in horizontal branch stars and (b) a sketch of the Asymptotic Giant Branch (AGB).

High mass stars can fuse carbon/oxygen under non-degenerate conditions. In these objects:

(1) Carbon fuses to a number of products, such as Ne, Na, and Mg. This phase lasts for a few years.
(2) Oxygen fuses to a number of products, such as S, P, and Si. This lasts for about a year.
(3) Photo-disintegration becomes important. The temperatures are so high that photons are capable of "ionizing" nuclei. Equilibrium among the various elements is reached, with the most tightly bound nuclei becoming more abundant. This phase is called "silicon burning", and it lasts for about a day. The principal product of these fusion processes is iron, the most tightly bound nucleus. Note that because iron is the most tightly bound nucleus, any reaction involving iron is endothermic. Once iron is produced, there is no more energy available in the star. The situation is represented in Fig. 7.18.

What eventually happens to a star? Stars with initial masses $M < 5M_\odot$ end they live as C/O (and possibly Ne) stars, see Fig. 7.17(b). They end up as white dwarfs. There aren't many white dwarfs because the hydrogen burning lasts very long (Serius B, the companion of Serius is such an example). Gravity is balanced by the pressure of degenerate electrons. When

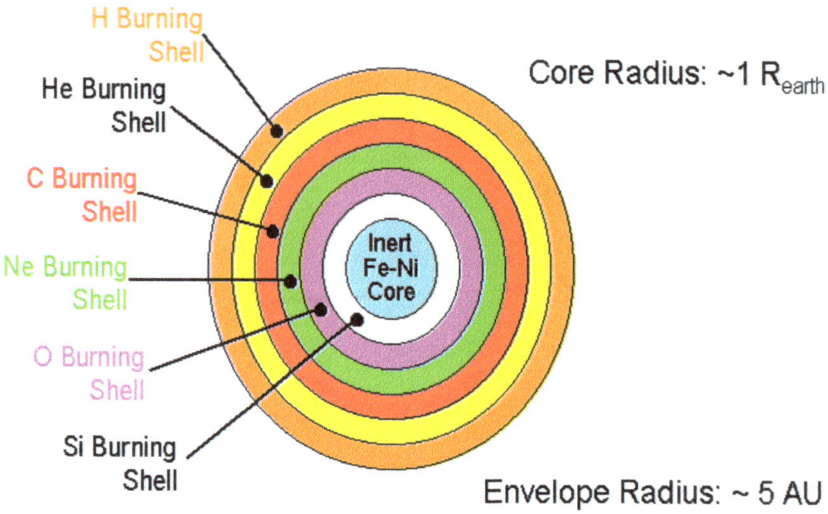

Fig. 7.18: A very massive star reaches the level of heavy element burning.

the fuel for burning is exhausted, if their mass is greater than the Chandrasekhar limit of $1.4M_\odot$, the star explodes and a supernova follows. Most of the energy is carried away by neutrinos. If the star's mass is less than $M < 1.4M_\odot$, it becomes a black star, i.e. its temperature is the same with its environment. Sometimes even a small white dwarf can explode into a supernova. This happens because many such stars are members of a binary system. Due to cannibalism their mass is increasing, lost by their partner, usually a red giant. When the Chandrasekhar limit is reached, a supernova follows, **supernova type Ia**. This supernova is extremely useful in cosmology because the energy output is exactly known and such supernovae serve as a **standard candle** indexstandard candle.[1] In other words by measuring the energy flux in the laboratory, we can determine the distance of the star from us. This is how the large cosmological distances are measured![2]

If the initial mass of the star is $5M_\odot < M < 25M_\odot$, the full nucleosynthesis cycle goes on and eventually the read giant explodes, while its core is left behind as a neutron star. It is supported be the high pressure of the degenerate neutron Fermi gas. If the mass is higher than $25M_\odot$, a black

[1]Astronomers can distinguish this particular type of supernova by its characteristics (light curve etc.).

[2]See Section 10.9.1 of the companion book.

hole is left behind. The mass of the black hole can even reach the mass of an entire galaxy (it can even exceed $10^6 M_\odot$).

7.5 Neutron stars

The most important compact stellar objects in the universe are White Dwarfs, Neutron Stars and Black Holes. Many interesting reviews are available, discussing these topics. The reader interested in all three is referred to [Shapiro and Teukolsky (1983)] and [Glendenning (2000)].

Neutron stars (NS) are the most interesting compact stellar objects, a fact that makes them extraordinary astronomical laboratories for studying the physics of dense nuclear matter. Very recently, the detection of gravitational waves from the merger of two neutron stars, in a binary neutron-star system, opened a new powerful window to the exploration of the physics of NS. In this book we will supply some simple ideas and facts about NS and for a more advanced exposition we refer to specialized books, like those mentioned above and in particular [Haensel *et al.* (2007)].

We will begin by summarizing some basic facts:

We have seen that if the star burns up its fuel it is doomed. It can become a white dwarf, a neutron star, a black hole or it can become completely disassembled. Of particular interest to us is the case of a neutron star. A neutron star is formed, if the initial star mass is $5M_\odot < M < 25M_\odot$. At some point the nuclei melt down and the electrons fuse with the protons to form neutrons. The neutron stars, which result, are fairly compact objects with a radius $R \approx 10$–14 km and a density $\rho \approx 10^{14}$ g/cm^3.

From a knowledge of the density, the composition of the star is known. The star behaves like a hydrostatic fluid with the pressure given in terms of the density, the composition and the temperature, which defines the equation of state. When the pressure is very low the various nuclear processes we have seen above take place, which are fast compared to the evolution rate. The star is in equilibrium composed of a nuclear soup with electrons mostly bound in their atoms. Eventually weak processes take place, generating neutrinos, which carry away energy, but a new equilibrium is established. As we have seen, at low pressure the ground state of the system is composed of electrons and Fe nuclei. At still higher densities $\rho \approx 10^4$ g/cm^3 the atoms are fully ionized, the electrons are no longer bound. These electrons acting as a degenerate Fermi gas provide the needed pressure to prevent the gravitational collapse of the star. We can estimate the energy and the pressure from the formulas we derived in the Fermi gas model, but we should use the correct relativistic formula in relating the energy and momentum $E = pc$. Thus

$$E = \left(\frac{\pi}{2}\right)^{2/3} \hbar c \left(\frac{N_e^{4/3}}{V^{1/3}}\right) \tag{7.31}$$

$$P = -\frac{\partial E}{\partial V} = \frac{1}{3}\left(\frac{\pi}{2}\right)^{2/3} \hbar c \left(\frac{N_e}{V}\right)^{4/3} . \tag{7.32}$$

At densities above $\rho \approx 10^{11}$ g/cm^3 $^{56}_{26}$Fe is no longer the ground state of the system. Electrons have acquired enough energy so that a series of electron captures takes place, with the emitted neutrinos carrying away energy. Eventually the neutron rich nucleus ^{112}Kr is formed. At densities about $\rho \approx 4 \times 10^{11}$ g/cm^3 neutrons begin to leak out forming a degenerate Fermi gas. At still higher densities more neutron rich nuclei are formed. At about $\rho \approx 2.5 \times 10^{14}$ g/cm^3 these nuclei almost fuse to form a continuous fluid of protons, neutrons and electrons, with the protons less than 4% and an equal number of electrons. Neutrons can no longer decay to protons, since the produced electrons must be below the Fermi surface and they are Pauli blocked.

As a fairly massive star evolves, it is natural to assume that:

- The outer region of the star contains predominantly nuclei, electrons and neutrons.
- At intermediate depths neutron, proton and electron fluids.
- In the interior (core) region a neutron fluid.

A reasonable picture is given in Fig. 7.19.

After the supernova explosion all material but the core is blown away. A neutron star is born. Neutron stars were discovered a long time ago [Baade and Zwicky (1934)]. A few years later objects emitting periodic radio signals were discovered, named pulsars [Hewish *et al.* (1968)]. The pulsars were found to be neutron stars [Gold (1968)].

We will continue with a more quantitative discussion. Many of the static properties as well as some dynamical processes of neutron stars are sensitively dependent on the employed equation of state. However, the knowledge of the equation of state, especially at high densities, is very uncertain and, consequently, the relevant predictions and estimations suffer from uncertainties. On the other hand, for low densities (close to the saturation density of symmetric nuclear matter) the equation of state (EoS) is well constrained and the relevant predictions are more reliable.

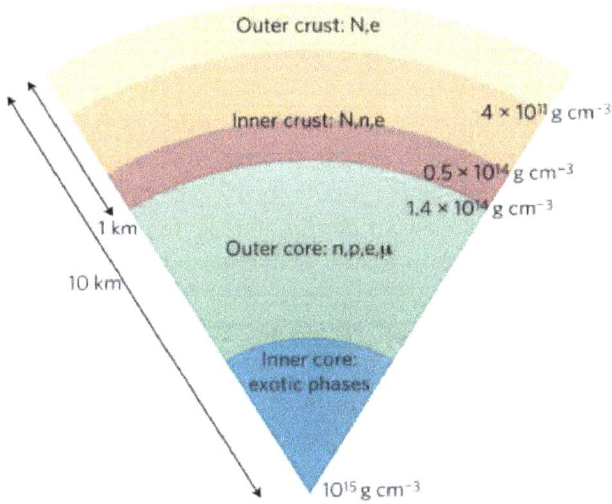

Fig. 7.19: We exhibit the profile of a typical neutron star, composition and density expected as a function of the distance from its center.

The starting point for determining the mechanical equilibrium of neutron star matter is the well known Tolman-Oppenheimer-Volkoff (TOV) equations [Oppenheimer and Volkoff (1939); Tolman (1939)]. This set of differential equations describes the structure of a neutron star. These equations were derived assuming that for a static spherically symmetric system, the metric can be written as follows:

$$ds^2 = -e^{\nu(r)}c^2dt^2 + e^{\lambda(r)}dr^2 + r^2\left(d\theta^2 + \sin^2\theta d\phi^2\right). \tag{7.33}$$

The TOV equations have the form

$$\frac{dP(r)}{dr} = -\frac{G\mathcal{E}(r)M(r)}{c^2r^2}\left(1 + \frac{P(r)}{\mathcal{E}(r)}\right)\left(1 + \frac{4\pi P(r)r^3}{M(r)c^2}\right)\left(1 - \frac{2GM(r)}{c^2r}\right)^{-1}, \tag{7.34}$$

$$\frac{dM(r)}{dr} = \frac{4\pi r^2}{c^2}\mathcal{E}(r). \tag{7.35}$$

To solve the above set of equations for $P(r)$ and $M(r)$, one can integrate outwards from the origin ($r = 0$) to the point $r = R$ where the pressure becomes zero. This point defines R as the coordinate radius of the star.

To do this, one needs an initial value of the pressure at $r = 0$. The radius R and the total mass of the star, $M \equiv M(R)$, depend on the value of the pressure at the center of the star. To be able to perform the integration, one also needs to know the energy density $\mathcal{E}(r)$ (or the mass density $\rho(r)$) in terms of the pressure $P(r)$. This relationship is the equation of state for neutron star matter. It should also be noted that, besides the stellar radius and mass, other global attributes of a neutron star are potentially observable, like the moment of inertia and the binding energy.

The interior of a neutron star is divided into the outer core and the inner one. It has a radius of approximately 10–14 km and contains most of the star's mass. The crust, with a thickness of about 10% of the total radius and containing only a few percent of the total mass, can also be divided into an outer and an inner part. The equation of state of neutron-rich nuclear matter is the important ingredient in the study of both the core and the crust, all the bulk properties of a NS. More specifically, the implementation of the EoS predicts the location of the inner edge of a neutron star crust. The inner crust comprises the outer region, from the density at which neutrons drip out of nuclei to the inner edge, separating the solid crust from the homogeneous liquid core. At the inner edge, in fact, a phase transition occurs from the high density homogeneous matter to the inhomogeneous one at lower densities. It was found that the transition density is related to some properties of finite nuclei including neutron-skin, dipole polarizability etc.

7.5.1 *Nuclear symmetry energy*

The symmetry energy plays an important role on the bulk properties of neutron star matter and consequently is a key quantity to explain in general many neutron star properties and dynamical processes. We consider that the energy per particle of nuclear matter $E_b(n, I)$ can be expanded around the asymmetry parameter I as

$$E_b(n,I)=E_b(n,I=0)+E_{\rm sym,2}(n)I^2+E_{\rm sym,4}(n)I^4+\cdots+E_{\rm sym,2k}(n)I^{2k}+\cdots \tag{7.36}$$

where $I = (n_n - n_p)/n = 1 - 2x$ (x is the proton fraction n_p/n). The coefficients of the expansion (7.36) are given by the expression

$$E_{\rm sym,2k}(n) = \frac{1}{(2k)!}\frac{\partial^{2k} E_b(n,I)}{\partial I^{2k}}\bigg|_{I=0}. \tag{7.37}$$

The nuclear symmetry energy $E_{\rm sym}(n)$ is defined as the coefficient of the quadratic term, that is

$$E_{\rm sym}(n) \equiv E_{\rm sym,2}(n) = \frac{1}{2!}\frac{\partial^2 E_b(n,I)}{\partial I^2}\bigg|_{I=0} \tag{7.38}$$

and the slope of the symmetry energy L at the nuclear saturation density n_s, which is an indicator of the stiffness of the EoS, is defined as

$$L = 3n_s \left.\frac{dE_{\rm sym}(n)}{dn}\right|_{n=n_s}. \tag{7.39}$$

In the framework of the parabolic approximation (PA) the energy per particle is given by the expression

$$E_b(n,x) \simeq E_b\,(n, I=0) + I^2 E^{PA}_{sym}(n), \tag{7.40}$$

where $E^{PA}_{\rm sym}(n)$ is simply defined as

$$E^{PA}_{\rm sym}(n) = E_b(n, I=1) - E_b\,(n, I=0)\,. \tag{7.41}$$

7.5.2 *An application*

In order to construct an equation of state of neutron star matter, one can consider a schematic equation for the symmetric nuclear matter energy, which is given by the expression

$$E(n, I=0) = m_n c^2 + \frac{3}{5}E^0_F u^{3/2} + V(u), \quad u = n/n_s \tag{7.42}$$

where $E^0_F = (\hbar k^0_F)^2/2m_n$ is the Fermi energy per baryon in equilibrium state and n_s is the saturation density of nuclear matter ($n_s \simeq 0.16$ fm^{-3}).

The density dependent potential energy per nucleon, $V(u)$, of the symmetric nuclear matter, can be parameterized as follows

$$V(u) = \frac{1}{2}Au + \frac{Bu^\sigma}{1+B'u^{\sigma-1}} + 3\sum_{i=1,2} C_i \left(\frac{\Lambda_i}{p^0_F}\right)^3 \left[\frac{p_F}{\Lambda_i} - \arctan\left(\frac{p_F}{\Lambda_i}\right)\right] \tag{7.43}$$

where p^0_F is related to the Fermi momentum p_F by $p_F = p^0_F u^{1/3}$. The parameters Λ_1 and Λ_2 parameterize the finite forces between nucleons and for example, to a very good approximation, $\Lambda_1 = 1.5p^0_F$ and $\Lambda_2 = 3p^0_F$. The rest of the parameters, A, B, B', σ, C_1 and C_2, are determined by the constraints provided by the properties of nuclear matter saturation. In particular, these parameters are selected in order that $E(n_s, I = 0) = -16$ MeV, $n_s =$ fm^{-3} and the incompressibility $K_0 = 240$ MeV $\left(K_0 = 9n_s^2 \left.\frac{d^2(E(n,I=0))}{dn^2}\right|_{n=n_s}\right)$.

Now the total energy per particle of neutron star matter $E(n,x)$ will be given by the sum of the energy per baryon and electron energy, that is

$$E(n,x) = E_b(n,x) + E_e(n,x), \tag{7.44}$$

where the fraction x is determined, in general, by Eq. (7.54). The electrons are considered as a non-interacting Fermi gas and consequently [Shapiro and Teukolsky (1983)]

$$E_e(n,x) = \frac{3}{4}\hbar c \left(3\pi^2 x^4 n^4\right)^{1/3}. \tag{7.45}$$

The total energy density $\mathcal{E}$ needed to construct the EOS is given by

$$\mathcal{E}(n) = E(n)/n. \tag{7.46}$$

Accordingly, the total pressure is decomposed also into baryon and lepton contributions

$$P(n,x) = P_b(n,x) + P_e(n,x), \tag{7.47}$$

where by definition

$$P_b(n,x) = n^2 \left(\frac{\partial E_b}{\partial n}\right)_x. \tag{7.48}$$

The contribution of the electrons to the total pressure is equal to

$$P_e(n,x) = \frac{1}{12\pi^2}\frac{\mu_e^4}{(\hbar c)^3} = \frac{\hbar c}{12\pi^2}\left(3\pi^2 x n\right)^{4/3}. \tag{7.49}$$

From Eqs. (7.46) and (7.47) we can construct the equation of state in the form $\mathcal{E} \equiv \mathcal{E}(P)$. Finally one can proceed with the solution of the TOV equations (7.34) and (7.35) using as input the above equation of state. What remains is the calculation of the proton fraction x. This calculation is based on β-stable nuclear matter where the following processes take place simultaneously

$$n \to p + e^- + \bar{\nu}_e, \quad p + e^- \to n + \nu_e \tag{7.50}$$

and, considering that neutrinos generated in these reactions have left the system, the chemical equilibrium condition takes the form

$$\mu_n = \mu_p + \mu_e. \tag{7.51}$$

It is easy to show that, after some algebra, one gets

$$\mu_n - \mu_p = \left(-\frac{\partial E_b}{\partial x}\right)_n. \tag{7.52}$$

Now, considering that the chemical potential of the electrons is given by

$$\mu_e = \hbar c(3\pi^2 xn)^{1/3} \tag{7.53}$$

we find that

$$\left(\frac{\partial E_b}{\partial x}\right)_n = -\hbar c(3\pi^2 xn)^{1/3}. \tag{7.54}$$

Equation (7.54) is the most general relation that determines the proton fraction of β-stable matter. In the specific case of the parabolic approximation, Eq. (7.54) leads to the equation

$$4(1-2x)E_{\text{sym}}(n) = \hbar c(3\pi^2 nx)^{1/3}. \tag{7.55}$$

Equation (7.55) determines the equilibrium proton fraction x as a function of the baryon number density n, once the density dependent symmetry energy $E_{\text{sym}}(n)$ is known. After a straightforward algebra one gets

$$x(n) = \frac{1}{2} - \frac{1}{4}\left([2\beta(\gamma+1)]^{1/3} - [2\beta(\gamma-1)]^{1/3}\right) \tag{7.56}$$

where

$$\beta = 3\pi^2 n\left(\frac{\hbar c}{4E_{\text{sym}}(n)}\right)^3, \qquad \gamma = \left(1+\frac{2\beta}{27}\right)^{1/2}.$$

To a very good approximation the symmetry energy can be written in a more instructive form by separating the kinetic and the potential contributions, that is

$$E_{\text{sym}}(u) \simeq 13u^{2/3} + 17F(u). \tag{7.57}$$

The function $F(u)$ is used in order to produce the results of more elaborate calculations and, in the most applications, it takes the following three forms (which correspond to soft, medium stiff and very stiff symmetry energy correspondingly):

$$F(u) = \sqrt{u}, \qquad F(u) = u, \qquad F(u) = \frac{2u^2}{1+u}. \tag{7.58}$$

It is worthwhile to notice that the present model satisfies the relativistic causality. That means the speed of sound c_s (in units of speed of light c) as defined by the relation,

$$\frac{c_s}{c} \equiv \sqrt{\frac{dP}{d\mathcal{E}}} = \sqrt{\frac{dP/dn}{d\mathcal{E}/dn}} \tag{7.59}$$

does not exceed the speed of light for any value of the baryon density. This is a basic demand for any realistic EOS and play important role on the estimation of the possible maximum mass of a neutron star.

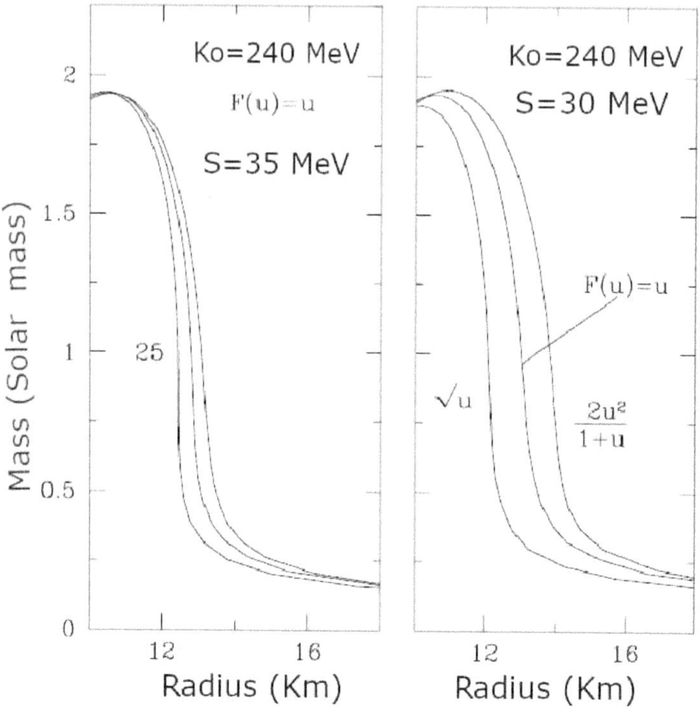

Fig. 7.20: Left: Mass-radius diagram for the potential contribution to the symmetry energy $F(u)$, for $K_0 = 240$ MeV and for the three choices of the value of the symmetry energy at saturation density, that is $S = 25, 30, 35$ MeV. Right: Mass-radius diagram for fixed values $K_0 = 240$ MeV and $S = 30$ MeV and for the three different choices of the potential contribution to the symmetry energy $F(u)$ [Lattimer and Prakash (2001)] (for more details see text).

In Fig. 7.20 we present the solution of the TOV equations for the three different parametrizations given in Eq. (7.58), while Table 7.1 summarizes the basic neutron star properties for the three cases, [Prakash *et al.* (1988)], [Lattimer and Prakash (2001)].

Finally, it is worth pointing out that, according to recent observations, the most massive neutron stars include PSR J1614-2230 ($M = 1.97 \pm 0.04\ M_\odot$) and PSR J0348+0432 ($M = 2.01 \pm 0.04\ M_\odot$).

Table 7.1: The maximum mass $M_{\max}$ (in units of solar mass $M_\odot$), the corresponding radius and the central baryons density n_c (in units of the saturation density $n_s = 0.16\ \mathrm{fm}^{-3}$) for the three different choices of the potential contribution to the symmetry energy $F(u)$ for incompressibility $K_0 = 240$ MeV and symmetry energy at saturation density $S = 30$ MeV.

$F(u)$	$M_{\max}/M_\odot$	R(km)	n_c/n_s
$\sqrt{u}$	1.895	10.112	7.740
u	1.935	10.570	7.269
$2u^2/(1+u)$	1.952	10.933	6.953

7.6 Problems

(1) Calculate the mass, density and pressure of a neutron star as a function of the radius r by employing power-series solution on the differential TOV equations [Schutz (2009)].

Solution:
We set $c = G = 1$ and the TOV equation are read

$$\frac{dP}{dr} = -\frac{(\rho + P)(m + 4\pi r^3 P)}{r(r - 2m)} \tag{7.60}$$

$$\frac{dm}{dr} = 4\pi r^2 \rho. \tag{7.61}$$

Employing a power-series solution for each of the main quantities we get

$$m(r) = \sum_{i=1}^{n} m_i r^i \tag{7.62}$$

$$P(r) = \sum_{i=1}^{n} P_i r^i \tag{7.63}$$

$$\rho(r) = \sum_{i=1}^{n} \rho_i r^i. \tag{7.64}$$

In addition we consider that near to the central density $\rho_c = \rho(0)$ the equation of state $P = P(\rho)$ has the expansion

$$P = P(\rho_c) + \frac{P_c \Gamma_c}{\rho_c}(\rho - \rho_c) + \dots \tag{7.65}$$

where Γ_c is the adiabatic index defined as

$$\Gamma = \frac{d \ln P}{d \ln \rho}$$

evaluated at ρ_c.

Substituting the power series (7.62) and (7.64) into Eq. (7.61) we find

$$\begin{aligned}\frac{dm}{dr} &= m_1 + 2m_2 r + 3m_3 r^2 + 4m_4 r^3 + 5m_5 r^4 + \text{hot} \\ &= 4\pi r^2 \rho = 4\pi(\rho_0 r^2 + \rho_1 r^3 + \rho_2 r^4 + \text{hot})\end{aligned} \tag{7.66}$$

where the index "hot" means higher order terms. Equating like powers in r we get

$$m_1 = 0,\ m_2 = 0,\ m_3 = 4\pi\rho_0/3,\ m_4 = 4\pi\rho_1/4,\ m_5 = 4\pi\rho_2/5.$$

Moreover, writing Eq. (7.60) in a more convenient form

$$-(r^2 - 2mr)\frac{dP}{dr} = (\rho + P)(m + 4\pi r^3 P) \tag{7.67}$$

the left-hand side, using the corresponding power series expansion is written

$$\begin{aligned}&-(r^2 - 2mr)\frac{dP}{dr} \\ &= -(r^2 - 2m_0 r - 2m_3 r^4 + \cdots) \\ &\quad (P_1 + 2P_2 r + 3P_3 r^2 + 4P_4 r^3 + \cdots) + \text{hot} \\ &= 2m_0 P_1 r - (P_1 - 4m_0 P_2) r^2 - (2P_2 - 6m_0 P_3) r^3 \\ &\quad -(-2m_3 P_1 + 3P_3 - 8m_0 P_4) r^4 + \text{hot}.\end{aligned} \tag{7.68}$$

The right-hand side of Eq. (7.67) gives

$$\begin{aligned}&(\rho + P)(m + 4\pi r^3 P) \\ &= \left(\rho_0 + P_0 + (\rho_1 + P_1) r + (\rho_2 + P_2) r^2 + \cdots\right) \\ &\quad \times \left(m_0 + (m_3 + 4\pi P_0) r^3 + (m_4 + 4\pi P_1) r^4\right) \\ &= (\rho_0 + P_0) m_0 + (\rho_1 + P_1) m_0 r + (\rho_2 + P_2) m_0 r^2 \\ &\quad + [(\rho_0 + P_0)(m_3 + 4\pi P_0) + (\rho_3 + P_3) m_0] r^3 \\ &\quad + [(\rho_0 + P_0)(m_4 + 4\pi P_1) + (\rho_1 + P_1)(m_3 + 4\pi P_0) \\ &\quad + (\rho_4 + P_4) m_0] r^4 + \text{hot}.\end{aligned} \tag{7.69}$$

Equating Eqs. (7.68) and (7.70) we found that

$$(\rho_0 + P_0)m_0 = (\rho_c + P_c)m_0 = 0 \to m_0 = 0$$

and also

$$(P_1 - \rho_1) = 0,\ P_1 = 0,\ 2P_2 = -(\rho_0 + P_0)(m_3 + 4\pi P_0),$$
$$3P_3 = -m_4(\rho_0 + P_0).$$

Combining the above equations we find that

$$P_2 = -\frac{1}{2}(\rho_c + P_c)(m_3 + 4\pi P_c) = -\frac{2}{3}\pi(\rho_c + P_c)(\rho_c + 3P_c). \quad (7.70)$$

The coefficient ρ_2 is defined as

$$\rho_2 = \frac{1}{2}\frac{\partial^2 \rho}{\partial r^2}\bigg|_{r=0} = \frac{\partial}{\partial r}\left(\frac{\partial \rho}{\partial P}\frac{\partial P}{\partial r}\right)\bigg|_{r=0} = \left(\frac{\partial \rho}{\partial P}\right)\left(\frac{\partial^2 P}{\partial r^2}\right)\bigg|_{r=0}$$
$$= \frac{2}{c_s^2}P_2 = \frac{\rho_c}{\Gamma_c P_c}P_2 \quad (7.71)$$

where we taken account that $P_1 = 0$ and the definition of the speed of sound in nuclear matter $c_s = \sqrt{\dfrac{\partial P}{\partial \rho}}$. The expansion of the bulk quantities are given now by the expressions

$$m(r) = \frac{4}{3}\pi\rho_c r^3 + \frac{4}{5}\pi\rho_2 r^5 + \text{hot} \quad (7.72)$$
$$= \frac{4}{3}\pi\rho_c r^3 - \frac{8}{15}\pi^2\frac{\rho_c}{\Gamma_c P_c}(\rho_c + P_c)(\rho_c + 3P_c)r^5 + \text{hot}$$

$$P(r) = P_c - \frac{2}{3}\pi(\rho_c + P_c)(\rho_c + 3P_c)r^2 + \text{hot} \quad (7.73)$$

$$\rho(r) = \rho_c - \frac{2}{3}\frac{\rho_c}{\Gamma_c P_c}\pi(\rho_c + P_c)(\rho_c + 3P_c)r^2 + \text{hot}. \quad (7.74)$$

(2) In a crude approximation, a neutron star is a sphere which consists almost entirely of neutrons which form a nonrelativistic degenerate Fermi gas. The pressure of the Fermi gas is counterbalanced by gravitational attraction.

(a) Estimate the radius of such a star to within an order of magnitude if the mass is 10^{33} g. Employ uniform density $\bar{\rho}$ for the star and estimate integrals you cannot easily evaluate, etc. (Knowing the answer is not enough here; you must derive it.)

(b) In the laboratory, neutrons are unstable, decaying according to $n \to p + e^- + \nu$ with a lifetime of 1000 s. Explain briefly why we can consider the neutron star to be made up almost entirely of neutrons, rather than neutrons, protons, and electrons.

Solution:

(a) Consider that the radius of the star is R then the gravitational potential energy is given by the integral

$$U_{\text{grav}} = -\int_{r=0}^{R} \frac{GM(r)}{r}\rho(r)4\pi r^2 dr \tag{7.75}$$

where $M(r)$ is the mass inside a shell of radius $r < R$. Obviously for a uniform density distribution $\rho(r) = \bar{\rho} = \frac{3M}{4\pi R^2}$ and $M(r) = \frac{4}{3}\pi r^3\bar{\rho}$, where M is the total mass of the star. Now, we find easily

$$U_{\text{grav}} = -\frac{3}{5}\frac{GM^2}{R}.$$

When R increases by ΔR the pressure of the Fermi gas does an external work given by

$$\Delta W = P\Delta V = 4\pi R^2 P\Delta R.$$

Now, since $\Delta W = -U_{\text{grav}}$ we find that the pressure needed to compensate the gravitation forces is

$$P = \frac{3GM^2}{20\pi R^4}.$$

Moreover, the pressure of a completely degenerate Fermi gas is given by

$$P = \frac{2}{5}NE_{\text{F}}$$

where $N = \bar{\rho}/m_n$ is the neutron number density, m_n is the neutron mass and E_{F} the Fermi energy given by

$$E_{\text{F}} = \frac{\hbar^2}{2m_n}\left(\frac{9\pi}{4}\frac{M}{m_n R^3}\right)^{2/3}.$$

Equating the two expression for the pressure and after some algebra we find

$$R = \left(\frac{9\pi}{4}\right)^{2/3}\frac{\hbar^2}{Gm_n^3}\left(\frac{m_n}{M}\right)^{1/3} \approx 16 \text{ Km}. \tag{7.76}$$

(b) We consider that if d s the distance between neighboring neutrons then to a crude approximation since $M/m_n \approx (2R/d)^3$ then

$$d \approx 2R\left(\frac{m_n}{M}\right)^{1/3} = 4 \times 10^{-15} \text{ m}.$$

If electrons exist in the star, the magnitude of their mean free path would be of order of d and the order of magnitude of their kinetic energy $E \approx cp \sim c\hbar/d \sim 50$ MeV. Now, since each neutron decay gives out 1 MeV and their kinetic energy is less than $E_F = 21$ MeV, it is unlikely that there could be electrons inside the neutron star. Moreover, because the neutrons are so close electrons and protons from a beta decay would immediately recombine to give neutrons. Thus, there would be no proton in a neutron stars. Of course more accurate calculations gives a small proton fraction (around 10%) to exist in the interior of neutron stars.

Chapter 8

Applications of nuclear science

The basic goal of physics is understanding the laws of nature, explaining the observed phenomena and making predictions about new ones to be observed. To achieve this goal, proper instruments have to be invented. In doing so, it affects all other physical sciences, mainly by providing the key physical concepts as well as the instruments they need in their own investigations. Furthermore physics feeds technology and, thus, has a strong impact on the social and economic development of mankind. More specifically, during the last one hundred years, nuclear physics has been a great contributor. We are not going to discuss here the great variety of its applications, but only focus on those that have affected the development of other sciences or had a large social impact.

8.1 Dating of old samples

We have in Section 13 that certain nuclei undergo beta decay with long life time. The life time depends on the nuclear system, it could be between thousands and billions of years. So by knowing the number of radioactive nuclei in the sample initially and measuring the amount left after some time later we can determine the elapsed time. This method can be used in determining the age of the rock, e.g. the age of the earth, or, by measuring the amount of radioactive ${}^{14}_{6}\mathrm{C}$ in an organic sample, we can determine the sample's age. This time must be of the order of the half life of the radioactive nucleus.

The radio-dating equation is, in general, written as

$$N_D(t) = N_D(0) + N(t)\left(e^{\lambda t} - 1\right) \tag{8.1}$$

where $N_D(t)$ and $N_D(0)$ are the number of daughter nuclei at time t (present) and initially (past) and $N(t)$ the number of the parent nuclei (present), λ is the decay constant $\lambda = \frac{1}{\tau}$ with τ given by Eq. (13.28). We recall that $N(t) = N_0 e^{-\lambda t}$, where N_0 the number of parent nuclei at $t = 0$. One measures $N(t)$ and must know $N_D(0)$, usually assumed to be zero. λ is known once the half life of the nuclide is known.

The above equation is simple, but the actual determination is not a trivial procedure, since the radiometric dating requires that neither the parent nuclide nor the daughter product can enter or leave the material after its formation. The possible confounding effects of contamination of parent and daughter isotopes have to be considered, as do the effects of any loss or gain of such isotopes since the sample was created. It is therefore essential to have as much information as possible about the material being dated and to check for possible signs of alteration. Precision is enhanced, if measurements are taken on multiple samples from different locations of the rock body. Alternatively, if several different minerals can be dated from the same sample and are assumed to be formed by the same event and were in equilibrium with the reservoir when they formed, they should form an isochron. This can reduce the problem of contamination.

Accurate radiometric dating generally requires that the parent has a long enough half-life that it will be present in significant amounts at the time of measurement, the half-life of the parent is accurately known, and enough of the daughter product is produced to be accurately measured and distinguished from the initial amount of the daughter present in the material. The procedures used to isolate and analyze the parent and daughter nuclides must be precise and accurate. This normally involves isotope-ratio mass spectrometry.

8.1.1 *Dating of organic samples*

In this case the radio-dating equation employed is:

$$N(t) = N(0) e^{-\frac{t}{\tau}} \tag{8.2}$$

where $N(0)$ the number of parent nuclei at time $t = 0$, which are known, $N(t)$ the number of nuclei left at time t, which are measured. It is useful to define the half life $t_{1/2}$ of the decaying nucleus, i.e. the time such that $N(t_{1/2}) = \frac{N_0}{2}$. Then

$$N(t) = N(0) e^{-\ln 2 \frac{t}{t_{1/2}}}. \tag{8.3}$$

From the above equation, one determines the age t.

The precision of a dating method depends in part on the half-life of the radioactive isotope involved. For instance, carbon-14 has a half-life of 5,730 years. After an organism has been dead for 60,000 years, so little carbon-14 is left that accurate dating cannot be established. On the other hand, the concentration of carbon-14 falls off so steeply that the age of relatively young remains can be determined precisely to within a few decades. This is called carbon-14 dating.

Carbon-14 has a relatively short life time compared to the age of the earth. Therefore the amount found today is produced by cosmic rays through collisions of neutrons generated by cosmic rays with nitrogen in the upper atmosphere and, thus, it remains at a near-constant level on Earth. The carbon-14 ends up as a trace component in atmospheric carbon dioxide (CO_2). So the initial concentration is known. A carbon-based life form acquires carbon during its lifetime. Plants acquire it through photosynthesis, and animals acquire it from consumption of plants and other animals. If an organism retreats from the actual atmosphere, say inside a cave, and dies buried there it ceases to take in new carbon-14, the amount of C-14 is not regenerated and therefore gets depleted through its beta decay. The proportion of carbon-14 left when the remains of the organism are examined provides an indication of the time elapsed since its death. This makes carbon-14 an ideal dating method to date the age of bones or the remains of an organism.

The earliest known European figurative cave paintings are those of the Chauvet Cave in France. These paintings date to earlier than 30,000 BCE (Upper Paleolithic) according to radiocarbon dating. Some researchers believe the drawings are too advanced for this era and question this age obtained by radiocarbon dating. However, more than 80 radiocarbon dates had been obtained by 2011, with samples taken from various spots (torch marks, from the paintings themselves, animal bones, charcoal, etc.) found on the cave floor. The paintings themselves are the most reliable, because they were there all the time. Radiocarbon dates obtained from such samples indicate that there were two periods of creation in Chauvet, one 35,000 years ago and the other 30,000 years ago. It was indeed a great surprise to find that many of the paintings were modified repeatedly over thousands of years, possibly explaining the confusion. It was even more surprising to find that the finer paintings seemed to date earlier than cruder ones [Zorich (2011)].

Many such artistic findings, in all continents of the earth, have been dated this way, becoming, thus, a useful tool in the study of global history.

The carbon-14 dating limit lies around 58,000 to 62,000 years [Plastino *et al.* (2001)]. Upper Paleolithic pieces of art, older than 40,000 years, have been dated this way.

8.1.2 *Dating of old glaciers*

In our days people are observing melting of the polar icebergs and worry whether this is induced by human activities unfavorable to the environment or it is a part of the a natural cycle. So interest has increased in understanding the age of the old icebergs, their movements in the past, climatic histories and, thus, possibly infer glacier flow dynamics and past glacier coverage. Organic materials such as plant or wood, found inside glaciers, can be used to determine the age of a deep glacier. This method has, however, certain problems, which can be somewhat overcome by taking more samples from various positions in the glacier. Another complementary method has recently been found. It is based on extracting the μg-amounts of the water-insoluble organic carbon (WIOC), a fraction of carbonaceous aerosols embedded in the ice matrix for subsequent ^{14}C dating, see e.g. [Uglietti *et al.* (2016)]. Meanwhile this new approach has improved considerably, thereby reducing the measurement time and improving the overall precision. Samples with 10 μg WIOC mass can now be dated with reasonable uncertainty of around 10–20% (depending on sample age). This requires about 100 to 500 g of ice considering the WIOC concentrations typically found in mid- and low-latitude glacier ice. Dating polar ice with satisfactory age precision is still not possible since WIOC concentrations are around one order of magnitude lower. The accuracy of the ^{14}C WIOC method was validated by applying it to independently dated ice. With this method the deepest parts of the ice cores from Colle Gnifetti and Mt. Ortles glacier in the European Alps, Illimani glacier in the Bolivian Andes, Tsambagarav ice cap in the Mongolian Altai, and 25 Belukha glacier in the Siberian Altai have been dated. In all cases a strong annual layer thinning towards bedrock was observed and the oldest ages obtained were in the range of 10000 yrs. ^{14}C WIOC-dating was not only crucial for interpretation of the embedded environmental and climatic histories, but additionally gave a better insight into glacier flow dynamics close to bedrock and past glacier coverage. For this the availability of multiple dating points in the deepest parts was essential, which is the strength of the presented WIOC ^{14}C-dating method, allowing determination of absolute ages from a large number of locations, essentially from every major piece of ice.

8.1.3 *Dating of old rocks*

Uranium-lead radiometric dating involves using uranium-235 or uranium-238 to date a substance's absolute age. This scheme has been refined to the point that the error margin in dates of rocks can be as low as less than two million years in two-and-a-half billion years. Error margins of 2–5% has been achieved on younger Mesozoic rocks.

One of the great advantages of uranium-lead radiometric dating is that any sample provides two clocks, one based on uranium-235's decay to lead-207 with a half-life of about 700 million years, and one based on uranium-238's decay to lead-206 with a half-life of about 4.5 billion years, providing a built-in crosscheck that allows accurate determination of the age of the sample, even if some of the lead has been lost. This can be seen in the concordia diagram. Further discussion on such issues is beyond the scope of this book and we refer the reader to specialized sources [Wetherill (1975)].

Anyway in uranium-lead dating, the concordia diagram is used which also decreases the problem of nuclide loss. Finally, correlation between different isotopic dating methods may be required to confirm the age of a sample. For example, the age of the Amitsoq Gneisses from western Greenland was determined to be 3.6 ± 0.05 million years ago (MA) using uranium-lead dating and 3.56 ± 0.10 MA using lead-lead dating, results that are consistent with each other [Dalrymple (1994)].

Uranium-lead isotope dating has been used for the determination of the age of earth taking samples from meteorites or in rocks in very deep oceans. Thus an age of 4.55 ± 0.07 billion years, is derived, in part, from the Canyon Diablo meteorite. It has been determined by Clair Cameron Patterson using uranium-lead isotope dating (specifically lead-lead dating) on several meteorites including the Canyon Diablo meteorite [Patterson (1956)].

It should, be mentioned, however, that Most geological samples from earth are unable to give a direct date of the formation of earth from the solar nebula because earth has undergone differentiation into the core, mantle, and crust, and, thus, it has undergone a long history of mixing and unmixing of these sample reservoirs, by plate tectonics, weathering and hydrothermal circulation etc. Anyway it seems an essential agreement exists, between dating with the samples from meteorites and with deep ocean rocks. This is considered as evidence that the earth was formed from an accumulation of objects like meteorites.

8.2 Radioisotopes in medicine

Radioisotopes are unstable nuclides produced mostly in nuclear reactors or in some cases in linear accelerators or cyclotrons. Chemically they behave similarly with the stable isotopes and thus, once they are inserted into the body, they follow the same metabolism. Every organ in our bodies acts differently from a chemical point of view. Doctors and chemists have identified a number of chemicals which are absorbed by specific organs. The thyroid, for example, takes up iodine, whilst the brain consumes quantities of glucose. With this knowledge, radio-pharmacists are able to attach various radioisotopes to biologically active substances. Once a radioactive form of one of these substances enters the body, it is incorporated into the normal biological processes and excreted in the usual ways. Since they can easily be identified, one can follow their actual path after they enter the body. Thus they can provide information about the functioning of a person's specific organs and/or treat them. In most cases, the information is used by physicians to make a quick diagnosis of the patient's illness. The thyroid, bones, heart, liver, and many other organs can be easily imaged, and disorders in their function revealed. In some cases radiation can be used to treat diseased organs, or tumors. Five Nobel Laureates have been closely involved with the use of radioactive tracers in medicine.

Radioisotopes find extensive use in medicine. Nuclear Medicine is a major branch of Medical Physics. So discussing in detail such issues is beyond the scope of this book. For the benefit of the students, however, we will examine some nuclear processes, which are relevant to medicine.

8.2.1 *Radio-diagnosis*

One of the great contributions of physics to other sciences and, in particular, to biology and medicine, is the fact that physics provides the instruments they use. Thus radioisotopes, in combination with imaging devices, which register the gamma rays emitted from within, can aid in studying the dynamic processes taking place in various parts of the body. The earliest technique developed uses single photons detected by a gamma camera which can view organs from many different angles. The camera builds up an image from the points from which radiation is emitted; this image is enhanced by a computer and viewed on a monitor for indications of abnormal conditions. Single photon emission computerized tomography (SPECT) is the current major scanning technology to diagnose and monitor a wide range of medical conditions.

A more recent development is positron emission tomography (PET) which is a more precise and sophisticated technique using isotopes, which can be produced in machines like a cyclotron. A positron-emitting radionuclide is introduced, usually by injection, and accumulates in the target tissue. As it decays it emits a positron, which promptly combines with a nearby electron resulting in the simultaneous emission of two identifiable gamma rays moving in almost opposite directions. These are detected by a PET (positron emitting tomography) camera and give very precise indications of their origin. PET's most important clinical role is in oncology, with fluorine-18 as the tracer, since it has proven to be the most accurate non-invasive method of detecting and evaluating most cancers. It is also well used in cardiac and brain imaging. New procedures combine PET with computed X-ray tomography (CT) scans to give co-registration of the two images (PET-CT), enabling 30% better diagnosis than with a traditional gamma camera alone. It is a very powerful and significant tool which provides unique information on a wide variety of diseases from dementia to cardiovascular disease and cancer.

The radioisotope most widely used in medicine is ^{99}Tc or Tc-99, employed in some 80% of all nuclear medicine procedures. It is an isotope of the artificially-produced element technetium and it has almost ideal characteristics for a nuclear medicine scan. These are:

- It has a half-life of six hours which is long enough to examine metabolic processes yet short enough to minimize the radiation dose to the patient.
- It decays by an 'isomeric' process, which involves the emitting of gamma rays and low energy electrons. Since there is no high-energy beta emission the radiation dose to the patient is low.
- The low-energy gamma rays emitted easily escape the human body and are accurately detected by a gamma camera.
- Its chemistry is so versatile that it can form tracers by being incorporated into a range of biologically-active substances that ensure it concentrates in the tissue or organ of interest.

Its logistics also favor its use. Technetium generators, a lead pot enclosing a glass tube containing the radioisotope, are supplied to hospitals from the nuclear reactor where the isotopes are made. They contain molybdenum-99 (Mo-99), with a half-life of 66 hours, which progressively decays to Tc-99. The Tc-99 is washed out of the lead pot by saline solution when it is required. After two weeks or less the generator is returned for recharging. A similar generator system is used to produce rubidium-82 for PET imaging from strontium-82, which has a half-life of 25 days.

8.2.2 *Radiotherapy*

Radiotherapy can be used to treat some kinds of cancer, using radiation to weaken or destroy the targeted malignant cells. For this to be useful some conditions must be met:

- The interaction of radiation in the body must be localized. The aim is to confine the radiation to well-defined target volumes of the patient. It should not damage cells in its path towards the tumor, if coming from the outside. It should not damage healthy cells, while leaving the tumorous area.
- The interaction of radiation must be selectively against the cancerous cells.
- The dose of radiation required for the treatment must be below a certain risk level.

External irradiation can be carried out using a gamma beam from a suitable source, e.g. radioactive cobalt-60 source. In developed countries, however, the much more versatile linear accelerators are now being used as high-energy X-ray sources. An external radiation procedure is known as gamma knife radiosurgery, and involves focusing gamma radiation from sources of Co-60 on a precise area of the brain with a cancerous tumor. This type of therapy is effective in some kinds of tumors rather than their removal. Internal radionuclide therapy is administered by planting a small radiation source, usually a gamma or beta emitter, in the target area. Short-range radiotherapy, known as brachytherapy, is becoming the main means of treatment.

Iodine-131 is commonly used to treat thyroid cancer, probably the most successful kind of cancer treatment. It is also used to treat non-malignant thyroid disorders.

Iridium-192 implants are used especially in the head and breast. They are produced in wire form and are introduced through a catheter to the target area. After administering the correct dose, the implant wire is removed to shielded storage.

Permanent implant seeds of about 40 to 100 of iodine-125 or palladium-103 are used in the early stage prostate cancer. Brachytherapy procedures give less overall radiation to the body, they are more localized to the target tumor and not very expensive.

Treating leukemia may require a bone marrow transplant, in which case the defective bone marrow will first be killed off with a massive, but otherwise lethal, dose of radiation before being replaced with healthy bone marrow from a donor. Many such procedures can be selected, usually with regard to relieving pain. For instance, strontium-89 and, more recently samarium-153, are used for the relief of cancer-induced bone pain. Rhenium-186 has recently been used for this purpose.

Lu-177 is essentially a low-energy beta-emitter together with some gamma emission. The carrier attaches to the surface of the tumor.

A new field is targeted alpha therapy (TAT) or alpha radioimmunotherapy, especially for the control of dispersed (metastatic) cancers. The short range of very energetic alpha emissions in tissue means that a large fraction of that radiative energy goes into the targeted cancer cells, once a carrier such as a monoclonal antibody has taken the alpha-emitting radionuclide such as bismuth-213 to the areas of concern. Clinical trials for leukemia, cystic glioma, and melanoma are underway. TAT using lead-212 is increasingly important for treating pancreatic, ovarian, and melanoma cancers.

A new experimental development is the Neutron Capture Enhanced Particle Therapy (NCEPT), which involves injecting a patient with a neutron capture agent, shortly before irradiation with protons or heavy ions. This approach boosts the target dose without increasing the dose to healthy tissue and delivers a significant dose to secondary lesions outside the primary treatment area. It uses boron-10 or gadolinium-157, which are known to concentrate in malignant brain tumors. The patient is, then, irradiated with thermal neutrons or protons which are strongly absorbed by the boron, producing high-energy alpha particles which kill the cancer. This, however, requires the patient to be brought to a nuclear reactor, rather than the radioisotopes being taken to the patient.

Radionuclide therapy has progressively become more successful in treating persistent disease and doing so with low toxic side-effects. With any therapeutic procedure the aim is to confine the radiation to well-defined target volumes of the patient.

8.3 Nuclear power

The two major processes for nuclear power is fission, Section 6.5, and fusion of hydrogen to Helium, Section 7.2.1. Furthermore the potential energy that can be extracted from such reservoirs is tremendous. The question is whether these sources can be used to solve the energy needs of humanity.

8.3.1 *Energy production in reactors*

Nuclear reactors have a long history. They are now extensively used in physics research. We have seen above that, in addition to research, they are used for the production of medical and industrial isotopes.

The first commercial nuclear power stations started operation in the 1950's. After a lot of research activity we now know that:

- Nuclear energy now provides about 10% of the world's electricity from about 450 power reactors.
- Nuclear energy is the world's second largest source of low-carbon power (29%of the total in 2017).
- Over 50 countries utilize nuclear energy in about 225 research reactors.

We see that the nuclear power production competes from an economic standpoint with the known fuel sources.

There remain, however, two main problems to be overcome:

- How to safely store the exhaust radioactive particles.
- How do deal with the possibility of accidents.

The first is indeed a serious problem regarding the long lived products. One can perhaps separate the short lined ones and store them inside suitable shells so that their radioactivity is contained. The long lived ones can live thousands of years. Thus their decay rate is quite slow, but the radioactivity lasts long. So special dumping sites must be constructed for them. Such sites cannot be existing caves or closed underground areas, e.g. old mine sites. The reason being that these cannot be water proof and they are subject seismic eruptions, which can result in radioactivity escape. Studies have been made for the possibility of digging holes in strong rocks inside solid mountains, far away from inhabited areas. It seems that to make them risk proof, the economic advantage is lost. Besides no country has undertaken to do this.

There is some laboratory evidence that, using a low energy accelerator, one can transform the long lived products into stable or short lived ones. This effort should continue, to make it commercially attractive.

The possibility of an accident is also a real issue. In a sense the relevant probability is not very high. In over seventy years we had only three major accidents:

- The Three Mile Island accident.
  It was a partial meltdown of reactor number 2 of Three Mile Island Nuclear Generating Station (TMI-2) in Dauphin County, Pennsylvania, USA, and subsequent radiation leak that occurred on March 28, 1979. The accident began with failures in the non-nuclear secondary system, followed by a stuck-open pilot-operated relief valve in the primary system, which allowed large amounts of nuclear reactor coolant to escape. The mechanical failures were compounded by the initial failure of plant operators to recognize the situation as a loss-of-coolant accident due to inadequate training and human factors, such as human-computer interaction design oversights in relation to ambiguous control room indicators in the power plant's user interface. In particular, a hidden indicator light led to an operator manually overriding the automatic emergency cooling system of the reactor, since the operator mistakenly believed that there was too much coolant water present in the reactor and causing the steam pressure release [Clarfield and Wiecek (194)].
- The Chernobyl disaster.
  It was a nuclear accident that occurred on 26 April 1986 at the No. 4 nuclear reactor in the Chernobyl Nuclear Power Plant, near the city of Pripyat in the north of the Ukrainian SSR. It is considered the worst nuclear disaster in history and is the first of only two nuclear energy disasters rated at seven-the maximum severity-on the International Nuclear Event Scale.

  The accident started during a safety test on an RBMK-type nuclear reactor, which was commonly used throughout the Soviet Union. The test supervisor failed to follow procedure, creating unstable operating conditions that, combined with inherent RBMK reactor design flaws and the intentional disabling of several nuclear reactor safety systems, resulted in an uncontrolled nuclear chain reaction [Eden (1999)]. This was immediately followed by an open-air reactor core fire that released considerable airborne radioactive contamination for about nine days that precipitated onto parts of the USSR and western Europe, before being finally contained on 4 May 1986.

- The Fukushima Daiichi nuclear disaster in Japan.
  It is the second accident rated at seven-the maximum severity-on the International Nuclear Event Scale, known as Fukushima March 2011 disaster. First commissioned in 1971, the plant consists of six boiling water reactors. These light water reactors drove electrical generators with a combined power of 4.7 GWe (i.e. Giga-Watt electrical), making Fukushima Daiichi one of the 15 largest nuclear power stations in the world. Fukushima was the first nuclear plant to be designed, constructed, and run in conjunction with the General Electric and the Tokyo Electric Power Company (TEPCO).

  The March 11, 2011 disaster, was the result of a huge tsunami following the Tohoku Earthquake at the 9 Richter scale. Both of them combined disabled the reactor cooling systems. It was indeed hard to manage the situation because the emergency power generators were ruined by the tsunami, and the reactor control system could not function to close the reactors down. The existing sea wall of a few meters in front of the reactor proved to be a toy for the 14 m high tsunami. This lead to huge release of radioactivity and triggering a 30 km (19 mi) evacuation zone surrounding the plant. On April 20, 2011, the Japanese authorities declared the 20 km (12 mi) evacuation zone a no-go area which may only be entered under government supervision [Lipscy *et al.* (2013)]. All four reactor plants were shut down about a year later and TEPCO decided not to reopen them again. The radioactivity releases continue to this day.

The Chernobyl disaster had another fallout: The economic and political toll hastened the end of the USSR and fueled a global anti-nuclear movement. The disaster has been estimated to cost some 235 billion USD in damages. What is now Belarus, which saw 23 percent of its territory contaminated by the accident, lost about a fifth of its agricultural land. At the height of disaster response efforts, in 1991, Belarus spent 22 percent of its total budget dealing with Chernobyl.

More than 2000 people died as a result of the Fukushima disaster, most of them after the earthquake and the tsunami (handling the earthquake itself was a simple matter for the excellent Japanese system). But the management of the reactor system was poor. The defective tsunami wall system was perhaps due to cost considerations, combined with the low expectation of the probability of such a huge tsunami occurring. The placement of the emergency power generators at the ground level was a terrible mistake.

In plants like these one would like to use affordable state of the art technology in building them. Because the plants are very expensive and the reactors are supposed to operate for many years, even if they start as state of the art, they do not possess this status a few years later. The Fukushima reactor was built in 1971, perhaps with state of the art technology, but, at the time of the disaster in 2011, reactors were being built with more advanced technology.

In recent years people are looking for different safe energy sources, which may replace in part the hydrocarbon fuels. It is doubtful, however, whether these may suffice after the hydrocarbon fuels are exhausted. If the society wants to continue consume energy at the current or higher levels, it should be prepared to deal with reactors, in case nuclear energy becomes unavoidable. The main issues involved are not for the experts to decide, but for the people involved. So:

- The governments must ensure the exhaust radioactive particles are safely stored (where and how). They should not be near borders without the consent of the neighboring countries, guaranteed under international law. The same applies for the location of the reactors themselves.
- It is essential that the decisions about reactor nuclear energy production should be decided democratically. The rule of applying state of the art technologies is enforced, even when the facilities are built under contract with private companies.
- Violation of the rules should constitute criminal as well compensation punishments.

It is clear that research to promote clean and safe nuclear power production should be supported at the appropriate level. These are issues for the experts to decide.

8.3.2 *Energy production by fusion*

Energy production by fusion, see Section 7.2.1, occurs naturally or for military purposes. The question is whether one can achieve energy production by controlled fusion. This means fuel and a confined environment with sufficient temperature, pressure and sufficient confinement time to create a plasma in which fusion can occur [Pfalzner (2006)]. In stars gravity provides extremely long confinement times that reach the conditions needed for fusion energy production.

Proposed fusion reactors generally use hydrogen isotopes such as deuterium and tritium, which react more easily than hydrogen to allow them to reach the Lawson criterion requirements with less extreme conditions. The relevant reaction is

$$^2_1\mathrm{D} +^3_1\mathrm{T} \rightarrow^4_2\mathrm{He}(3.5\ \mathrm{MeV}) +^1_0\mathrm{n}\ (14.1\ \mathrm{MeV}). \qquad (8.4)$$

Deuterium (D) is an abundant isotope of hydrogen, while tritium (T) is not so abundant. Its short life time 12.32y, makes hard to handle and keep. So it is usually lithium (Li) used, with tritium produced via the reactions:

$$^1_0\mathrm{n} +^6_3\mathrm{Li} \rightarrow^4_2\mathrm{He} +^3_1\mathrm{T},\ ^1_0\mathrm{n} +^7_3\mathrm{Li} \rightarrow^4_2\mathrm{He} +^3_1\mathrm{T} +^1_0\mathrm{n}. \qquad (8.5)$$

Isotope separation is no problem since the isotope mass ratios are larger. From Eq. (8.4) we see that a small fraction of the energy goes into charged particles. It is clear that for fusion to occur the interacting particles should come sufficiently close for the short ranged nuclear interaction to become operative. This is opposed by the repulsive coulomb force. So confinement can be achieved only if the interacting particles have enough energy, i.e. the plasma must be at high temperature as we know from fusion in the stars, see Section 7.2.1.

Fusion researchers have investigated various confinement concepts. The emphasis was initially on three systems, namely z-pinch, stellarator and magnetic mirror. The current leading designs are the tokamak and inertial confinement (ICF) by laser. Both designs are under research at very large scales, most notably the ITER tokamak in France, and the National Ignition Facility laser in the United States. Other designs may offer cheaper approaches. Among these alternatives there is increasing interest in magnetized target fusion and inertial electrostatic confinement, and a new variation of the stellarator.

Most designs aim to heat their fuel to tens of millions of degrees, which presents a major challenge in producing a successful design. Plasma clouds lose energy through conduction and radiation. Conduction occurs when ions, electrons or neutrals impact other substances, typically a surface of the device, and transfer a portion of their kinetic energy to the other atoms. Radiation is energy that leaves the cloud as light in the visible, UV, IR, or X-ray spectra. Radiation increases with temperature. Fusion power technologies must overcome these losses.

As a source of power, nuclear fusion enjoys some advantages not possessed by fission. These include reduced radioactivity in operation and little high-level nuclear waste, ample fuel supplies, and increased safety. However, achieving the above mentioned conditions has proven to be difficult, especially in connection with producing energy in a practical and economical manner. Research into fusion reactors began in the 1940s, but, to date, no design has produced appreciably more fusion power output than the electrical power input, defeating the purpose. A second issue that affects common reactions, is managing neutrons that are released during the reaction, which over time degrade many common materials used within the reaction chamber.

In summary: Nuclear fusion power is ecologically preferred to hydrocarbon burning, since it does not produce CO_2. In spite of the research efforts, commercially exploitable nuclear fusion power still remains elusive. Research on nuclear fusion power is well financed in many developed counties. Let us hope it works and energy problem is solved.

Chapter 9

Introductory particle physics

In this chapter we are going to discuss the most important theoretical and experimental tools needed for a basic understanding of particle physics.

9.1 The natural system of units

In basic physics we consider a system of three elementary units the length (L), the time (T) and the mass (M). From these the dimensions of the other physical quantities can be determined via definitions ($v = s/t$ etc.) and the laws of nature, (e.g. $F = ma$). This way any physical quantity dimensionally can take the form:

$$[\Pi] = L^{\lambda} T^{\mu} M^{\nu} \tag{9.1}$$

where λ, μ, ν rational numbers. So, if units are defined for (L), (T) and (M), the units of all the other quantities can be obtained. Traditionally we select the meter (m), the second (s) and the kilogram (Kg) as the units of length, time and mass respectively. This system is called the International System of Units (IS). In a electromagnetism we introduced one more unit for the charge via Coulomb's law:

$$F = k_1 \frac{q_1 q_2}{r^2}, \quad k_1 = \frac{1}{4\pi\epsilon_0}, \quad \epsilon_0 = \frac{1}{\mu_0} c^2, \quad \mu_0 = 4\pi \times 10^{-7}, \tag{9.2}$$

$$\epsilon_0 = 8.85 \times 10^{-12}, \quad \frac{1}{4\pi\epsilon_0} = 8.99 \times 10^9 m/F (\text{values used in calculations}),$$

with significant digits 8.854187817 or 8.987551787 (exactly) and dimensions

$$[k_1] = [\epsilon_0^{-1}] = Q^{-2} L^3 M T^{-2}.$$

The dimensionless quantities have the same value in all systems:

$$\alpha = \frac{e^2}{4\pi\epsilon_0} \frac{1}{\hbar c} = \frac{1}{137.035599911} \approx \frac{1}{137}, \text{ (MKSA)},$$

$$\alpha = \frac{e^2}{4\pi}\frac{1}{\hbar c} = \frac{1}{137.035599911}, \text{ (Heaviside-Lorentz).}$$

In hindsight it might have been better had we chosen a more a unambiguous system of units like the mass m_0 of an elementary particle (e.g. the mass of the electron) as the unit of mass, a given wavelength as a unit of length and the period of a simpler microscopic system as the unit of time.

In particle physics we often encounter the constants $\hbar = \frac{h}{2\pi}$, with h Planck's constant, and c the velocity of light. It would be nice, if we devise a system where these constants do not show up. So we use a system whereby

$$\hbar = 1, c = 1 \text{ and as unit of mass the mass } m_0 \text{ of a particle.} \tag{9.3}$$

This way:

- Length and time have the same dimension.
- The velocity is dimensionless.
- Energy has the dimension of mass.
- $E = \hbar\omega = E = \hbar/(2\pi T) \Rightarrow [T] = [E^{-1}] \Rightarrow [T] = [L] = [m_0^{-1}]$.
- For any quantity Π

$$[\Pi] = m_0^{\rho}, \quad \rho = \text{rational}.$$

The parameters λ, μ, ν and ρ for various needed variables are exhibited in Table 9.1.

For the needed conversions we will use:

$$\hbar c = 1.97 \times 10^{-13} \text{ MeV-m } = 3.15 \times 10^{-15} \text{ J-m.} \tag{9.4}$$

Thus for the length we get:

$$\ell_0 = \frac{\hbar c}{m_0 c^2} = \frac{\hbar c}{m_p c^2}\frac{m_p}{m_0} = \frac{1.97 \times 10^{-13} \text{ MeV-m}}{939 \text{ MeV}}\frac{m_p}{m_0}$$
$$\approx 2.10 \times 10^{-16}\frac{m_p}{m_0} \text{ m,} \tag{9.5}$$

while for the time:

$$t_0 = \frac{\hbar}{m_0 c^2} = \frac{\hbar}{m_p c^2}\frac{m_p}{m_0} = \frac{2.10 \times 10^{-16} \text{ m}}{3.00 \times 10^8 \text{ m/s}} = 7.00 \times 10^{-25}\frac{m_p}{m_0} \text{ s.} \tag{9.6}$$

In both cases the left hand side is in m_0^{-1}. Thus selecting as unit the proton mass $m_p = 939$ GeV/c^2, as a basis (unit) we find:

$$m_p^{-1} = 2.10 \times 10^{-16} \text{ m or } 1 \text{ m} = 4.76 \times 10^{15} m_p^{-1}, \tag{9.7}$$

$$m_p^{-1} = 7.00 \times 10^{-25} \text{ s or } 1 \text{ s} = 1.43 \times 10^{24} m_p^{-1}. \tag{9.8}$$

Table 9.1: The dimensions of some physical quantities in various systems.

quantity	symbol	MKSA			NSI
		λ	μ	ν	ρ
Action	$\hbar$	2	-1	1	0
velocity	v	1	-1	0	0
mass	m	0	0	1	1
length	ℓ	1	0	0	-1
time	t	0	0	1	-1
momentum	p	1	-1	1	1
Energy	E	2	-2	1	1
Fine structure constant	α	0	0	0	0
Newton's constant	G_N	3	2	-1	-6
Fermi's constant	G_F	5	-2	1	-2

Sometimes we use GeV for a unit of energy. Then

$$\begin{aligned} &\text{GeV}^{-1} = 1.97 \times 10^{-16} \text{ m} = 6.57 \times 10^{-25} \text{ s}, \; 1 \text{ m} = 5.06 \times 10^{15} \text{ GeV}^{-1}, \\ &\quad 1 \text{ s} = 1.52 \times 10^{24} \text{ GeV}^{-1}, \end{aligned} \tag{9.9}$$

$$G_F = 1.16 \times 10^{-5} \hbar c \left(\frac{\hbar}{m_p c} \right)^2 \Rightarrow G_F = 1.16 \times 10^{-5} m_p^{-2}. \tag{9.10}$$

The converse can perhaps be found by some tricks.

We will, however, exhibit the general method which is a bit harder. What we should do is to find a combination of $\hbar, c$ and m_p needed to express G_F in the SI system. What we need to know is that its dimension is energy times length^3, that is

$$G_F = \text{energy} \times (\text{length})^3 = (L/T)^2 M L^3 = L^5 M T^{-2}. \tag{9.11}$$

We try

$$G_F = \hbar^p c^q m_0^r. \tag{9.12}$$

Then from Table 9.1 we find:

$$[\hbar] = L^2 T^{-1} M \ , \ [c] = L T^{-1} \ , \ [m_0] = M. \tag{9.13}$$

Thus

$$G_F = (L^2 T^{-1} M)^p (L T^{-1})^q M^r = L^{2p+q} T^{-p-q} M^{p+r}. \tag{9.14}$$

Comparing (9.11) and (9.14) we get:

$$2p+q=5\ ,\ \ -p-q=-2\ ,\ p+r=1 \Rightarrow p=3\ ,\ q=-1\ ,\ r=-2, \tag{9.15}$$

$$G_F = \hbar^3 c^{-1} m_0^{-2}. \tag{9.16}$$

The numerical value is obtained by comparing it with the expression of G_F given by (9.10)).
Example 1: The life time of muon μ in natural units is given by:

$$\frac{1}{\tau} = \frac{G_F^2}{192\pi^3} m_\mu^5. \tag{9.17}$$

Let us find it in seconds.

$$\frac{1}{\tau} = \frac{(G_F m_p^2)^2}{192\pi^3} \frac{m_\mu^5}{m_p^5} m_p = \frac{(1.16\times 10^{-5})^2}{0.60\times 10^4} \left(\frac{0.106}{0.940}\right)^5 m_p = 3.37\times 10^{-19} m_p, \tag{9.18}$$

$$\tau = 3.10\times 10^{18} m_p^{-1} = 3.10\times 10^{18}\times 7.00\times 10^{-25}\ \mathrm{s} = 2.17\times 10^{-6}\ \mathrm{s}. \tag{9.19}$$

Example 2: Electromagnetism.
In the system MKSA the energy of the hydrogen atom is:

$$U = -\frac{e^2}{4\pi\epsilon_0}\frac{1}{r} = -\alpha\frac{\hbar c}{r}, \tag{9.20}$$

where α is the fine structure constant. In particle physics we choose $\epsilon_0 = 1$, so that it coincides with the Heaviside-Lorentz system. Then the unit of charge is:

$$e = \sqrt{4\pi\alpha} = 0.303. \tag{9.21}$$

So if q is a given charge in standard units and q_e is the charge of the electron in the same units, the charge q in natural units is

$$Q = \frac{q}{q_e}\sqrt{4\pi\alpha}\ \text{(natural units)}. \tag{9.22}$$

9.2 The Planck natural system of units

Up to now the mass unit has been arbitrary. In an effort to remove this arbitrariness, we write gravitational energy as:

$$U = -G_N\frac{m^2}{r} = -\alpha_G\frac{\hbar c}{r}\ ,\ \alpha_G = \frac{G_N m^2}{\hbar c}, \tag{9.23}$$

$$m = \sqrt{\frac{\alpha_G \hbar c}{G_N}}. \tag{9.24}$$

Choosing $\alpha_G = 1$ we obtain the so called Planck mass:

$$M_P = \sqrt{\frac{\hbar c}{G_N}}. \tag{9.25}$$

This mass from a particle point of view is tremendous, almost macroscopic. Taking $\hbar c = 1.97 \times 10^{-13} \times 10^6 \times 1.610^{-19}$ Kg m^{3}s^{-2}, $G_N = 6.67 \times 10^{-11}$ m^{3}Kg^{-1}s^{-2} we find

$$M_P = \left[\frac{3.15 \times 10^{-26}}{6.67 \times 10^{-11}}\right]^{1/2} \text{Kg} = 2.17 \times 10^{-8} \text{ Kg} \approx 0.02 \text{ mgr}. \tag{9.26}$$

This can be written:[1]

$$M_P = \frac{2.17 \times 10^{-8} \text{ Kg}}{1.67 \times 10^{-27} \text{ Kg}} m_p = 1.27 \times 10^{19}\ m_p = 1.20 \times 10^{19} \text{ GeV}. \tag{9.27}$$

Note also that, through the Eq. (9.25) for $M_P = 1$, in the Planck system get

$$G_N = 1. \tag{9.28}$$

We can verify that

$$\ell_P = \sqrt{\frac{\hbar G}{c^3}} = 1.62 \times 10^{-35} \text{ m}, \ t_P = \sqrt{\frac{\hbar G}{c^5}} = 5.39 \times 10^{-44} \text{ s}. \tag{9.29}$$

Furthermore Boltzmann's constant in the Planck scale can be set to unity, $k_B = 1$. Thus in this system the energy and absolute temperature have the same dimension. Furthermore a new unit of absolute temperature is defined by

$$T_P = \frac{M_P c^2}{k_B} = 1.42 \times 10^{32} \text{ K}. \tag{9.30}$$

In the Planck system we choose ϵ_0 so that

$$\epsilon_0 = \mu_0 = 1/4\pi \Rightarrow e = \sqrt{\alpha}. \tag{9.31}$$

Thus the charge will be:

$$q_P = \sqrt{4\pi\epsilon_0 \hbar c} = 1.88 \times 10^{-18} \text{ C}, \tag{9.32}$$

while the length and time units are:

$$\ell_P = \sqrt{\frac{\hbar G}{c^3}} = 1.62 \times 10^{-35} \text{ m}, \ t_P = \sqrt{\frac{\hbar G}{c^5}} = 5.39 \times 10^{-44} \text{ s}. \tag{9.33}$$

[1]There exist other definitions of the Planck mass. If we choose $\alpha_G = \alpha = 1/137$, we find $M_P = 1.02 \times 10^{18}$.

At energies in the Planck scale, gravity will be as strong as electromagnetism. Something like that happened in the very early Universe (at a time of the order of the Planck time).

Then the school books will define the three basic units as follows:

$$\hbar = 1, \quad c = 1, \quad G_N = 1. \tag{9.34}$$

Then the basic laws of nature will be written in simple, compact and symmetric form, i.e. as:

$$F = -\frac{M_1 M_2}{r^2} \text{ (Newton)}, \quad F = \frac{Q_1 Q_2}{r^2} \text{ (Coulomb)}, \quad E = T, \quad E = \omega. \tag{9.35}$$

The third and fourth equations apply in the case one particle and/or one quantum of energy.[2] Before concluding this section we present in Tables 9.2 and 9.3 some useful quantities in natural units.

9.3 Relativistic effects

Since the field of elementary particles is related to high energies, see Chapter 1, relativistic effects must be taken into account. The essential ingredients of the special theory have been given in Chapter 3. Here we are going to review some interesting results.

9.3.1 *Covariant and contravariant derivatives*

In general an invariant quantity constructed out of two 4-vectors A^μ and B^μ can be written on the form

$$AB = A_\mu B^\mu = A^\mu g_{\mu\nu} B^\nu = A_0 B_0 - \mathbf{A}.\mathbf{B}. \tag{9.36}$$

One such vector can the derivative with respect to the coordinates:

$$\partial_\mu = \frac{\partial}{\partial x^\mu} = \left(\frac{\partial}{\partial x^0}, \frac{\partial}{\partial x^1}, \frac{\partial}{\partial x^2}, \frac{\partial}{\partial x^3}\right) = \left(\frac{\partial}{\partial x_0}, -\frac{\partial}{\partial x_1}, -\frac{\partial}{\partial x_2}, -\frac{\partial}{\partial x_3}\right). \tag{9.37}$$

Note the upper and the lower indices. To remember this rule think as follows: A term ∂_μ (lower index) is generated by the derivative acting on an invariant $x^\mu x_\mu$. So clearly in the process x^μ must be eliminated. This is what the derivative $\partial/\partial x^\mu$ actually does.

[2] For actual physical systems one needs an expression of the form $E = fNT$, where N is the number of particles of the system and f a dimensionless parameter, which depends on the microscopic properties of the system.

Table 9.2: We exhibit some quantities of interest in natural as well as MKSA systems of units. Note $4\pi\epsilon_0 = 1$ is used in this table.

Quantity	Symbol	Natural units	MKSA
Length	l	1/eV	$1.9732705 \cdot 10^{-7}$ m $\approx 0.2\ \mu m$
Mass	m	1/eV	$1.7826627 \cdot 10^{-36}$ Kg
Time	t	1/eV	$6.5821220 \cdot 10^{-16}$ s ≈ 0.66 fs
Frequency	ν	1/eV	$1.5192669 \cdot 10^{15}$ Hz
Speed	u	1	$2.99792458 \cdot 10^{8}$ m/s
Momentum	p	1 eV	$5.3442883 \cdot 10^{-28}$ Kgm/s
Force	F	1 eV^2	$8.1194003 \cdot 10^{-13}$ N
Power	P	1 eV^2	0.24341350 mW
Energy	E	1 eV	$1.6021773 \cdot 10^{-19}$ J
Charge	q	1	$1.8755408 \cdot 10^{-18}$ C
Charge density	ρ	1 eV^3	244.10013 C/m^3
Current	I	1 eV	2.8494561 mA
Current density	J	1 eV^3	$7.3179379 \cdot 10^{10}$ A/m^2
Electric Field	E	1 eV^2	432.90844 V/mm
Potential	Φ	1 eV	85.424546 mV
Polarization	q	1 eV^2	$4.8167560 \cdot 10^{-5}$ C/m^2
Conductivity	σ	1 eV	$1.6904124 \cdot 10^{5}$ S/m
Resistance	R	1	29.979216 Ω
Capacitance	C	1/eV	$2.1955596 \cdot 10^{-17}$ F
Magnetic Flux	ϕ	1	$5.6227478 \cdot 10^{-17}$ Wb
Magnetic induction	B	1 eV^2	1.440271 mT
Magnetization	M	1 eV^2	$1.4440271 \cdot 10^{4}$ A/m
Inductance	L	1/eV	$1.9732705 \cdot 10^{-14}$ H

By the same token, one can define

$$\partial^\mu = \left(\frac{\partial}{\partial x_0}, \frac{\partial}{\partial x_1}, \frac{\partial}{\partial x_2}, \frac{\partial}{\partial x_3}\right), \tag{9.38}$$

we find

$$\partial^\mu \partial_\mu = \frac{\partial^2}{\partial x_0^2} - \nabla^2 = -\Box^2, \quad \Box^2 = \text{ D'Alembertian operator.} \tag{9.39}$$

Table 9.3: Table 9.2 (continued)

Quantity	Symbol	Natural units	MKSA
Planck's quantum	$\hbar$	1	$1.05457266 \cdot 10^{-34}$ Jc
$h = 2\pi\hbar$	q	2π	$6.6260755 \cdot 10^{-34}$ Jc
Charge of electron	e	$8.5424546 \cdot 10^{-2}$	$1.60217733 \cdot 10^{-19}$ C
Bohr radius, $\hbar^2/mc^2$	α_0	$2.6817268\ 10^{-4}/eV$	$5.29177249 \cdot 10^{-11}$ m
Energy 1 electron Volt	eV	1 eV	$1.60217733 \cdot 10^{-19}$ J
Rydlberg energy, $e^2/2\alpha_0$	E_{Ryd}	13.605698 eV	$2.1798711 \cdot 10^{-18}$ J
Hartree energy, e^2/α_0	E_h	27.211396 eV	$4.3597482 \cdot 10^{-18}$ J
Speed of light	c	1	$2.99792458 \cdot 10^{8}$ m/s
Permeability of vacuum	μ_0	4π	$4\pi \cdot 10^{-7}$ H/m
Permittivity of vacuum	ϵ_0	$1/4\pi$	$8.854187817 \cdot 10^{-12}$ F/m
Bohr magneton	μ_B	$8.3585815\ 10^{-8}$/eV	$9.2740154 \cdot 10^{-24}$ J/T
Mass of electron	m_e	510.99906 keV	$9.1093897 \cdot 10^{-31}$ kg
Mass of proton	m_p	938.27234 MeV	$1.6726231 \cdot 10^{-27}$ kg
Mass of neutron	m_n	939.56563 MeV	$1.6749286 \cdot 10^{-27}$ kg
Gravitation constant	G	$6.70711\ 10^{-57}/eV^2$	$6.67259 \cdot 10^{-11}$ Nm^2/Kg^2

9.3.2 *Some simple implications of relativity to particle physics*

We will consider the implications of Chapter 3 in particle physics. First the relation (3.16) between energy and momentum is very useful. Among other things it allows the determination of the identity of an unknown particle by extracting its mass after measuring its momentum end energy. On the other hand for a given particle the relation (3.18) in one dimension is represented by Fig. 9.1(a). We see that for large momenta this curve becomes a straight lime. In the classical limit it becomes a parabola. Whenever this relation holds we say that a particle is in the **mass shell**, since in the classical limit it looks like a shell, see Fig. 9.1(b).

Obviously the energy and momenta depend on the coordinate system they are measured. Suppose that in one system Σ their values are such that they correspond to the point a of the graph of Fig. 9.1(b). Their values in another system must also fall somewhere on the same graph, e.g. b.

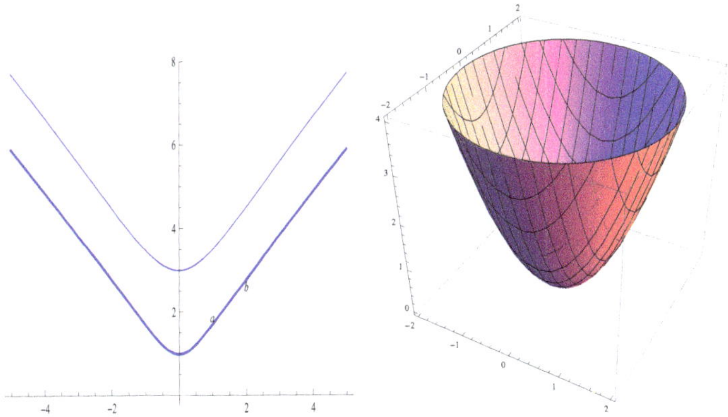

Fig. 9.1: We exhibit the dependence of the kinetic energy on the momentum of a particle, e.g. of a proton, moving in one dimension, bottom graph of panel (a). We also show the same graph for a hypothetical particle with mass 3 times larger (top). We also plot the classical expression for a particle of mass m_p moving in two dimensions (b). All quantities are given in the natural system of units ($c = 1$). The points a and b in panel (a) correspond to the energy and momentum of the proton as measured in two different inertial frames Σ and Σ'.

As a second example we will clarify the meaning of the life time of a particle like the μ. This is like the electron, but 210 times heavier. In the particle's rest frame the lifetime has been experimentally found to be $\tau_0 = 2.2 \times 10^{-6}$ s. It is clear, however, that in such a short time this particle could not have reached the Earth. Even if it were moving with the velocity of light, it could not have traveled more than $c\tau_0 = 0.66$ km. How come it was seen for the first time in cosmic rays?

We have seen in Chapter 3 that the invariant quantity, which is the same in all inertial frames, is $(\Delta s)^2 = c^2(\Delta t)^2 - (\Delta \mathbf{x})^2$. Thus in the system Σ', which we imagine to be attached to the particle μ, we have:

$$\Delta \mathbf{x}' = 0 \quad \Delta s = c(\Delta t') = c\tau_0,$$

while in the observer's system Σ we have

$$(\Delta s)^2 = c^2(\Delta t)^2 - (\Delta \mathbf{x})^2 = (c\Delta t)^2\left[1 - \frac{1}{c^2}\left(\frac{\Delta \mathbf{x}}{\Delta t}\right)^2\right] = c^2(\Delta t)^2(1 - \beta^2)$$

or

$$(\Delta s)^2 = c^2 \tau^2 \frac{1}{\gamma^2}.$$

That is

$$c\tau \frac{1}{\gamma} = c\tau_0 \quad \text{or} \quad \tau = \tau_0 \gamma.$$

Since $\gamma \geq 1$ we find that, relative to an observer on Earth, the particle lives longer. For relatively high velocities we obtain:

$$\begin{array}{lll} v = 0.99\ c & \tau \approx 10\ \tau_0 & c\tau = 6.6\ \text{km} \\ v = 0.999\ c & \tau \approx 22\ \tau_0 & c\tau = 14.5\ \text{km} \\ v = 0.9999\ c & \tau \approx 70\ \tau_0 & c\tau = 46.6\ \text{km} \quad \text{etc.} \end{array}$$

So the fast moving particles indeed have a high probability to reach the Earth. The higher the velocity the longer the lifetime!

9.4 Invariants in kinematics

Consider the elastic collision of two particles A and B with masses m_A and m_B respectively.

We define the invariant quantity:

$$s = (p_A + p_B)^2 = (E_A + E_B)^2 - (\mathbf{p}_A + \mathbf{p}_B)^2. \tag{9.40}$$

We will evaluate s in some frames of special interest.

(1) The center of momentum system $\mathbf{p}_A + \mathbf{p}_B = 0$.
In this system the meaning of s will become simpler

$$s = (E_A^* + E_B^*)^2 \to \sqrt{s} = E_A^* + E_B^*. \tag{9.41}$$

In the other words $\sqrt{s}$ is the energy in the center of momentum frame.

(2) In the laboratory frame in which, let us say, the particle B is at rest.

We have

$$s = (E_A + m_B)^2 - \mathbf{p}_A^2.$$

But

$$\mathbf{p}_A^2 = E_A^2 - m_A^2$$

i.e.

$$s = 2m_B E_A + m_B^2 + m_A^2. \tag{9.42}$$

Thus

$$\sqrt{s} = \sqrt{2m_B E_A + m_A^2 + m_B^2}. \tag{9.43}$$

Consider now two processes of historical interest, namely Compton scattering and anti-proton production.

9.4.1 *Compton scattering*

This problem can be handled traditionally, i.e. by the conservation of energy and momentum, but we will approach it by using the invariant quantity s.

Consider the collision of a massless particle of momentum k with a particle of mass m at rest. In the laboratory we have initially

$$s = (k+m)^2 - k^2 = 2km + m^2.$$

If the momentum of the outgoing massless particle is k' and the momentum of the massive particle is $\mathbf{p}$, we have

$$s = (k' + E)^2 - (\mathbf{p} + \mathbf{k}')^2 = 2Ek' - 2\mathbf{k}' \cdot \mathbf{p} + m^2,\ E = \sqrt{p^2 + m^2}.$$

Conservation of energy momentum yields $E = k + m - k'$ and $\mathbf{p} = \mathbf{k} - \mathbf{k}'$, thus we get

$$s = 2mk' + 2kk'(1 - \cos\theta) + m^2$$

where θ is the scattering angle, i.e. the angle between $\mathbf{k}$ and $\mathbf{k}'$. Equating the two expressions we obtain

$$m(k - k') = kk'(1 - \cos\theta) \text{ or } \frac{1}{k'} - \frac{1}{k} = \frac{1}{m}(1 - \cos\theta)$$

in the case of the photon, scattered off electrons, one uses the wavelength instead of the momentum, $k = 1/\lambda$. Converting it into conventional units we find

$$\Delta\lambda = \lambda' - \lambda = \frac{h}{m_e c}(1 - \cos\theta), \text{ (Compton scattering formula).} \tag{9.44}$$

The outgoing photon is shifted to higher wavelengths, by an amount which is independent by the wavelength, as observed by the Compton experiments. This fact, by augmenting Einstein's arguments for the explanation of the photoelectric effect, established the corpuscular nature of light. The quantity $\lambda_c = h/(m_e c)$ is called Compton wavelength.

Strictly speaking there is a tiny deviation from this formula due to the binding of the electrons.

9.4.2 *Production of anti-protons in proton on proton collision*

Consider the production of anti-protons via the reaction

$$p + p \to p + p + p + \bar{p}$$

where one of the initial photons is at rest. Initially in the laboratory frame we have:

$$s = (m_p + E_p)^2 - \mathbf{p}^2 = 2m_p^2 + 2m_p E_p.$$

For the final system we will evaluate this quantity in the center of momentum system. Obviously the minimum energy of the oncoming protons (threshold energy) occurs when all 4 momenta of the final system are zero (this is possible in the center of momentum system, but not possible in the lab frame). Thus for the final system $s = (4m_p)^2$ (the anti-proton has the same mass with the proton). On the other hand the kinetic energy proton is $T = E_p - m_p$. Thus the threshold energy T is given by

$$16m_p^2 = 2m_p^2 + 2m_p(m_p + T) \Rightarrow 12m_p^2 = 2m_p T \Rightarrow T = 6m_p = 5.64 \text{ GeV}.$$

This is the reason why the anti-proton, expected to exist in the Dirac theory, could only be discovered when a machine capable of achieving this energy was constructed (Bevatron, Berkeley 1954).

9.5 Colliding beams

Consider two particles moving in opposite directions, $\mathbf{p}_A = -\mathbf{p}_B$.

An example is the LINAC accelerator at Stamford, USA, or the LEP at CERN, Geneva, employing electron positron beams and HERA in Hamburg, Germany, accelerating protons and electrons (or positrons) in opposite directions.

i) The LEP accelerator.

The two particles have equal mass and energy. The quantity s is given by:

$$s = (E_A + E_B)^2 - (p_A - p_B)^2 = E_A^2 + E_B^2 + 2E_AE_B - p_A^2 - p_B^2 + 2p_Ap_B$$
$$= m_A^2 + m_B^2 + 2E_AE_B\left(1 + \sqrt{1 - \frac{m_A^2}{E_A^2}} + \sqrt{1 - \frac{m_B^2}{E_B^2}}\right)$$
$$s \approx 4E_AE_B. \tag{9.45}$$

The last relation is valid in the limit in which the masses can be neglected in front of the corresponding energies.

Let us compare this to the case of the fixed target experiment. We express $\sqrt{s}$ in terms of kinetic energies. From Eq. (9.43).

$$K_{cm} = K_A^* + K_B^* = \sqrt{2m_BE_A + m_A^2 + m_B^2} - m_A - m_B \text{ or}$$
$$K_{cm} = \sqrt{2m_BK_A + (m_A + m_B)^2} - m_A - m_B. \tag{9.46}$$

At relativistic energies the energy available in the center of momentum frame becomes:

$$K_{cm} \approx \sqrt{2m_bK_A}, \tag{9.47}$$

i.e. it increases as the square root of the energy of the beam. Thus, to obtain an energy of 2 GeV in the center of momentum system, the beam energy for electrons should be:

$$K_A \approx \frac{K_{cm}^2}{2m_e} \approx \frac{4\text{ GeV}^2}{0.511 \times 2\text{ MeV}} \approx 4000\text{ GeV!}$$

ii) The HERA accelerator.

Now the leptons with mass $m_A = 0.511 \times 10^{-3}$ GeV are accelerated to kinetic energy $K_A = 27.5$ GeV, while the protons with mass $m_B = 0.940$ GeV to kinetic energy $K_B = 920$ GeV. Both are relativistic.

The energy in the center of mass system is:

$$E_{cm}^2 = (E_A^* + E_B^*)^2 = (E_A + E_B)^2 - (\mathbf{p}_A + \mathbf{p}_B)^2.$$

Noting that the momenta are in opposite directions we get

$$E_{cm}^2 = m_A^2 + m_b^2 + 2E_AE_B + 2p_A2p_B.$$

We re-express it now in terms of the kinetic energies of the particles, $E_i = K_i + m_i$, $p_i = \sqrt{K_i(K_i + 2m_i)}$, $i = 1, 2$, and get:

$$E_{cm}^2 = 2\,(K_A + m_A)\,(K_B + m_B)$$
$$+2\sqrt{K_A\,(K_A + 2m_A)}\sqrt{K_B\,(K_B + 2m_B)} + m_A^2 + m_B^2. \tag{9.48}$$

Table 9.4: Typical cross sections σ for weak, electromagnetic and strong interactions. (μb $= 10^{-6}$b. b $= 10^{-24}$ cm^2.)

	weak	Electromagnetic	strong
σ	$10^{-8}\mu$b	10μb	$10^4\mu$b

The center of mass kinetic energy is:

$$K_{cm} = \Big(2\,(K_A + m_A)(K_B + m_B) + 2\sqrt{K_A\,(K_A + 2m_A)}\sqrt{K_B\,(K_B + 2m_B)} + m_A^2 + m_B^2\Big)^{1/2} - m_A - m_B \approx 2\sqrt{K_A K_B} \tag{9.49}$$

where the last expression holds in the completely relativistic limit. Using the above data we find $K_{cm} = 318$ GeV, while the exact expression yields $K_{cm} = 317.3$ GeV.

In the HERA detector, if the protons are fixed target, proceeding as above we find:

$$K_A \approx \frac{K_{cm}^2}{2m_p} \approx \frac{318\ \text{GeV}^2}{0.940 \times 2\ \text{GeV}} \approx 5.410^4\ \text{GeV}.$$

The situation becomes worse, if the electrons are fixed target. Now the resulting proton energy becomes 10^8 GeV. In both cases the energies required are not feasible

Clearly accelerating a particle this way is not a practical for achieving high energy in the center of momentum system.

In reactions the energy in the center of momentum counts, so that in such an arrangement we do not lose most of it as kinetic energy of the fragments.

COLLIDING BEAMS IS THE SECRET TO HIGH ENERGIES

TWO BEAMS, ONE MACHINE AND ONE CHANNEL SUFFICE

9.6 Cross sections and luminosities

The cross section in particle physics is expressed in units

$$1\ \text{b} = 1\ \text{barn} = 10^{-24}\ \text{cm}^2 = 100\ \text{fm}^2 \tag{9.50}$$

typical cross sections at energies of 100 GeV are given in Table 9.4.

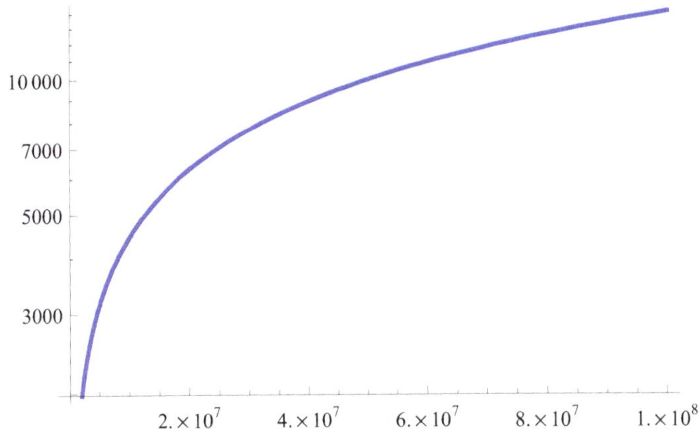

Fig. 9.2: The kinetic energy of two particles of the same mass in the center of momentum frame versus the energy of one of them in a fixed target set up. Both energies are given in units of the rest energy of each particle.

9.6.1 *Luminosity for fixed target experiments*

The number of events produced during the collision is proportional to the flux of the particles in the beam (see Fig. 9.3). Indeed, if n is the particle density in volume dV we will have:

$$dN = ndV.$$

But during the time dt only the particles which are υdt way will be able to hit the surface dS. Thus $dV = \upsilon dtdS$, i.e. $dN = n\upsilon dtdS$ or

$$\text{particle flux } \Phi = \frac{dN}{dSdt} = n\upsilon. \tag{9.51}$$

The number of particles going through an area dS is going to be:

$$\frac{dN}{dt} = n\upsilon dS.$$

For the process $A + B \to C + D$, (A beam, B target) we have $S = N_B\sigma$, i.e.

$$\frac{dN}{dt} = n_A\upsilon_A N_B\sigma = L\sigma, \quad \sigma = \sigma(A + B \to C + D). \tag{9.52}$$

The quantity L given by

$$L = n_A\upsilon_A N_B \tag{9.53}$$

is called **fixed target luminosity** of the accelerator and is measured cm^{-2} s^{-1}.

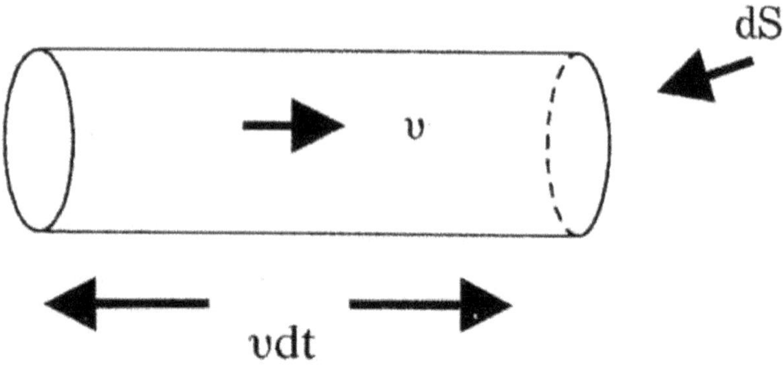

Fig. 9.3: The geometry relevant for obtaining the particle flux, $\Phi = dN/dtdS = nv$, where n is the number of particles per unit volume.

9.6.2 *Luminosity for colliding beam experiments*

For colliding beams (C.B.):

$$n_A = \frac{N_A}{V_A} = \frac{N_A}{\ell S}, \tag{9.54}$$

where ℓ is the length of the accelerator and S cross sectional area of the beam. Now the quantity of interest is the relative velocity, $v \to v_{AB} = |v_A - v_B|$. Thus for a colliding beam accelerator

$$L = \frac{N_A N_B}{S} \frac{v_{AB}}{\ell} \quad \text{(C.B. Luminosity)}. \tag{9.55}$$

In the case of a circular C.B. accelerator we have

$$\ell = 2\pi R, \quad v_{AB} = 2v_A = 2\omega R = 4\pi f R$$

that is

$$L = 2\frac{N_A N_B}{S} f. \tag{9.56}$$

If we have k bunches accelerated simultaneously

$$L = 2k\frac{N_A N_B}{S} f. \tag{9.57}$$

Usually instead of N_A and N_B one uses the beam intensities I_A and I_B, i.e. the number of particles going through the unit area in unit time. Thus in the interval of one period $T = 1/f$ we have:

$$N_A = I_A S T, \quad N_B = I_B S T,$$

$$L = 2kI_A I_B \frac{S}{f}. \tag{9.58}$$

Usually $I_A = I_B$ (we do not want any particles not to find partners). Then

$$L = 2kI^2 \frac{S}{f}, \quad I = I_A = I_B.$$

Unfortunately the luminosities that can be achieved in colliding beam accelerators typically are:

$$L \approx 10^{31} \text{ cm}^{-2}\text{s}^{-1}(e^+, e^-), \quad L \approx 10^{32} \text{ cm}^{-2}\text{s}^{-1}(\bar{p}, p) \text{ or } (p, p), \tag{9.59}$$

i.e. they relatively small compared to those of a fixed target accelerator. Indeed for fixed target:

$$n_A v_B = \frac{1}{S} \frac{dN_A}{dt}.$$

Furthermore, if the target has mass M_B, $M_B = N_B A_B m_p$ with A_B the molecular weight of the substance of B and m_p the proton mass, we find

$$n_B = \frac{N_B}{V_B} = \frac{N_B}{(M_B/\rho_B)} = \frac{N_B \rho_B}{m_p A_p},$$

$$L = \frac{dN_A}{dt} \left(\frac{\rho_B}{m_p A_B} \right) \ell_B. \tag{9.60}$$

As an example consider a synchrotron of fixed target so that 2×10^{12} protons fall per second on liquid hydrogen with thickness of 1 m and density $\rho = 0.084$ gr/cm^3. Then for $A_B = 1$, $m_p = 1.7 \times 10^{-24}$ gr, the achieved luminosity is:

$$L = 2 \times 10^{12} \text{ s}^{-1} \left(\frac{0.065 \text{ s gr/cm}^3}{1.7 \times 10^{-24} \text{ gr}} \right) \times 10^2 \text{ cm} = 7.73 \times 10^{36} \text{ cm}^{-2} \text{ s}^{-1}.$$

Much larger than that of a colliding beam accelerator.

The small luminosity is the price one has to pay to achieve high energies.

9.7 Kinematics of particle decay

We will consider the decay of a particle into two or three other particles.

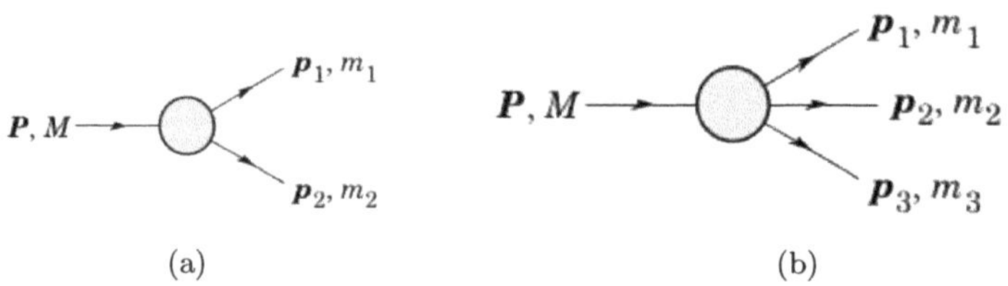

Fig. 9.4: The decay of one particle into two particles (a) and three particles (b).

9.7.1 *The decay of one particle into two*

This situation is exhibited in Fig. 9.4(a), with a slight different notation.

This is the simplest case of one particle decay. Pictorially we have the following:

$$A \to B + C.$$

Four momentum conservation implies

$$p_A = p_B + p_C.$$

In the rest frame of the decaying particle we have

$$p_A = (m_A, 0), \quad p_B = (E_B, \mathbf{p}_B), \quad p_C = (E_C, \mathbf{p}_C), \quad \mathbf{p}_C = -\mathbf{p}_B.$$

Combining them with the relations:

$$\begin{aligned} E_B^2 &= \mathbf{p}_B^2 + m_B^2 = \mathbf{p}^2 + m_B^2, \\ E_C^2 &= \mathbf{p}_C^2 + m_C^2 = \mathbf{p}^2 + m_C^2, \\ m_A &= E_B + E_C, \end{aligned}$$

we find

$$m_A = \sqrt{p^2 + m_B^2} + \sqrt{p^2 + m_C^2}, \quad p = p_B = p_C. \tag{9.61}$$

It thus follows:

$$p = \frac{\sqrt{\lambda(m_A^2, m_B^2, m_C^2)}}{2m_A}, \tag{9.62}$$

where

$$\lambda(x, y, z) = x^2 + y^2 + x^2 - 2xy - 2xz - 2yz$$

is the triangular function. Then it is easy to show:

$$\begin{aligned} E_B &= \frac{m_A^2 + m_B^2 - m_C^2}{2m_A}, \\ E_C &= \frac{m_A^2 - m_B^2 + m_C^2}{2m_A}. \end{aligned}$$

9.7.2 *The decay of one particle into three particles*

This situation is exhibited in Fig. 9.4(b). Now the kinematics is not fully determined and there exist many possibilities. In the rest frame of the initial particle momentum conservation forces the products to be in a plane. Thus the orientation of this plane relative to some vector $\hat{n}$ must be specified. This, e.g., can be chosen to be $\hat{n} = \hat{p}_1 \times \hat{p}_2$ with $\mathbf{p}_1$ and $\mathbf{p}_2$ the momenta of two of the three particles, provided they are not collinear. In addition, one needs two angles, say θ and ϕ, i.e. one needs, in general, three parameters to specify the configuration in the plane. The decay is specified by two more variables, which, e.g., could be an angle θ_3 and the energy of one of the particles, e.g. E_1 or E_2. Thus a set of 5 parameters is needed.

One can show, see, e.g., the companion book [Vergados (2017)], that the decay width (probability) is given by:

$$d\Gamma = \frac{1}{M}\int\left(\frac{1}{(2\pi)^3}\right)^3 \bar{\mathcal{M}}^2 \frac{d^3\mathbf{p}_1}{2E_1}\frac{d^3\mathbf{p}_2}{2E_2}\frac{d^3\mathbf{p}_3}{2E_3}(2\pi)^4\delta(P-p_1-p_2-p_3) \quad (9.63)$$

where P is the 4-momentum of the initial particle and p_1, p_2 and p_3 the 4 momenta of the final particles and $\bar{\mathcal{M}}^2$ is the invariant amplitude, which depends on the specific process. Integrating over the directions of the momenta we obtain:

$$\begin{aligned} d\Gamma &= \frac{1}{M}\frac{1}{(2\pi)^3}\int \bar{\mathcal{M}}^2 \frac{p_1 dp_1}{2E_1}\frac{p_2 dp_2}{2E_2}\frac{p_3 dp_3}{2E_3}\delta(M-E_1-E_2-E_3) \\ &= \frac{1}{M}\frac{1}{(2\pi)^3}\frac{1}{8}\int \bar{\mathcal{M}}^2 dE_1 dE_2 dE_3 \delta(M-E_1-E_2-E_3). \end{aligned} \quad (9.64)$$

Or using the δ function we can cast it, e.g., in the form

$$d\Gamma = \frac{1}{(2\pi)^3}\frac{1}{8M^3}\bar{\mathcal{M}}^2 dE_1 dE_3. \quad (9.65)$$

How many parameters one will need for the description of the experimental data, will depend on the nature of the particles, e.g. their spin, and the interaction responsible for the decay. The set of parameters of phase space, used in the analysis of experimental data is given by a Dalitz plot [Dalitz (1953)], [Kamal (2014)] for a detailed exposition, [Link *et al.* (2004)]. First we define the invariant quantities $m_{ij}^2 = (p_i + p_j)^2$ and then one can easily show that:

$$m_{12}^2 + m_{13}^2 + m_{23}^2 = M^2 + m_1^2 + m_2^2 + m_3^2.$$

Thus only two of them are independent. Furthermore

$$m_{12}^2 = (p_1 + p_2)^2 = (P - p_3)^3 = M^2 + m_3^2 - 2ME_3$$

$$m_{23}^2 = (p_2 + p_3)^2 = (P - p_1)^3 = M^2 + m_3^2 - 2ME_1$$

where E_3 and E_1 are the energies of particles 3 and 1 respectively (in the rest frame of the initial particle). Thus

$$d\Gamma = \frac{1}{(2\pi)^3}\frac{1}{32M^3}\bar{\mathcal{M}}^2 dm_{12}^2 dm_{23}^2. \tag{9.66}$$

This is the standard form of the Dalitz plot, which is a scatter plot in the (m_{12}^2, m_{23}^2) plane. Energy-momentum conservation defines a boundary in that plane, which depends on the particle masses (see Fig. 9.5). If there are no angular correlations between the decay products, then Eq. (9.66) implies a uniform distribution of the data, i.e. the distribution of these variables is flat. However symmetries may impose certain restrictions on the distribution. Furthermore, three-body decays are often dominated by resonant processes, in which the particle decays into two decay products, with one of those decay products immediately decaying into two additional ones. In this case, the Dalitz plot will show a non-uniform distribution, with a peak around the mass of the resonant decay. In this way, the Dalitz plot provides an excellent tool for studying the dynamics of three-body decays. A typical example of a Dalitz plot of D^+ decaying to a final state involving $\pi^+K^-\pi^+$ [Link *et al.* (2007)] (the D mesons are genuinely charmed, i.e. they involve a single charmed quark or anti-quark).

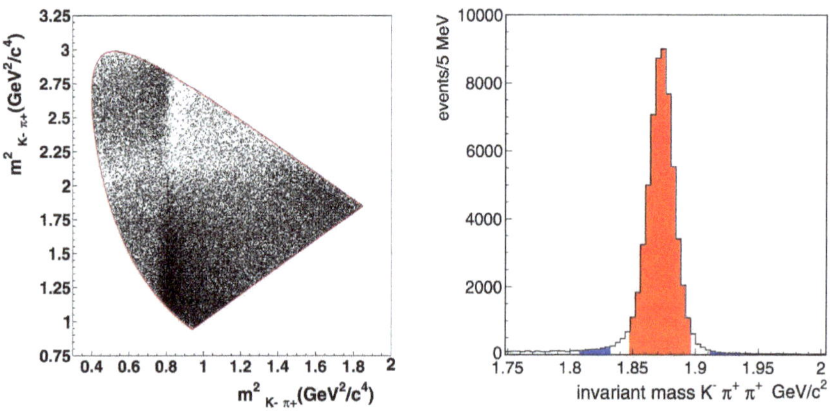

Fig. 9.5: The Dalitz plot of D^+ decaying to a final state involving $\pi^+K^-\pi^+$. Also shown is the invariant mass. Courtesy of the FOCUS Collaboration [Link *et al.* (2007)].

9.8 Invariant quantities in the scattering of two particles

Invariants are quantities which are the same in all frames. Consider the reaction

$$A + B \to C + D.$$

If, for the moment, we ignore the spin and the internal quantum numbers, the cross section will depend on p_A, p_B, p_C and p_D. The possible invariant quantities arising from these momenta are of the form:

$$p_i^\mu p_{j,\mu} \quad i, j = A, B, C, D.$$

But

$$p_A^2 = m_A^2, \; p_B^2 = m_B^2, \; p_C^2 = m_C^2, \; p_D^2 = m_D^2,$$

in other words 4 of these quantities are trivial. We are thus left with the six quantities:

$$p_A \cdot p_B, \quad p_A \cdot p_C, \quad p_A \cdot p_D, \quad p_B \cdot p_C, \quad p_B \cdot p_D, \quad p_C \cdot p_D. \tag{9.67}$$

They are not all independent, however, since we have four relations between them arising from energy-momentum conservation:

$$p_A^\mu + p_B^\mu = p_C^\mu + p_D^\mu \quad \mu = 0, 1, 2, 3.$$

The two independent variables can be any of the following three:

$$s = (p_A^\mu + p_B^\mu)^2 = (p_C^\mu + p_D^\mu)^2 \tag{9.68}$$

$$t = (p_A^\mu - p_C^\mu)^2 = (p_D^\mu - p_B^\mu)^2 \tag{9.69}$$

$$u = (p_A^\mu - p_D^\mu)^2 = (p_C^\mu - p_B^\mu)^2, \tag{9.70}$$

where $\sqrt{s}$ is the center of mass energy and $\sqrt{-t}$ is the 4-momentum transfer. The quantities s, t and u are called Mandelstam variables. They correspond to the square of the propagator in the three possible topologies (channels) shown in Fig. 9.6. In this figure the flow of momenta is indicated. They are not properly drawn Feynman diagrams. The last are discussed Section 9.12.

One can easily show that:

$$s + t + u = m_A^2 + m_B^2 + m_C^2 + m_D^2. \tag{9.71}$$

In a scattering experiment we can express the kinematical variables in terms of s, t and u. Consider, e.g., the reaction

$$A + B \to C + D$$

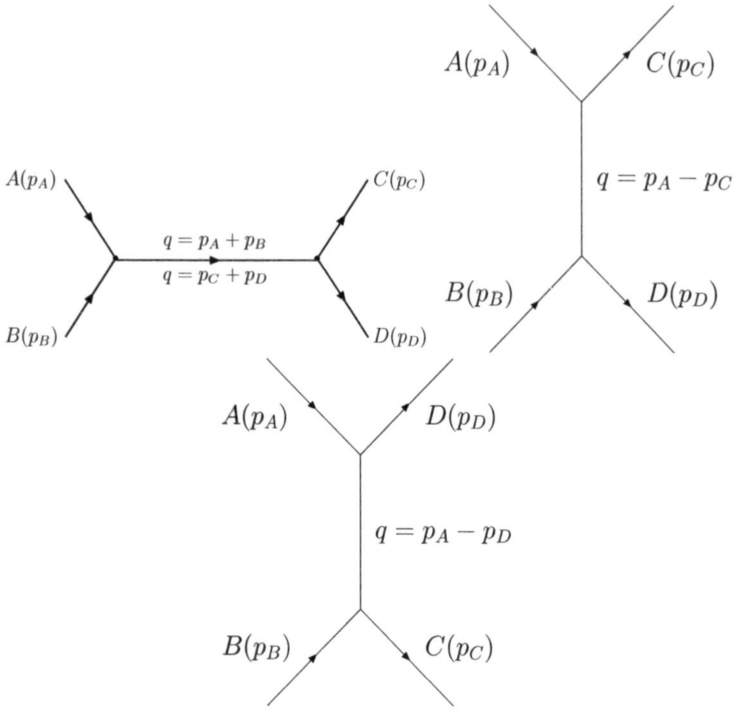

Fig. 9.6: The three available channels of the reaction $A+B \to C+D$. The arrows indicate the oncoming and outgoing particles.

in the center of momentum system. The sum of the three momenta before and after the reaction is zero. Thus:

$$\mathbf{p}_A^* + \mathbf{p}_B^* = \mathbf{p}_C^* + \mathbf{p}_D^* = 0. \tag{9.72}$$

The four momenta are given by

$$p_A = (E_A^*, \mathbf{p}), \quad p_B = (E_B^*, -\mathbf{p}), \tag{9.73}$$
$$p_C = (E_C^*, \mathbf{p}'), \quad p_D = (E_D^*, -\mathbf{p}'), \tag{9.74}$$

with

$$\mathbf{p} = \mathbf{p}_A^* = -\mathbf{p}_B^*, \quad \mathbf{p}' = \mathbf{p}_C^* = -\mathbf{p}_D^*,$$

from these we can find s, t and u. Thus, if θ is the scattering angle, i.e. the angle between $\mathbf{p}$ and $\mathbf{p}'$ we have

$$\begin{aligned} s &= (E_A^* + E_B^*)^2 = \left[\sqrt{m_A^2 + \mathbf{p}^2} + \sqrt{m_B^2 + \mathbf{p}^2}\right]^2 \\ &= (E_C^* + E_D^*)^2 = \left[\sqrt{m_C^2 + \mathbf{p}'^2} + \sqrt{m_D^2 + \mathbf{p}'^2}\right]^2, \\ t &= (E_A^* - E_C^*)^2 - (\mathbf{p} - \mathbf{p})'^2 = m_A^2 + m_C^2 - 2E_A^* E_C^* + 2|\mathbf{p}||\mathbf{p}'|\cos\theta, \\ u &= (E_A^* - E_D^*)^2 - (\mathbf{p} + \mathbf{p})'^2 = m_A^2 + m_D^2 - 2E_A^* E_D^* - 2|\mathbf{p}||\mathbf{p}'|\cos\theta. \end{aligned}$$

Solving them in terms of p, p' and $\cos\theta$ we find:

$$|\mathbf{p}|^2 = \frac{1}{4s}\left[s - (m_A + m_B)^2\right]\left[s - (m_A - m_B)^2\right], \tag{9.75}$$

$$|\mathbf{p}'|^2 = \frac{1}{4s}\left[s - (m_C + m_D)^2\right]\left[s - (m_C - m_D)^2\right]. \tag{9.76}$$

$$\cos\theta = \frac{s(t-u) + (m_A^2 - m_B^2)(m_C^2 - m_D^2)}{[(s-(m_A^2+m_B^2))^2 - 4m_A^2 m_B^2]^{1/2}[(s-(m_C^2+m_D^2))^2 - 4m_C^2 m_D^2]^{1/2}}. \tag{9.77}$$

Solving them for the energy we get:

$$E_A^* = \frac{1}{2\sqrt{s}}(s + m_A^2 - m_B^2), \tag{9.78}$$

$$E_B^* = \frac{1}{2\sqrt{s}}(s + m_B^2 - m_A^2), \tag{9.79}$$

$$E_C^* = \frac{1}{2\sqrt{s}}(s + m_C^2 - m_D^2), \tag{9.80}$$

$$E_D^* = \frac{1}{2\sqrt{s}}(s + m_D^2 - m_C^2). \tag{9.81}$$

In the special case of elastic scattering:

$$A + B \to C + D,$$

we have

$$|\mathbf{p}| = |\mathbf{p}'|, \quad \cos\theta = 1 + \frac{t}{2p^2}.$$

If we consider the laboratory frame with particle B as target we get the picture of Fig. 9.7. The 4-momenta in this system are given by

$$\begin{aligned} p_A &= (E_{AL}, \mathbf{p}_L), \quad p_B = (m_B, \mathbf{0}), \\ p_C &= (E_{CL}, \mathbf{p}'_L), \quad p_D = (E_{DL}, \mathbf{p}_L - \mathbf{p}'_L). \end{aligned}$$

Fig. 9.7: The scattering $A + B \to C + D$ in a frame in which B is at rest (laboratory frame).

Obviously

$$E_{DL} = E_{AL} + m_B - E_{CL}.$$

The s, t and u are given as follows:

$$s = (E_{AL} + m_B)^2 - \mathbf{p}_L^2 = m_A^2 + m_B^2 + 2E_{AL}m_B, \tag{9.82}$$

$$t = 2||\mathbf{p}||_L||\mathbf{p}'||_L \cos\theta_L + m_A^2 + m_C^2 - 2E_{AL}E_{CL}, \tag{9.83}$$

$$u = m_A^2 + m_B^2 + m_C^2 + m_D^2 - s - t. \tag{9.84}$$

Following the above procedure the lab frame, we find:

$$E_{AL} = \frac{1}{2m_B}(s - m_A^2 - m_B^2), \tag{9.85}$$

$$|\mathbf{p}_L| = \frac{1}{2m_B}(s^2 + m_A^4 + m_B^4 - 2s + m_A^4 - 2s + m_B^4 - 2 + m_A^2 + m_B^2)^{1/2}, \tag{9.86}$$

$$\begin{aligned} &\cos\theta_L = \cos\theta_{AC} \\ &= \frac{(s - m_A^2 - m_B^2)(s - m_C^2 - m_D^2) + t(s + m_B^2 - m_A^2) - 2m_B^2(m_A^2 + m_C^2)}{\left[(s + m_A^4 + m_B^4 - 2sm_A^2 - 2sm_B^2 - 2m_A^2m_B^2)\right]^{1/2}} \\ &\times \frac{1}{\left[(u + m_B^4 + m_C^4 - 2um_B^2 - 2um_C^2 - 2m_B^2m_C^2)\right]^{1/2}}. \end{aligned} \tag{9.87}$$

All the above can be generalized in more complicated processes.

We will not consider such cases here. We will, instead, consider a simple application involving the kinematics for resonance production in pion-proton scattering:

$$\pi + p \to \pi + N^*$$

with N^* a particle with mass $m_{N^*} > m_p > m_\pi$. In the center of momentum

$$\mathbf{p}_\pi^* + \mathbf{p}_p^* = 0. \tag{9.88}$$

The energy of the resonance is known, once the pion energy is known. The minimum energy required in the center of momentum is:

$$E_{\rm tot}^* = m_\pi + m_{N^*}.$$

Obviously the initial particles must have energies such that:

$$E_\pi^* + E_p^* = m_\pi + m_{N^*}.$$

The question is: What is the kinetic energy of the π particles in the lab so that the resonance is produced? We have

$$\begin{aligned}(E_\pi^* + E_p^*)^2 &= (E_\pi + E_p)^2 - (\mathbf{p}_\pi + \mathbf{p}_p)^2 = m_p^2 + E_\pi^2 + 2E_\pi m_p - E_\pi^2 + m_\pi^2 \\ &= 2E_\pi m_p + m_p^2 + m_\pi^2, \end{aligned} \tag{9.89}$$

that is

$$E_\pi = \frac{(E_{\rm tot}^*)^2 - (m_\pi^2 + m_p^2)}{2m_p}.$$

The π kinetic energy is $K_\pi = E_\pi - m_\pi$.

The threshold energy is $E_{\rm tot}^* = m_\pi + m_{N^*}$, i.e.

$$E_\pi^{th} = \frac{(m_\pi + m_{N^*})^2 - (m_\pi^2 + m_p^2)}{2m_p} = \frac{2m_\pi m_{N^*} + m_{N^*}^2}{2m_p} - \frac{m_p}{2}. \tag{9.90}$$

We, thus, see that the mass of the resonance can be computed, once E_π^{th} is known. Indeed from Eq. (9.90) we obtain:

$$m_{N^*} = \sqrt{2m_p E_\pi^{th} + m_p^2 + m_\pi^2} - m_\pi.$$

How this is done experimentally will be discussed in Section 9.10.1.

9.9 The resonance in elastic scattering

A simple picture describing a resonance has been given in Chapters 4 and 5. Here we will continue our discussion of how to extract a resonance in elastic scattering.

Let us suppose the amplitude encountered in a elastic scattering takes the form

$$f = \frac{1}{2i} e^{i\delta} \left(e^{i\delta} - e^{-i\delta}\right), \; \delta = \delta(E) \tag{9.91}$$

where δ is the phase shift, which is a function of the energy of the particle E in the center of momentum system for angular momentum $\ell = 0$, see Eq. (5.7).

$$f = e^{i\delta}\sin\delta = \frac{\sin\delta}{e^{-i\delta}} = \frac{\sin\delta}{\cos\delta - i\sin\delta} = \frac{1}{\cot\delta - i}.$$

Suppose further that there exists an energy $E = E_R$ such that the phase shift becomes $\pi/2$. Then, expanding around the energy E_R, we get:

$$\cot\delta(E) = \cot\delta(E_R) + (E - E_R)\left.\frac{d\cot\delta(E)}{dE}\right|_{E=E_R} + \frac{1}{2}(E-E_R)^2\left.\frac{d^2\cot\delta(E)}{dE^2}\right|_{E=E_R} + \cdots.$$

But the first term vanishes by hypothesis and the second derivative is small. Then we get:

$$\cot\delta(E) = -(E-E_R)\frac{2}{\Gamma}, \quad \frac{2}{\Gamma} \equiv -\left.\frac{d\cot\delta(E)}{dE}\right|_{E=E_R}.$$

It can be shown that the condition for ignoring higher order terms is $|E - E_R| \approx \Gamma$, $\Gamma << E_R$. Then we get

$$f(E) \approx \frac{\Gamma/2}{-(E-E_R) - i\Gamma/2}.$$

Thus the elastic cross section takes the form:

$$\sigma_{el} \propto |f(E)|^2 \approx \frac{\Gamma^2/4}{(E-E_R)^2 + \Gamma^2/4} \tag{9.92}$$

which is the celebrated form of a Breit-Wigner resonance we discussed in Chapter 4.

As an illustration we will consider the hard sphere scattering discussed in Section 5.2. In this case

$$\delta(E) = -\frac{\sqrt{2mE}}{\hbar} + n\pi, \; n = 0, \pm1, \pm2, \cdots \tag{9.93}$$

$$E_R(n) = \frac{\pi^2\hbar^2}{2ma^2}(2n-1)^2, \; n = 0, \pm1, \pm2, \cdots. \tag{9.94}$$

Furthermore if

$$d(n) = \left.\frac{d\cot(\delta(E))}{dE}\right|_{E=E_R} \tag{9.95}$$

we find

$$d(0) = \frac{b^2}{\pi}, d(1) = \frac{b^2}{\pi}, d(n) = \frac{b^2}{\pi}\frac{1}{2n-1}, \; n = 2, 3, \ldots,$$

$$d(n) = \frac{b^2}{\pi} \frac{1}{-2n+1}, \; n = -1, -2, \ldots.$$

Then

$$\Gamma(n) = -\frac{2}{d(n)}. \tag{9.96}$$

We obtain a number of resonances

$$f(n) = \frac{1}{(E - E_R(n))} \frac{2}{\Gamma(n)} - i = -\frac{\frac{\Gamma(n)}{2}}{E - E_R(n) + i\frac{\Gamma(n)}{2}}. \tag{9.97}$$

The cross section is proportional to

$$|f(n)|^2 = \frac{\left(\frac{\Gamma(n)}{2}\right)^2}{(E - E_R(n))^2 + \left(\frac{\Gamma(n)}{2}\right)^2}. \tag{9.98}$$

This expression is valid, however, when terms higher than 1 can be neglected. We find

$$\frac{\Gamma(n)}{E_R(n)} = \frac{8\pi}{2n-1} << 1 \text{ holds for } n > 2 \tag{9.99}$$

while for $n < 0$

$$\frac{\Gamma(n)}{E_R(n)} = \frac{8\pi}{2|n|+1} << 1 \text{ holds for } |n| > 1. \tag{9.100}$$

These correspond to the high energy peaks of Fig. 5.2.

9.10 Detection of resonances

Suppose we consider the reaction:

$$p + \pi^- \rightarrow n + \pi^- + \pi^+. \tag{9.101}$$

This could have proceeded directly producing the the above three particles or indirectly via a very short lived particle indicated by ρ^0, i.e.

$$p + \pi^- \rightarrow n + \rho^0, \quad \rho^0 \rightarrow \pi^- + \pi^+. \tag{9.102}$$

Could one tell? We will see that one can.

9.10.1 *The signature of a resonance*

Suppose we measure the invariant mass of the two produced pions.

$$s = (E_1 + E_2)^2 - (\mathbf{p}_1 + \mathbf{p}_2)^2 . \tag{9.103}$$

This quantity can easily be measured by measuring the energies and momenta of the two produced pions. If the reaction proceeds as in Eq. (9.102) s must be very sharp and equal to the mass of the intermediate particle. If, on the other hand, it proceeds as in Eq. (9.101) this quantity must take continuous values between a minimum and a maximum determined by the energy and momenta of the two initial particles and the energy and momentum of the produced neutron. So given enough energy, one expects the spectrum indicated in Fig. 9.8.

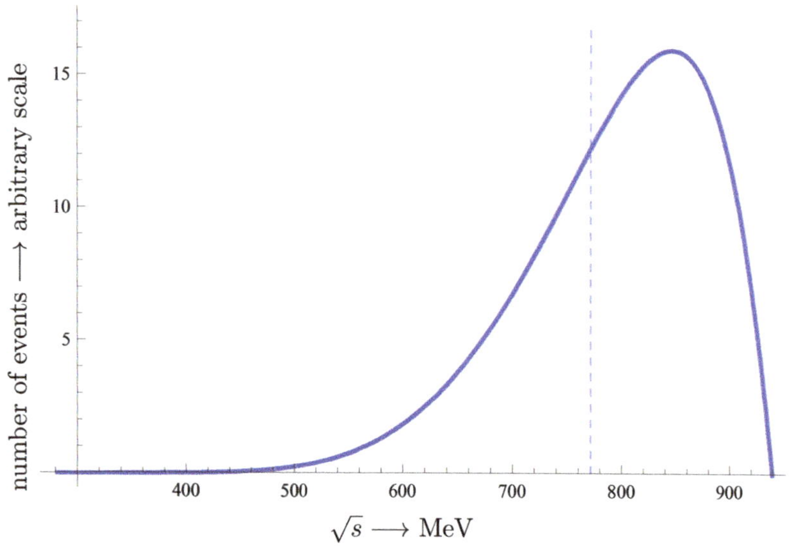

Fig. 9.8: The spectrum of invariant mass of the final particles π^+ and π^-. If the reaction proceeds through the the decay of an intermediate very short lived particle ρ^0, one expects a single line (the dotted in the figure). If the two pions are produced directly the spectrum is continuous as expected from of three particles in the final state. The scale for the number of events is arbitrary. The energy is in MeV.

9.10.2 *The experimental determination of a resonance*

The experimental situation is given in Fig. 9.9. The question is: how does one find the mass of the resonance? Clearly in the center of momentum we have

$$\sqrt{s} = m_n + m_\rho = E_\pi^* + E_p^* = \sqrt{(E_\pi + E_p)^2 - (\mathbf{p}_\pi + \mathbf{p}_p)^2}.$$

Since the proton is at rest we find

$$s = m_\pi^2 + m_p^2 + 2E_\pi m_p \Rightarrow m_\pi^2 + m_p^2 + 2E_\pi m_p = \left(E_\pi^* + E_p^*\right)^2.$$

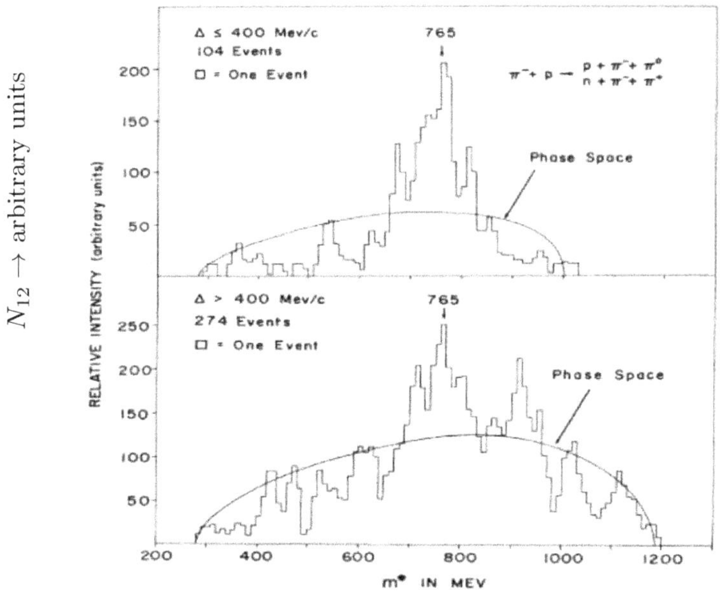

The two pion invariant mass $m_{12} \longrightarrow \text{MeV/c}^2$

Fig. 9.9: The spectrum, i.e. the number of observed events (104 total), top panel, as a function of the invariant mass m_{12} of the two pions produced in the reaction $p + \pi^- \rightarrow n\pi^+\pi^-$ in MeV. The phase space line indicates the expected continuous spectrum in the absence of a resonance. The resonance is clearly seen in the top panel, $\Delta \leq 400$ MeV/c, Δ being the four-momentum transfer to the nucleon, with a peak at 765 MeV/c. In the ideogram for $\Delta > 400$ MeV/c the peak is still present but seems to be smeared to higher values of the di-pion mass m_{12}. The figure has been adopted from Ref. [Erwin *et al.* (1961)].

We thus get:

$$E_\pi = \frac{\left(E_\pi^* + E_p^*\right)^2 - m_\pi^2 - m_p^2}{2m_p}.$$

The minimum (threshold) pion energy obviously happens when

$$E_\pi^* + E_p^* = m_\rho + m_n.$$

We thus get

$$(E_\pi)_{th} = \frac{\left(m_\rho + m_n\right)^2 - m_\pi^2 - m_p^2}{2m_p} \to m_\rho = \sqrt{2m_p\left(E_\pi\right)_{th} + m_\pi^2 + m_p^2} - m_n. \tag{9.104}$$

Experimentally it is found that the pion threshold energy is 1068 MeV. This leads to 765 MeV for the ρ^0 mass, which is in agreement with Fig. 9.9. From this figure we see that the resonance has a width $\Gamma = 125$ MeV. From this we conclude that the lifetime of this particle, Eq. (4.11), is:

$$\tau = \frac{\hbar}{\Gamma} \approx 6 \times 10^{-24}! \tag{9.105}$$

Yes, such a short lived particle gave its signature in our apparatus. Since then numerous narrow resonances (short lived) particles have been discovered this way, see, e.g., Chapter 10. Thanks to the ingenuity of the experimentalists!

9.11 Transformation properties of some non-invariant quantities

One very useful quantity is the cross section σ. It has the meaning of an effective surface of the target particles. Since the lengths, which are perpendicular to the velocity, do not Lorentz contract, the cross section is the same in two inertial frames. Consider, e.g., the process:

$$A + B \to C + D,$$

then $\sigma = \sigma(s,t)$.

What do we do, if a quantity is not invariant? Such a quantity is, e.g., the differential cross section $d\sigma/d\Omega$, Ω = the solid angle. How can we transform it from one system to the next? We can multiply such a quantity with some suitable factor to make it invariant. One can show, e.g., that the quantity $(d^3\mathbf{p})/E$ is invariant. Thus for two inertial frames we have:

$$\frac{E}{p^2dp}\frac{d\sigma}{d\Omega} = \frac{E'}{(p')^2dp'}\frac{d\sigma'}{d\Omega'} \quad \text{or} \quad \frac{d\sigma}{pdEd\Omega} = \frac{d\sigma'}{p'dE'd\Omega'} \tag{9.106}$$

(we used the relation $pdp = EdE$).

From the known relation between the two momenta and after integrating Eq. (9.106) over the momenta, we find the desired relation.

Before concluding this chapter we should say a few words about the Feynman diagrams, which are very useful in particle physics. We will not make extensive use of them, but they may be needed in understanding various physically processes or in reading relevant published work.

9.12 The Feynman diagrams and rules

The Feynman diagram is an ingenious way of summarizing pictorially the results of Field Theory (see, e.g., [Majore (2015)] and the classic books by S. Weinberg, [Weinberg (1995)], [Weinberg (1996)]). It is an essential tool in understanding the various particle physics processes. It has become an economic way to obtain via a perturbative expansion the relative amplitudes, which lead to the determination of of the experimental observables, see e.g., [Vergados (2017)].

We are not going to derive these rules and diagrams, only list them and give some heuristic arguments. We are not going to use these diagrams for calculating cross sections and decay rates. It is only going to be a bookkeeping device to give us an idea of the type of interactions between various particles.

The Feynman diagrams contain lines, which may or may not intersect. The point of intersection is called a knot and will be indicated by a point or full circular disc. The diagrams contain both external lines, with one open end, and internal lines. Out of the many possible diagrams we only need consider the topologically non-equivalent, i.e. those that cannot by brought to each other by a continuous deformation, which leaves the external lines unaffected.

9.12.1 *Propagators*

We will begin with lines indicating the propagation of particles, **propagators**, for fermions, bosons with spin one and scalars (see Fig. 9.10). We should mention that the arrow of the fermion line does not indicate the direction of motion, but the flow of charge. It coincides with the direction of motion of particles, but it is opposite to that of antiparticles.

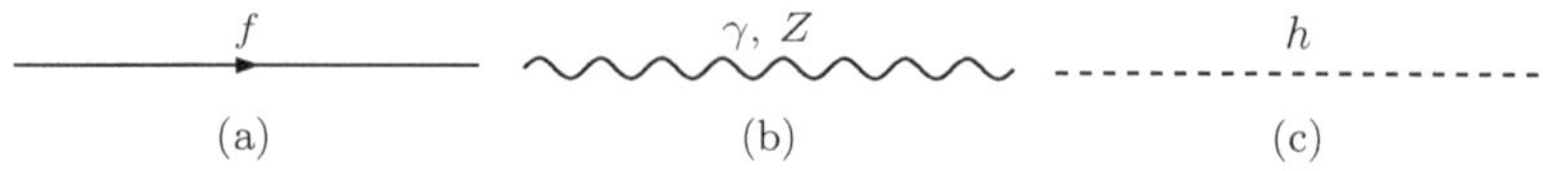

Fig. 9.10: (a) The propagating line for a fermion. The direction is important. (b) and (c) the the propagating lines for vector boson and scalar particle respectively.

9.12.2 *The simplest diagrams in quantum electrodynamics; the photon-charge interaction*

The interaction of the photon with a charge is shown by the diagrams of Fig. 9.11. There exist two possibilities corresponding to Figs. 9.11(a) and 9.11(b) respectively. This process by itself, as it stands, is not allowed by conservation of energy and momentum. More than one photon must be produced. The arrows on the fermions do not indicate the direction of momentum. The convention is different for particles and antiparticles. In this case time is assumed running upwards. So in the case of particles in the vertex both have direction upwards. In the case of antiparticles the direction is reversed.

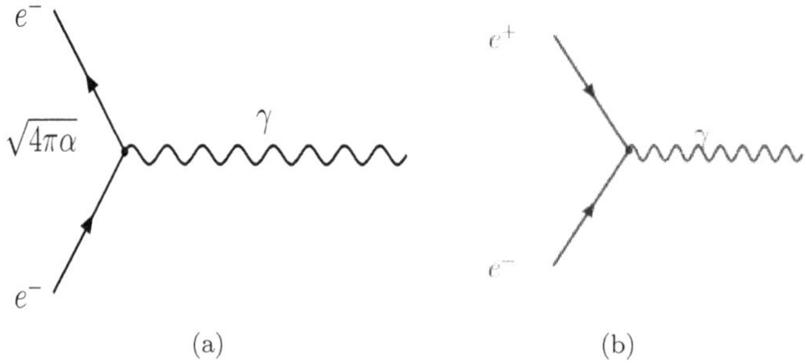

Fig. 9.11: (a) An oncoming electron interacts with a photon and exits (outgoing electron). (b) An electron and a positron collide to produce a photon. In the vertex we indicate the strength of the interaction $e = \sqrt{4\pi\alpha}$. Note the direction of the arrow. All lines are "external".

9.12.3 *Quantum electrodynamics; the interaction between two charges*

We will consider two cases:
i) the simple case of electron scattering the situation is shown in Figs. 9.12(a) and 9.12(b).

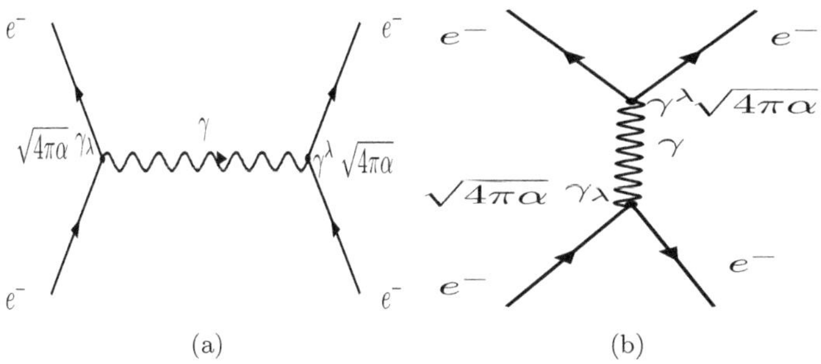

Fig. 9.12: The interaction between two electrons (a) and (b) to second order in the interaction. The fermion lines are external, the photon line internal.

In this book, strictly speaking, we will not use Feynman diagrams for calculations. So we may draw Feynman diagrams as in (a) or as in (b), provided that any conservation laws are obeyed.

ii) The electron positron collision: $e^- + e^+ \rightarrow e^- + e^+$. The Feynman diagrams are shown in Figs. 9.13(a) and 9.13(b). Given sufficient energy in an electron positron accelerator the produced outgoing particles can also be μ^-, μ^+ or τ^-, τ^+. Figure 9.13(b) with μ's at the top is allowed.

As mentioned above for our purposes Feynman diagrams can be drawn as in (a) or as in (b).

One can draw similar diagrams, beyond those of QED. The procedure is similar, but the relevant interactions must be incorporated in the appropriate vertices as dictated by the standard model or its extensions.

9.13 Quantum chromodynamics *QCD*

We have seen in Chapter 1 that the quarks are characterized by a quantum number, called color, which can take three values, r, g, b for red, green and blue. The strong interaction acts on the color degrees of freedom and the resulting theory that governs the dynamics is called **Quantum**

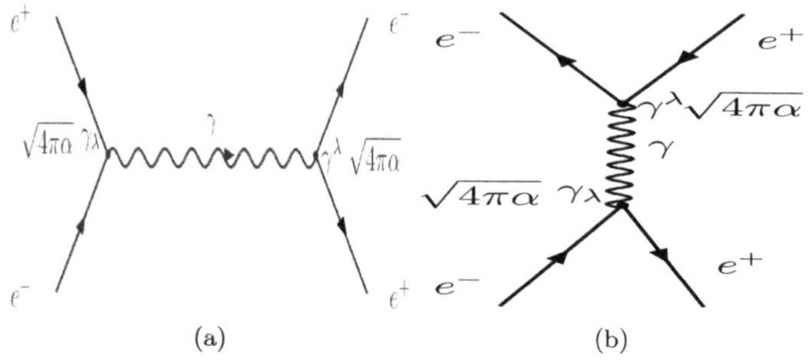

Fig. 9.13: The diagrams involved in electron-positron collisions to second order in the interaction. The fermion lines are external, the photon line internal.

Chromodynamics. It is governed by the group $SU_c(3)$ of the standard model, see, e.g., the companion group [Vergados (2017)] and a summary in Section 15.6.3. Associated with this symmetry we have a set of $3^2 - 1 = 8$ gauge bosons[3] G^a_λ, with λ a Lorentz index and $a = 1 \cdots 8$, the **gluons**. The interactions mediated by the gluons is given in Fig. 9.14.

A comparison of the gluon exchange and the W exchange interaction is given in Fig. 9.15.

The gluons remain massless even after the spontaneous symmetry breaking. Thus this symmetry remains unbroken. Since, however, the symmetry group involved is not Abelian, it does not lead to any long range force, like electromagnetism. The emerging theory is called **Quantum Chromodynamics**

Note that, in contrast to the other unbroken Abelian $U_{EM}(1)$ symmetry, the presence of the SU(3) term leads to three and four point gluon self couplings. This effect leads to some technical difficulties in QCD as compared to the electrodynamics.

- The gluons tend to polarize the medium. In other words it becomes energetically cheaper to create quark anti-quark pairs out of the vacuum (see Fig. 9.16). So we have the production of hadron jets (see Fig. 9.17).

[3]Sometimes the Lorentz index is understood and omitted and the gluons are written in the form of a 3×3 matrix G^μ_λ with μ and λ color degrees of freedom taking values, e.g. r, g, b. The two diagonal matrices have the form diag$(1, -1, 0)$ and $(1/\sqrt{3})$diag$(1, 1, -2)$.

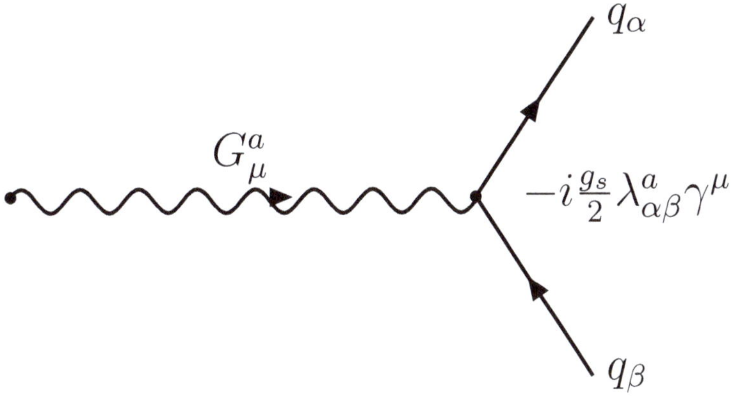

Fig. 9.14: A Feynman diagram indicating the gluon mediated interaction. Here $g = g_s$ is the strong coupling constant and $\lambda^a_{\alpha,\beta}$, α, $\beta = r, g, b$, $a = 1, 2, \ldots, 8$ are connected with the $SU(3)_c$ generators.

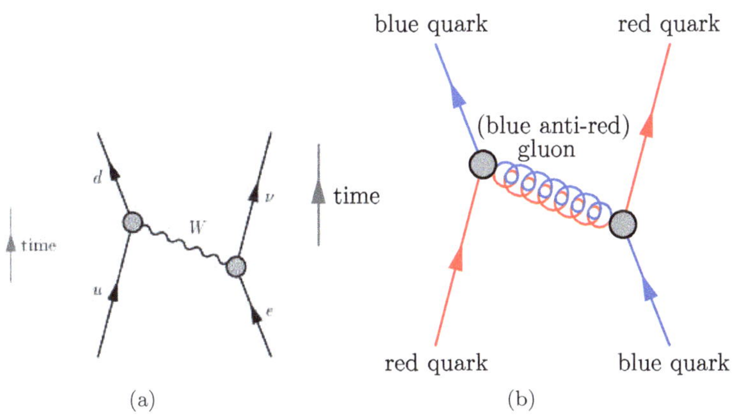

Fig. 9.15: A comparison of a W exchange (weak interaction) (a) and a gluon exchange (strong interaction) (b). Note the propagation of color via the gluons. Other color combinations are possible, $3^2 - 1 = 8$ color combinations in total.

- The polarization of the vacuum is not like the dipole creation familiar from dielectrics, which tends to decrease the interaction between two charges. Instead it behaves more like quadrupoles, which in dielectrics is very hard to make in macroscopic scale. In QCD such quarupoles tend to increase the interaction between the colored quarks (see Fig. 9.18). In other words at high energies, or small distances, the quarks behave almost like being free, i.e. we have **asymptotic freedom**. On the other hand at low energies (large distances of the order of fm) the interaction becomes very strong. Thus we encounter **confinement**, perpetual quark slavery.

- Since the interaction becomes strong at low energies, one cannot invoke perturbation theory. So multi-gluon exchange or pair creation out of the vacuum diagrams become important (see Fig. 9.16).

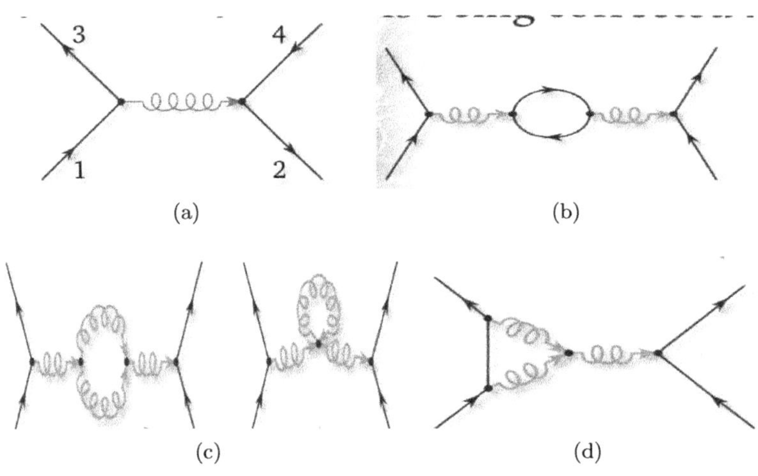

Fig. 9.16: The one gluon exchange (a). At some point it becomes cheaper to create quark-antiquark pairs out of the vacuum (b). As a result the interaction between quarks is screened and decreased. This is analogous to QED and leads to a decrease in the interaction between to charged particles in dielectrics. In QCD, however, we also have virtual gluon creation out of the vacuum. For some cubic and quartic couplings, which are present in a non Abelian theory, this is exhibited in (c). Another diagram involving a cubic coupling is shown in (d). Such virtual gluons lead to anti-screening, i.e. the interaction increases, "color dielectric". This has dramatic effects at low energies (large distances). As a result we have confinement of colored particles.

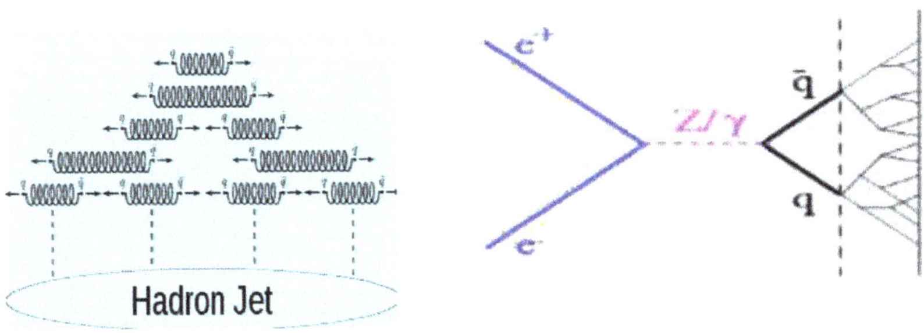

(a) (b)

Fig. 9.17: Hadron jets produced via the strong interaction (a). Hadron jets produced in an e^+, e^- (electron-positron) collider (b).

The strength of the interaction depends on the energy E, sometimes indicated as energy scale Q, and the number n_f of active quark pairs:

$$\alpha_s(E) = \frac{4\pi}{\left(-33 + 2n_f \ln \frac{E^2}{\mu^2}\right)}, \tag{9.107}$$

$$\text{where } \begin{cases} n_f = \text{the number of quarks active pair production (up to 6)} \\ \mu = \text{experimentally determined scale} \approx 0.2 \text{ GeV} \end{cases}.$$

Another common parametrization is given by

$$\alpha_s(q^2) = \frac{\alpha_s(\mu^2)}{1 + \beta\alpha_s(\mu^2)\left(\ln \frac{q^2}{\mu^2}\right)}, \; \beta = \frac{33 - 2n_f}{12\pi}. \tag{9.108}$$

These formulas yield $\alpha_s = 0.12$ at $q^2 = (100 \text{ MeV})^2$.

We have seen that at high enough energies the strong coupling is weak and one can apply perturbation theory. This is not possible at low energies. There exist, however,e general features. The strong interaction acts on the color, it cannot change flavor. This the reason that strange particles could not be generated strongly by the strong interactions, but always in pairs (one strange and one anti-strange). So if $N(f)$ is the number of quark flavor and $N(\tilde{f})$ that of anti-quarks present, the quantity $N_f = N(f) - N(\bar{f})$ is conserved in strong interactions. The electromagnetic interaction cannot change flavor or color, while the weak interaction can change flavor, but it cannot change color.

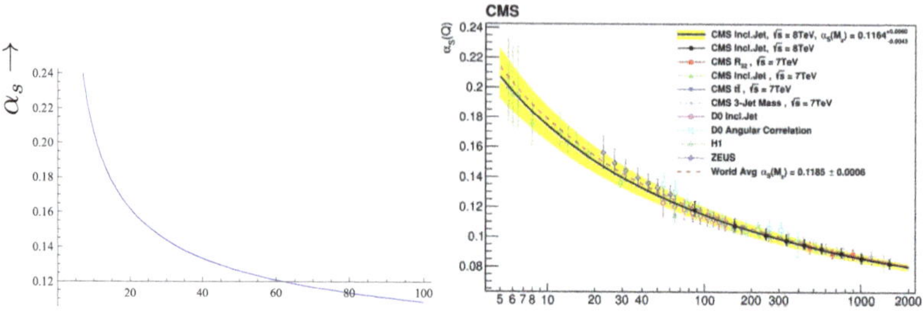

(a) (b)

$Q \to$ GeV

Fig. 9.18: The strong coupling strength decreases as a function of the energy scale. At large scale it becomes very weak (asymptotic freedom), while at small scales it becomes very strong (imposed slavery). (a) Shown is the prediction of the simple formula (Eq. (9.107)) and (b) the available experimental data courtesy of the CMS collaboration [Khachatryan *et al.* (2017)].

Thus the quarks q are characterized by the following states:

$$\begin{pmatrix} u_r & u_g & u_b \\ d_r & d_g & d_b \\ s_r & s_g & s_b \end{pmatrix}.$$

Asymptotic freedom implies that the quarks and the gluons cannot be free. The only hadrons found free are colorless, color singlets. We will see in Section 15.6.3 that the color symmetry for baryons is the colorless antisymmetric combination of three quarks. Omitting the flavor indices we write:

$$(0,0)_c = [1^3] = \frac{1}{\sqrt{6}}\left((rg - gr)b + (gb - bg)r + (br - rb)g\right), \qquad (9.109)$$

while for mesons the colorless singlet is combination of all colors in equal proportion, that is:

$$(0,0)_c = \frac{1}{\sqrt{3}}\left(r\bar{r} + g\bar{g} + b\bar{b}\right). \qquad (9.110)$$

9.13.1 *Approximations at low energies-interaction potentials between quarks*

We know that the quarks do not appear free and all the observed hadrons are colorless. So all experimental information regarding quarks is necessarily indirect and a complicated manifestation of chromodynamics. It looks as though we are trying to find out about electrodynamics with our only access to it coming from the Van der Waals forces between molecules.

We have already seen that due to anti-screening the interaction between quarks becomes very strong at low energies so perturbative techniques are not going to be effective. So some approximations have to be made. It is common to assume that the quarks have a mass, which cannot be directly determined, since they are never free. Their masses are extracted from fits of the appropriate spectra of hadrons viewed as bound states of quarks. This is achieved by assuming a confining potential, which is attractive, with a strength that increases with the distance between the interacting quarks. The most popular are the linear, $V(r) \propto r$, logarithmic $V(r) \propto \ln(r/a)$ and quadratic $V(r) \propto r^2$ [Vergados and Strottman (2018)]. Here we are going to derive the effective potential between quarks in the one gluon approximation in the non relativistic limit up to including terms of second order in the quark momenta, i.e. of order of p^2/m_q^2. To this end we express the 4-spinor forms into matrix elements involving two component wave functions. Clearly this approximation will be applicable to heavy quarks or perhaps for light quarks assuming constituent masses, about a third of the nucleon mass.

i) One gluon exchange potential in a process involving only baryons
The non relativistic reduction of the one gluon exchange amplitude (see Fig. 9.15; see also Fig. 15.2 for the relevant color structure at the vertices) leads to the effective 2-body operator:

$$\begin{aligned}\tilde{V} = &-\frac{1}{(2\pi)^3}\boldsymbol{\lambda}_1.\boldsymbol{\lambda}_2\delta(\tilde{\mathbf{q}}_1+\tilde{\mathbf{q}}_2)\frac{4\pi\alpha_s}{\tilde{\mathbf{q}}_1^2}\\ &\Big[1-\frac{1}{(2m_q^2}\Big(\tilde{\mathbf{Q}}_1.\tilde{\mathbf{Q}}_1+i\boldsymbol{\sigma}_1.(\tilde{\mathbf{q}}_1\times\tilde{\mathbf{Q}}_2)+i\boldsymbol{\sigma}_2.(\tilde{\mathbf{q}}_2\times\tilde{\mathbf{Q}}_1)\\ &-(\boldsymbol{\sigma}_1\times\tilde{\mathbf{q}}_1).(\boldsymbol{\sigma}_2\times\tilde{\mathbf{q}}_2))\Big],\end{aligned} \tag{9.111}$$

where $\boldsymbol{\lambda}_1.\boldsymbol{\lambda}_2$ the SU(3) invariant,[4] see Section 15.6.3 and for more details [Vergados (2016)]. Furthermore α_s is the strong coupling constant, $\tilde{\mathbf{q}}_i = \mathbf{p}'_i - \mathbf{p}_i$ and $\tilde{\mathbf{Q}}_i = \mathbf{p}'_i + \mathbf{p}_i$, $i = 1, 2$. It is traditional to perform a Fourier transform and go to the coordinate space. Thus we get

$$V = -\alpha_s\boldsymbol{\lambda}_1.\boldsymbol{\lambda}_2\left[\frac{1}{r}+\frac{1}{(2m_q)^2}\left(\frac{2}{3}V_S+\frac{1}{3}V_T+\frac{1}{r}V_{QQ}+\frac{1}{r^3}V_{qQ}\right)\right], \tag{9.112}$$

[4]Here, of course, we talk about the color group $SU_c(3)$, see Section 15.6.3. The physics is different, but the mathematical structure is the same with that of the flavor $SU_f(3)$. For the latter see Section 15.6.1.

with α_s treated as a parameter to be fitted to the spectra. It is assumed to be of order 1, i.e. about five times larger than a typical value used in high energy physics (see Fig. 9.18). Furthermore

$$V_S = 4\pi\delta(\mathbf{r})\boldsymbol{\sigma}_1.\boldsymbol{\sigma}_2,$$

$$V_T = 3\boldsymbol{\sigma}_1.\hat{\mathbf{r}}.\boldsymbol{\sigma}_2.\hat{\mathbf{r}} - \boldsymbol{\sigma}_1.\boldsymbol{\sigma}_2,$$

$$V_{QQ} = \nabla_{r_1}(\leftarrow).\nabla_{r_2}(\leftarrow) + \nabla_{r_1}(\rightarrow).\nabla_{r_2}(\rightarrow) - \nabla_{r_1}(\leftarrow).\nabla_{r_2}(\rightarrow)$$
$$-\nabla_{r_1}(\rightarrow).\nabla_{r_2}(\leftarrow).$$

$$V_{qQ} = (\sigma_1 + \sigma_2).(\ell(\rightarrow) - \ell(\leftarrow))$$
$$-\frac{1}{2}\left((\boldsymbol{\sigma}_2 - \boldsymbol{\sigma}_1).(i\, r \times \nabla_R(\leftarrow) - (\boldsymbol{\sigma}_2 - \boldsymbol{\sigma}_1).(i\, r \times \nabla_R(\rightarrow)\right).$$

In the above expressions the arrows in parenthesis indicate the direction (bra or ket) on which the non local operator acts. As usual $\ell = \ell_2 - \ell_1$ is the relative orbital angular momentum, $\mathbf{r} = \mathbf{r_2} - \mathbf{r_1}$ and $\mathbf{R} = \frac{1}{2}(\mathbf{r_2} + \mathbf{r_1})$. The relative orbital angular momentum gives no contribution since the relevant matrix element is diagonal (this part of the operator does not depend on the CM coordinates).

To leading order we obtain for quark interaction inside baryons:[5]

$$V_B = -\alpha_s \boldsymbol{\lambda}_1.\boldsymbol{\lambda}_2 \frac{1}{r} \tag{9.114}$$

(see Section 15.6.3). The two quark system can be of in color states the type

$$(2,0) \leftrightarrow \underline{6} \text{ (sextet) or } (0,1) \leftrightarrow \underline{3}^* \text{ (antitriplet).}$$

The quantity $\boldsymbol{\lambda}_1.\boldsymbol{\lambda}_2$ is related to the Casimir invariants of $SU(3)$ via the expression [Vergados and Strottman (2018)] (see Section 15.6.3)

$$\boldsymbol{\lambda}_1.\boldsymbol{\lambda}_2 = \begin{cases} \frac{2}{3} & \text{(for sextet)} \\ -\frac{4}{3} & \text{(for antitriplet)} \end{cases}. \tag{9.115}$$

ii) One gluon exchange potential in processes involving meson spectra

The non relativistic reduction of the one gluon exchange amplitude (see Fig. 9.19) in this case leads to the effective potential:

$$\tilde{V} = -\frac{1}{(2\pi)^3}\boldsymbol{\lambda}_1.\boldsymbol{\lambda}_2 \delta(\tilde{\mathbf{Q}}_1 - \tilde{\mathbf{Q}}_2)\frac{4\pi\alpha_s}{\tilde{\mathbf{Q}}_1^2}$$

[5] Had we including the V_s spin dependent term we would have obtained in this case:

$$\boldsymbol{\lambda}_1.\boldsymbol{\lambda}_2 = \begin{cases} \frac{2}{3}, \text{ (for sextet and) } S = 1 & -\frac{4}{3}, \text{ (for antitriplet and) } S = 1 \\ -2, \text{ (for sextet and) } S = 0 & 4, \text{ (for antitriplet and) } S = 0 \end{cases}. \tag{9.113}$$

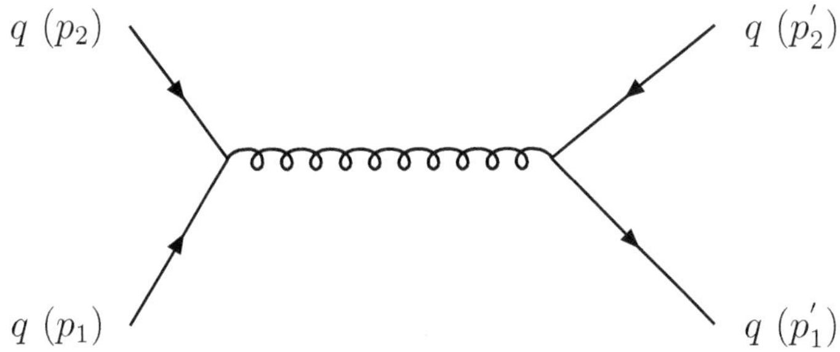

Fig. 9.19: The one gluon effective potential relevant for meson spectra.

$$\left\{ \boldsymbol{\sigma}_1.\boldsymbol{\sigma}_2 + \frac{1}{2}\frac{1}{(2m_q)^2}\left[\boldsymbol{\sigma}_2.(\tilde{\mathbf{q}}_1 \times \tilde{\mathbf{Q}}_1) + \boldsymbol{\sigma}_1.(\tilde{\mathbf{q}}_2 \times \tilde{\mathbf{Q}}_1)\right] \right.$$
$$\left. + \frac{1}{4}\frac{1}{(2m_q)^2}\left[2\tilde{Q}_1^2 - \tilde{q}_1^2 - \tilde{q}_2^2 + (\tilde{\mathbf{q}}_1 - \tilde{\mathbf{q}}_2) \times \tilde{\mathbf{Q}}_1).(\boldsymbol{\sigma}_1 \times \boldsymbol{\sigma}_2)\right\}. \quad (9.116)$$

We notice the appearance of the spin interaction to leading order.

In coordinate space, to leading order we obtain for the quark-antiquark interaction in mesons

$$V_m = -\alpha_s \boldsymbol{\lambda}_1.\boldsymbol{\lambda}_2 \boldsymbol{\sigma}_1.\boldsymbol{\sigma}_2 \frac{1}{r}. \quad (9.117)$$

In this case the $q\tilde{q}$ states can be

$$(0,0) \leftrightarrow \underline{1}\ \text{(singlet) or } (1,1) \leftrightarrow \underline{8}\ \text{(octet)}.$$

In proceeding as above we find [Vergados and Strottman (2018)], see Section 15.6.3

$$\boldsymbol{\lambda}_1.\boldsymbol{\lambda}_2 = \left\{ \begin{array}{ll} \frac{1}{3}, \text{(for octet and) } S = 1 & -\frac{8}{3}, \text{(for singlet and) } S = 1 \\ -1, \text{(for octet and) } S = 0 ' & 8, \ \text{(for singlet and) } S = 0 \end{array} \right. . \quad (9.118)$$

Again one does not know what to use for α_s, but it should be much larger than 0.1, see Fig. 9.18, e.g. $\alpha_s = 1$.

In order to proceed further with calculations like those mentioned in Chapter 10 one must use a confining potential. The simplest one to use is a harmonic version:

$$V_c = -\boldsymbol{\lambda}_1.\boldsymbol{\lambda}_2 k(\mathbf{r_1} - \mathbf{r_2})^{\mathbf{2}}. \quad (9.119)$$

Linear potentials of the form $k_L\sqrt{(\mathbf{r_1} - \mathbf{r_2})^{\mathbf{2}}}$ or even logarithmic $k_{Lg}\left(\log{(\mathbf{r_1} - \mathbf{r_2})^{\mathbf{2}}}/a^2\right)$ have been employed. The parameters of these potentials can be determined by suitable fits [Vergados (2017)].

One mast include the one body terms like the energy $m_q + p^2/(2m_q)$, with m_q being the constituent mass, one third of the nucleon mass, $m_p/3$, for quarks u, d, s or their actual mass for the heavier quarks.

9.14 Some concluding remarks on QCD

A complete understanding of strong interaction at low energies in terms of QCD, i.e. based in the standard model, is not a trivial problem, especially in a presentation aimed at the undergraduate level, which is the aim of this book. Perhaps the interested reader can benefit, if he undertakes the task to work out the related problems of Section 9.15, illustrated, e.g., by Figs. 9.21, 9.22, 9.23, 9.24 and 9.25.

One can say that QCD had quite a bit of success in quite a number of problems, but, in a way, it still remains elusive. This is best summarized by one of the main contributors in this field, A. De Rujula, who also closely witnessed the various steps in its evolution, in an excellent recent review [Rujula (2019)]. He gives a pictorial presentation of the situation, see Fig. 9.20.

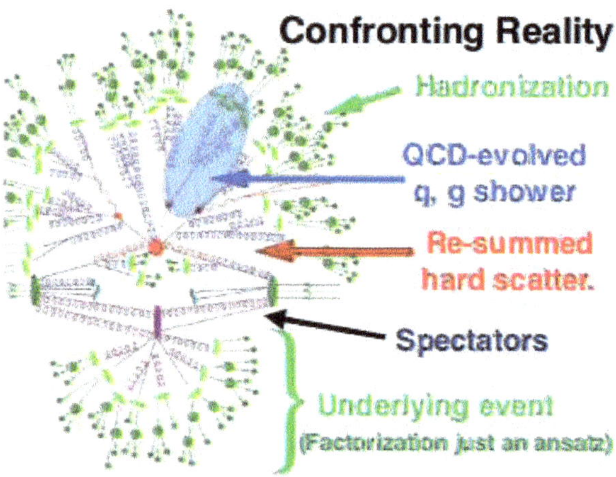

Fig. 9.20: QCD confronting reality; not quite a chaos.

9.15 Problems

(1) Adopting a mass $m_0 = m_e$ express in natural units the expressions:
i) The Bohr radius $a_B = \frac{\hbar}{m_e}$.
ii) The Bohr magneton $\mu_B = e\frac{\hbar}{2m_e c}$ (we use here $4\pi\epsilon_0 = 1$).
Check your answer with the expected value of 8.356×10^{-8} eV^{-1},

which corresponds to $\approx 9.274 \times 10^{-24}$ J/T of Table 9.3.
iii) The classical electron radius $r_e = \frac{e^2}{4\pi}\frac{1}{m_e c^2} = \alpha\frac{\lambda_e}{2\pi}$, where λ_e the Compton wavelength.

(2) A particle of mass m decays at rest into two particles with masses m_1 and m_2, $m > m_1 + m_2$. Find the kinetic energy of each final particle. Are the momenta of the final particles fully specified, magnitude and direction?

(3) An astrophysical source at 5000 light-years distance emits neutrons. Find the minimum energy that neutrons must have to reach the Earth.

To make the estimate, assume that neutrons decay after a mean lifetime and the half-life is about 10 min.

(4) Consider the decay

$$\pi^- \to \mu^- + \nu_\mu.$$

Find the neutrino energy in the pion's rest frame in terms of the mass of the particles involved. Do the same in the center of momentum system. Comment on the results.

(5) The previous decay occurs in flight with $\gamma = \frac{E_\pi}{m_\pi c^2} > 1$.
i) Repeat the procedure of the previous problem. For a given γ is the neutrino energy uniquely determined? If not, find its maximum and its minimum.
ii) What is the distance the pion travels for $\gamma = 10000$ before its decay?

(6) In experimental particle physics an important concept is that of rapidity defined by:

$$y = \ln\frac{E + p_{||}}{E - p_{||}} \tag{9.120}$$

where E is the energy of the particle and $p_{||}$ the projection of its momentum along its trajectory. In the pp scattering at colliders, this component is usually identified as the projection onto the beam directions in the interaction point.
(i) Show that under Lorentz transformation, the rapidity is only shifted by a quantity that depends on the β of the transformation. Therefore the difference between the rapidities of two particles is invariant with respect to Lorentz boosts along the z-axis. This is the key reason why rapidities are so crucial in accelerator physics. **Rapidity differences are invariant with respect to Lorentz boosts along the beam axis.**

(ii) Calculate the maximum and minimum rapidities in the case of LHC (pp colliding beams at 14 TeV in the center-of-momentum reference system).
(Assume that the maximum (minimum) rapidity is obtained for the elastic scattering, $pp \to pp$, at $\theta = 0^0$, ($\theta = 180^0$).
(iii) Show that in the relativistic limit ($E \approx p$) the rapidity can be approximated by

$$y \approx \eta \text{ with } \eta = -\ln \tan\frac{\theta}{2},\ \theta = \cos^{-1}\left(\frac{p_{||}}{p}\right),$$

which is sometimes called pseudorapidity. This is important since pseudorapidity can be determined most simply experimentally.
(iv) Compare the rapidities and the pseudorapidities at 90^0 and 1^0 in the case of LHC.
Hints:

$$E = \gamma(E' + \beta p'_{||}),\ p_{||} = \gamma(p'_{||} + \beta E')$$

$$\frac{E + p_{||}}{E - p_{||}} = \frac{1+\beta}{1-\beta}\frac{E' + p'_{||}}{E' + -p'_{||}} \Rightarrow$$

$$y = y' + \ln\frac{1+\beta}{1-\beta} = y' - \tanh^{-1}(\beta) \Rightarrow y_1 - y_2 = y'_1 - y'_2$$

$$\left(p_{||}\right)_{max} = 7 \text{ TeV},\ y_{max} = -y_{min} \approx 9.6$$

$$y = \frac{1}{2}\ln\frac{E(1+\cos\theta)}{E(1-\cos\theta)} = -\ln\tan\frac{\theta}{2} = \eta.$$

(7) Consider the weak decay $\Xi^0 \to \Sigma^+ + e^- + \tilde{\nu}_e$. Assuming that the Ξ^0 baryon is at rest with mass $M_{\Xi^0} = 1315$ MeV/c^2, calculate the maximum and minimum energies of the electron.
($M_{\Sigma^+} = 1189$ MeV/c^2, $M_e = 0.511$ MeV/c^2.)
(8) The annihilation of anti-protons with momentum $p_{\bar{p}} = 1$ GeV/c against protons at rest, in which $K^- - K^+$ pairs are produced.
i) In the LS examine the symmetric production, i.e. kaons having the same energy. Find the the relevant momenta of the two kaons.
ii) How is the most symmetric production defined in the CM system? What are now the Kaon energies and momenta in this system?
Assume $M_K = 0.49$ GeV/c^2.

(9) Consider the reaction

$$p + \pi \to p + \pi.$$

Let p_π and p'_π be the magnitude of the pion momenta in the laboratory (proton at rest) and the center of momentum frame respectively. Show that they are related by the expression $p_\pi = \sqrt{s/m_p^2} p'_\pi$.
Hint: You may ignore the pion mass in front of the proton mass.

(10) Consider the reaction:

$$e^- + e^+ \to W^- + W^+$$

where $W^\pm$ are the charged vector bosons with a mass of about 80 GeV. This reaction has been employed in the colliding beam LEP accelerator for the detection of these bosons. Find the minimum energy required for the detection of these bosons. What would be the minimum positron energy for their detection in a fixed target experiment? Comment on the obtained result.

(11) The HERA accelerator in DESY uses colliding beams of protons and electrons. The proton energy is 820 GeV and the electron energy 30 GeV.

- Find the total energy in the center of momentum system.
- What energy would be required to achieve this in fixed target experiments i) If the proton is the target and ii) if the electron is the target.
- Find the energy of each beam in a standard electron-positron collider to achieve the same goal.
- Do the same for a proton-proton colliding beam accelerator.
- Comment on the feasibility of each proposal.

(12) The following reactions are proposed:

$$\pi^- + p \to K^0 + \Lambda^0,\ K^- + p \to \pi^0 + \Lambda^0$$

in a fixed target experiment. What is the threshold energy in each case? From an experimental point of view, which one should be preferred?

Can these be done in a colliding beam set up?

(13) Consider the process:

$$A + B \to C + D.$$

Suppose k_i, $i = A, B, C, D$ are the corresponding four-momenta. Show that:

$$(k_A - k_B)(k_C - k_D) = u - t,\ (k_A - k_D)(k_B - k_C) = t - s,$$

$$(k_A - k_D)(k_B - k_C) = u - s,$$

where u, s, t are the Maldestam variables.

(14) Investigate the prospect of the existence of protonium, a bound state of a proton with an antiproton, $p\bar{p}$, or $\bar{p}$H.
i) Suppose very slow antiprotons of a storage ring could combine with protons to do this. Assume that the elecctromagnetic interaction is responsible for this a la hydrogen atom.
In the experiment ATHENA (ApparaTus for High precision Experiment on Neutral Antimatter) [Zurlo *et al.* (2006)], it is claimed that about 1 protonium has been made out of $10^4\bar{p}$, with major quantum number up to $n = 70$. Find possible binding energies and the range of Bohr radii.
ii) In an earlier experiment of the Bess Collaboration [Bai *et al.* (2003)] it was found that a narrow resonance ξ appeared in the process:

$$J/\psi \to \gamma + \xi,\ \xi = N\bar{N} = \frac{1}{\sqrt{2}}(p\bar{p} \pm n\bar{n}), \tag{9.121}$$

the $\pm$ sign depending on the isospin of $N\bar{N}$, with en energy about 2 m_p and a width less than 30 MeV. This was interpreted as a nucleon-antinucleon resonance with $J^\pi = 0^-$ (s-wave) or $J^\pi = 1^+$ (p-wave).
Analyze it as a Breit-Wigner resonance and obtain its expected life time. For simplicity assume it is an s-wave resonance.
iii) Subsequently this resonance was observed the Belle Collaboration [Abe *et al.* (2002)] in

$$B^+ \to K^+ + p\bar{p},\ B^0 \to D^+ + p\bar{p}.$$

This requires further work to establish that, what is observed here, is indeed a $p\bar{p}$ resonance. Certain models, however, give results consistent with this interpretation, in agreement with the interpretation of ii) above.

(15) Consider an electron-electron collider in which two e^- beams hit each other moving in opposite directions with respective beam energies are $E_1 = 12$ GeV and $E_2 = 5$ GeV.

i) Find the total energy in the center of mass system (CMS).
ii) Find the electron momenta in the CMS.
iii) Evaluate β_{CM} and γ_{CM} of the Lorentz transformation from the LS (laboratory frame) to the CMS.

(16) In pion-nucleon scattering many fermion resonances were found, e.g. the $I = 3/2$, $s = 3/2$ resonances with a mass 1250 GeV denoted by Δ^{++}, Δ^{+}, Δ^{0} and Δ^{-}, which were discovered by Fermi and Anderson in 1940. Consider the following reactions:

$$\pi^+ + p \to \pi^+ + p \text{ elastic process}$$

$$\pi^+ + p \to \pi^+ + p + \pi^0$$

$$\pi^- + p \to \pi^+ + n + \pi^-.$$

- Indicate which resonance was found in which reaction. Are there any resonances observed in two reactions?
- Find the threshold pion kinetic energy required for the production of each resonance (the initial proton is at rest).
- Obtain the life time for each resonance, if their width is 140 MeV.
- Give the quark content of each resonance. Does it appear reasonable in the case of Δ^{++}?

Data: proton mass 938 MeV, neutron mass 940 MeV, charged pion mass 140 MeV, neutral pion mass 135 MeV.

(17) The elastic scattering in the previous problem was very useful in determining the angular momentum of the resonance. Indeed the p-wave of the initial pion dominates. Choosing the direction of the momentum of the pion as a quantization axis, its wave function is a state of the form $|\ell = 1, m = 0\rangle$, since the pion is spinless. The resonance must have an angular momentum $|j, m = \alpha\rangle$, with α the spin projection of the target nucleon and $j = 3/2$ or $1/2$. Assume, e.g., that $\alpha = 1/2$. The resonance decays to a pion and proton with angular momenta $|\ell = 1,\ m = -\beta + 1/2\rangle$ and $|s = 1/2,\ m = \beta\rangle$ respectively according to the rule:

$$|j, m = 1/2\rangle = C_1^j |\ell = 1, m = 1\rangle |1/2, -1/2\rangle +$$

$$C_0^j |\ell = 1, m = 0\rangle |1/2, 1/2\rangle$$

where C_β^j are the relevant angular momentum coupling (Clebsch-Gordan) coefficients. Furthermore we know that, if the outgoing pion is scattered at an angle θ, its wave function can be expressed in terms of spherical harmonics, i.e. we have

$$|\ell = 1, m = 1\rangle = Y_1^1 = -\sqrt{\frac{3}{8\pi}} \sin\theta e^{i\phi},$$

$$|\ell = 1, m = 0\rangle = Y_0^1 = \sqrt{\frac{3}{4\pi}} \cos\theta.$$

The angular distribution of the outgoing pions is going to be proportional to a function $f(\theta)$ given by

$$f(\theta) = \frac{3}{4\pi}\left\{\left|C_1^j\right|^2 \frac{1}{2}\sin^2\theta + \left|C_0^j\right|^2 \cos^2\theta\right\}.$$

From tables we find $C_1^{3/2} = \frac{1}{\sqrt{3}}$ and $C_0^{3/2} = \sqrt{\frac{2}{3}}$, while $C_1^{1/2} = \sqrt{\frac{2}{3}}$ and $C_0^{1/2} = -\frac{1}{\sqrt{3}}$. Thus:

$$f(\theta) = \frac{3}{4\pi}\left(\frac{1}{6}\sin^2\theta + \frac{2}{3}\cos^2\theta\right) = \frac{3}{4\pi}\frac{1}{6}\left(1 + 3\cos^2\theta\right)$$

for a $j = \frac{3}{2}$ resonance and

$$f(\theta) = \frac{3}{4\pi}\left(\frac{1}{3}\sin^2\theta + \frac{1}{3}\cos^2\theta\right) = \frac{3}{4\pi}\frac{1}{3}, \text{ for a } j = \frac{1}{2} \text{ resonance,}$$

i.e. isotropic.

The same result holds if the outgoing nucleon is in the state $|1/2, -1/2\rangle$.

The experiment revealed that the distribution was not isotropic, but of the form $\left(1 + 3\cos^2\theta\right)$, consistent with the spin assignment of 3/2.

(18) The purpose of this problem is to understand that the old Yukawa type strong interaction, with the mesons been its carriers between nucleons, can be viewed as a Van der Waals type interaction originating from QCD.

- Since the gluons can interact with $q\bar{q}$ the diagram of Fig. 15.2 one obtains a new diagram by replacing a a $q - q$ line by a $q - \bar{q}$ pair as in Fig. 9.21.
- Consider the above diagram, combined with a number of "spectator" (non interacting) quarks.

  This combination can lead to processes involving only color singlet states (see Fig. 9.22), as, e.g., baryon decay into a

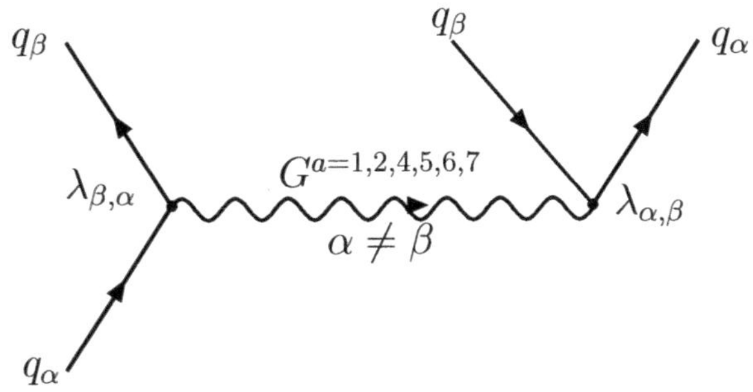

Fig. 9.21: A typical diagram involving the creation of a $q - \bar{q}$ pair by a gluon.

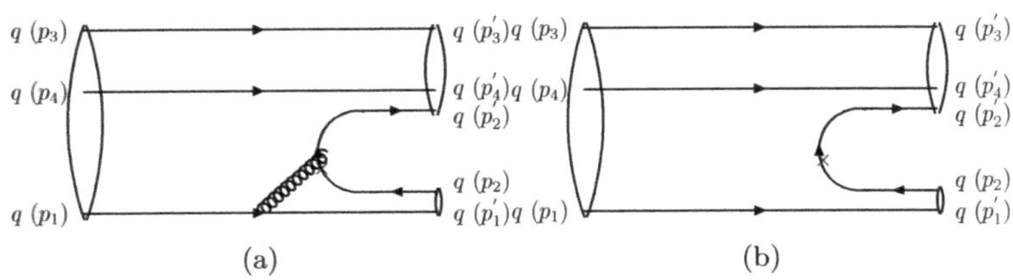

Fig. 9.22: The 1 gluon exchange potential acting between the quarks labeled $q(p_1)$, $q(p_1')$, $q(p_2)$ and $q(p_2')$, while the other quarks do not participate in the interaction (they are spectators) (a). The same effect can be accomplished by creating a $q\bar{q}$ pair out of the vacuum, marked by $\times$, (b), in which case the pair is normally in an $S = 1$, $\ell = 1$, $J = 0$ state ($^3P_0$ model). The process exhibited describes the baryon meson coupling. If the initial baryon is heavier than the final, it provides the decay amplitude $B_1 \to B_2 M$ (the lens like curves indicate a baryon or meson bound state). The same diagram is applicable in any process in which the interaction causes an $1q \to 2q - 1\bar{q}$ transition. For example, if the middle horizontal line is missing and the top arrow is reversed, it describes the decay of one meson into two mesons.

baryon and a meson, if energy allows it [Henley *et al.* (1988)]. Another approach is to consider the $^3P_0$, which is derived from the quark string model [Perkins (1971)], yielding a $q - \bar{q}$ pair, "created out of the vacuum".

- The above diagram can be considered as an effective nucleon-pion coupling, with the coupling constant dependent on the nucleon and pion structure, provided that the quarks in each vertex of Fig. 9.21 are restricted to be u or d.

Find the quark combinations that lead to

$$N \rightarrow N + \pi^0,\ n \rightarrow p + \pi^-,\ p \rightarrow n + \pi^+$$

(see Fig. 9.23). Show that i) Fig. 9.23(a) results from Fig. 9.22(a) with both $q(p_2)$, $q(p_1')$ being u or d quarks and ii) Fig. 9.23(b) results from Fig. 9.22(a) with the $q(p_2)$, $q(p_1')$ being u in the case of π^- and d in the case of π^+.

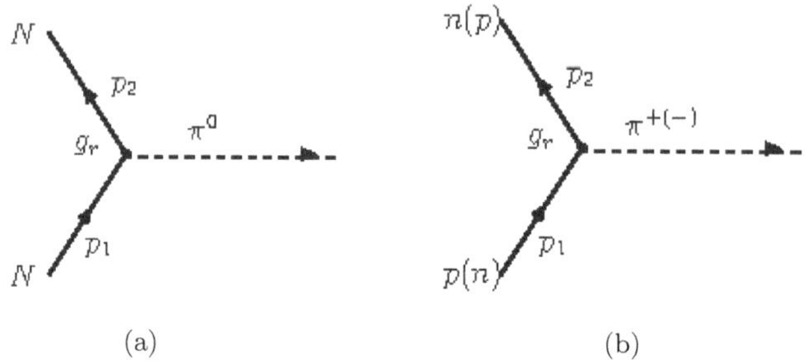

Fig. 9.23: The diagram of Fig. 9.22 re-written as pion nucleon coupling in the standard notation of a scalar particle. (a) Involving neutral pion and (b) charged pion.

- Combining the previous couplings one can get the effective Yukawa nucleon interaction shown in Fig. 9.24.

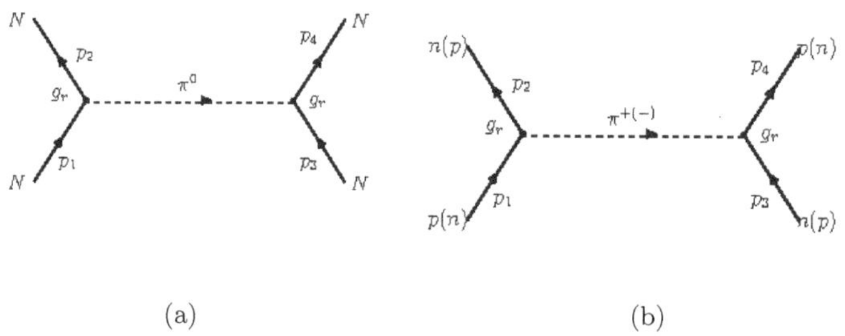

Fig. 9.24: The one pion exchange diagram between two nucleons, leading to the Yukawa interaction between nucleons. (a) neutral pion exchange and (b) charged pion exchange.

(19) Proceed as far as you can generalizing the above to other meson exchanges used in nuclear physics.

(20) Show that the interaction in coordinate space obtained in the one meson exchange between nucleons, Fig. 9.24, is of the form

$$V_m(r) = \frac{g_r^2}{4\pi r} e^{-mr}, \; r = |\mathbf{r}_i - \mathbf{r}_j|, \; m = \text{the meson mass.} \quad (9.122)$$

(21) It is suggested that in a $p-p$ collider process $p+p \to pp+D^-+D^+$, shown Fig. 9.25(a), can occur.

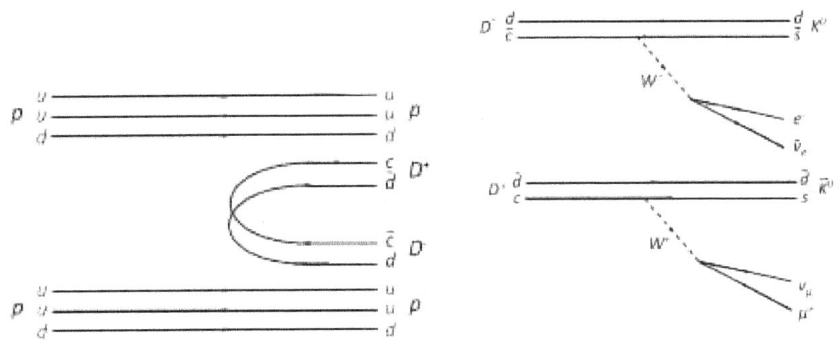

(a) (b)

Fig. 9.25: (a) Heavy meson production at LHC. (b) Possible decay mechanisms for D mesons.

i) Give arguments to support this view. What interaction is responsible?
ii) Examine the feasibility of this process occurring in the context of strong interactions. If feasible, add in the diagram the missing gluon lines responsible for the binding of the quarks into hadrons and those that create $q\bar{q}$ pairs out of the vacuum, including color indices at the vertices.
iii) Examine the feasibility of the processes shown in Fig. 9.25(b). Which interaction is responsible for the decay? For the hadron binding? Indicate flavor and color indices at the vertices.

(22) We have seen that an $N\bar{N}$ resonance has been observed, see Eq. (9.121). This resonance can decay under the strong interaction into a number of mesons. Thus:
i) Show that after rearrangement the strong interaction can lead to decays into three mesons, e.g. into 3 pions:

$$p(u_1, u_2, d_3)\bar{p}(\bar{d}_1\bar{u}_2\bar{u}_3) \rightarrow \text{rearrange}$$
$$\rightarrow (u_1\bar{d}_1)(u_2\bar{u}_2)(d_1\bar{u}_1) = \pi^+\pi^0\pi^-$$

(the numbers were put just to guide the eye). Also into $\pi^+\eta\pi^-$. In the same spirit, consider other three meson combinations.
ii) Show that, since he strong interaction can create $q\bar{q}$ pairs out of the vacuum, it can lead into decays of 4 meson combinations etc, depending on the available energy.

Chapter 10

The quark model for hadrons

In this chapter we will briefly discuss the quark model for hadrons, i.e. baryons, which are relatively heavy particles like the proton and the neutron, and mesons which have mass intermediate between the baryons and the electron. For more technical aspects the reader is referred to more specialized books, e.g. [Lichtenberg (1978)].

10.1 Brief history of hadrons

As we have seen in Chapter 1 experiments involving cosmic rays revealed the existence of particles that had strange behavior. Namely they were produced copiously, a signature that they interact strongly, but they lived very long, suggesting that they die under the influence of weak interaction. In 1957 the very high energy accelerator, the Cosmotron, started operating in Berkeley. This machine confirmed the properties of the known particles and discovered new ones. In this zoo of strange particles theorists were called upon to put some order. Initially Pais [Pais (1952)], [Pais and Jost (1952)] and later in a superior manner Gell-Mann [Gell-Mann (1953)] and Nishijima [Nagado and Nishijima (1953)] provided an explanation of the strange behavior, introducing an additive internal quantum number named *strangeness*. The strange particles are produced in pairs, a particle and its antiparticle, so strangeness is conserved under the strong interaction. The probability of them dying together, after they are produced and separated, is practically zero. So they die alone!

10.1.1 *The notion of strangeness*

Theorists tried as hard as they could, to explain things with the introduction of isotopic spin and strangeness quantum numbers. A sample is given in Table 10.1.

Reactions which are allowed by the strong interaction are:

$$\begin{aligned}
&p+p \to p+p+K^+ + K^- \\
&p+\pi^+ \to p+K^+ + \bar{K}^0 \\
&p+K^- \to \Lambda^0 + \pi^0 \\
&p+\pi^- \to K^0 + \Lambda^0 \\
&p+\pi^- \to K^+ + \Sigma^- \\
&p+p \to p+\Sigma^0 + K^+ \\
&p+K^- \to K^+ + \Xi^- \\
&p+K^- \to K^0 + \Xi^0 \\
&p+K^- \to \Omega^- + K^0 + K^+ \\
K^+ + d \to\ &\bar{\Omega}^- + \Lambda^0 + \Lambda^0 + p + \pi^+ + \pi^- .
\end{aligned} \tag{10.1}$$

Conservation of strangeness in strong interactions implies the conditions:

$$\begin{array}{ll}
S(K^-) = -S(K^+) & S(K^-) = S(\Sigma^+) \\
S(\bar{K}^0) = -S(K^+) & S(\Xi^-) = S(K^-) - S(K^+) = -2S(K^+) \\
S(\Lambda^0) = S(K^-) = -S(K^+) & S(\Xi^0) = S(K^-) - S(K^0) = -2S(K^+) \\
S(K^0) = -S(\Lambda^0) = S(K^+) & S(\Omega^-) = S(K^-) - S(K^+) - S(K^+) = -3S(K^+) \\
S(\Sigma^-) = -S(K^+) & S(\bar{\Omega}^-) = S(K^+) - 2S(\Lambda^0) = 3S(K^+) \\
S(\Sigma^0) = -S(K^+) & S(\Sigma^-) = S(\Sigma^+) \text{ (it is derived differently).}
\end{array}$$

The strangeness violating decay modes of mesons are given in Table 10.2.

Note that the life time of K^0 and $\bar{K}^0$ is not defined (they are not antiparticles of each other). One can define:

$$\begin{aligned}
|K_1^0\rangle &= \frac{|K^0\rangle + |\bar{K}^0\rangle}{\sqrt{2}} \\
|K_2^0\rangle &= \frac{|K^0\rangle - |\bar{K}^0\rangle}{\sqrt{2}}
\end{aligned} \tag{10.2}$$

which may decay, but with life times that are quite different, the first is short lived with a life time $(0.8953 \pm 0.0005) \times 10^{-10}$ s and the second $(5.116 \pm 0.020) \times 10^{-8}$ s. This interesting behavior will not be explained here, the interest reader can find it elsewhere, e.g. in the companion book [Vergados (2017)].

The Sigma baryons, Σ^+, Σ^0 and Σ^-, have spin 1/2 and a mass $\approx$ 1190 MeV. The charged ones, Σ^+ and Σ^- have a mean lifetime $\approx 10^{-10}$,

Table 10.1: Some of the strange particles. Particles are accompanied by their antiparticles, with opposite batyon number (B), strangeness (S) and isotopic spin projectionI_3, but with the same mass, lifetime etc. Antibaryons, however, have not been included here.

particle		lifetime (s)	mass MeV/c^2	strangeness (S)	Isotopic spin I	Projection of isotopic spin I_3
mesons	K^+	$(1.2369 \pm 0.0032) \times 10^{-8}$	493.69	1	1/2	1/2
	K^0	not defined	497.63	1	1/2	- 1/2
	K^0	not defined	497.63	-1	1/2	1/2
	K^-	$(1.2371 \pm 0.0029) \times 10^{-8}$	493.69	-1	1/2	-1/2
Baryons	Λ^0	$(2.632 \pm 0.20) \times 10^{-10}$	1115.6	-1	0	0
	Σ^+	$(0.799 \pm 0.004) \times 10^{-10}$	1189.36	-1	1	1
	Σ^0	$(7.4 \pm 0.7) \times 10^{-14}$	1192.46	-1	1	0
	Σ^-	$(1.4799 \pm 0.11) \times 10^{-10}$	1197.34	-1	1	-1
	Ξ^0	$(2.90 \pm 0.09) \times 10^{-10}$	1314.9	-2	1/2	1/2
	Ξ^-	$(1.639 \pm 0.015) \times 10^{-10}$	1321.32	-2	1/2	-1/2
	Ω^-	$(0.822 \pm 0.012) \times 10^{-10}$	1672.22	-3	0	0

while the neutral Σ^0 has a life time $\approx 10^{-14}$ s. They decay as follows:

$$\Sigma^+ \to p + \pi^0 \ (51.6\%)$$
$$\Sigma^+ \to n + \pi^+ \ (48.4\%)$$
$$\Sigma^0 \to \Lambda^0 + \gamma \ (100\%)$$
$$\Sigma^- \to n + \pi^- \ (100\%).$$

The particle Ω^- is an electrically charged baryon with spin 3/2 and a mass of 1672 MeV. As we will see below, its existence had been predicted before its discovery. Its life time is relatively long, 0.82×10^{-10} s, if one takes into account its large mass. It decays as follows:

$$\Omega^- \to \Lambda^0 + K^- \ (68.6\%)$$
$$\Omega^- \to \Xi^0 + \pi^- \ (23.4\%)$$
$$\Omega^- \to \Xi^- + \pi^0 \ (8\%).$$

The neutral Λ^0 is a baryon with a mass ≈ 1116 MeV, spin $s = 1/2$ and a mean lifetime $\approx 2.6 \times 10^{-10}$ s, decaying as follows:

$$\Lambda^0 \to p + \pi^- \quad (64.2\%)$$
$$\Lambda^0 \to n + \pi^0 \quad (35.8\%).$$

Table 10.2: Decays of strange mesons. The μ-particles are leptons like the electrons, but 210 times heavier. They are accompanied by their own neutrinos (antineutrinos).

Reaction	Percentage
$K^+ \to \mu^+ + \nu_\mu$	(63.3%)
$K^+ \to \pi^+ + \pi^0$	(21.2%)
$K^+ \to \pi^+ + \pi^- + \pi^+$	(5.6%)
$K^- \to \mu^- + \bar{\nu}_\mu$	(63.6%)
$K^- \to \pi^- + \pi^0$	(21.2%)
$K^- \to \pi^- + \pi^+ + \pi^-$	(5.6%)
$K_S^0 \to \pi^- + \pi^+$	(68.6%)
$K_S^0 \to \pi^0 + \pi^0$	(31.4%)
$K_L^0 \to \pi^\pm + e^\mp + \bar{\nu}_e(\nu_e)$	(38.7%)*
$K_L^0 \to \pi^\pm + e\mu^\mp + \bar{\nu}_\mu(\nu_\mu)$	(27.1%)**
$K_L^0 \to \pi^0 + \pi^0 + \pi^0$	(21.5%)
$K_L^0 \to \pi^+ + \pi^- + \pi^0$	(12.4%)

*In reality $K^0 \to \pi^- e^+ \nu_e, \quad \pi^- \mu^+ \nu_\mu$.
**In reality $\bar{K}^0 \to \pi^+ e\, \tilde{\nu}_e, \quad \pi^+ \mu\, \tilde{\nu}_\mu$.

It is produced together with its antiparticle $\bar{\Lambda}^0$ in proton antiproton collisions:

$$p + \bar{p} \to \Lambda^0 + \bar{\Lambda}^0.$$

There exist two Ξ-baryons, Ξ^0 and Ξ^- (with their antiparticles $\bar{\Xi}^0$ and Ξ^+). They have masses ≈ 1315 MeV and 1321 MeV and a mean life time $\approx 2.9 \times 10^{-10}$ s and $\approx 1.6 \times 10^{-10}$ s respectively. They have spin 1/2 and they decay yielding a Λ^0 as follows:

$$\Xi^+ \to \pi^+ + \Lambda^0 \; (100^0/_0)$$
$$\Xi^- \to \pi^- + \Lambda^0 \; (100^0/_0)$$
$$\Xi^0 \to \pi^0 + \Lambda^0 \; (100^0/_0).$$

Taking into account the above facts, especially their mass, we can classify them as follows:

- Two groups of mesons:
  (K^+, K^0), $(K^-, \bar{K}^0)$.

- Four groups of baryons:
  (Λ^0), $(\Sigma^+, \Sigma^0, \Sigma^-)$, (Ξ^0, Ξ^-) and (Ω^-).

The particles of each group have the same strangeness quantum number. They form an isopin multiple with isospin I, containing $(2I+1)$ members, distinguished from each other be the isospin projection quantum number I_3 (see Table 10.1).

10.1.2 *Resonances*

At about the same time two types of resonances were discovered (see Section 4.1):

i) Meson resonances.
As we have seen the first such resonance to be observed was the family indicated by $\rho^{\pm}$, ρ^0 (see Section 4.1). This resonance was observed in pion nucleon scattering at a pion kinetic energy of 295 MeV. It is characterized by a mass of 775 MeV and a width $\Gamma = 125$ MeV, which yields a life time of 4.5×10^{-24} s. This resonance was easy to accommodate as the sister of the pion family, but with spin one. Other similar meson resonances followed, ϕ^0, ω^0, η^0, etc., a whole army of them. All these, with the exception of η^0, have spin 1 and they can be neutral or charged, while the η^0 is neutral with spin zero.

ii) Baryon resonances.
The first baryon resonance observed was a resonance with spin 3/2, which could be accommodated with isospin 3/2. This was indicated as $\Delta(3/2, 3/2)$, with its 4 members denoted by Δ^{++}, Δ^+, Δ^0 and Δ^-. The resonance members have a mass about 1240 MeV and approximately the same life time, 5.5×10^{-24} s.

The resonance is characterized by two parameters (see Fig. 10.1): the energy E_0 at which it peaks and its width Γ. The width gives the uncertainty to assign a definite energy to it. As we have seen from the uncertainty principle $(\Delta E)(\Delta t) \approx \hbar$, we may obtain the lifetime τ of this new particle:

$$\tau = (\Delta t) = \frac{\hbar}{(\Delta E)} = \frac{\hbar}{\Gamma}. \tag{10.3}$$

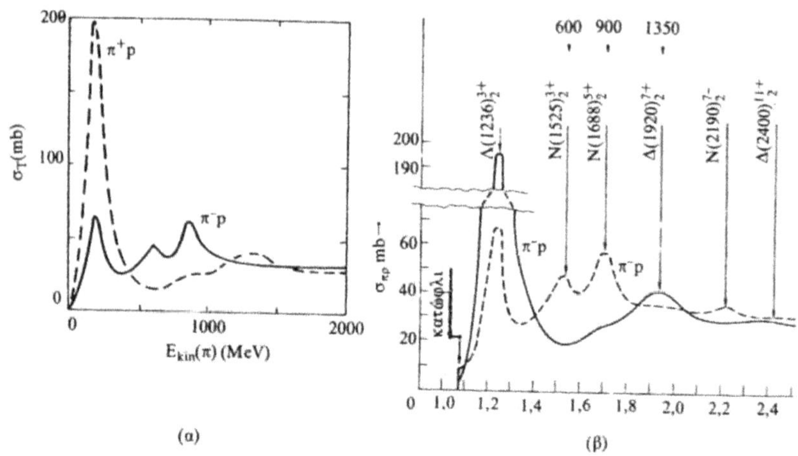

Fig. 10.1: (α) The cross section of the scattering π^+ and π^- mesons on a proton target as a function of the energy of mesons. (β) The invariant mass in GeV/c^2, while above the invariant mass the corresponding pion kinetic energy T_π is indicated. Δ indicates $I = 3/2$ and N $I = 1/2$ resonances.

The most probable decay schemes are as follows:

$$\begin{aligned}
\Delta^{++} &\to p + \pi^+ \\
\Delta^{+} &\to p^+ + \pi^0 \\
\Delta^{0} &\to n + \pi^0 \\
\Delta^{-} &\to n + \pi^- .
\end{aligned}$$

As indicated in Fig. 10.1 other resonances have been observed, called N resonances, even though it was harder to get the corresponding data and analyze them. Above the mass of 1450 MeV/c^2, there exists another family of resonances indicated by a * (star) over their name.[1] Exotic resonances like Σ^{*0}, Σ^{*-}, Ξ^{*0}, Ξ^{*-}, Ω^{*0}, etc. were discovered. We could easily build for them a table like Table 10.1, but it is pointless. We have a real zoo of elementary particles!

Faced with such a zoo with more than 150 hadrons, one asks: Can one put an order to it? Can we recover the beauty characterizing the world of particle physics, reminiscent of that of the late nineteen thirties?

[1]It is common to indicate the resonance with * (star) above the corresponding particle, e.g. $N^*(N^{*+}, N^{*0})$, K^{*+}, Σ^{*+}, etc.

10.2 The SU(3) symmetry in the hadron zoo

The first obvious extension of the $SU(2) \times U_1(S)$ symmetry was the group $SU(3)$. This is a classical symmetry first considered in particle physics in 1962 by the American Physicist Murray Gell-Mann and, independently, by the Israeli physicist Yuval Ne'eman, as a theory of organizing subatomic baryons and mesons. This group contains both isospin and strangeness as its subgroups. Among its simplest representations, they considered only the octet $\underline{8}$ and the decuplet (or decaplet) $\underline{10}$. Into these they put particles with the same spin and parity and, approximately, the same mass. If the symmetry were exact the particles should be degenerate in mass, but this approximation was a good starting point for strong interactions. The other simple representations $\underline{1}$ and $\underline{6}$ did not serve any purpose and they were not considered. Even the fundamental $\underline{3}$, from which all the other representations can mathematically be obtained, was missing.

A rigorous method of reduction of these representations into $SU(2) \subset SU(3)$ is beyond the scope of this book, see, however, a summary in Section 15.6.1. The reader interested in rigorous derivations is referred to more specialized books, e.g. [Vergados (2016)]. The weight diagram associated with the quantum numbers I_3 and the operator Y defined by the relation $Q = I_3 + Y/2$ was also known. This is exhibited for the decuplet in Fig. 10.2. The usual baryons were put in the octet, $\underline{8}$, see Fig. 10.3. The reason for calling this root diagram will be explained in Section 15.6.1, see Fig. 15.1.

The agreement with experiment is remarkable. It should be noted that the particle Ω^- with a strangeness -3 was not known at the time. It was a prediction based on $SU(3)$ in 1964, which was an ingredient for the award of Nobel price to Gell-Man in 1969. This picture of the octet motivated Gell-Mann to coin for $SU(3)$ the name the *Eightfold Way* alluding to the Noble Eightfold Path of Buddhism.

There was no room for the decuplet for mesons. But the octet was found useful. The two interesting root diagrams are exhibited in Fig. 10.4. In the case of vector bosons both ω (782) and ϕ (1020) have been included, since it could not be decided which belongs to the octet on the basis of their mass, since they are close. The ω decays predominantly into three pions, while the ϕ into two K mesons. One may get a better picture by studying their internal structure, e.g. their quark content, see below.

In spite of the above success the $SU(3)$ model did not solve the fundamental problem of too many elementary particles. It did not really explain why there appear so many elementary particles by giving reasons for their

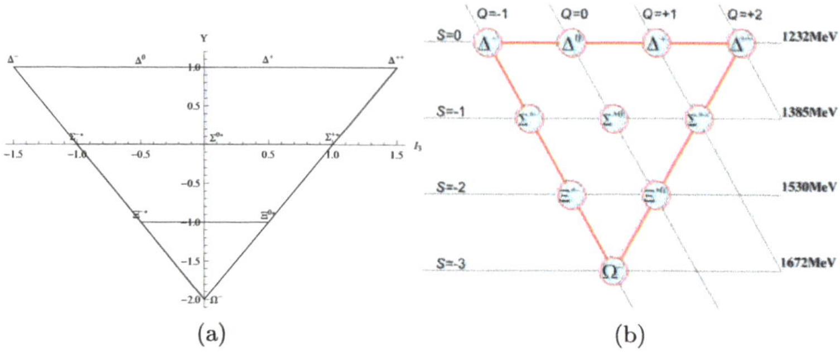

Fig. 10.2: (a) The weight diagram (I_3, Y) associated with the decuplet with hadrons indicated. (b) The same diagram in the (I_3, S) plane. The masses associated with each I_3 are indicated, the split between the isospin members is quite a bit smaller.

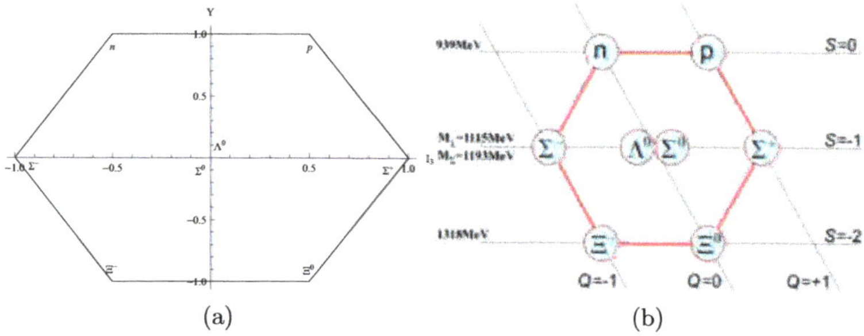

Fig. 10.3: (a) The root diagram (I_3, Y) associated with the baryon octet. (b) The same diagram in the (I_3, S) plane. In this presentation the strangeness is decreasing downwards. The masses associated with each I_3 are indicated, the split between the isospin members is quite a bit smaller.

existence. The parton model proposed by Feynman employed in the analysis of experiments had already given some evidence that the protons and the neutron possessed internal structure. $SU(3)$, however, paved the way to find the fundamental building blocks that particle physics is made of, which began with the unsuccessful proposal of three particles, proton, neutron and Λ proposed by Sakata in 1959, $\pi^+ = p\bar{n}$ etc. It culminated with the successful quark model proposed Sweig and Gell-Mann in 1964. The

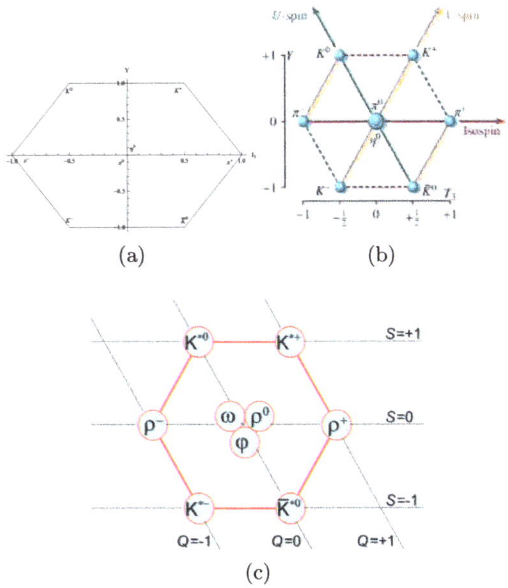

Fig. 10.4: (a) The root diagram (I_3, Y) associated with the pseudoscalar (spin 0) meson octet. (b) The same diagram in the (I_3, S) plane. The Kaons are almost degenerate (see Table 10.1). The pions are around 140 MeV and the η^0 is a bit out of place with 550 MeV. The split between the isospin members is quite small. (c) The (I_3, S) root diagram for vector (spin 1) mesons. Here the mass of K* is around 892 MeV. Both ω (782) and ϕ (1020) have been included, since in addition to the octet there exists an SU(3) singlet. It could not be decided which one belongs to the octet on the basis of their mass, since they are close. The name root diagram is understood by comparing this figure to Fig. 15.1.

strange name was coined by Gell-Mann:

Three quarks for Muster Mark!

Sure he has not got much of a bark

And sure any he has it's all beside the mark.

-James Joyce, Finnegans Wake

10.3 The quark model

In this context the known hadrons are bound states of quarks and one needs three of them. The game is not as simple as in the case of atoms, molecules and nuclei. The structures with quarks are complicated and rather tricky [Veltman (2003)]. The main reason is that the mediators of the interaction between quarks, called gluons, unlike the photons, interact very strongly among themselves. Big globs of gluons are needed to keep the quarks bound. Even though the gluons are massless, the gluon globs contain a large amount of energy, which contributes to the binding energy of the system. The quarks are embedded in the gluon blog and their mass is a tiny fraction of the total. The mass of the light quarks making up the proton are expected to have a mass less than 20 MeV. Thus the gluon blob contains an energy of $938 - 20 = 918$ MeV. It is thus hard to even speak of a binding energy under these circumstances. Another reason is that the strength of the interaction between the quarks increases fast as you try to pull them apart and separate the hadron into its constituents. The blog gets bigger and bigger and glues the quarks together (hence their name). Anyway this type of binding does not appear elsewhere and as a result the quarks never become free. So it is very difficult to determine their mass, especially in the case of the lightest ones. The situation gets somewhat improved for the heavier hadrons. At the moment, however, this is the only hadron picture we have.

In its simplest context the quark model contains three elementary particles, two of them indicated indicated by u (up) and d (down) with isospin quantum numbers $(I, I_3) = (1/2, 1/2)$ and $(1/2, -1/2)$ respectively and strangeness $S = 0$, while the third s has strangeness[2] $S = -1$ and isospin 0. These can be considered as the three states of an object, the quark q with three flavors, spin 1/2 and positive parity.

Out of these building blocs one should construct all the known particles with the rules:

- The known mesons as a combination of a quark with an antiquark with the known rules of spin and isospin.
- The known baryons as a combination of three quarks with the known rules of spin and isospin.

The baryon construction assigns a baryon number of 1/3 to the quarks and

[2]The negative sign is dictated by the fact that the strangeness was defined to +1 for the K^+, which contains a strange antiquark.

constraints their charge. One requires that any combination of three quarks should lead to charges of Ne, $N = \pm 2, \pm 1, 0$. This constrains the quarks to have charges:

$$Q = \pm 2/3,\ \pm 1/3$$

(in units of the elementary charge e). The above analysis showed, however, that there exist doubly charged baryons with positive charge, so the quark u should have a charge of 2/3. The proton has a charge +1 and it involves u and d quarks so the charge of the d quark should be $-1/3$. The is consistent with the structure of the neutron. Furthermore the $S = -1$ baryon has a charge -1 and, thus, the charge of the s quark should also be be $-1/3$. Thus the only possibility is

$$Q = \frac{2}{3}, -\frac{1}{3}, -\frac{1}{3} \text{ for } u, d, s \text{ respectively.} \tag{10.4}$$

At this point we remark that with three flavors there exist $3^3 = 27$ baryon states. The resulting irreducible representations Of $SU(3)$ must have dimensions less than 27. The only way these 27 states can be accommodated as irreducible $SU(3)$ representations is via the reduction $\underline{1} + 2 \times \underline{8} + \underline{10}$. So the pre-quark $SU(3)$ enthusiasts hit right on target considering the baryon decuplet and octet. These irreducible representations will be of interest to us here.

10.3.1 *Construction of some simple states of a given symmetry*

In the Appendix B, see Chapter 15, using the method of Young tableaux, we show that all representations, which occur in the 3-q (baryon) and $q - \bar{q}$ (meson) systems can be generated by the fundamental representation.

$$(1,0) \otimes (1,0) \to (0,1) + (2,0) \leftrightarrow \underline{3}^* + \underline{6}. \tag{10.5}$$

From the formulas of Section 15.6.2 we find:
The (0,1) contains $I = 1/2$, $-Y = 1/3$ identified with the anti-quarks of up and down quarks and $I = 0$, $Y = 2/3$ identified with the antiquark of the strange quark.
The (2,0) contains $I = 0$, $Y - 4/3$, $I = 1$, $Y - 1/3$ and $I = 1, Y = 2/3$

$$(0,1) \otimes (1,0) \to (0,0) + (1,1) \leftrightarrow \underline{1} + \underline{8}. \tag{10.6}$$

The (0,0) contains $I = 0$, $Y = 0$.
The (1,1) contains $I = 0$, $Y = 0$, $I = 1/2$, $Y = -1$, $I = 1$, $Y = 0$, $I = 1/2$, $Y = 1$. The representation (1,1) is self conjugate,

$$(2,0) \otimes (1,0) \rightarrow (3,0) + (1,1) \leftrightarrow \underline{10} + \underline{8}. \tag{10.7}$$

The (1,1) has already been decomposed.
The (3,0) contains $I = 0$, $Y = -2$, $I = 1/2$, $Y = -1$, $I = 1$, $Y = 0$, $I = 3/2$, $Y = 1$.

The antiquark is associated with the conjugate of the previous representation, i.e. $(\lambda, \mu) = (0, 1)$. Proceeding as above, Section 15.6.1, Eq. (15.36), we get $q = 0$, 1 and $p = 0$. For $q = 1$ we get $I = 1/2$, $Y = 2/3$ to be identified with $\bar{s}$ (s-antiquark), while for $q = 0$ we get $I = 1/2$, $Y = -1/3$ leading to $I = 1/2$, $I_3 = -1/2$, $Y = -1/3$ to be identified with $\bar{u}$ (u-antiquark) and $I = 1/2$, $I_3 = 1/2$, $Y = -1/3$ to be identified with $\bar{d}$ (d-antiquark). As expected the additive quantum numbers are the opposite of the previous ones. The root diagram is shown in Fig. 10.5.

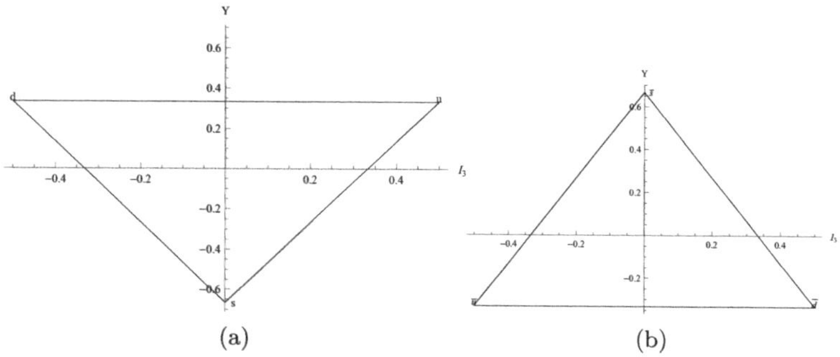

Fig. 10.5: The root diagram (I_3, Y) associated fundamental representation (1,0) ($\underline{3}$, triplet or q) (a) and its adjoined (0,1) ($\underline{3}^*$, antitriplet or $\bar{q}$) (b).

Let us now consider the regular representation (1,1) or $\underline{8}$ in more detail.

$$p = 0, q = 0 \rightarrow Y = -1, I = \frac{1}{2}, I_3 = \pm\frac{1}{2} \Leftrightarrow ssu, ssd$$

$$p = 0, q = 1 \rightarrow Y = 0, I = 0, I_3 = 0 \Leftrightarrow \Lambda_0 = \frac{1}{\sqrt{2}}(ud - du)$$

$$p = 1, q = 0 \rightarrow Y = -1, I = 1, I_3 = 0, \pm 1 \Leftrightarrow uus, \frac{1}{\sqrt{2}}(ud + du)s, dds$$

$$p = 1, q = 1 \rightarrow Y = -2, I = \frac{1}{2}, I_3 = \pm\frac{1}{2} \Leftrightarrow uss, dss.$$

For the representation (3,0) or $\underline{10}$.

$$p = 0, q = 0 \rightarrow Y = -2, I = 0, I = 0 \Leftrightarrow sss$$

$$p = 1, q = 0 \rightarrow Y = -1, I = \frac{1}{2}, I_3 = \pm\frac{1}{2} \Leftrightarrow uss, dss$$

$$p = 2, q = 0 \rightarrow Y = 0, I = 1, I_3 = 0, \pm 1 \Leftrightarrow \frac{1}{\sqrt{2}}(ud - du)s, uus, dds$$

$$p = 3, q = 0 \rightarrow Y = 1, I = \frac{3}{2},$$

$$I_3 = \pm\frac{3}{2} \pm \frac{1}{2} \Leftrightarrow \Delta^{++}(uuu), \Delta^{-}(ddd), \Delta^{+}(uud), \Delta^{0}(udd).$$

The root diagrams associated with the symmetric (3,0), decuplet, and the regular, octet, representation (1,1) are shown in Fig. 10.6. The equation relating the charge to Y and S is given as follows:

$$Q = I_3 + \frac{Y}{2}, \ Y = B - S. \tag{10.8}$$

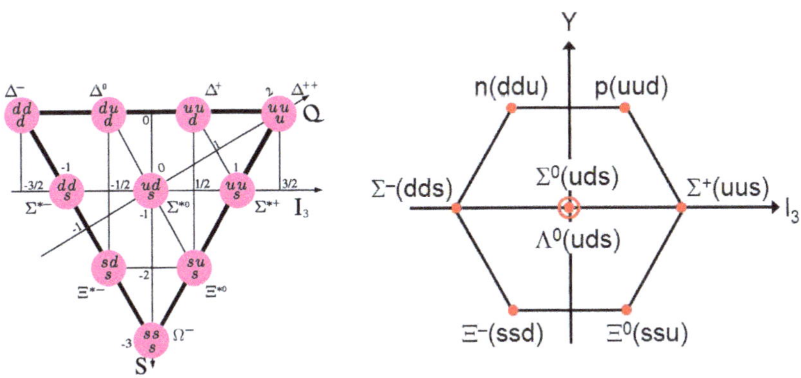

(a) (b)

Fig. 10.6: The root diagram (I_3, S) associated the symmetric representation (3,0) ($\underline{10}$ or decuplet) (a) and diagram (I_3, Y) of the regular representation (1,1) ($\underline{8}^*$ or octet) (b).

10.3.2 *The symmetry of the wave function under S_n*

In constructing the wave function of multiquark system of identical particles we have to make sure that it has the proper symmetry under the exchange of any two particles. The wave function must change sign, if the system is composed of particles with angular momentum half integral, called fermions. In this case it obeys the Pauli exclusion principle. On the other hand the wave function should have same sign if the system is composed of particles with integral angular momentum, called bosons.

The singlet $SU(3)$ representation is completely antisymmetric, corresponding to a Young tableau with one column. If a system is composed of u, d and s quarks from Eq. (15.12) of Chapter 15 we find that this state can easily be written as

$$\psi^{[1^3]}(u,d,s) = \frac{1}{\sqrt{6}}\left[(u(1)d(2) - u(2)d(1))s(3)) + (u(2)d(3) - u(3)d(2))s(1) + (u(3)d(1) - u(1)d(3))s(2)\right]. \tag{10.9}$$

A mnemonic rule is starting from the first term in the square bracket and proceeding by cyclic permutation. We notice, however, that this does not apply to baryons that do not contain a strange quark, like the proton and neutron. Furthermore flavor is not the only quantum number. The constituents have a spin, and this dictates whether a particle is a fermion or boson. The procedure outlined in Appendix B, Chapter 15, applies when we consider the maximum symmetry of the system, in our case flavor and spin.

In this context in the case of the nucleon we have the 4 possibilities:

$$|1\rangle = u_\uparrow, |2\rangle = u_\downarrow, |3\rangle = d_\uparrow, |4\rangle = d_\downarrow$$

where the arrows indicate the spin orientation. Thus the maximum symmetry is $U(4)$. From Eq. (15.2) and the discussion following it there is an anti-symmetric representation which for $n = 4$ has dimension 4, i.e. it can accommodate the nucleon. Now let us consider the proton with spin up. Since its state must be anti-symmetric, the two up quarks must have opposite spin orientations due to Pauli principle. Thus the Young operator can operate only on the state $u_\uparrow(1)u_\downarrow(2)d_\uparrow(3)$. Thus from Eq. (15.12) of Appendix B for $\alpha = u_\uparrow, \beta = u_\downarrow, \gamma = d_\uparrow$ we obtain:

$$\begin{aligned} p_\uparrow &= \psi^{[1^3]}(u_\uparrow, u_\downarrow, d_\uparrow) \\ &= \frac{1}{\sqrt{6}}\left[u_\uparrow(1)u_\downarrow(2)d_\uparrow(3) - u_\uparrow(2)u_\downarrow(1)d_\uparrow(3) - u_\uparrow(3)u_\downarrow(2)d_\uparrow(1)\right. \\ &\quad \left. - u_\uparrow(1)u_\downarrow(3)d_\uparrow(2) + u_\uparrow(2)u_\downarrow(3)d_\uparrow(1) + u_\uparrow(3)u_\downarrow(1)d_\uparrow(2)\right]. \end{aligned} \tag{10.10}$$

The situation for the spin down and the neutron is analogous.

The enthusiasm for this success did not, however, last long, as will see below.

10.3.3 *The snug of antisymmetry — The color is born!*

Soon after the quark model was introduced an examination of the decouplet showed that there is a violation of the Pauli principle. The doubly charged member of the (3/2, 3/2) is of the type uuu, which is flavor symmetric. The spin part of the wave function $s = 3/2$, $m_s = 3/2$ is also symmetric, i.e. $\uparrow\uparrow\uparrow$. This yields a total three fermion wave function, which is totally symmetric (the orbital part is composed of three quarks, each in an $\ell = 0$ state, and, therefore, is also symmetric). Many ideas were proposed, but the one that survived all tests was the one proposed by Greenberg (1964) and independently and more completely by Han and Nambu (1965). The quarks, in addition to flavor, contain another quantum number called color, which also takes three values with poetic names r (red), g (green) and b (blue). This is done in the context of a new symmetry $SU_c(3)$, see Section 15.6.3, contrasted with the flavor symmetry $SU_f(3)$, but with both having the same mathematical structure. Thus the quarks q are characterized by the following states:

$$\begin{pmatrix} u_r & u_g & u_b \\ d_r & d_g & d_b \\ s_r & s_g & s_b \end{pmatrix}.$$

This important symmetry characterizes the strong interaction and will be discussed in detail elsewhere. For the moment we can say that we impose the constraint that all observed hadrons are colorless, i.e. the transform as singlets under the group $SU_c(3)$. In the case of baryons the color part of the wave function is antisymmetric. Thus the total wave function of the decuplet becomes antisymmetric. Thus the $S_3 = 3/2$ Δ^{++} wave function is:

$$\Delta^{++} = \frac{1}{\sqrt{6}}(u_{r\uparrow}u_{g\uparrow}u_{b\uparrow} - u_{r\uparrow}u_{b\uparrow}u_{g\uparrow} - u_{b\uparrow}u_{g\uparrow}u_{r\uparrow}$$

$$+u_{g\uparrow}u_{b\uparrow}u_{r\uparrow} - u_{g\uparrow}u_{r\uparrow}u_{b\uparrow} + u_{b\uparrow}u_{r\uparrow}u_{g\uparrow}), \qquad (10.11)$$

which is clearly antisymmetric. In writing the last equation we assume the standard convention of the permutation group, i.e. the first particle is labeled 1, the second 2 and the last one 3. The other spin components are obtained as usual by the lowering spin operator. All members of the decuplet can be dealt this way.

The situation with the $I = 1/2$, $S = 1/2$ objects like the proton and the neutron is less obvious. We have seen above that the appropriate group to consider is $SU(4)$, which contains as subgroup the $SU_I(2) \otimes SU_s(2)$. Without the color we saw above that the proton and the neutron can be accommodated in the anti-symmetric representation of $SU(4)$, yielding a completely anti-symmetric wave function in spin-isospin. This is not acceptable after the introduction of color, since now all hadrons must be colorless (singlets), i.e. anti-symmetric in color space. Thus an anti-symmetric spin color component would lead to a symmetric wave function, which is contrary to the Pauli principle for fermions. From the Kronecker product in Eq. (15.3) of Chapter 15, we find, however, that the [3] of SU(4) is 20-dimensional and contains $(I, s) = (3/2, 3/2)$ and $(1/2,\ 1/2)$. The first is identified with the Δ multiplet, discussed above, and the second with the nucleon. This is just fine, a completely antisymmetric color wave function combined with a completely symmetric spin-isospin component.

To write down a symmetric $(I = 1/2,\ s = 1/2)$ we follow an analogous procedure that lead to Eq. (10.10), but for the symmetric case. Since now the Young operator is completely symmetric, in the case of the proton spin up it can act on $u_\uparrow(1)u_\downarrow(2)d_\uparrow(3)$ and $u_\uparrow(1)u_\uparrow(2)d_\downarrow(3)$ (the second did not contribute when the Young operator was antisymmetric). This procedure is straightforward, but a bit more tedious and we will skip the derivation. The result for the nucleon spin up, written out explicitly in spin, flavor and color spaces, has many components! In fact one finds that the spin flavor proton wave function with spin up is conveniently written the form:

$$
\begin{aligned}
p(\uparrow) = \frac{1}{9\sqrt{2}} & \{2\left[d(\downarrow)u(\uparrow)u(\uparrow) + u(\uparrow)u(\uparrow)d(\downarrow) + u(\uparrow)d(\downarrow)u(\uparrow)\right] \\
& - \left[d(\uparrow)\left(u(\uparrow)u(\downarrow) + u(\downarrow)u(\uparrow)\right) + \left(u(\uparrow)u(\downarrow) + u(\downarrow)u(\uparrow)\right)d(\uparrow)\right] \\
& - \left[u(\downarrow)d(\uparrow)u(\uparrow) + u(\uparrow)d(\uparrow)u(\downarrow)\right]\},
\end{aligned}
$$

(the particle indices 1, 2, 3 labeling the quarks are understood). One can verify that this function is symmetric. One must include the color degree of freedom in the colorless antisymmetric combination

$$
[1^3] = \frac{1}{\sqrt{6}}\left((rg - gr)b + (gb - bg)r + (br - rb)g\right),
$$

where again we have omitted the particle indices 1, 2, 3.

Fortunately this tedious procedure is not needed in practice. The sophisticated group theorists only need to know what we call the coefficients of fractional parentage (CFP) for each of the symmetry groups involved. These express the irreducible representations of the groups involved in terms of the symmetries $[f]$ of the symmetric group S_n and those of S_{n-1} and S_{n-2} symbolically written as $\langle f_{n-1}; n | \} f_n \rangle$ (one particle CFP), $\langle f_{n-2}; [f_2] | \} f_n \rangle$ (two-particle CFP), where $[f_2]$ is the symmetric or antisymmetric combination of the the last two particles. Such topics, however, are beyond the scope of this book and the reader is referred to specialized group theory books, e.g. [Chen *et al.* (2002)], [Vergados (2016)].

10.3.4 *The meson octet in the quark model*

As we have mentioned the mesons are combination of a quark and antiquark. The antiquark transforms like the conjugate representation of the quark, namely as (0,1) or $\underline{3}^*$. One can see that

$$(0,1) \otimes (1,0) = (0,0) \text{ or } \underline{1} + (1,1) \text{ or } \underline{8}, \tag{10.12}$$

see Chapter 15, Section 15.6.1.

The mesons are combination of 1 quark and 1 anti-quark. The flavor structure of the meson octet states follows from the above decomposition of the octet in terms of Y, I, I_3 and the fundamental representation of the group, i.e. with the identification u and $\bar{d}$ as $I = 1/2$, $I_3 = 1/2$, while d and $\bar{u}$ as $I = 1/2, I_3 = -1/2$ respectively. Furthermore the s and $\bar{s}$ are characterized by $I = 0$.

The state $u\bar{d}$ has $I = 1, I_3 = 1$. Then acting with the isospin lowering operator we get $|I = 1, I_3 = 0\rangle = \frac{1}{\sqrt{2}}\left(u\bar{u} + d\bar{d}\right)$ and $|I = 1, I_3 = -1\rangle = |d\bar{u}\rangle$. Similarly the $|Y = 0 (S = 0), I = 0, I_3 = 0\rangle = \frac{1}{\sqrt{2}}\left(u\bar{u} - d\bar{d}\right)$.

The color singlet is colorless, i.e. it is combination of all colors in equal proportion, that is:

$$(0,0)_c = \frac{1}{\sqrt{3}}\left(r\bar{r} + g\bar{g} + b\bar{b}\right). \tag{10.13}$$

So all we have to do adjoin the color singlet to each flavor state combined to the desired spin, i.e. zero (for the pseudoscalar) and one (for the vector) bosons. E.g. for the π^+ we get

$$\pi^+ = \frac{1}{\sqrt{3}}\left[u_r\bar{d}_r + u_g\bar{d}_g + u_b\bar{d}_b\right]^{s=0} = \frac{1}{\sqrt{3}}$$
$$\left[\frac{(u_r \uparrow \bar{d}_r \downarrow - u_r \downarrow \bar{d}_r \uparrow)}{\sqrt{2}} + \frac{(u_g \uparrow \bar{d}_g \downarrow - u_g \downarrow \bar{d}_g \uparrow)}{\sqrt{2}} + \frac{(u_b \uparrow \bar{d}_b \downarrow - u_b \downarrow \bar{d}_b \uparrow)}{\sqrt{2}}\right] \tag{10.14}$$

and ρ^+ is given by a similar expression:

$$\rho^+ = \left[u_r\bar{d}_r + u_g\bar{d}_g + u_b\bar{d}_b\right]^{s=1} \tag{10.15}$$

is given by a similar expression as the previous one by merely replacing the − sign with a + sign. The other components are obtained by operating with the isospin lowering operator. One finds e.g.:

$$\pi^0 = \frac{1}{\sqrt{3}}\left[\frac{\left(u\bar{u}+d\bar{d}\right)_r}{\sqrt{2}} + \frac{\left(u\bar{u}+d\bar{d}\right)_g}{\sqrt{2}} + \frac{\left(u\bar{u}_+d\bar{d}\right)_b}{\sqrt{2}}\right]^{s=0} \tag{10.16}$$

$$\rho^0 = \frac{1}{\sqrt{3}}\left[\frac{\left(u\bar{u}+d\bar{d}\right)_r}{\sqrt{2}} + \frac{\left(u\bar{u}+d\bar{d}\right)_g}{\sqrt{2}} + \frac{\left(u\bar{u}_+d\bar{d}\right)_b}{\sqrt{2}}\right]^{s=1} \tag{10.17}$$

with the spin decomposition as a above.

In a similar fashion one proceeds with the other mesons not containing the strange quarks (anti-quarks).

The positive strangeness was associated with K^-, which must have the structure $s\bar{u}$, which means $|K^+\rangle = |u\bar{s}\rangle$. Then, since the strangeness of $K^0 = -1$, it must be of the form $d\bar{s}$ to be identified as $I = 1/2$, $I_3 = -1/2$ and the $s\bar{d}$ as the $I = 1/2$, $I_3 = 1/2$.

In addition we need the spin structure of the octet. Instead of simply adjoining the spin quantum number, different names are used for the corresponding mesons. One thus obtains the corresponding root diagram (see Fig. 10.7).

10.4 The extension of the hadronic sector — The charm

The quark model discussed above was quite successful, except for the fact that net color has not been observed. Another concern has been the lack of symmetry between leptons and hadrons. There existed only three quarks, (u, d, s) and four leptons (e^-, ν_e, μ^- and ν_μ). For this reason Glashow and Bjorken proposed in 1964 the existence of a fourth quark named charmed (c), which has the same charge as the u-quark, carrying the extra quantum number, c or charm. Now we have:

$$\begin{pmatrix}\nu_e\\ e^-\end{pmatrix} \leftrightarrow \begin{pmatrix}u\\ d\end{pmatrix}$$

and

$$\begin{pmatrix}\nu_\mu\\ \mu^-\end{pmatrix} \leftrightarrow \begin{pmatrix}c\\ s\end{pmatrix}. \tag{10.18}$$

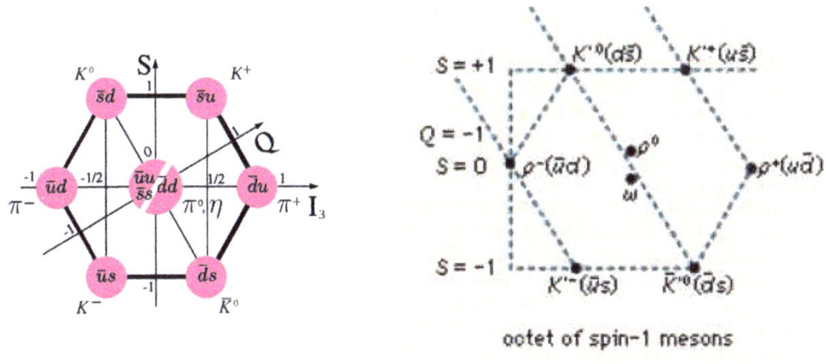

(a) (b)

Fig. 10.7: The root diagram (I_3, S) associated the pseudoscalar octet (a) and the vector meson octet (b) in the quark model. The color index is omitted.

The arrangement into doublets has been according to their behavior under the weak interaction, acting vertically between its members. The corresponding quantum number is denoted as weak isospin, not to be confused with the strong isospin discussed above. We used the standard convention putting the higher electric charge on top. For the quark doublets

$$u \leftrightarrow d \text{ and } c \leftrightarrow s. \tag{10.19}$$

The strong interaction acts, of course, among all hadrons as dictated by the color symmetry $SU_c(3)$.

10.4.1 *The need for a fourth quark — The GIM mechanism*

The above arguments in favor of the c quark were a mere conjecture. The need for the additional quark became imperative in 1970, when Iliopoulos, Glashow and Maiani showed that the three quarks do not explain all the facts. With only three quarks one cannot explain, e.g., why the decay rate for the processes:

$$\bar{K^0} \rightarrow \mu^- + \mu^+$$
$$K^+ \rightarrow \pi^+ + \nu + \bar{\nu}$$

is one million times smaller than the theoretical prediction. These authors proposed a mechanism, known as GIM, derived from their initials, which

Table 10.3: The possible mesons from the quarks u, d, s and c as well their antiparticles. These are pseudoscalar (0^-) and vector bosons (1^-). Experimentally it is found that the ω and ϕ can be better represented as $\phi = s\bar{s}$ and $\omega = \frac{1}{\sqrt{2}}(u\bar{u} + d\bar{d})$.

	meson	composition		charm		
0^-	1^-	quark- antiquark	S	c	I	I_3
π^+	ρ^+	$u\bar{d}$	0	0	1	1
π^0	ρ^0	$(u\bar{u} - d\bar{d})/\sqrt{2}$	0	0	1	0
π^-	ρ^-	$d\bar{u}$	0	0	1	-1
K^+	K^{*+}	$u\bar{s}$	1	0	1/2	1/2
K^-	K^{*-}	$s\bar{u}$	-1	0	1/2	-1/2
K^0	K^{*0}	$d\bar{s}$	1	0	1/2	1/2
$\bar{K}^0$	$\bar{K}^{*0}$	$s\bar{d}$	-1	0	1/2	-1/2
η	ω	$(u\bar{u} + d\bar{d} - 2s\bar{s})/\sqrt{6}$	0	0	0	0
η'	ϕ	$(u\bar{u} + d\bar{d} + s\bar{s})/\sqrt{3}$	0	0	0	0
η_c	Ψ/J	$c\bar{c}$	0	0	0	0
D^+	D^{*+}	$c\bar{d}$	0	1	1/2	1/2
D^-	D^{*-}	$d\bar{c}$	0	-1	1/2	-1/2
D^0	D^{*0}	$c\bar{u}$	0	1	1/2	-1/2
$\bar{D}^0$	$\bar{D}^{*0}$	$u\bar{c}$	0	-1	1/2	1/2
F^+	F^{*+}	$c\bar{s}$	1	1	0	0
F^-	F^{*-}	$s\bar{c}$	-1	-1	0	0

explains this reaction with the introduction of the fourth quark carrying the quantum number $c = +1$. This quark combined, of course, with the other quarks can give a plethora of new charmed particles, even some without net charm quantum number, e.g. Ψ/J. Examples of mesons containing this quark or its antiquark are included in Table 10.3. Some such possibilities for baryons are given in Table 10.4.

Here are some meson examples:

$$\begin{aligned} D^+ &= c\bar{d} \to B = 0,\ Q = 1,\ S = 0,\ \underline{c = 1},\ I = \tfrac{1}{2},\ I_3 = \tfrac{1}{2} \\ D^- &= d\bar{c} \to B = 0,\ Q = -1,\ S = 0,\ \underline{c = -1},\ I = \tfrac{1}{2},\ I_3 = -\tfrac{1}{2} \\ \eta_c &= c\bar{c} \to B = 0,\ Q = 0,\ S = 0,\ \underline{c = 0},\ I = 0,\ I_3 = 0 \text{ etc.} \end{aligned} \quad (10.20)$$

Table 10.4: The possible baryons with spin parity $1^+/2$ or $3^+/2$ resulting from a combination of the four u, d, s, c. For simplicity the multiplet $I = 3/2$, $J = 3/2$ (Δ^{++} etc.) has been omitted. We mention that at the time this nomenclature had not been widely accepted (see, e.g., Gailard [Gailard *et al.* (1975)]).

baryons $(1/2)^+$, $(3/2)^+$	structure from quark	S	c	I	I_3
p	uud	0	0	1/2	1/2
n	udd	0	0	1/2	-1/2
Σ^+	uus	-1	0	1	1
Σ^0	$1/\sqrt{2}(ud+du)s$	-1	0	1	0
Σ^-	dds	-1	0	1	-1
Λ^0	$1/\sqrt{2}(ud-du)s$	-1	0	0	0
Ξ^0	uss	-2	0	1/2	1/2
Ξ^-	dss	-2	0	1/2	-1/2
Ω^-	sss	-3	0	0	0
C_1^{++} or Σ_c^{++}	uuc	0	1	1	1
C_1^+ or Σ_c^+	$1/\sqrt{2}(ud+du)c$	0	1	1	0
C_1^0 or Σ_c^0	ddc	0	1	1	-1
C_0^+ or Λ_c^+	$1/\sqrt{2}(ud-du)c$	0	1	0	0
S^+	$1/\sqrt{2}(us+su)c$	-1	1	1/2	1/2
S^0	$1/\sqrt{2}(ds+sd)c$	-1	1	1/2	-1/2
A^+	$1/\sqrt{2}(us-su)c$	-1	1	1/2	1/2
A^0	$1/\sqrt{2}(ds-sd)c$	-1	1	1/2	-1/2
T^0	ssc	-2	1	0	0
X_u^{++} or Ξ_c^{++}	ucc	0	2	1/2	1/2
X_d^+ or Ξ_c^+	dcc	0	2	1/2	- 1/2
X_s^+	scc	-1	2	0	0
Ω_c^{++}	ccc	0	3	0	0

and some baryons:

$$S^+ = usc \rightarrow B = 1,\ Q = 1,\ S = -1,\ \underline{c = 1},\ I = \tfrac{1}{2},\ I_3 = \tfrac{1}{2}$$

etc. (10.21)

The flavor group must now be extended, naturally to $SU(4)$, but we will not elaborate more on this symmetry.

10.5 Charm is discovered — A new era begins

The advocates of charm did not have to wait long for their celebration. The first discovery came in November 1974 in two different experiments. The first experiment at BNL (Brookhaven National Laboratory) headed by the MIT professor Ting investigation lepton production in the reaction

$$p +_2^4 He \rightarrow \underbrace{e^+ + e^-}_{J/\psi} + X \tag{10.22}$$

where X stands for anything else. He observed a very narrow resonance at 3.1 GeV. At about the same time and before the BNL results were announced, a resonance appeared in $e^+ - e^-$ collisions in the Stamford Linear Accelerator (SLAC) in SPEAR collaboration with much better resolution at center of mass energy 3.095 GeV, as seen in Fig. 10.8, by measuring the cross sections:

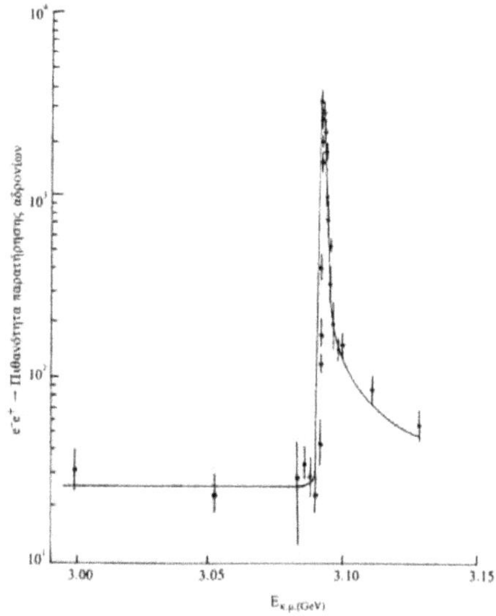

Fig. 10.8: The fine resonance that appeared in $e^+ - e^-$ collisions at SPEAR involving a particle named J/Ψ. As it can be seen the resonance is very narrow. It was determined that it is a meson, a very long lived bound state of the fourth quark c and $\bar{c}$.

$$\sigma(e^-, e^+ \to \text{anything}),$$
$$\sigma(e^-, e^+ \to e^-, e^+),$$
$$\sigma(e^+, e^- \to \mu^+, \mu^-).$$

It was determined that the new particle is a meson, a very long lived bound state of the fourth quark c and $\bar{c}$, named J/Ψ. Its width could not be determined directly, since it was narrower than the experimental resolution of 2 MeV. The way it was extracted is a bit complicated and we will not present it here. Anyway it was found that the total width is

$$\Gamma_{\text{tot}} = (69 \pm 15) \text{ KeV}, \tag{10.23}$$

while the electromagnetic width is

$$\Gamma_{\text{el}} = (4.8 \pm 0.6) \text{ KeV}. \tag{10.24}$$

In other words the electromagnetic width, due to electromagnetic interactions, competes with the total width, which is mainly hadronic involving the strong interaction. In fact the widths of other vector bosons, like $\rho(770)$ and $\omega(784)$, are $\Gamma(\rho) = 100$ MeV and $\Gamma(\omega) = 11$ MeV, i.e. much larger. So, somehow, its decay due to the strong interaction must be suppressed.

At the beginning the particle's nature was not clear.[3] The fact that its width is very narrow compared to its mass,

$$\frac{\Gamma_{\text{tot.}}}{mc^2} = \frac{70 \text{ KeV}}{3.1 \text{ GeV}} = 2.2 \times 10^{-5} = 0.0022\% \tag{10.25}$$

was a manifestation of new physics. The story of the strange particles repeated itself. It was produced copiously, but it chose to die weakly, suggesting a new quantum number. Since it was produced in e^-, e^+ collisions, it had to have the quantum numbers spin and parity similar to those of the photon 1^-, i.e. a structure[4] $n\,{}^{2s+1}\ell_J = 1^3S_3$, suggestive of a similarity with that of the positronium, a bound state of e^- and e^+ ($n = 1$, $l = 0$, $s = 1$, $J = 1$). A further indication that the Ψ has the same quantum numbers as the photon comes the asymmetry exhibited in Fig. 10.8. This asymmetry can be explained as the result of the two interfering amplitudes exhibited in Fig. 10.9. The photon-Ψ mixing occurs, if the two particles have the same quantum numbers.

This assignment was firmly established when similar states with different radial quantum numbers $n = 2, 3, 4$ were soon observed at SLAC, see

---

[3]Its life time is:
$\tau = \frac{h}{\Gamma_{\text{el}}} = \frac{6.6\times10^{-19} \text{ KeV}\cdot s}{69\text{KeV}} = 10^{-18}$ s.

[4]$n\,{}^{2s+1}\ell_J$ is the atomic physics notation. The orbital angular momentum is indicated $s, p, d, f \cdots (S, P, D, F \cdots)$ for $\ell = 0, 1, 2, 3 \cdots$

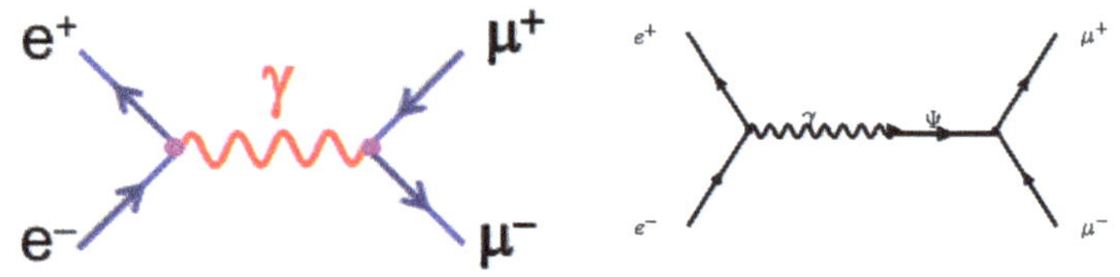

(a) (b)

Fig. 10.9: The (μ^-, μ^+) production the electron positron collider can proceed via (a) a virtual photon or (b) via the photon-Ψ mixing.

Fig. 10.10, with the rest of quantum numbers the same, $l = 0$, $s = 1$, $J = 1$. Quite naturally they were given the names Ψ', Ψ'', Ψ''' etc. In fact the Ψ' with m $= 3.684$ GeV/c^2 and width $\Gamma_{\text{tot.}} = 226$ KeV, was discovered only 10 days after J/Ψ. It decayed to J/Ψ via γ de-excitation. A new spectroscopy begins involving bound states of quarks!

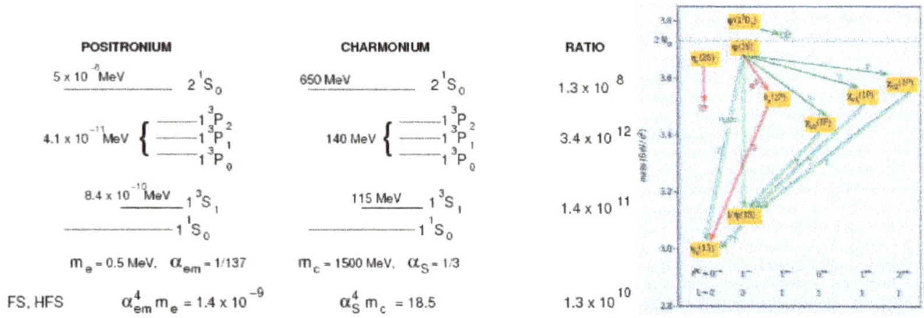

(a) (b)

Fig. 10.10: (a) The spectra of positronium, the atom of the positron (e^+), and charmonium, i.e. a meson which contains the bound charmed quark without itself carrying a total charm, i.e. of the form $c\bar{c}$. The similarity is impressive, only the scale differs. Even the symbols employed are those of atomic physics. Mesons which have the quantum numbers of the photon were produced first and more copiously in electron-positron collisions. (b) The charmonium was produced again much more beautifully many years later during the first runs of LHC at CERN.

The situation of Fig. 10.10 does not completely prove the existence of charm. This will be definitely established after the truly charmed mesons D^+, D^0 of Table 10.3 are found. Since the $J/\Psi, \Psi'$ etc. have a small width it is clear that their mass does not allow the opening of the channels $c\bar{d}+d\bar{c}$ or $c\bar{u}+u\bar{c}$. Consequently the mass of the charmed mesons was correctly estimated to be between $m \approx 1.85$ and 2.00 GeV. Thus the minimum required center of mass energy for their production was estimated to be between 3.7 and 4.0 GeV. Once produced, due to the special requirement imposed by the GIM mechanism, such charmed mesons should preferentially decay to strange mesons. Anyway more than two mesons are produced and this process is suppressed by the Sweig rule [Perkins (1972)], Section 5.15.

The discovery of such mesons followed within a year at SLAC from the decays:

$$\begin{aligned} &D^0 \to K^- + \pi^+ \text{ or } D^0 \to K^- + \pi^+ + \pi^- + \pi^+ \\ &\bar{D^0} \to K^+ + \pi^- \text{ or } \bar{D^0} \to K^+ + \pi^- + \pi^- + \pi^+. \end{aligned} \tag{10.26}$$

The name given is not essential. The important thing is that c decays to s and $\bar{c}$ to $\bar{s}$. Thus the particle which decays toK^- ($s = 1$ by convention) must contain the c quark, i.e. it must be of the form $c\bar{u}$ and we agree to call it D^0. The charged mesons D^+ and D^- followed soon with the structure $D^- = d\bar{c}$ and $D^+ = c\bar{d}$. They were detected via the reactions:

$$\begin{aligned} D^+ &\to K^- + \pi^+ + \pi^+ \\ D^- &\to K^+ + \pi^- + \pi^-. \end{aligned} \tag{10.27}$$

According to the preferred mode ($c \to s$, $\bar{c} \to \bar{s}$), one does not expect the reaction:

$$D^+ \to K^+ + \pi^+ + \pi^-. \tag{10.28}$$

The real triumph came with the discovery of the charmed baryons C_1^{++}, C_1^+, C_1^0 (see Table 10.4) with a mass around 2.26 GeV. We note the observation of Λ_c and the excitations D^{*0}, $\bar{D}^{*0}$, D^{*+}, D^{*-} (see Fig. 10.11). The equation relating the charge to Y is still valid but Y is modified as follows:

$$Q = I_3 + \frac{Y}{2}, \; Y = B - S + c. \tag{10.29}$$

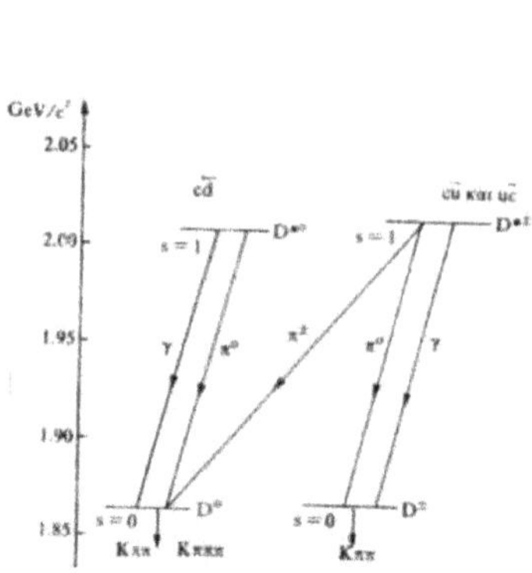

Fig. 10.11: The simple charmed mesons and their decay modes. The lower mesons have spins antiparallel ($s = 0$), while the upper have spin parallel ($s = 1$).

10.6 New surprises — New quarks and leptons

The order observed above was shaken in August 1976 when in the colliding beams in SLAC a new particle was discovered by SPEAR, which turned to be a spin 1/2 fermion with mass 1.8 GeV/c^2 and otherwise similar to muon and electron. It was named τ^- and, in spite of the fact that it is twice as heavy as the proton, it belongs to the lepton family. It is produced via the mechanism:

$$e^+ + e^- \to \tau^+ + \tau^- + X \tag{10.30}$$

where, as usual, X stands for whatever else. This was confirmed in other experiments, e.g. DORIS in Hamburg. This particle is accompanied by its own neutrino ν_τ, directly detected 20 years after its charged partner at FERMILAB. So three generations of leptons but two for quarks! This asymmetry did not last long.

Almost a year later in the summer of 1977 at Fermilab Lederman and collaborators discovered a new particle in the reaction

$$p + \bar{p} \to \mu^+ + \mu^- + X \text{ (with X whatever else)}, \tag{10.31}$$

with mass 9.4 GeV/c^2 and fine narrow width ($<$ 100 MeV), which was named Υ (upsilon) from the Greek letter Υ. A detailed analysis showed that we have a similar story with that of J/Ψ. In other words a new quark is born named b, carrying a new quantum number named b for beauty, i.e.

$$\Upsilon = b\bar{b}. \tag{10.32}$$

This quark is found to carry charge $-1/3$, like the d and s. Soon afterwards its radial excitations Υ' and Υ'' with masses 10.0 GeV/c^2 and 10.4 GeV/c^2 were found (see Fig. 10.12). The obtained spectrum is called bottomonium and shows a perfect similarity with that of the charmonium (see Fig. 10.13). Not surprisingly soon after mesons carrying net beauty were found

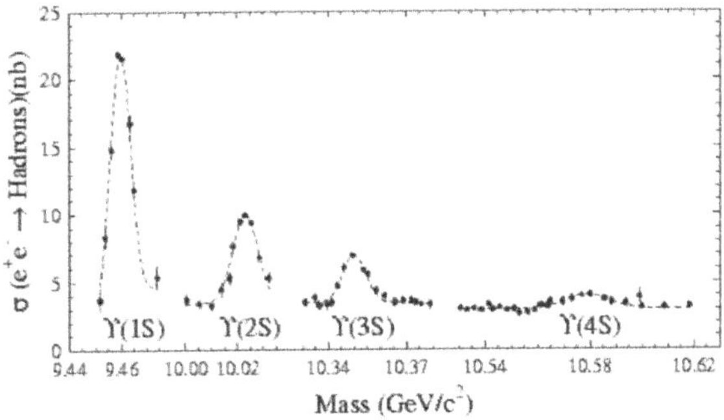

Fig. 10.12: Shown are three resonances Y, Y', Y'' produced in $e^+ - e^-$ collisions at center of mass energies 9.4 GeV, 10.0 GeV and 10.4 GeV respectively. They are considered to be bound states of b and $\bar{b}$.

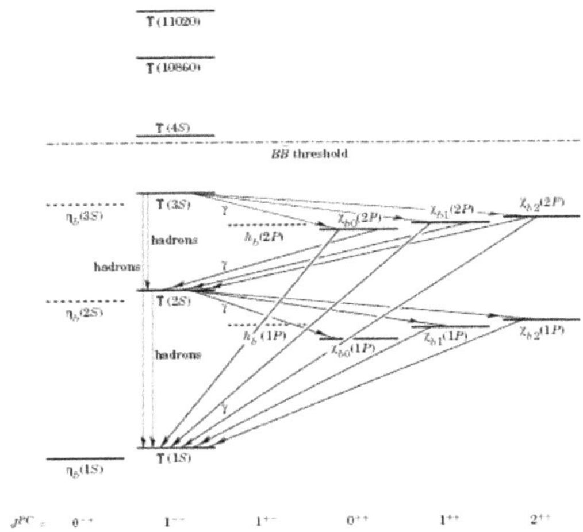

Fig. 10.13: The spectrum of a $b\bar{b}$. The similarity with that of charmonium is impressive (see Fig. 10.10).

$$B^0 = d\bar{b},\ \bar{B^0} = b\bar{d},\ B^+ = u\bar{b},\ B^- = b\bar{u} \tag{10.33}$$

(other mesons, containing the s and c quarks, more difficult to be produced were also found later). On this basis we expect predominantly the decays:

$$B^0 \to B^- + e^+ + \nu_e,\quad \bar{B^0} \to B^+ + e^- + \bar{\nu}_e, \tag{10.34}$$

with lifetimes of 10^{-8} s (typical of weak interaction) initially expected. This however does not occur since the neutral partner is only barely heavier than the charged, by 0.33 MeV. So the process is forbidden. Their actual life times are 1.6×10^{-12} decaying to many complicated channels.

One could now expand the flavor symmetry to $SU_f(5)$, but it is complicated and not terribly useful. One cannot draw the corresponding root diagrams. It is amusing to see that, if one considers baryons which do not involve any of the charmed quarks, one may obtain root diagrams similar to $SU_f(4)$ (see, e.g., Fig. 10.14 with b taking the place of c). The same applies to the 16-plets for the pseudoscalar and vector mesons (Fig. 10.15 with b taking the place of c) made of the u, d, s, and c quarks.

Theoretical prejudices insisted that there must be a third quark named t (top) with charge 2/3 similar to u and c, but its observation was hindered by its large expected mass. Its discovery, however, followed as soon as the appropriate center of mass energy of 1.8 TeV = 1.8 $\times 10^3$ GeV was reached at Tevatron (FNL) in proton anti-proton collisions. After a sophisticated analysis of the data, obtained with the detector CDF in 1994, its mass was found to be (173.34 $\pm$ 0.27 (stat) $\pm$ 0.71 (sys)) Gev/c^2, quite a bit heavier than expected.

In summary we now have three generations of leptons and three generations of quarks. Each of the latter appearing in three colors. Since, however, we now have a whole new list of non elementary hadrons, we extend the earlier Table 10.3 making in Table 10.5 a list of the mesons, which existed soon after the Υ was discovered. This list is presented for historical reasons only. The reader is encouraged to update and extend it using the PDG, e.g. [Patrignani *et al.* (2016)]. In understanding this table we note that the intrinsic parity of the antiquark is opposite to that of the quark. This means that the parity of mesons is $(-1)^{\ell+1}$ where ℓ is the relative angular momentum involved. The G parity of a meson with relative angular momentum ℓ, isotopic spin I and spin s is given by $G = (-1)^{\ell+s+I}$.

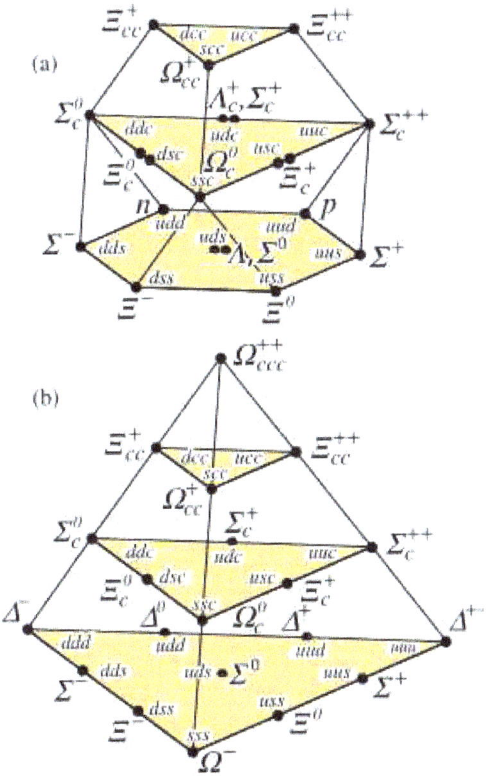

Fig. 10.14: SU(4) multiplets of baryons made of u, d, s, and c quarks. (a) The spin 1/2 20-plet with an SU(3) octet. (b) The spin 3/2 20-plet with an SU(3) decuplet. Courtesy of PDG [Tanabashi *et al.* (2018b)].

10.7 Exotic hadrons

The possibility of gluons creating $q\bar{q}$ pairs out of the vacuum, see Fig. 9.21, gives rise to the possibility of tetra quark bound messons of the form $q^2\bar{q}^2$ or pentaquark baryons of the $q^4\bar{q}$. The detection of such bound states is not easy. Occasionally claims of such exotic hadrons have been presented. Most of them proved spurious. Among them two seem to to have passed all the tests.

i) Theoretically predicted tetraquarks of the form $T(bb\bar{u}\bar{d})$ with $J^\pi = 1^+$

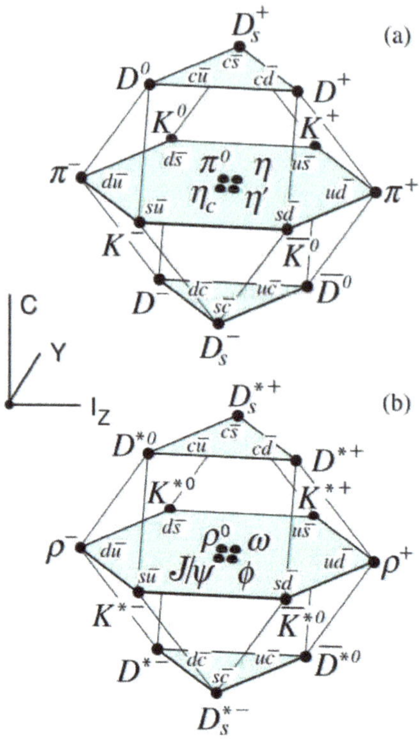

Fig. 10.15: SU(4) weight diagram showing the 16-plets for the pseudoscalar (a) and vector mesons (b) made of the u, d, s, and c quarks as a function of isospin $\mathrm{I_z}$, charm C, and hypercharge $Y = B + S\frac{C}{3}$. The nonets of light mesons occupy the central planes to which the cc states have been added. Courtesy of PDG [Tanabashi *et al.* (2018b)].

and a mass $M = 10389 \pm 12$ MeV [Karliner and Rosner (2017)]. This is only a theoretical prediction motivated in part by the recent LHCb discovery of the first doubly-charmed baryon $\Xi^{++} = ccu$ with a mass of 3621.40 ± 0.78 MeV, which is very close to their theoretical prediction of such a state. So using the same methods they predict a doubly-bottom tetraquark of the type $T(bb\bar{u}\bar{d})$ at $10,389 \pm 12$ MeV, i.e. 215 MeV below the $B - B^0$ threshold and 170 MeV below threshold for decay to $BB^0\gamma$. Thus they claim that $T(bb\bar{u}\bar{d})$ is stable under strong and electromagnetic (EM) interactions and can only decay weakly, the first exotic hadron with such a property. This needs experimental confirmation.

Table 10.5: Some of the mesons resulting from a combination of the u, d, s, c, b quarks and anti-quarks, containing at least one of the type s, c, b. In addition there exists an $I = 0$ state of the type $b\bar{c}, \bar{b}c$ with quantum numbers $1,^1 S_0\, 0^{-+}$ named $B^{\pm}c$. The first column shows the relevant quantum numbers n, ℓ, s, J involved in atomic physics notation.

$n\,^{2s+1}\ell_J$ J^{PG}	$I = 0$ $c\bar{c}$	$I = 0$ $b\bar{b}$ $\bar{c}u, \bar{c}d$	$I = \frac{1}{2}$ $c\bar{u}, c\bar{d}$,	$I = 0$ $c\bar{s}, \bar{c}s$	$\frac{1}{2}$ $b\bar{u}, b\bar{d}$, $\bar{b}u, \bar{b}d$	$I = 0$ $b\bar{s}, \bar{b}s$
$1,^1 S_0\, 0^{-+}$	$\eta_c(1S)$	$\eta_b(1S)$	D	$D_s^{\pm}$	B	B_s^0
$1,^3 S_1\, 1^{--}$	$J/\psi(1S)$	$\Upsilon(1S)$	D^*	$D_s^{*\pm}$	B^*	B_s^*
$1,^1 P_1\, 1^{+-}$	$\chi_c(1P)$	$\chi_b(1P)$	D_1	$D_{s1}^{\pm}$	B_1	B_{s1}
$1,^3 P_0\, 0^{++}$	$\chi_{c0}(1P)$	$\chi_{b0}(1P)$	D_0^*	$D_{s0}^{*\pm}$		
$1,^3 P_1\, 1^{++}$	$\chi_{c1}(1P)$	$\chi_{b1}(1P)$	D_1	$D_{s1}^{\pm}$		
$1,^3 P_2\, 2^{++}$	$\chi_{c2}(1P)$	$\chi_{b2}(1P)$	$D_2(3460)$	$D_{s2}^{*\pm}$	B_2^*	B_{s2}^*
$1,^3 D_1\, 1^{++}$	ψ			$D_s^{*\pm}$		
$2,^1 S_0\, 0^{-+}$	$\eta_c(2S)$	$\Upsilon(2s)$				
$2,^3 S_1\, 1^{--}$	$\psi(2S)$		D			
$1,^1 P_1\, 1^{+-}$		h	$h_b(2P)$			
$1,^3 P_{0,1,2}$ $,0^{++}1^{++}$ $,2^{++}$	$\chi_{c2}(2P)$	$\chi_{b0,1,2}(2P)$				

ii) The LHCb collaboration has announced the discovery of a pentaquark particle as a resonance in the decay $\Lambda_c^0 \to J/\Psi p K^-$. The particle, named $P_c(4312)^+$, is of the form $\Sigma_c\bar{D}^0$ or $\Sigma_c\bar{D}^{0*}$ (see Tables 10.3 and 10.4 for the relevant definitions). It is formed in the decay of Λ_c^0 and it immediately decays to $J/\Psi p K^-$. Recall that J/ψ is composed of a charm quark and an anticharm quark. This latest observation has a statistical significance of 7.3σ, passing the threshold of 5 σ traditionally required to claim a discovery of a new particle [Aaij *et al.* (2019)].

10.8 Problems

(1) Comment on the content of Fig. 10.10. Is it surprising that phenomena repeat themselves?

(2) Explain the nature of the transitions implied by Fig. 10.11.
(3) It is suggested that the production and decay of charmonium occurs as in Fig. 10.16. If you agree, explain each of the steps involved. If the charmonium is produced in an electron-positron accelerator and the final black arrows represent fermion lines, could there be an odd number. What is the most interesting combination for them?

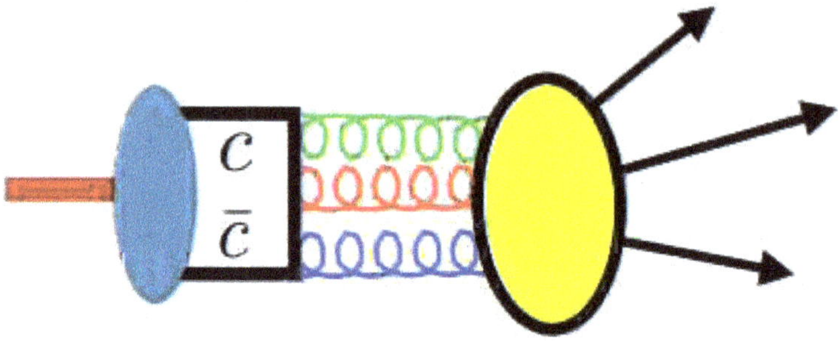

Fig. 10.16: A pictorial way of charmonium production and decay.

(4) Explain the spectrum and the nature of the transitions exhibited in Fig. 10.13.
(5) A very spectacular event was found Brookhaven bubble-chamber group in 75:

$$\nu_\mu + p \to \Sigma_c^{++} + \mu^-,\ \Sigma_c^{++} \to \Lambda_c^+ + \pi^+,\ \Lambda_c^+ \to \Lambda + \pi^+ + \pi^+ + \pi^-. \tag{10.35}$$

Two charmed baryons discovered in one shot! This was a source of delight for everyone especially for the experimentalists involved.

Draw a Feynman diagram for each step. Comment of the interaction responsible for the transition in each step.
(6) In the older days one read "It seems very likely to us that four-quark molecules involving a cc pair do exist, and have a rich spectroscopy. Our conjecture that the 4.028 GeV and perhaps de 4.4 GeV peaks in e + e annihilation are indeed due to the production of thesemolecules is a bit speculative". By looking at Fig. 10.17 would you agree? Do you think it is real?

See Section 10.7 and keep in mind the case of the atomic spectra before the electron was discovered.

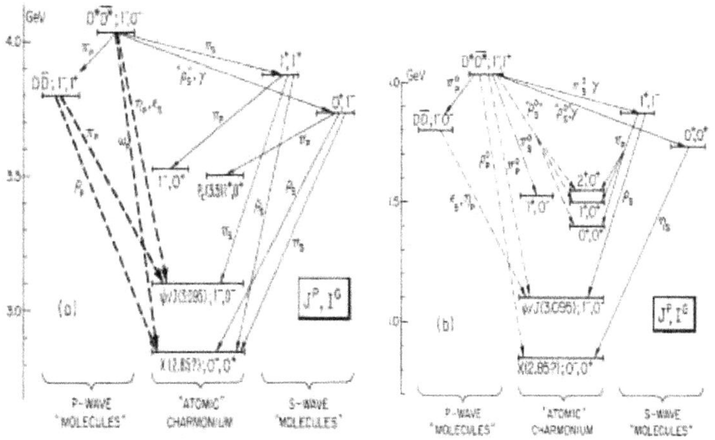

(a) (b)

Fig. 10.17: Are the states at 4.028 GeV and 4.4 GeV molecular or bound states of four quarks?

(7) Imagine that a baryon pentaquark of the form $q^4\bar{q}$ with J^π exists. Suppose that it is composed of quarks u and d and it has the same quantum numbers as the nucleon, i.e. it is a color singlet with isospin 1/2 and spin 1/2.

- What are the allowed quantum numbers, color, spin and isospin of the q^4 subsystem? You can use the tables found in [SO and Strottman (1976)].
- Indicate how the above results are modified, if one of the quarks and the antiquark are strange, i.e. of the form $q^3 s\bar{s}$.

(8) Suppose the tetraquark $T(bb\bar{u}\bar{d})$ exists. Draw Feynman diagrams for its weak decay to known particles.
Hint: Consider the possibilities:
i) Via the standard process $bb\bar{u}\bar{d} \to cb\bar{u}\bar{d} + W^-$, leading e.g. to $D^0\bar{B}^0$, $D^+B^-\pi^-$ etc.
ii) Via the W-exchange one of the b's undergoing the transition $b\bar{d} \to c\bar{u}$, e.g. $bb\bar{u}\bar{d} \to D^0B^-$.

(9) Suppose the pentaquark baryon $P_c(4312)^+$ exists and is of the form $\Sigma_c\bar{D}^0$ or $\Sigma_c\bar{D}^{0*}$ (see Tables 10.3 and 10.4 for the relevant definitions). Draw a Feynman diagram for its production from the decay of Λ_c^0. Do the same for its decay to $J/\Psi pK^-$. What type of interactions are involved?

Chapter 11

Experimental devices

11.1 Introduction

The relevance of experimental physics in settling the various theoretical issues has become pretty obvious from our previous discussions. The experiment is the best shift in sorting out the correct theories from the false ones. In this chapter we will discuss in some detail the various devices, which are useful in nuclear and particle experimental physics.

In Chapter 1 we have seen that, if we want to be able to see at very small distances, we need probes of very high energy. The history of microscopes exhibits many good examples. Thus with the electronic microscope we can see large chemical molecules, not seen by the optical ones. New devices are needed to observe smaller objects (atoms, nuclei, elementary particles). Such devices have been developed, but they have the disadvantage that the observation is no longer direct. In other words the needed information (mass, electric charge, size, etc.) can be obtained after a detailed analysis of a number of experimental data. Typically such data are the results of scattering of two or more particles. You recall that this way Rutherford obtained the structure of the atom. It still serves as a prototype. The scattering process plays in our days the role of the microscope, which has the ability to provide an image on the screen of our computers.

A scattering experiment can be separated into two stages. In the first stage the interacting particles must be accelerated at sufficient energies to approach each other. This can only be achieved with the aid of electric and magnetic fields and, as a result, only charged particles can be accelerated. The set up forms the **accelerator**. To get a large number of events a large number of particles must be accelerated, so that a sufficient number of them can interact. This can happen if the particles to interact approach each other. Thus a focusing device, a **lens**, is needed. The second stage, i.e. after the scattering has taken place, involves the detection of the produced particles. The needed device forms the **detector**. The detector must be sufficiently sophisticated to distinguish between the various products and it must respond quickly, since some of the produced particles may be unstable.

So when we talk about experimental devices we have in mind the accelerators, the focusing set up and the detector system.

11.2 Direction and quality if the beam

A charged particle in the presence of electric and magnetic fields feels a force known as **Lorentz force**:

$$\mathbf{F} = q\Big(\mathbf{E} + \frac{\boldsymbol{v}}{c} \times \mathbf{B}\Big). \tag{11.1}$$

In the above expression we assume a system of units so that the electric field **E** and the magnetic field **B** have the same dimensions, e.g. like in the Gaussian system. The time derivative of momentum equals the force **F**, i.e.:

$$\frac{d\mathbf{p}}{dt} = q\Big(\mathbf{E} + \frac{\boldsymbol{v}}{c} \times \mathbf{B}\Big). \tag{11.2}$$

The magnetic field yields a force which is perpendicular to the velocity, i.e. perpendicular to the displacement of the moving particle and, as a result, it produces no work. The electric field can change the energy of the particle or as we normally say it can accelerate it. In other words, if a particle of charge q is moving in a region where there exists an electric field, it gains an energy E given[1] by:

$$\Delta E = q \cdot \Delta V \tag{11.3}$$

where ΔV is the potential difference between the initial and final position of the charge.

In addition to the acceleration process, the moving charged particles must be organized into very fine beams. Thus the main features of the beam are:

- The direction of the beam.
- The focusing of the beam.

[1]Not to confuse ΔE (energy change) with **E** (the electric field).

11.2.1 *The directionality of the beam*

As we have already mentioned the force due to the magnetic field is perpendicular to the orbit, see Eq. (11.1). Thus under the influence of the magnetic field **B**, the particle is accelerated in such a way that the direction of its momentum changes, while its magnitude remains the same. Thus if $\hat{e}$ is the unit in the direction of momentum, the change of momentum in the direction perpendicular to the orbit is:

$$\hat{e}_r \cdot \frac{d\mathbf{p}}{dt} = \hat{e}_r \cdot \frac{d}{dt}(p\hat{e}) = p\hat{e}_r \frac{d\hat{e}}{dt} = -p\frac{d\theta}{dt} = \frac{-p}{\rho}\frac{\rho d\theta}{dt} = -\frac{p}{\rho}v.$$

Finally

$$\hat{e}_r \cdot \frac{d\mathbf{p}}{dt} = -\frac{p}{\rho}v \tag{11.4}$$

where ρ is the radius of curvature of the orbit. If we suppose now that the magnetic field **B** is perpendicular to the plane of the orbit we find:

$$\frac{p}{\rho}v = q\frac{v}{c}B.$$

Solving this equation for ρ we find

$$\rho = \frac{pc}{|q|B}, \quad \leftrightarrow p = \rho\frac{|q|B}{c} \text{ (Gaussian system of units).} \tag{11.5}$$

The radius of curvature ρ depends only on the ratio $|q|/p$. Thus, Eq. (11.5) can be used for the determination of momentum p once the magnetic field B is known and the curvature ρ of the orbit is measured. It is also used for the determination of the beam direction in a circular accelerator by adjusting the magnetic field according to the momentum. We note from its derivation that Eq. (11.5) is valid relativistically. It is useful to write Eq. (11.5) in natural units and even in GeV for the variable $p\ c$. The quantity $B\rho$ has dimensions of potential and is measured in stat-volt when B is measured in Gauss and ρ in cm (1 stat-volt = 300 Volt). Usually one prefers the expression in Volt whenever B and ρ are measured in Gauss and cm respectively. We find it useful to define the quantity

$$K = 300 \text{ Volt } B \text{ (Gauss) } \rho \text{ (cm)}$$

which is called magnetic hardness. One finds

$$(pc) = K|q| = K\frac{|q|}{e}e.$$

In any case

$$pc \text{ (GeV)} = 3 \times 10^{-7}\frac{|q|}{e}\rho \text{ (cm)} B \text{ (Gauss)}. \tag{11.6}$$

Thus, if $|q| = e$, $B = 1$ Tesla $= 10^4$ Gauss and $\rho = 0.47$ cm, we find $pc = 1.42$ MeV. On the other hand, Eq. (11.6), in MKS units is written as follows:

$$pc\ (\text{GeV}) = 3 \times 10^{-1} \frac{|q|}{e} \rho\ (\text{m})\ \text{B(Tesla)}. \tag{11.7}$$

Typical magnetic fields required for the determination of the beam direction are

$B = 1.8$ T in SPS, suitable for energies 450 GeV

$B = 0 \cdot 135$ T in LEP, for energies 100 GeV

$B = 8 \cdot 4$ T in LHC for energies $7 + 7$ TeV $= 1.4 \times 10^4$ GeV.

One achieves a practically uniform magnetic field in the region the beam is moving employing two solenoids with n_p coils and current I each at a distance h (gap between the two solenoids). Then

$$B_0 h = (n_p + n_p)\mu_0 I \quad \text{or} \quad B \cong B_0 = n_\rho \mu_0 \frac{I}{h/2}.$$

The beam is moving vertically to the field B_0.

11.2.2 *Focusing of the beam — Magnetostatic lenses*

If a particle beam is left to itself, it diverges! In order not to diverge, a focusing device is required.

Let us consider a magnetic quadrupole, i.e. 4 magnets arranged as in Fig. 11.1. Since the space between the magnets we have no currents ($\mathbf{J} = 0$) and no change in the electric field ($\partial \mathbf{E}/\partial t = 0$), the following relations hold:

$$\vec{\nabla} \times \vec{B} = 0 \quad \text{and} \quad \vec{\nabla} \cdot \vec{B} = 0. \tag{11.8}$$

Equations (11.8) imply that there exists a scalar function ψ such that:

$$\nabla^2 \psi = 0. \tag{11.9}$$

Now by mere inspection we see that, since the field B is practically perpendicular to the boundary surface, the function ψ must obtain a constant value on the surface of the magnet. As a result the boundary is characterized by a constant potential. In other words the the functions of constant potential in the plane of the figure are the hyperbolas $xy =$ constant. Thus

$$\psi = \lambda xy, \quad \text{where} \quad \lambda = \text{constant} \Rightarrow \tag{11.10}$$

$$\mathbf{B} = \nabla \psi = \lambda (y \hat{e}_x + x \hat{e}_y). \tag{11.11}$$

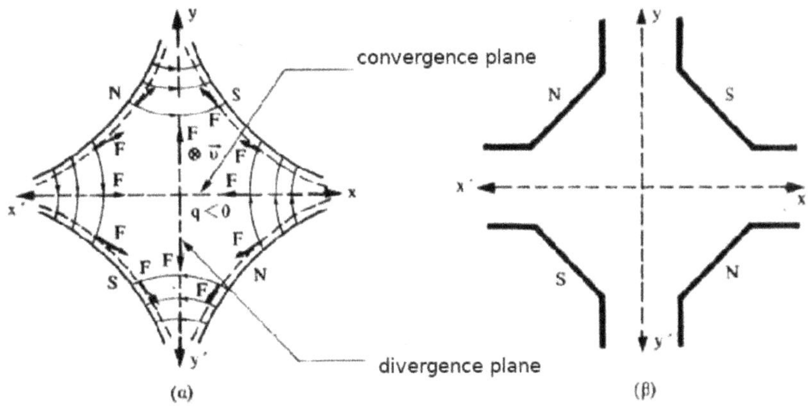

Fig. 11.1: A cross section of a magnetostatic lens: (α) ideal case and (β) a realistic one. The planes of convergence and divergence are indicated.

In other words the field lines are hyperbolas perpendicular to the equipotential lines $xy =$ constant. Furthermore the field B is fully specified once the quantity λ is specified.

In Fig. 11.1 we draw the forces acting on a charge $q < 0$ with velocity $\boldsymbol{v}$ inwardly normal to the plane with direction indicated by $\otimes$. We note that we have convergence along x axis and divergence the y axis. In other words not all particles are focused. Focusing can be achieved, however, if we combine this quadurpole with another one rotated by $\pi/2$ with respect to the first (crossed quadupoles). This combination results in focusing in both directions. Indeed the beam focused by the first quadrupole diverges a little bit in the second. The divergence caused by the second quadrupole is, however, substantially smaller than the convergence in the first, the reason being that the magnetic field is weaker in the region of divergence. For the same reason, the beam which had been diverging in the first quadrupole converges stronger in the second (the magnetic field in the second is now stronger in the relevant region). We will see that this can, also, be proved quantitatively. Such focusing devices are indeed very useful not only in the accelerators themselves but also in the beam lines leading from the accelerators to the experimental areas. With suitable such devices the beam can be transferred retaining its intensity.

Application: Suppose we want to find the divergence a high energy particle feels after going through a lens of thickness h.

We will suppose that while the particle is found into the lens, its velocity does not change appreciably. In such a case

$$\mathbf{F} = q\frac{\boldsymbol{\upsilon} \times \mathbf{B}}{c} = \lambda q \frac{\upsilon}{c}(-x\hat{e}_x + y\hat{e}_y).$$

Since

$$B_x = \lambda y,\ B_y = \lambda x,$$

supposing further that the dependence on x and y for relatively large x and y is weak, we find

$$\lambda y \approx B,\ \lambda x \approx B,\ B \approx \text{constant}.$$

Thus

$$\frac{dp_x}{dt} = F_x = -q\lambda\frac{\upsilon}{c}x = -q\frac{\upsilon}{c}B,\ \frac{dp_y}{dt} = F_y = q\lambda\frac{\upsilon}{c}y = q\frac{\upsilon}{c}B,\ \frac{dp_z}{dt} = 0. \tag{11.12}$$

In other words with a judicious choice of the axes we obtain:

$$\upsilon = \frac{z}{t},\ \frac{dp_x}{dz} = -q\frac{B}{c},\ \frac{dp_y}{dz} = q\frac{B}{c}$$

or

$$p_x = -q\frac{B}{c}z,\quad p_y = q\frac{B}{c}z,\quad p_z = \text{independent of time} \tag{11.13}$$

furthermore using Eqs. (11.12) we find:

$$p_x = m\gamma\frac{dx}{dt} = -q\frac{B}{c}z,\quad \gamma = \left(1 - \frac{\upsilon^2}{c^2}\right)^{-1/2}$$

or

$$\upsilon\gamma m \underbrace{\frac{dx}{\upsilon dt}}_{dz} = -q\frac{B}{c}z$$

i.e.

$$p\frac{dx}{dz} = -qz\frac{B}{c} \quad \text{where} \quad p = m\gamma\upsilon. \tag{11.14}$$

Integrating Eq. (11.14) over z we obtain:

$$x = -\frac{qB}{2pc}z^2.$$

Thus the divergence along the x axis is:

$$\Delta x = \frac{|q|B}{2pc}h^2. \tag{11.15}$$

Supposing now that $h = 2$ m, $B = 20$ k, Gauss $= 2$ Tesla and that the particle has momentum $p = 400$ GeV/c, we find $\Delta x = 0.3$ cm. We thus confirm that with the assumed geometry the divergence is very small.

The solution can be found with higher precision, assuming that the velocity v does not appreciably change due to the presence of the quadrupoles. Then the equations of motion take the form

$$\gamma m \frac{dv_x}{dt} = -q\frac{v}{c}x\frac{dB_y}{dx}, \quad \gamma m \frac{dv_y}{dt} = q\frac{v}{c}y\frac{dB_x}{dy}, \tag{11.16}$$

where

$$\frac{dB_y}{dx} \cong \frac{dB_x}{dy} \cong \lambda.$$

Thus

$$\frac{d^2x}{dz^2} = -\frac{q}{pc}x\frac{dB_y}{dx}, \quad \frac{d^2y}{dz^2} = \frac{q}{pc}y\frac{dB_x}{dy}, \quad z = vt.$$

Defining now

$$k = \frac{1}{Bp}\frac{dB_y}{dx} \cong \frac{1}{Bp}\frac{dB_x}{dy} \quad \text{(focusing intensity)}, \tag{11.17}$$

we get

$$\frac{d^2x}{dz^2} = -k\sin(q)x, \quad \frac{d^2y}{dz^2} = k\sin(q)y. \tag{11.18}$$

From these for $q > 0$ we find

$$x = x(0)\cos\sqrt{k}z + \frac{x'(0)}{\sqrt{k}}\sin\sqrt{k}z \tag{11.19}$$

$$x' = -x(0)\sqrt{k}\sin\sqrt{k}z + x'(0)\cos\sqrt{k}z \tag{11.20}$$

$$y = y(0)\cosh\sqrt{k}z + \frac{y'(0)}{\sqrt{k}}\sinh\sqrt{k}z \tag{11.21}$$

$$y' = y(0)\sqrt{k}\sinh\sqrt{k}z + y'(0)\cosh\sqrt{k}z. \tag{11.22}$$

For a lens of thickness ℓ we find in tabular form

$$\begin{pmatrix} x(\ell) \\ x'(\ell) \end{pmatrix} = \begin{pmatrix} \cos\sqrt{k}\ell & \frac{\sin\sqrt{k}\ell}{\sqrt{k}} \\ -\sqrt{k}\sin\sqrt{k}\ell & \cos\sqrt{k}\ell \end{pmatrix} \begin{pmatrix} x(0) \\ x'(0) \end{pmatrix}$$

$$\begin{pmatrix} y(\ell) \\ y'(\ell) \end{pmatrix} = \begin{pmatrix} \cosh\sqrt{k}\ell & \frac{\sinh\sqrt{k}\ell}{\sqrt{k}} \\ \sqrt{k}\sinh\sqrt{k}\ell & \cosh\sqrt{k}\ell \end{pmatrix} \begin{pmatrix} y(0) \\ y'(0) \end{pmatrix}.$$

Introducing now

$$f_x = \frac{1}{k\ell}, \; f_y = -\frac{1}{k\ell}, \tag{11.23}$$

we obtain

$$\begin{pmatrix} x(\ell) \\ x'(\ell) \end{pmatrix} = \begin{pmatrix} \cos\sqrt{\frac{\ell}{f_x}} & \sqrt{f_x\ell}\sin\sqrt{\frac{\ell}{f_x}} \\ -\frac{1}{\sqrt{f_x\ell}}\sin\sqrt{\frac{\ell}{f_x}} & \cos\sqrt{\frac{\ell}{f_x}} \end{pmatrix} \begin{pmatrix} x(0) \\ x'(0) \end{pmatrix},$$

$$\begin{pmatrix} y(\ell) \\ y'(\ell) \end{pmatrix} = \begin{pmatrix} \cosh\sqrt{-\frac{\ell}{f_y}} & \sqrt{f_y\ell}\sinh\sqrt{\frac{\ell}{f_y}} \\ \frac{1}{\sqrt{-f_y\ell}}\sinh\sqrt{-\frac{\ell}{f_y}} & \cosh\sqrt{-\frac{\ell}{f_y}} \end{pmatrix} \begin{pmatrix} y(0) \\ y'(0) \end{pmatrix}.$$

The above equations take the form:

$$\begin{pmatrix} u(\ell) \\ u'(\ell) \end{pmatrix} = \begin{pmatrix} 1 & 0 \\ -1/f & 1 \end{pmatrix} \begin{pmatrix} u(0) \\ u'(0) \end{pmatrix},$$

where $u = x$ or y and $f = f_x$ or $-f_y$, provided that for $\ell \ll |f|$, i.e. $\ell \to 0$ with $k\ell \to$ finite.

The angular spread of the beam is

$$\Delta\theta = u'(\ell) = -\frac{u(0)}{f} \cong -\frac{u}{f}, \tag{11.24}$$

i.e. $\Delta\theta$ is proportional to the initial position of the beam with convergence in the x-direction and divergence in the y-direction.

Suppose now that we have two lenses with focal distances f_1 and f_2 and thicknesses ℓ_1 and ℓ_2 separated by a distance $d << |f_1|, |f_2|$. Then:

$$\begin{pmatrix} u(d+\ell_1+\ell_2) \\ u'(d+\ell_1+\ell_2) \end{pmatrix} = \begin{pmatrix} 1 & 0 \\ -1/f_2 & 1 \end{pmatrix} \begin{pmatrix} u(d+\ell_1) \\ u'(d+\ell_1) \end{pmatrix}.$$

But

$$u(d+\ell) \cong u(\ell) + du'(\ell)$$
$$u'(d+\ell) \cong u'(\ell),$$

hence

$$\begin{pmatrix} u(d+\ell_1) \\ u'(d+\ell_1) \end{pmatrix} = \begin{pmatrix} 1 & d \\ 0 & 1 \end{pmatrix} \begin{pmatrix} u(\ell_1) \\ u'(\ell_1) \end{pmatrix} = \begin{pmatrix} 1 & d \\ 0 & 1 \end{pmatrix} \begin{pmatrix} 1 & 0 \\ -1/f_1 & 1 \end{pmatrix} \begin{pmatrix} u(0) \\ u'(0) \end{pmatrix}.$$

Thus finally

$$\begin{aligned} \begin{pmatrix} u(d+\ell_1+\ell_2) \\ u'u(d+\ell_1+\ell_2) \end{pmatrix} &= \begin{pmatrix} 1 & 0 \\ -1/f_2 & 1 \end{pmatrix} \begin{pmatrix} 1 & d \\ 0 & 1 \end{pmatrix} \begin{pmatrix} 1 & 0 \\ -1/f_1 & 1 \end{pmatrix} \begin{pmatrix} u(0) \\ u'(0) \end{pmatrix} \\ &= \begin{pmatrix} 1-d/f_1 & d \\ -\frac{1}{f_1} & 1-d/f_2 \end{pmatrix} \begin{pmatrix} u(0) \\ u'(0) \end{pmatrix} \\ &\cong \begin{pmatrix} 1 & 0 \\ -1/f_{\text{tot}} & 1 \end{pmatrix} \begin{pmatrix} u(0) \\ u'(0) \end{pmatrix}, \end{aligned}$$

with

$$\frac{1}{f_{\text{tot}}} = \frac{1}{f_1} + \frac{1}{f_2} - \frac{d}{f_1 f_2}. \tag{11.25}$$

We see that the system of two lenses behaves like a composite lens with a focal distance f_{tot} given by Eq. (11.25). The similarity with the usual optics is impressive! In the special case $f_1 = -f_2 = f$, which means a combination of two quadrupoles with fields vertical to each other, we find

$$\frac{1}{f_{\text{tot}}} = \frac{d}{f^2} \quad \text{or} \quad f_{\text{tot}} = \frac{f^2}{d} > 0. \tag{11.26}$$

This way we achieve focusing in both directions.

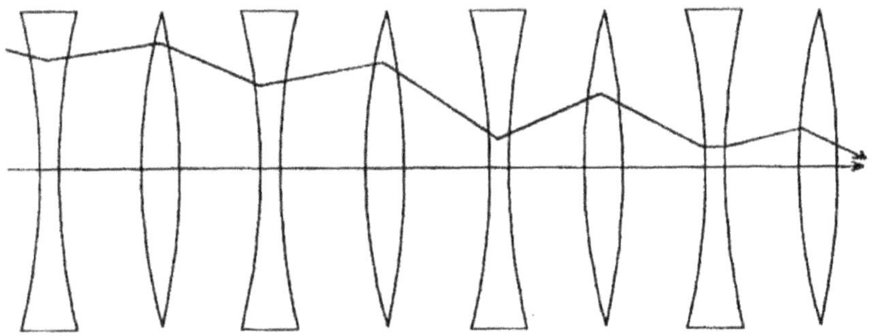

Fig. 11.2: A system of crossed quadrupoles leads to convergence in both directions.

Combining n such lenses one can easily show that

$$\begin{pmatrix} u(s) \\ u'(s) \end{pmatrix} = M(s/s_0) \begin{pmatrix} u(s_0) \\ u'(s_0) \end{pmatrix},$$

where

$$M(s/s_0) = M_n(s_n = s) \cdots M_2(s_2) M_1(s_1) \begin{pmatrix} u(0) \\ u'(0) \end{pmatrix}.$$

The quantities $M_i(s_i)$ are of the above form with $z \to s_i$, $k \to k_i$.

11.3 Accelerators

We have seen above that in order to give energy to charged particles we must set up an electric field. Particles moving in this electric field gain energy equal to $q\Delta V$, where ΔV is the potential difference between the initial and final position of the particle (see Eq. (11.3)). Clearly the particles must be moving in an area with high vacuum, since, otherwise, as a result of collisions with the other particles present, the source particles may lose more energy than that acquired via the electric field.

Furthermore it is obvious that there should be an appropriate source of the charged particles to be accelerated.

Thus the main components of an accelerator are: α) the particle source, β) the set up of the accelerating field and γ) a vacuum pump.

11.3.1 *TANDEM accelerators*

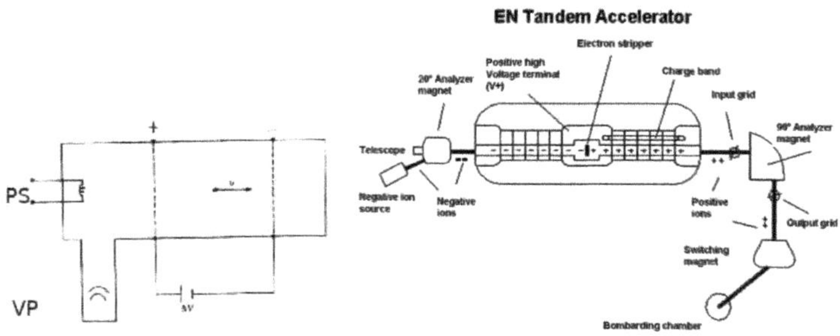

(a) (b)

Fig. 11.3: (a) The simplest possible accelerator. VP indicates the vacuum pump and PS stands for particle source. (b) A typical TANDEM accelerator employed in modern nuclear physics experiments.

It is possible to generate a beam of kinetic energy of up to 20 MeV with the simple device of Fig. 11.3. Everyone who has played with high voltage knows the difficulties involved. Even at voltages of a few thousand Volts, terrible electric sparks begin and only with special skill we can reach 100 kV. The above energy can be achieved in electrostatic accelerators known as TANDEM. In such devices, usually of cylindrical shape, we put the negative electrode in each base and the positive in the middle. The maximum terminal voltage achieved is $V = 18$ MV. Suppose that a negative ion with n extra electrons, charge $q = -n|e|$, is produced at the source. Then in the middle of it attains an energy $n|e|V$. At that point the atom is stripped of all its electrons. Then it continues to be accelerated and, when it reaches the other negative electrode, it has an energy $(Z+n)|e|V$.

For a ${}^{12}_{6}$C ion, $n = 1$, $Z = 6$ and $V = 5$ MV, at the TANDEM at the National Center for Scientific Research "Demokritos" (NCSRD), Athens, Greece, energies of 35 MeV have been reached.

Such devices, sketched in Fig. 11.3(b), are still employed in nuclear physics experiments.

Returning back to the subject of accelerators, to-day there exist three such types: α) Electrostatic accelerators (like TANDEM), β) linear accelerators and γ) synchrotrons. The particles that are accelerated are ions, electrons, positrons (anti-electrons), protons and antiprotons.

The most important standard (fixed target) accelerators are shown in Table 11.1.

11.3.2 *Linear accelerators*

To overcome the difficulties encountered with electrostatic accelerators mentioned above, lead to a new idea by Widoroe (1928). Instead of the particle accelerated once, why not accelerate it a little bit but many times in succession? The successive acceleration sites were placed along a straight line (liner accelerators). A series of cylindrical cavities are connected with an electric oscillator, a device operating in the area of radio frequencies (Fig. 11.4). We now consider a charged particle with charge e, moving along the axis of the device. We arrange matters so that while the particle is in the cavity the electric field **E** is zero. In the gap between the cylinders a field is created, which accelerates the particle, taking care so that the field attains its maximum, while the particle is there. By the time the particle reaches the next gap, the polarity becomes appropriate for another acceleration. In other words the length of the cylinder must be appropriate so that:

$$L = \frac{1}{2} v T, \tag{11.27}$$

with v the particle's velocity and T the oscillation period. It is clear that, as the velocity of the particle increases, the period T must be appropriately modified. Thus, if the potential difference is V_0, after n accelerations the particle attains an energy:

$$E = neV_0. \tag{11.28}$$

This has, indeed, been a great invention. Many years passed, however, before linear accelerators (LINAC) became truly useful. The essential reason was that there appeared many technical difficulties in their design.

Table 11.1: We list here the most well known common accelerators. The parameters of the colliding beam accelerators will be given below, see Table 11.2.

Accelerators	date of first operation	accele- rated particles	Energy beam (GeV)	Energy at the center mass (GeV)	Intensity (particles per bunch)
PS CERN Geneva	1959	protons	29	7.4	
AGS Brookhaven New York	1961	protons	33	8	8×10^{12}
SLAC (Linear accelerator) Stanford	1966	electrons	22	6,5	
Synchrotron protons Serpukkov Russia	1967	protons	78	12	5×10^{12}
Synchrotron center ring Fermi Batavia	1972	protons	500	30,7	2×10^{13}
DESY Hamburg	1974	electrons	7	3,8	
KEK Japan	1975	protons	12	5	
SPS CERN	1976	protons	500	30,7	10^3
Tevatron Laborayory Fermi	1982	protons	1000	43	2×10^{13}
Synchrotron protons Beijing	1985	protons	50	9,0	10^{13}

One problem had been the precise coordination between the length L of every cylinder and the period T. The main problem, however, had been the requirement of high powered radio frequencies (RF) in the region of ≥ 100 MHz. Thus up to 1946 the LINAC's were essentially used for accelerating heavy ions in nuclear physics experiments. Soon after that, however, high frequency generators utilized RF devices, like the Magnetron,

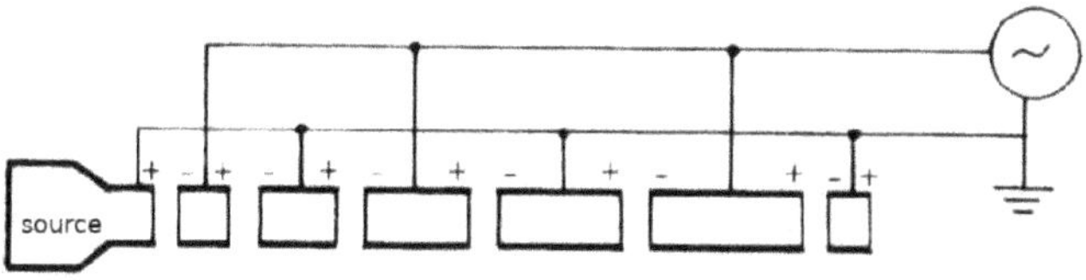

Fig. 11.4: A sketch of a linear accelerator. Inside the cavity the electric field **E** is zero. In the gap between the cavities the potential has the shown polarity, just at the moment the charged particle exits the cavity and, as a result, each time gets accelerated along the axis of the cylinder.

Klystron etc., initially employed in Radars for military purposes. These were utilized in making high energy LINAC's. One good example was the LINAC at Stanford, which has a length of 3 km and accelerates electrons up to 20 GeV in energy. It employs hundreds of Klystrons at frequencies of thousands of MHz and supplies power to the accelerated particles of some hundreds of kW. Another such machine accelerating protons up to 800 MeV at Los Alamos, USA, has been used to produce intensive meson beams (meson factory).

Note that at high energies the velocity of the accelerated particles approaches the speed of light and practically $L \approx (1/2)cT$, i.e., it becomes independent of the velocity of the particles. To simplify the design one may use a pre-accelerator, were the velocity dependence is taken into account, before the main high energy accelerator.

11.3.3 *Circular accelerators — Synchrotrons*

In spite of the progress made in the development of the LINAC accelerators, some practical problems remain at high energies. The use of a linear accelerator, analogous to that operating at Stanford (L = 3 km energy 20 GeV), to achieve 500 GeV will require a length of 75 km! It is clear that this is not a realistic prospect. We need a system such that the accelerated beam passes through a given acceleration site many times. This generated the idea of a circular accelerator.

Circular accelerators have, in fact, been operating since 1930, when Lawrence proposed them, known as cyclotrons. They played an important role in the development of nuclear physics and they are still used for that purpose even today. We are not going to discuss them in detail here, since we are going to discuss the more advanced sister accelerator, known as synchrotron, which is more appropriate at high energies. It has been proposed independently by Mcmillan and Veksler in 1945. The essential ingredients of the operation of a synchrotron are exhibited in Fig. 11.5.

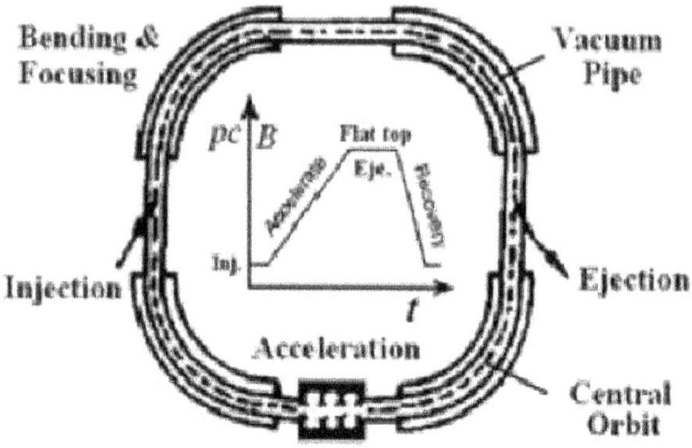

Fig. 11.5: The essential elements of a synchrotron accelerator.

We note that the particles leave the source with energy E_i before entering the synchrotron ring. The particles are repeatedly accelerated inside a number of cavities fed with RF of frequency ω. Once they enter the acceleration cavity they are focused with a set of quadrupoles into a fine beam, a necessary requirement for an economic construction. The actual particle orbit is composed of linear segments in the cavities of acceleration, focusing etc., and circular in the area of bending magnets (dipoles). As a result the geometric radius R of the ring is larger than the radius of curvature ρ (Eq. (11.5)).

At the moment the particles enter the ring they have momentum

$$p_i = \sqrt{\left(\frac{E_i}{c}\right)^2 - (mc)^2} \quad i = 1, \ldots, N.$$

If the RF power source is not connected, the particles will follow an almost circular orbit with velocity:

$$v_i = \frac{p_i}{m\gamma} = \frac{p_i c^2}{E_i},$$

with period

$$T = \frac{2\pi R}{v_i} = \frac{2\pi R E_i}{p_i c^2} \tag{11.29}$$

and corresponding circular frequency Ω:

$$\Omega = \frac{2\pi}{T} = \frac{p_i c^2}{R E_i}. \tag{11.30}$$

The required magnetic field is given by Eq. (11.5)

$$B = \frac{p_i c}{|q|\rho}. \tag{11.31}$$

Once we connect the RF supply the situation becomes quite different. In order to always have acceleration it necessary to have.

$$\omega = n\Omega = n\frac{c}{r}\frac{pc}{E}, \quad n = \text{positive integer}. \tag{11.32}$$

In other words the frequency ω should increase as the energy increases till we reach a completely relativistic situation ($p_c/E \approx 1$). Then:

$$\omega \approx n\frac{c}{R} \text{ and } B = \frac{pc}{|q|\rho}, \quad n = \text{positive integer}. \tag{11.33}$$

Thus, provided Eqs. (11.32) and (11.33) are obeyed, acceleration is guaranteed.

The whole process is summarized as follows: At $t = 0$ the particle beam enters the accelerator with energy E_i. The magnetic field is then B_i and the RF frequency is ω. Afterwards they take the values B_f and ω_f respectively, but always obeying the relations (11.32) and (11.33). The resulting particle energy is E_f (the time required for this depends on the machine).

Synchrotons are primarily used for protons and electrons. For low mass particles one has large losses due to EM radiation emitted while they are accelerated. These Losses can be computed in the context of Classical Electrodynamics, see e.g. [Jackson (1062)].

A particle moving in a circular orbit with a relatively low velocity radiates a power given by

$$P = \frac{2}{3}\frac{e^2}{4\pi\epsilon_0}\frac{\dot{\mathbf{v}}^2}{c^2} = \frac{2}{3}\frac{e^2}{4\pi\epsilon_0}\frac{\mathbf{p^2}}{m^2c^2}, \quad \dot{\mathbf{v}} = \frac{d\mathbf{v}}{dt}. \tag{11.34}$$

This can be written in relativistic form as

$$P = \frac{2}{3}\frac{e^2}{4\pi\epsilon_0}\frac{1}{m^2c^2}\frac{dp_\mu}{d\tau}\frac{dp^\mu}{d\tau}. \tag{11.35}$$

For an arbitrary motion one finds:

$$P = \frac{e^2}{4\pi\epsilon_0}\frac{2}{3c}\gamma^6\left[|\dot{\boldsymbol{\beta}}|^2 - |\boldsymbol{\beta}\times\dot{\boldsymbol{\beta}}|^2\right], \quad \dot{\boldsymbol{\beta}} = \frac{d\boldsymbol{\beta}}{dt}. \tag{11.36}$$

For a linear accelerator $\boldsymbol{\beta}\times\dot{\boldsymbol{\beta}} = 0$, while for a circular accelerator $|\boldsymbol{\beta}\times\dot{\boldsymbol{\beta}}|^2 = \beta^2|\dot{\boldsymbol{\beta}}|^2$. Thus for a circular accelerator

$$P = \frac{e^2}{4\pi\epsilon_0}\frac{2}{3c}\gamma^4|\dot{\boldsymbol{\beta}}|^2, \quad \dot{\boldsymbol{\beta}} = \frac{1}{c}\frac{v^2}{R} \Rightarrow \tag{11.37}$$

$$P = \frac{e^2}{4\pi\epsilon_0}\frac{2}{3c^3}\gamma^4\frac{v^4}{R^2}. \tag{11.38}$$

But

$$E = pc = \gamma mvc,\; v = \frac{E}{\gamma mc} \Rightarrow P = \frac{e^2}{4\pi\epsilon_0}\frac{2}{3}\frac{c}{R^2}\left(\frac{E}{mc^2}\right)^4. \tag{11.39}$$

Looking at this equation we notice that the synchrotron radiation losses are:

- Proportional to the fourth power of the energy achieved. An unavoidable price to pay, if one wants to achieve high energies.
- Inversely proportional to the fourth power of the mass of the accelerated particle. For the same energy an electron beam consumes 10^{13} more power than a proton one. Electron accelerators usually aim at lower energies, since their energy can be more efficiently used.
- Inversely proportional to the second power of the radius. A larger radius is preferred, but then the accelerator becomes costly to build and difficult to maintain.

By looking at Eqs. (11.32) and (11.33) one notices another feature of large accelerators: *the particles cannot begin accelerating from zero in the same engine.* The RF frequencies and the magnetic fields required would have to vary in a very large domain, which would result in an impossible construction. So all machines are characterized by a number of stages. The ones encountered, e.g., in the case of FERMILAB, USA are: (a) One Cockroft-Walton accelerator, which yields a proton beam of 0.75 MeV. (b) A LINAC, which raises the energy to 200 MeV. (c) A Booster Synchrotron, which raises the energy to 8 GeV. After these stages the task of the large machine begins, see Fig. 11.6. It is, thus, clear that a modern accelerator is essentially a laboratory town, see Fig. 11.7.

11.3.4 *Colliding beam accelerators*

In the type of accelerators we discussed so far one of the colliding particles is accelerated, while the other is fixed. This arrangement is the simplest

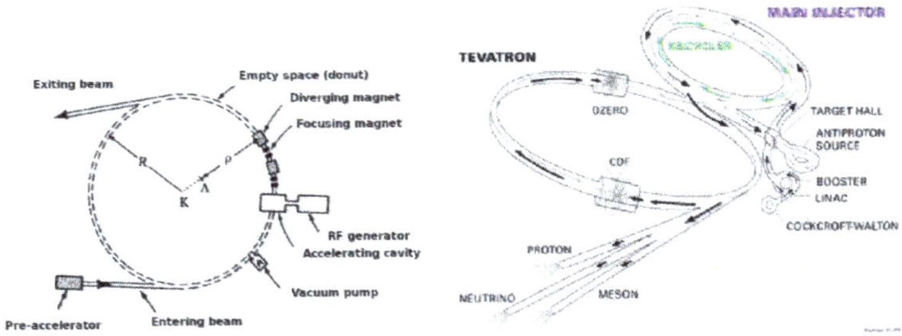

(a) (b)

Fig. 11.6: (a) A sketch of a typical circular accelerator like the one at Fermi Laboratory. (b) Its actual lay out as a colliding beam accelerator (dark arrow $\leftrightarrow$ protons, light arrow $\leftrightarrow$ antiprotons).

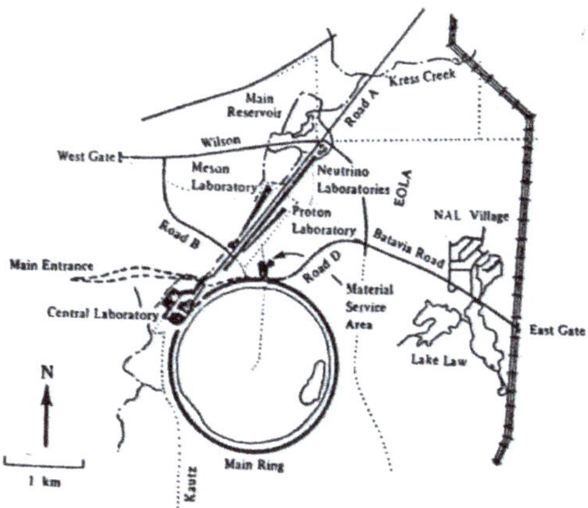

Fig. 11.7: Map of the accelerator NAL in Batavia, Illinois, U.S.A.

possible, but, as we have seen in Section 9.5, it is not the most efficient. The usable energy E^* in the center of mass is

$$E^* \approx \sqrt{2m_2 E_1}, \tag{11.40}$$

with E_1 the energy of the accelerated particle m_2 the mass of the target particle. We thus see that, if the energy of the accelerator is increased 100 times, the usable energy is increased only 10 times. It was clear that a new idea was needed, and in fact it was conceived when, in 1956, the physicists Kerst and O'Neil and their collaborators unveiled the proposal of a colliding beam accelerator. Thus the colliding 100 GeV LEP accelerator is equivalent to a fixed target accelerator of energy

$$E_1 \approx \frac{(E^*)^2}{2m_2} \approx 10^4 \text{ (GeV)!} \tag{11.41}$$

In Fig. 11.8 we present the energy in the center of mass as a function of the proton energy in a fixed target as well as for a colliding beam accelerator.

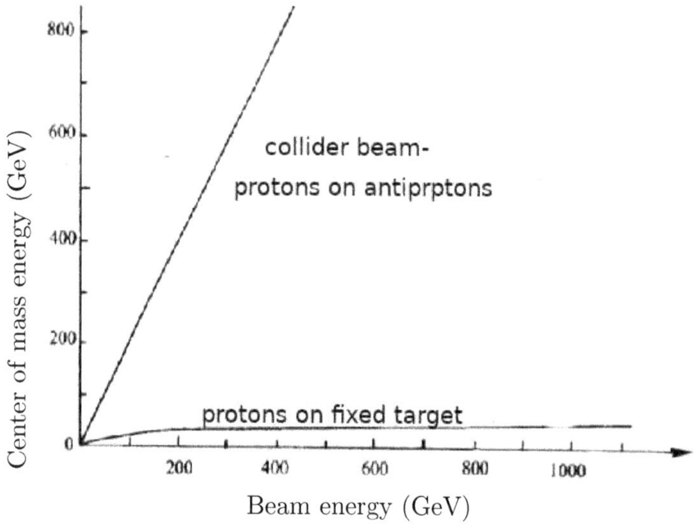

Fig. 11.8: The dependence of the obtained energy in the center of mass as a function of the proton energy in a fixed target accelerator. We also present the same quantity for a colliding beam accelerator.

Another important aspect of an accelerator is its luminosity, which has been discussed in Section 9.6. The two most important parameters of the various accelerators, namely the energy and luminosity, are presented in Table 11.2.

11.3.5 *Storage rings*

It is clear that the luminosity of a colliding beam is limited by the fact that the number of available antiparticles (antiprotons and positrons) is limited. Those particles, however, are pretty stable once they are found in an area

Table 11.2: The main features of the most well known colliding beam accelerators in various parts of the world.

accelerators	date of first obser-vation	stored particles	energy per beam (GeV)	energy in the center of momentum (GeV)	luminosity $(\mathrm{cm})^{-2}\ \mathrm{s}^{-1}$
storage rings Stanford	1954	e^+e^-	0.56	1.1	10^{27}
ISR CERN Geneva	1971	pp	3.1	62	4×10^{31}
CEA-ByPass Cambridge USA	1967	e^+e^-	3.5	7	2×10^{29}
SPEAR LINAC Stanford	1972	e^+e^-	4.2	8.4	10^{32}
DORIS DESY Hamburg	1974	e^+e^-	4.5	9	10^{32}
DSI Orsay Paris	1975	e^+e^-	3.7	7.4	10^{32}
VEPP-4 Novosibirsk Russia	1978	e^+e^-	7	14	10^{31}
PETRA DESY	1978	e^+e^-	19	38	10^{32}
CESR Cornell	1979	e^+e^-	8	16	10^{32}
ISR CERN	1980	$p\bar{p}$	31	62	10^{29}
PEP LiNAC Stanford	1980	e^+e^-	18	36	10^{32}
SPS CERN	1981	$p\bar{p}$	270	540	10^{30}
Fermi Batavia	1982	$p\bar{p}$	1000	2000	10^{30}
LEP CERN	1989	e^+e^-	48	98	10^{30}
HERA DESY	1992	ep	e:30 p:92	318	8×10^{32}
Tevatron Fermilab	1987	$p\bar{p}$	980	1960	3×10^{32}
LHC CERN	2008	pp	70000	14000	10^{34}

free of matter. Once they are separated, they can be stored in a ring and kept in a high vacuum space, orbiting in fine circular beam with a suitable magnetic field (dipoles). This is called *storage ring*. Thus, while only 10^8 antiprotons per second are produced at CERN or FERMILAB one can have 10^{11} antiprotons in 10^3 s in a storage ring, a number adequate to achieve the luminosities listed in Table 11.2.

The luminosity can increase, if we can make the particles of each beam encounter each other many times. So in the storage ring one can send many pulses of particles. In the CERN proton accelerators, known as ISR (Intersecting Storage Rings) each beam has about 200 pulses and is characterized by a current 20 A. The maximum center of mass energy is 60 GeV, i.e. 1900 GeV of a traditional fixed target accelerator. At the points of the two rings meet the interaction can take place

It is clear that, for such a storage arrangement to be effective, there should not be large dispersion in the antiproton momenta. Thus these particles must be "cooled". For antiprotons this can be achieved by bombarding them with low energy electrons, which steal energy from them. In the end the antiprotons can be separated by a magnetic field, even though they have the sane charge with the electrons, since they travel in paths with different curvature.

All particles can be cooled by a method invented by der Meer in 1970 and is called **stochastic cooling**, but we will not elaborate further here, see the classic paper on the subject [Mohl *et al.* (1980)].

11.3.6 *Future accelerators*

LHC at CERN, the highest energy colliding beam accelerator so far, lead to the discovery of the Higgs particle, a crucial missing element for the establishment of the standard model, but, unfortunately, it gave no signal of any other particle beyond the standard model. In particular it provided no candidate for dark matter. After some needed upgrades, it will continue to provide useful data during the next decade. The high energy community, however, has already been thinking bout the idea of a next generation accelerator. Based on past experience, indicated by the Livingston plot, see Fig. 11.9, and, in particular, the successful debate that lead to the construction of LHC, the prospects that such a quite expensive project may be undertaken are quite good. It may be necessary for the construction to have a consortium formed, involving the European Union, USA, China and Japan. Based on past experience the site should be chosen near existing accelerator networks. The future machine, in spite of designs for linear electron/positron colliders (ILC and CLIC), will probably by a circular collider, which, of course, will benefit from existing designs. The Future Circular Collider (FCC) is a conceptual study that aims to develop designs for a post-LHC particle accelerator with an energy significantly above that of previous circular colliders (SPS, Tevatron, LHC) see, e.g., [Benedikt and Zimmermann (2018)]. After injection at 3.3 TeV, each beam would have a total energy of 560 MJ. At collision energy of 100 TeV this increases to 16.7 GJ. These total energy values exceed the present LHC by nearly a factor of 30.

The following options are being considered

- FCC-ee (electron/positron), with centre-of-mass collision energies between 90 and 350 GeV.
- FCC-hh (proton/proton) with 50 TeV proton beam, which could also operate as ion/ion.
- FCC-ep (electron/proton), with 50 TeV proton beam and an electron beam with energy of the order above 100 GeV.

The above accelerator strategies no doubt will involve improving and refining existing technologies. Building accelerators at very high energies, however, unless a really new idea is put forward, may be proved to be too expensive. In other words the society may be unwilling to provide the needed support. In such a case one may have to use the high energy accelerators that nature provides, such as the most energetic supernova explosions and the ultra high energy particles found in cosmic rays, see Section 7.1.

These natural accelerators cannot be used as efficiently as the man made accelerators, but, still, they may help us unravel any secrets beyond the standard model. Also some of the remaining secrets of nature may be retrieved not from the energy frontier, but from the intensity frontier, i.e. with the observation of extremely rare events. Especially those involved in detecting in the laboratory the dominant components of the universe, namely dark matter and dark energy.

11.4 Detectors

As **detectors** we imagine devices, which enable us to measure all the elements characterizing the identity of a particle, so that we can recognize

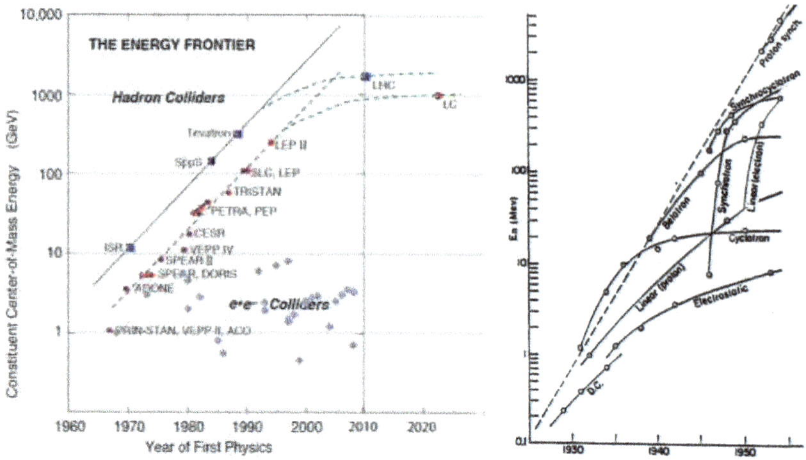

(a) (b)

Fig. 11.9: (a) A Livingston plot of the history of the accelerators. An illustration of how history of discovery on the energy frontier has been enabled by the history of invention (red arrows) in accelerator science and technology. (b) This plot shows how the energy of accelerators has been increasing as a function of time. Note the exponential increase in the energy indicated by the straight line, a sort of covering of the various curves.

it. It thus appears that the detectors are even more important than accelerators. As we have already mentioned, without accelerators something can be done utilizing the laboratories of nature (radioactivity, cosmic rays, rare events like dark matter etc.). Without detectors, since the human eye is no longer reliable, only telepathy remains! The information we need for the identification of a particle are its momentum and energy, which, if combined, can yield the mass of the particle and the distribution of events in space-time, which can lead to the determination of the spin of the particle.

In this section we will consider the main types of detectors, namely:

- Time of arrival detectors.
- Scintillation detectors (counters).
- Semiconductor detectors.
- Spark chamber detectors.
- Bubble chamber detectors.

- Wire detectors.
- Time projection counters (TPC).
- Cherenkov detectors.
- Micromegas.
- Calorimeters.

The above detectors can be used in physics experiments performed at the sites of the accelerators discussed above. They can also be used in underground searches for the detection of rare events produced in nature (cosmic and solar neutrinos, dark matter, proton decay (if it occurs) etc.).

A full understanding of detector performance presupposes some knowledge of he way the particles are produced as well as the way they loose energy due to their collisions with matter etc. A detailed discussion of such topics is beyond the goals of the present text. We will, nevertheless, provide a brief introduction of such topics.

The detectors are usually placed quite far from the interaction area. During the experiment the position of the particles and the time of the measurement are already specified. The quantities which must be determined are the 4-momentum of every particle, its charge, its spin etc. Once the 4-momentum of a particle is known, its mass can be obtained via the relation:

$$m_i = \sqrt{E_i^2 - \mathbf{p}_i^2}. \tag{11.42}$$

In the above equation we have employed natural units.

We also know that, if the charge of the particle is known, the measurement of momentum is reduced to the measurement of the radius of curvature in a magnetic field. We also know that, if the momenta of two particles are known, their mass can be determined from the time of flight between two detectors that are a distance L apart. Indeed

$$\Delta T = \frac{L}{\beta_1} - \frac{L}{\beta_2} \tag{11.43}$$

and

$$\beta_1 = \frac{p_1}{E_1} \text{ and } \beta_2 = \frac{p_2}{E_2} \text{ (natural units)},$$

i.e.,

$$\Delta T = L(\frac{E_1}{p_1} - \frac{E_2}{p_2}) = L\frac{p_2\sqrt{p_1^2 + m_1^2} - p_1\sqrt{p_2^2 + m_2^2}}{p_1 p_2}. \tag{11.44}$$

From this relation, if one of the masses is known, the other can be determined.

Table 11.3: Common materials used in scintillation detectors with some of their properties.

fluorescence material	pulse order (relative to Anthracene)	decay time ns	θ_{max} (A)	density gr/cm^3
Polysterine + p-terphenol	0.28	3	3550	0.9
+ Tetraphenyloboutadiene	0.38	4.6	4800	
Na-oxide (+ thalium)	2.1	250	4100	3.7
Anthracene	1.0	32	4100	3.7
Tolumenium	0.7	<3	4300	0.9

Furthermore, if the particles have the same momenta $p_1 = p_2 = p$ and they are relativistic $m_i << p$, $i = 1, 2$, the above equation becomes:

$$\Delta T = \frac{L}{p}\frac{(m_1^2 - m_2^2)}{\sqrt{p_1^2 + m_1^2} + \sqrt{p_2^2 + m_2^2}} = \frac{L}{2p^2}(m_1^2 - m_2^2) \tag{11.45}$$

or in standard units:

$$\Delta T = \frac{L}{c}\frac{(m_1^2 - m_2^2)c^2}{2p^2}. \tag{11.46}$$

This method cannot, of course, be used for particles that live for a very short time. Then for the determination of the particle's mass, one applies the method discussed in Section 4.1.

11.4.1 *Scintillation detectors*

The principle of operation of scintillation detectors (counters) depends on the fact that, when a charged particle passes through matter, it excites the atoms and molecules present and, thus, radiation (fluorescence) is produced, following their de-excitation. The produced light is transformed into an electric signal with the aid of a number of photon multipliers (PMP). Neutral particles may be detected this way, if they produce charged particles. Thus γ rays can be detected via the e^+, e^- pairs they produce, after interacting with matter. A very simple scintillation detector is drawn in Fig. 11.10 and some characteristic properties of some useful materials are given in Table 11.3.

We should mention that the first such detector was completed in 1903 by Sir William Crookes using ZnS as fluorescence material.

Typical photomultipliers contain up to 14 levels of amplification and thus achieve amplification factors of the order of 10^9. This way even a small

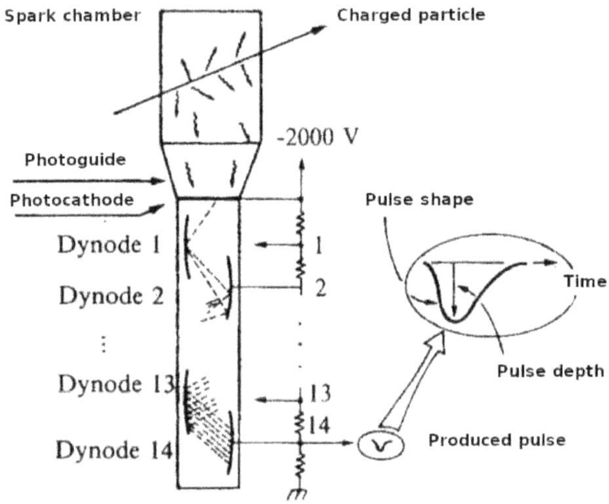

Fig. 11.10: A schematic drawing of a scintillation counter.

number of initial photons can lead to a measurable signal (see Fig. 11.10). Assume now that the particle looses all its energy by producing each time photons of little energy, typically $E_{ph} \cong 3eV (\cong 4.000 A^{\circ})$. If ϵ_{ph} is the efficiency of the detector in producing photons, the final number of the photons produced is:

$$N_{ph} = \frac{E_p}{E_{ph}} \epsilon_{ph}. \tag{11.47}$$

Not all these photons reach the photomultiplier. If ϵ_{col} is the efficiency of collection and ϵ_{cath} the probability a photon to kick out an electron as it hits the cathode, the number of photoelectrons produced is

$$n_e = \frac{E_p}{E_{ph}} \epsilon_{ph} \epsilon_{col} \epsilon_{cath}. \tag{11.48}$$

The efficiency ϵ_{col} can get very large employing special plastics or photoguiding air. Typical values are $\epsilon_{ph} = 0.02$, $\epsilon_{col} = 0.1$ and $\epsilon_{cath} = 0.1$.

In a standard scintillation detector with thickness 1 cm a particle of high energy, e.g. a μ with energy of a few hundred of MeV, looses energy about 1.5 MeV. With the above data we find

$$n_e \cong 100 \quad \text{photoelectrons.}$$

For a NaI (Tl) crystal $\epsilon_{ph} \approx 0.1$ one gets $n_e \approx 500$.

It is clear that, since all processes entering Eq. (11.48) are statistical in nature, the values of n_e are subject to variations. These lead to a large width of the signal we get. Another source of width broadening comes from the statistical fluctuations in the photomultipliers. Classical scintillation detectors are characterized by pulses with widths $\tau \approx$ 3–30 ns (1 ns = 10^{-9} s) and about 250 ns for those involving NaI. Small widths can be obtained with special care, a feature which is important for attaining a small error in the determination of ΔT (see Eq. (11.46)).

The main problem of these detectors is that they cannot be activated at will. They cannot distinguish between one event involving two particles, one with energy E_1 and the other with energy E_2, and another event with only one particle but with energy $E_1 + E_2$. In other words the signal depends on the total energy. They can, of course, be used, combined with an equipment that measures momentum (Eq. (11.7)) as we discussed in previous section, for the determination of a particle's mass, once its charge is known.

11.4.2 *Semiconductor detectors*

The scintillation detectors have been replaced by the semiconductor detectors. The principle operation of the latter is quite simple. As a charged particle passes through matter it ionizes the atoms and eventually it produces ion pairs. The number of pairs is given by:

$$n_{\text{ion}} = \frac{E}{W},$$

where W is the energy required to produce one pair. As a result the energy E is measured once W is known and n_{ion} is measured. n_{ion} can be determined, if the ions are separated in a magnetic field and the total charge is counted.

If a gas ionization is used, the quantity W is large, 42 eV for He, 22 eV for Xe, 34 eV for the air. Thus, for a given energy, few ion pairs are produced. In a semiconductor, however, W is much smaller, 2.9 eV for Ge and 3.5 eV for Si. This happens because ionization does not occur from the atomic levels to the continuum, but to the conduction zone.

Furthermore the semiconductors have the advantage of higher density plus the fact that they do not require cooling at low temperature. On the other hand, they have the disadvantage that their volume rarely can exceed 100 cm^3, while spark detectors can be much larger.

11.4.3 *Cherenkov counters*

The principle of operation of Cherenkov counters is based on the **Cherenkov phenomenon** (1934), i.e. when a particle passes through a dielectric, it produces light. This happens, since the EM field accompanying the particle excites the dipoles of the dielectric, which then emit radiation. The waves produced this way by he dipoles under some conditions interfere constructively and, thus, even a weak initial radiation can be observed. The necessary condition for this to happen is

$$v_p > \frac{c}{n} = v_{ph}, \tag{11.49}$$

where v_p is the particle velocity and n the index of refraction of the medium. In other words the particle velocity must be higher than the velocity v_{ph} of light in the dielectric. This can be seen by applying Huyghens (Fig. 11.11).

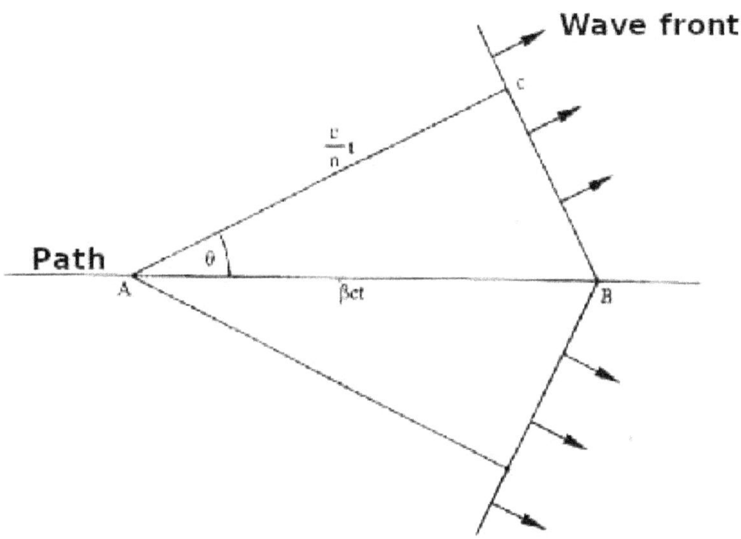

Fig. 11.11: Huyghens principle for the production of Cherenkov light. AB is the particle's path.

We have

$$\cos\theta = \frac{\frac{c}{n}t}{\beta ct} = \frac{1}{n\beta}. \tag{11.50}$$

Obviously $\cos\theta \leq 1 \rightarrow \beta \geq 1/n$ or $v \geq c/n$. The critical velocity is $v = c/n$ and, then, $\theta \approx 0$. The maximum angle is obtained in the extreme relativistic case $\beta = 1$, that is

$$\cos\theta_{max} = \frac{1}{n}. \tag{11.51}$$

For water, $n = 1.33$ (visible spectrum), the condition $v \geq c/n$ for radiation Cherenkov in water becomes $\beta \geq 0.75$.

In general, however, since the index of refraction is a function of the frequency, the critical velocity and the maximum angle are frequency dependent.

The impressive blue color seen in reactors of open tank is due to the Cherenkov radiation. In this case it is caused by electrons resulting from nuclear β decay. The energy lost by a relativistic particle $v \approx c$ in water, associated Cherenkov light in the visible spectrum (4000–8000°A), is about 400 eV/cm. This is a small fraction of a typical ionization energy, which is of the order of 2 MeV/cm. A standard Cherenkov counter is exhibited in Fig. 11.12.

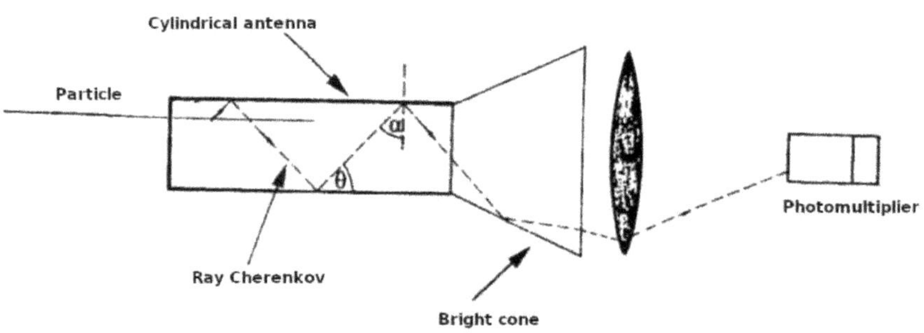

Fig. 11.12: The basic elements of a Cherenkov counter.

For a given n and for a particular choice of observation angle θ, one can measure the particle velocity v. Thus, knowing the particle's momentum (Eq. (11.7)), we can determine its energy and, thus, measure its mass. This equipment can also be used as a mass selector. For this purpose we select a dielectric so that only one particle, with unknown mass, can satisfy the condition $v \geq c/n$ (see Fig. 11.13). For the measurement of a definite velocity (or of mass, if the momentum is known), what is needed is to perform the measurement in a definite angle. So photons coming out in other angles are not observed. Such a device is called differential Cherenkov counter. In this case the angular resolution is related to the mass resolution via Eq. (11.50). That is

$$-\sin\theta d\theta = -\frac{1}{n\beta^2}d\beta.$$

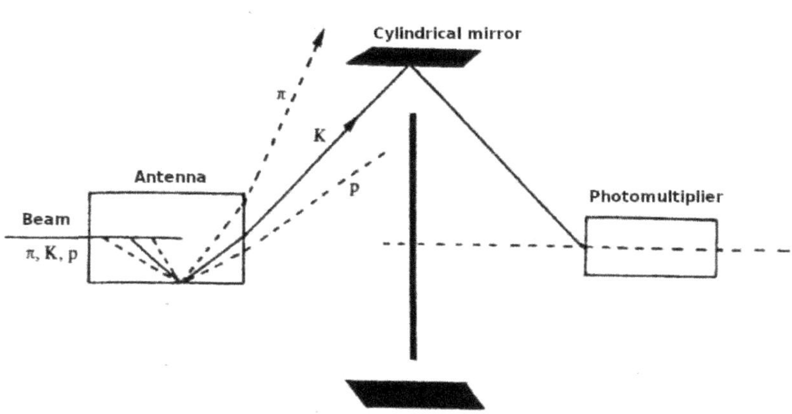

Fig. 11.13: A typical differential Cherenkov counter, designed to select only one beam component, here only K particles.

But

$$\beta = \frac{p}{E},$$

hence

$$d\beta = \frac{pmdm}{E^3}$$

and

$$\tan\theta d\theta = \frac{mdm}{E^2}. \tag{11.52}$$

In order to detect particles of very high velocities, one needs material with small n, $n \approx 1$, which corresponds to gases. In this case it is necessary to employ a detector with a size of a few meters.

11.4.4 *Spark chambers*

The main features of a spark chamber are exhibited in Fig. 11.14. It includes a series of parallel plates, which are about 1 cm apart, in an atmosphere of a noble gas, Ne or Ar, i.e. under a pressure in the range of $P = 1$ to 1.5 Atm. The presence in of a charged particle the chamber is manifested by the scintillation detectors S_1 and S_2, which yield a coincident signal. This signal excites a hydrogen thyratron lamp. The capacitor C_1, which is connected with a high voltage supply $V \approx 10$ kV, is discharged to the earth through a resistor producing a signal with a voltage $C_1/(C_1 + C_2)V$ on the plates, which is equivalent with a capacitor with capacitance C_2.

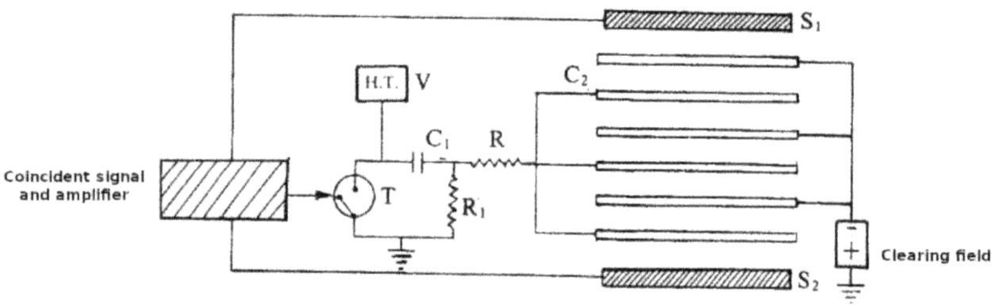

Fig. 11.14: The main features of a spark chamber with parallel plates. S_1 and S_2 are spark detectors. T is a thyratron lamp or a spark slit.

One can choose $C_2 << C_1$ in which case the duration of the pulse, i.e. of the voltage increase, in the spark chamber is $\tau \approx C_2 R \approx 10^{-8}$ s. If a particle passes now through the chamber, it produces electric discharges with the immediate effect the decrease the voltage in the plates with a time constant $\tau_1 \approx RC_1$. The time required for the development of the electron burst typically is of the order of μs, i.e. much larger than τ. Thus the same particle activates the device at time τ and manifests its presence at time τ_1.

On the plates one has an electric field produced by a potential difference $\Delta V \approx 10^3$ V. The purpose is to clear the space of any ions so that the chamber is ready to start operating again. With such a field all ions are cleared in 0.5 μs. If, during this period, another particle passes through the chamber, the device is not activated (the produced electrons have a high chance to recombine with the ions). Thus the chamber has a time resolution about 1 μs and can function with a rate of 10^5 particles/s. In other words the device selects an interesting event and is activated once. It can be reactivated again only after 10 ms $= 10^{-2}$ s, which is essentially equal to the time required for charging the capacitor C_1 to the voltage 10 kV. This inertia is not a great disadvantage, since other types of detectors have even longer inertial times. In optical spark chambers, e.g., the activation of the cameras takes longer.

One important feature of spark chambers is that they have good spatial resolution, i.e. they can locate the particle very accurately. The sparks can be located with an accuracy of 0.3 mm, while in scintillation detectors the particles cannot be located any better then the size of the detector. The sparks are registered photographically using two devices at a solid angle of 90° (for a typical picture see Fig. 11.15).

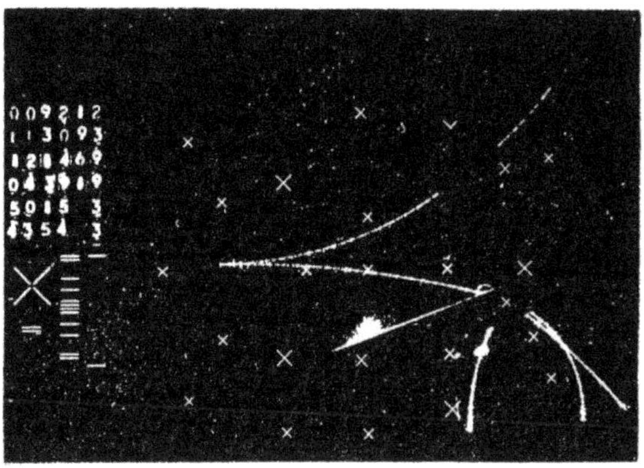

Fig. 11.15: This is a spark chamber photograph of a long lived neutral K_L meson. It is inferred by the decay $K_L \rightarrow \mu^+\pi^-\nu_\mu$. The upper path is that of μ^+ and the lower that of a π^-. The magnetic field is perpendicular to the plane of the figure. The picture was taken at the Stanford linear accelerator (SLAC), USA.

To-day there exist, of course, additional methods for registering the above information. Such are crystals, which transform the acoustic signal, accompanying the spark, into an electric signal. The latter can be registered electronically without the need of processing films etc.

In addition there exist spark chambers, which use, instead of plates, fine parallel metallic wires at a small distance apart. This way currents are produced only in the wires at small distances from the path of the particles. Thus one can determine quite accurately the particle coordinates in the plane of the chamber. Even better localization can be achieved by combining different chambers with wires perpendicular to each other. We will not, however, elaborate further on these rather technical subjects.

11.4.5 *Gas detectors*

11.4.5.1 *Bubble chambers*

The operation of bubble chambers is based on the same principle as that of spark chambers except that, in this case, one takes pictures of bubbles formed along the path of an ionizing particle, after it penetrates a space with super heated transparent liquid, which begins boiling. This is exhibited in Fig. 11.16. Bubble chambers suffer from the fact that they cannot be activated at will. It is like the policeman who inquires identification from everyone, not just from the suspects. They played a important role in the early elementary particle experiments, but they are no longer in use.

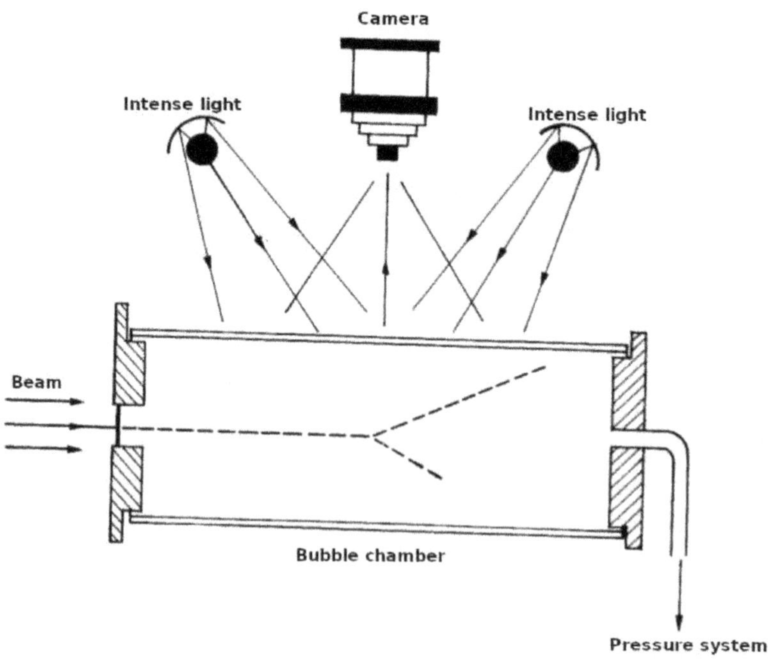

Fig. 11.16: A schematic presentation of a bubble chamber.

11.4.5.2 *Analog detectors*

The analog detectors aim at curing the problems encountered in the above devices, namely the fact that the scintillation detectors have poor spacial resolution and the spark champers remain inert for a long time. They are in fact a technological improvement of an old device known as Single Wire Proportional Counter (SWPC), invented by Thomson, Rutherford and Geiger at the beginning of the 20th century. It is essentially a wire acting as anode put along the axes of a metallic tube, which acts as a cathode.

In such devices one can arrange the gas pressure so that the ions produced by the charged particle produce more ions as a result of their collisions with the atoms present in the gas. The signal produced, when the electrons reach the anode, is proportional to the initial ionization. Since no spark is produced, the time the analog detector remains inactive is very short, 1 μs $= 10^{-6}$ s. Real progress was made in 1968 when Charpak [Charpak *et al.* (1968)] and collaborators invented the Multi Wire Proportional Counter (MWPC). For this work Charpak was awarded the Physics Nobel Prize in 1992.

MWPCs utilize a collection of thin parallel wires made of steel acting as one electrode placed between metallic plates or plane steel nets acting as the other electrode. This way the obtained information about the coordinates of the particles is more accurate. Typical potential difference used is 200 V with wires of diameter 50 μm placed at a distance 0.1 to 1.0 cm. The tubes are filled with a mixture of Ar/alcohol. Two such parallel detectors with wire directions at right angles can determine the particle position with an accuracy of 0.25 μm. At the same time these detectors have a good energy resolution. It should be mentioned that each wire acts as an analog detector and must be equipped with its own amplifier etc.

The performance of these devices was substantially improved with the use of MICRO STRIP. Thus achieving rates $10^3/(\text{mm}^2\text{s})$, particle location about 30 μm, separation of double tracks about 400 μm and, of course, very good energy resolution.

Such devices are used not only in particle physics experiments, but in many other applications as well (chromatography, radiography etc.).

11.4.5.3 *Micromegas*

A subsequent great step forward came with the invention of **micromegas**, MICROMEsh GAseous Structure. Micromegas has been invented in 1995 by Giomataris and collaborators [Giomataris *et al.* (1996)]. It is a two-stage parallel plate avalanche chamber with narrow amplification gap (few tens to one or two hundred microns). The gas volume is split in two regions by a thin micromesh which separates the amplification gap from the drift gap. The electric field is homogeneous both in the drift and amplification gaps. Due to the narrow amplification region in Micromegas, locally small variations of the amplification gap are compensated by an inverse variation of the amplification coefficient and therefore do not induce gain fluctuations [Giomataris (1998)].

Excellent spatial resolution: 12 m (RMS) accuracy (limited by the pitch of micromesh) is achieved in a beam test at CERN, using anode strips with a pitch of 100 m and low diffusion CF4/iC4H10 (80:20) mixture [Wotschack (2013)], [Alexopoulos *et al.* (2011)]. A time resolution of 600 picosecond (RMS) was achieved by the KABES beam spectrometer of the NA48 experiment. Micromegas with single electron capabilities have been studied in [Derre *et al.* (1999)], [Derre *et al.* (2000)], [Campbell *et al.* (2005)].

Recent developments which are currently under way using Cherenkov light converted in a CsI photocathode have shown the capability of reaching time resolution of about 10 ps. The energy resolution is 11% FWHM[2] with 5.9 keV photons, a result which is close to the limit imposed by statistical fluctuations. It shows that intrinsic fluctuations during the amplifications are quite small.

In the traditional way Micromegas is obtained by suspending a mesh over the anode strips or pads, the precise gap is obtained by using adequate insulating spacers (pillars) printed on top of the anode plane by lithography of a photo-sensitive film.

A big step in the direction of the industrial manufacturing of large-size detectors is the development of the bulk micromegas technology [Giomataris *et al.* (2006)]. The basic idea is to build the whole detector in a single process: the anode plane with copper strips, a photo-image-able polyamide film and the woven mesh are laminated together at a high temperature forming a single object.

Micromegas detectors for tracking purposes have been successfully used

[2]Full width at half maximum (FWHM) is the width of a line shape at half of its maximum amplitude.

in High Energy Physics experiment. The first example is the COMPASS detector at CERN with sizable dimension of 40 cm × 40 cm [Kunne *et al.* (2003)], [Magnon *et al.* (2002)]. Such devices are of use in specific experiments, like CAST [Abdon *et al.* (2007)], future accelerators [Colas *et al.* (2004)], underground experiments searching for dark matter [Mayet *et al.* (2009)] etc. For new capabilities see Ref. [Iguaz *et al.* (2012)].

The device is exhibited[3] in Figs. 11.17–11.18 and some of its advantages are summarized in Fig. 11.19.

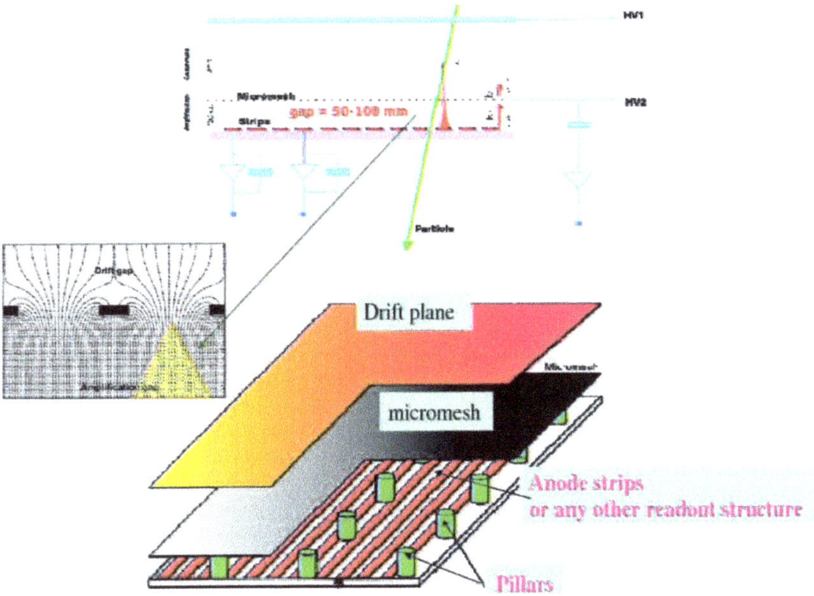

Fig. 11.17: A sketch of a micromegas detector [Giomataris *et al.* (1996)].

11.4.6 *Time projection counters (TPC)*

The detectors discussed in the previous section suffer from the fact that they provide limited information for the position of the particles in space, at best in two dimensions. For the determination of the particle's position in the third direction one needs more planes of wires. Then the detector becomes cumbersome and the angle of acceptance becomes smaller. The solution to this is provided by time projection counters (TPC), invented in 1974 by Nygren.

[3]We thank I. Giomataris and G. Fanourakis for supplying us with these figures.

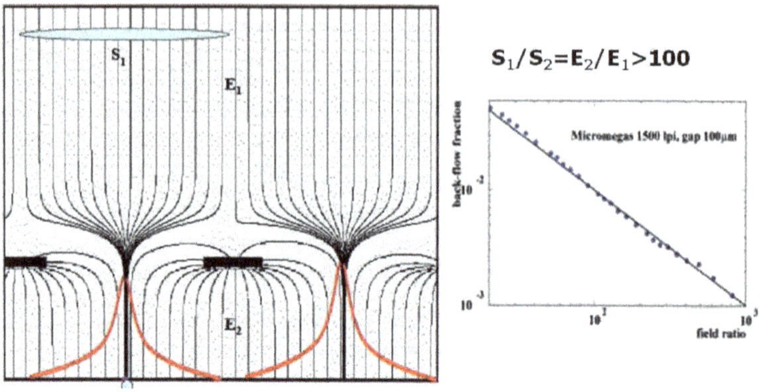

Fig. 11.18: The electric field and the suppression of ion reproduction in a typical micromegas detector.

11.4.6.1 *Cylindrical TPC*

A traditional cylindrical TPC is exhibited in Fig. 11.20. This way the electrons, with the combined action of the electric and magnetic fields, follow a helical path with axes along the axes of a cylinder. This way:

- The two coordinates are determined by the collection point on the end wires.
- The third coordinate is specified by the arrival time at the wires.
- The total energy is measured as above, i.e. by counting the total charge on the wires.
- Knowing the coordinates and the magnetic field the particle's momentum can be determined (see Section 11.2.1).
- If care is taken, so that the detector fully contains the beam, all events are registered.

It, thus, looks like an excellent detector and, if properly designed, ideal for high energy physics experiments.

11.4.6.2 *Spherical TPC*

The cylindrical detectors are very useful devices, but they suffer from the fact that their accuracy is limited by edge effects, which affect the shapes of the electric and magnetic fields near the bases of the cylinder. These are

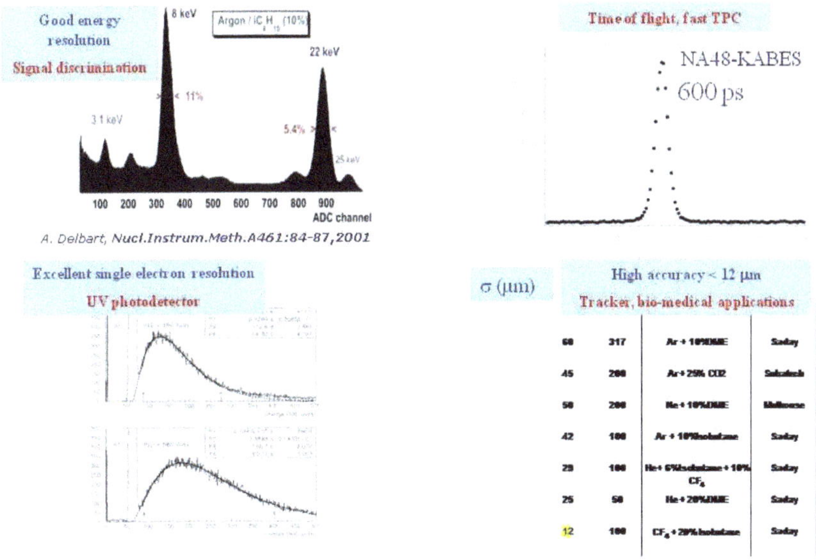

Fig. 11.19: Schematic exhibition of the advantages (high resolution etc.) of a micromegas detector.

cured by a spherical design, which, in addition, offers the advantage that only an electric field is needed.

In this case we consider two concentric spheres, one with a mall radius r_1, which acts as anode, and the other with a larger radius r_2 acting as a cathode. The potential in the space between them, $r_1 < r < r_2$, is a function of r only and is given by:

$$V = V_0 \frac{1/r - 1/r_2}{1/r_1 - 1/r_2}, \quad V_0 > 0. \tag{11.53}$$

The electric field is radial pointing outwards:

$$E(r) = \frac{V_0}{r^2} \frac{1}{1/r_1 - 1/r_2}, \quad V_0 > 0, \tag{11.54}$$

where V_0 is the applied voltage on the small sphere.

The space between the spheres is filled with a gas or a mixture of gases. If a charged particle or photon finds itself in that space, it ionizes the gases and the ions move towards the anode, while the electrons move to the small sphere. As soon as they approach the sphere, a few mm away, they feel a stronger electric field, $E = 10$ keV/cm and they produce more electrons

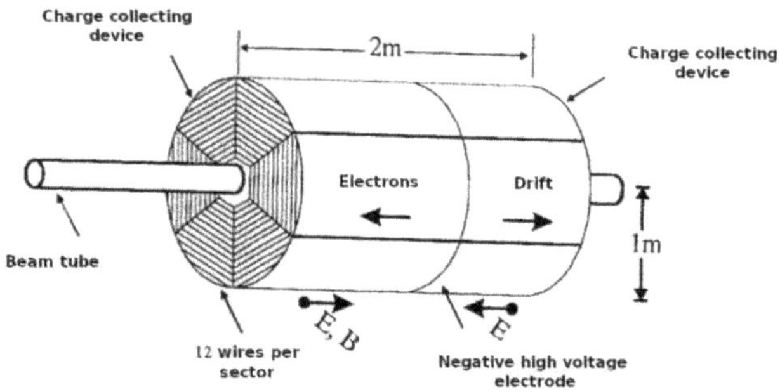

Fig. 11.20: A schematic presentation of a cylindrical TPC. The charged particles of the beam, after their collisions with the atoms of the gas, produce electrons and ions. These electrons are collected in the proper wires in the base of the cylinder with appropriate electric and magnetic fields parallel to the axes of the beam.

resulting in an avalanche. The positive ions induce a pulse in the charge preamplifier. Due to the fact that the ions travel a much larger distance than the electrons, this pulse is due to the ions with the electrons making a small contribution. One can show [Giomataris *et al.* (2008)] that the detector signal produced by an avalanche at $t = 0$ is:

$$V(t) = \frac{V_0}{(1/r_1 - 1/r_2)} \left[\left(\frac{3\mu V_0 t}{1/r_1 - 1/r_2} + r_2^3 \right)^{-\frac{1}{3}} - 1/r_2 \right], \quad (11.55)$$

where μ is the ion mobility, which, by definition, is the drift velocity through the electric field, that is:

$$\mu = \frac{v}{E(r)} = \frac{dr}{dt}\frac{1}{E(r)} \approx \text{constant}. \quad (11.56)$$

Equation (11.55) is found to be in good agreement with experiment. From the dispersion in this equation one can determine very accurately the point at which the interaction of the particle in question took place. Furthermore from the collected charge its energy can be determined.

The inner sphere acts as an analog detector and can be equipped with micromegas, as represented in Fig. 11.21. With this one would like to detect low energy neutrinos produced from a radioactive source inserted inside the small sphere. To this end, one detects recoiling electrons, produced as a

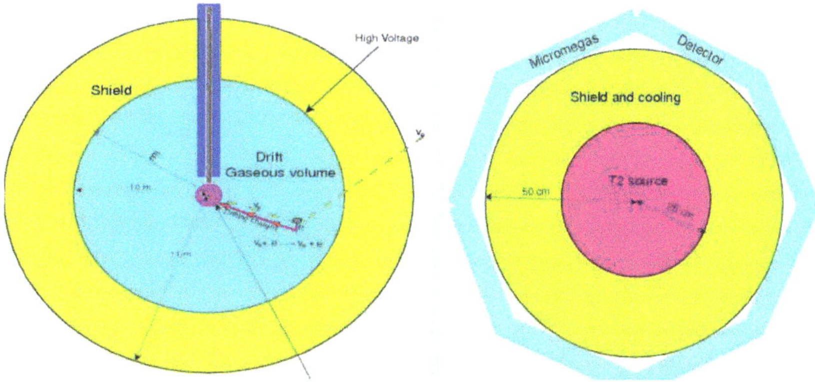

Fig. 11.21: A spherical time projection counter (TPC) detector. On the left we show the analog detector (interior sphere), the cathode and he space filled with a gas. For the application of a high voltage a proper connection of the two spheres, indicated by a bar, with a source characterized by a potential difference V_0. In this case the particle to be detected is a neutrino, which produces a recoiling electron, which drifts towards the anode producing an avalanche of electrons. On the right we show the small sphere, enlarged and equipped with micromegas. Inside this sphere a radioactive source may be inserted, if one wants to have neutrinos produced inside the device. The sphere is equipped with a cooling system and is properly shielded so that only neutrinos can go through.

result of the neutrino-electron interaction, with the goal of observing neutrino oscillations [Giomataris and Vergados (2004)]. In a sightly different set up the neutrinos are produced outside the detector, more specifically in a supernova explosion [Giomataris and Vergados (2006a)] or from a spallation source [Vergados *et al.* (2009)]. The neutrino energy in the last two cases is sufficiently high so that nuclear recoils can be observed.

In order to achieve the required accuracy the electric field must be spherically symmetric, especially in the area near the little sphere. The presence of the connecting wire, however, spoils the symmetry and, thus, some of the field lines end up on the wire and not on the little sphere, see Fig. 11.22(a). The problem was solved [Giomataris *et al.* (2008)] with a clever arrangement of coaxial cylindrical conductors with axes parallel to the connecting wire, see Fig. 11.22(b).

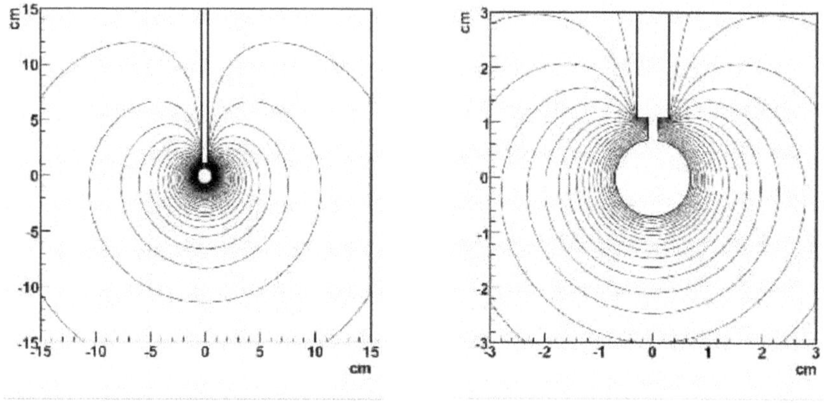

(a) (b)

Fig. 11.22: (a) The equipotential surfaces are exhibited near the little sphere (anode). Due to the presence of the conducting wire they deviate from spherical symmetry and not all of them end up on the surface of the sphere. (b) After the introduction of proper cylindrical conductors around the connecting wire, the equipotential surfaces become spherical and, more important, they end up at the anode.

11.4.7 *Calorimetric detectors*

The total energy of a particle or of a number particles can be obtained by methods of total absorption. This way the original particles of the beam, interacting with atoms of the target, produce secondary ones. The latter in a similar fashion produce tertiary and so on. This leads to the formation of a shower. Care is taken so that none of the particles in the shower escapes. So, knowing the ionization energy of the medium, one obtains the energy of the initial particle. Since such detectors measure the total energy of the particles, they are called **colorimeters**. They are, in particular, very useful for the detection of neutral particles. They also allow for the measurement of very fast, 10 ns, total energy signals and, thus, they can be used for diagnostic purposes. Typically, calorimeters are segmented: transverse segmentation provides information about the direction of the particle or particles, as well as the energy deposited, and longitudinal segmentation can provide information about the identity of the particle based on the shape of the shower as it develops.

The calorimeters are of two types: (a) electromagnetic calorimeters, which are specifically designed to measure the energy of particles that interact primarily via the electromagnetic interaction and (b) hadronic calorimeters, designed to detect particles that interact via the strong nuclear force. The response of a calorimeter can be described in terms of the e/h (electron to hadron) ratio. This is a measure of how well a calorimeter responds to leptons or photons versus hadrons. Ideally one would want a ratio e/h about 1, this condition is called compensation.

11.4.7.1 *Electromagnetic calorimeters*

High energy photons and electrons are the combined result of synchrotron radiation, bremsstrahlung, see Section 11.2, and electron positron pair production, e^+e^-, shown schematically in Fig. 11.23. From this figure we see that what happens depends on the parameter X_0, which in turn depends on the material, see Table 11.4.

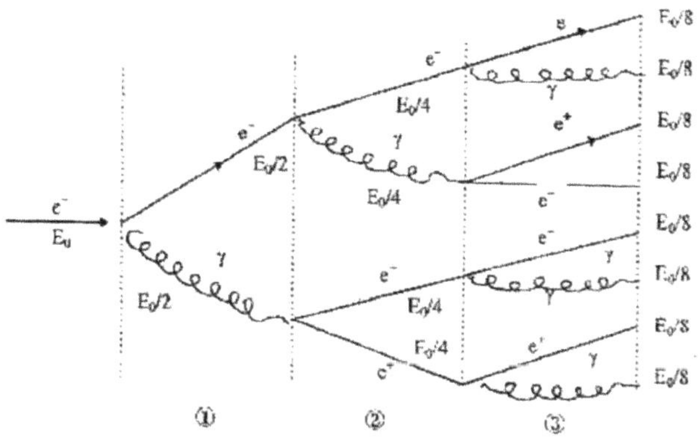

Fig. 11.23: At every stage $1, 2, \ldots, n$ corresponding to n radiation lengths, (nX_0), 2^1, 2^2, 2^3, $\cdots 2^n$ particles are produced, each having an energy $E_0/2$, $E_0/2^2, \cdots E_0/2^n$.

Note that in Table 11.4 the radiation length X_0 is given in units $(\mathrm{gr/cm^2})$. This is important since this length becomes different for materials of different density. In fact the relation between X_0 and the usual length ℓ_0 is given by:

$$\ell_0 = X_0/\rho, \tag{11.57}$$

with ρ the density of the material.

Table 11.4: The radiation length X_0 and the critical energy E_c are shown. At energies smaller than E_c, radiation (Bremsstrahlung) process stops and ionization dominates. Hydrogen is a liquid, He is a gas and the rest are solids.

element	Z	E_c(MeV)	X_0 (gr/cm^2)	ρ(gr/cm^3)
hydrogen	1	340	58	0,065
helium	2	220	85	-
carbon	6	103	42.5	2.3
aluminum	11	47	23.9	2.7
iron	26	24	13.8	7.7
lead	82	6.9	5.8	10.4

What happens to the shower depends on a critical energy E_c, which depends on the material. For energies $E > E_c$ the radiation Bremsstrahlung is dominant and the energy deposited in the detector increases exponentially. For smaller energies ionization is operative and the energy deposited per X_0 remains constant and, after some time, all of a sudden it drops to zero.

One can show that for a particle with energy E_0:

(1) The position of the maximum increases logarithmically with E_0/E_c.
(2) The number of particles at the maximum is proportional to E_0.
(3) The relative energy resolution is given by

$$\Delta E/E \approx 0.05\ (E/1\ \text{GeV})^{-1/2}, \qquad (11.58)$$

i.e. it increases with energy.
(4) Materials with high Z are preferred. A good choice are lead glasses, e.g. (S_i0_2 45% lead 55%).

For more details see, e.g., the review [Wigmans (2000)].

11.4.7.2 *Hadronic calorimeters*

In this section we will summarize the most essential features of hadronic calorimeters. For more details the reader is referred to the excellent books [Wigmans (2000)], [Fabjan (1987)].

In this case the reaction proceeds through inelastic nuclear reactions, which lead to the production of secondary, tertiary, etc. hadrons. They are characterized by longer absorption lengths in comparison with the EM calorimeters, e.g. 80 gr/cm^2, 130 gr/cm^2 and 210 gr/cm^2 for C, Fe and Pb respectively. Thus they are much bigger than the the EM ones. In addition a substantial fraction of the energy (30%) is lost due to nuclear decays. This, however, can be calculated and replaced by placing radioactive uranium in the detector.

The energy resolution is much worse than that of the EM detectors and is given by:

$$\frac{\Delta E}{E} = 0.5 \; (E/1 \text{ GeV})^{-1/2}. \tag{11.59}$$

11.5 Problems

(1) Consider the LEP accelerator, see Table 11.2. It is known that the current of the accelerated charges is 6 mA and that $2\pi R$ is 27 km.
i) What is the accelerated charge in one period?
(Hint assume that the charges are moving with the speed of light.)
ii) Find what is the energy radiated in one period, assuming that Eq. (11.39) holds.
iii) What is the number of the particles accelerated?
iv) For the luminosity given in the table, what is the area crossed by the beam?

(2) In the Positron Emission Tomography (PET), the e^+e^- annihilation is exploited to produce photons which are detected and measured in the apparatus. The positrons are produced in the organic material, where a suitable β^+ emitting radionuclide is introduced for this purpose. Subsequently e^+e^- annihilate at rest into pairs of photons. Assuming that photons are detected through the Compton scattered electrons, find their minimum and maximum energy.

(3) An electromagnetic calorimeter is calibrated using different particle beams. The calorimeter is a sandwich of lead (Pb) and plastic scintillators. It is made up of consecutive stacks of one scintillator and one Pb slab, both 1 cm thick. Beams of 5 GeV electrons, muons and photons are used for calibration.
i) Determine the mean energy deposited by electrons and muons in the fourth scintillator.
ii) Compare the results obtained using electron and photon beams.

iii) Explain how to discriminate between these two cases.
Data for Pb: $X_0 = 0.56$ cm, $\rho = 11$ g/cm^3.
Data for scintillator: $X_0 = 42$ cm, $\rho = 1.03$ g/cm^3.

(4) A typical experiment measures neutral pions through the detection of the photons emitted in the decay $\pi^0 \to \gamma\gamma$. Assume that the pion energies are around 1 GeV and that a Pb plate of 1 cm is used for their conversion to electrons (+ and −), which are measured in a downstream detector. The Pb radiation length is 5.6 mm. What is the pion detection efficiency?
Hint: Detection efficiency is the probability that both photons from π^0-decays are converted in one or more electrons.

(5) A charged pion beam having momenta ranging between 0.5 and 1.5 GeV/c is collimated by a narrow slit 1 cm wide after having crossed a region 1.1 m long, where a magnet generates a 0.2 Tesla uniform field (see Fig. 11.24(a)).

Find what is the distance between the slit and the end of the magnet to select a momentum $p_0 = (1 \pm 5\%)$ GeV/c.

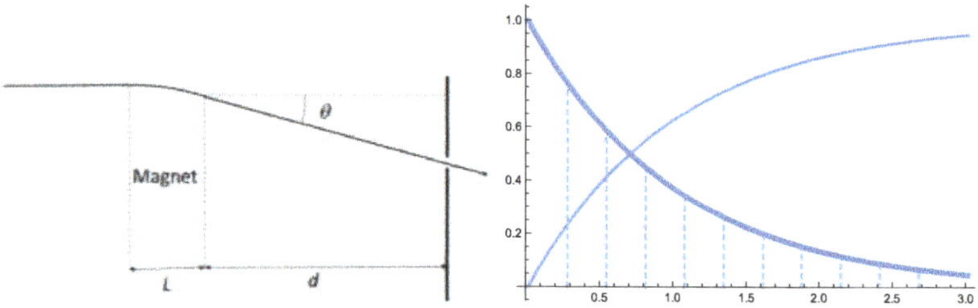

(a) (b)

Fig. 11.24: (a) The deflection of a pion beam in a magnetic field. (b) The fraction of muons (thick solid line) and the the corresponding number of appearing electrons (solid line) as a function of time in units of τ_μ. The dashed vertical lines guide the eye to find the fraction of moons left, at times which are multiples of the period.

(6) A muon beam is circulating in an accumulation ring of 28 m radius with a magnetic field of 0.5 Tesla. Knowing that $m_\mu = 106$ MeV/c^2 and $\tau_\mu = 2.2 \times 10^{-6}$ s find i) the muon momentum, ii) the time of one revolution (period), and iii) the fraction of muons lost in one turn.
Hints: $p = 4.2$ GeV/c, $T = 0.267\tau_\mu$, see Fig. 11.24(b)

(7) Two equal scintillation counters, S_1 and S_2, 1 cm thick, placed on a beamline at a relative distance $L = 3$ m, are used to measure the time-of-flight of crossing particles. All the particles in the beamline have the same momentum p. The scintillator radiation length is $X_0 = 40$ cm.

Suppose this set-up is used to separate $\pi^+(m_\pi = 139\ \mathrm{MeV/c^2})$ from $K^+(m_K = 493\ \mathrm{MeV/c^2})$ in a beamline with $p = 1$ GeV/c.

Find what is the time resolution needed to achieve a particle discrimination within 4 standard deviations.

Hint: Eq. (11.46) maybe useful.

(8) There seem to be two paths to high energy experiments:

i) To proceed with the next generation of a high energy accelerators. This way we may find new exotic particles, like supersymmetric particles, those that are constituents of the dominant matter component of the universe, which is dark, etc.

ii) To build high quality efficient detectors and exploit natural accelerators (cosmic rays, ultra high energy cosmic neutrinos etc.). At the same time one may detect rare events (dark matter searches etc.).

Suppose that some people find that option i) is too expensive to continue for ever. Give arguments in favor of one or the other option.

Chapter 12

Neutrino physics — Neutrino oscillations

12.1 The elusive neutrino

From its discovery neutrino has been elusive, but it does not really escape in neutron decay:

$$n \to p + e^- + (?)$$

its presence was inferred by Pauli (1930, 1932), since without it one has violation of energy and angular momentum conservation. Thus they proposed:[1]

$$n \to p + e^- + \tilde{\nu}_e.$$

Soon afterward Fermi (1934) proposed a theory of beta decay as a contact interaction.

Bethe (1939) found for the neutrino an important role in our world. Without it our sun and the stars *cannot shine*. Symbolically:

$$4p \to \text{He} + 2\nu_e + 2e^+ + \text{light}.$$

(For more details see Section 7.2.1.) Its direct detection was made much later by Reiness and Cowan (1953)

$$\tilde{\nu}_e + p \to n + e^+.$$

The fact the the neutrino produced in weak interaction is different from its anti-neutrino was shown by Davis (1955) looking at the reactions:

$$\nu_e +^{37}_{37} \text{C}\ell \to^{37}_{38} \text{Ar} + e^- (\text{yes!}),$$

$$\tilde{\nu}_e +^{37}_{37} \text{C}\ell \to^{37}_{38} \text{Ar} + e^- (\text{No!}).$$

[1]They actually wrote ν, since at that time they did not about the existence the antineutrino or of the presence three neutrino generations.

Neutrino has been a real revolutionary, it violates sacred laws: The fall of parity was first proposed [Lee and Yang (1956)] and experimentally found by [Wu *et al.* (1957)]. Furthermore the neutrino is very prejudiced: always left handed!, see Section 16.5, as shown by [Goldhaber *et al.* (1958)]. A new neutrino is born in 1947, named ν_μ, since it always accompanies μ^-, with its own antineutrino found at about the same time. This neutrino was different from its older brother since it was shown that $\nu_\mu \neq \nu_e$ at Brookhaven experiments (1962). One more neutrino, named ν_τ, was later born. A neutral fermion that appears always together with the charged lepton τ^- found at SPEAR and SLAC (1976). It was later shown that $\nu_\tau \neq \nu_\mu \neq \nu_e$ at Fermilab (1997).

The neutrino is the only known particle that does not interact electromagnetically or strongly. Its properties were the cornerstone in the development of the standard model by GSW [Glashow (1961)], [Weinberg (1967)], [Salam (1968)].

12.2 Lepton flavor conservation

In weak interactions it appears that the neutrino appears in 3 distinct generations as it is manifested by the charged parent. It is always accompanied by its charged partner. This behavior is understood if we associate with each lepton pair, neutrino and charged particle, a additive quantum number called flavor, i.e. (L_e, L_μ, L_τ) and postulate that it is conserved. The anti-neutrinos have an opposite lepton flavor number and, as we have seen above, they are distinct from the neutrinos. A consequence of lepton flavor conservation is that processes like:

$$\mu^- \to e^- + \gamma, \quad \tau^- \to e^- + \gamma, \quad \tau^- \to \mu^- + \gamma, \tag{12.1}$$

$$\mu^- \to e^- + e^- + e^+, \quad \tau^- \to e^- + e^- + e^+ \text{ etc.}, \tag{12.2}$$

$$\mu_b^- + (A, Z) \to e^- + (A, Z)^* \text{ (bound muon conversion in the nucleus)}, \tag{12.3}$$

are forbidden. Experimentalists have, in fact, searched for such processes long before gauge theories, but no such signals have been observed. So rather stringent experimental limits on the branching ratios (BR) exist:

$$BR(\mu^- \to e^- + \gamma) \leq 5.6 \times 10^{-13}, \text{ MEG (2012)}$$

$$BR(\tau^- \to \mu^- + \gamma) \leq 3.8 \times 10^{-8}, \text{ B-factories (2011)}$$

$$BR(\tau^- \to e^- + \gamma) \leq 4.8 \times 10^{-8}, \text{ B-factories (2011)}$$

$$BR(\mu_b^- + \text{Au} \to e^- + \text{Au}) \leq 7 \times 10^{-13} \text{ SINDRUM II (2006)}.$$

Lepton flavor conservation implies that the total lepton number $L = L_e + L_\mu + L_\tau$ is also conserved. So neutrinoless double beta decay

$$(A, Z) -> (A, Z+2) + e^- + e^-$$

is forbidden. In fact it has been searched for over 80 years, but it has not yet been observed. This important process will be discussed below.

Lepton flavor conservation implies that a neutrino of a given flavor cannot be transformed into one of a different flavor. Neutrino oscillation experiments persisted, however, and, as we will see below, they were finally successful.

12.3 History of neutrino oscillations

It is known from quantum mechanics that if a system is found in a state, which is not stationary, i.e. it is not an eigenstate of the Hamiltonian, it oscillates with time between the various eigenstates. Many such oscillations are known, e.g. in molecules like NH_3 with the possibility of different arrangements of the atoms in space, which are degenerate to zeroth order, but the degeneracy may be removed by some perturbation. Another system, known to exhibit oscillations since the middle of the 20th century, which is discussed e.g. in the companion book [Vergados (2017)], is that of the neutral kaons (see Section 1.5), K^0 and $\bar{K}^0$, which are eigenstates of the strong interactions, i.e. they are produced as such. However K^0 and $\bar{K}^0$ are not eigenstates of the symmetry CP. CP is the combined symmetry (see [Vergados (2017)] Section 12.5) of charge conjugation C, see Section 16.6 and parity P, see Section 16.4. The eigenstates of CP are the K_1 and K_2, i.e.

$$K_1 = \frac{1}{\sqrt{2}}\left(K^0 + \bar{K}^0\right), \quad K_2 = \frac{1}{\sqrt{2}}\left(K^0 - \bar{K}^0\right). \tag{12.4}$$

Thus we encounter the transitions:

$$|\langle K^0(t)|K^0(0)\rangle|^2 = \cos^2 \omega t, \quad |\langle \bar{K}^0(t)|K^0(0)\rangle|^2 = \sin^2 \omega t, \quad \omega = \frac{\delta m c^2}{2\hbar}. \tag{12.5}$$

Can a similar oscillation exist in another neutral system, like the neutrinos? The first concept of this idea is due to Pontecorvo (1958). The neutrinos, $(\nu_e, \nu_\mu, \nu_\tau)$ are produced in weak interactions, which are flavor conserving.

Thus they have a definite flavor associated with that of their charged partners, e^-, μ^-, τ^-. Flavor, however, is only a global symmetry and there is no reason to expect it to be absolutely conserved. If it is not, one expects to see neutrino oscillations, provided that the neutrinos have a non zero mass. In such a case we need to go beyond the Standard Model of Particle Physics, which predicts that the neutrinos are massless.

The first hints of neutrino oscillations came from the deficit of solar neutrinos, Davis (1960–1970). The first evidence of neutrino oscillations came from observations of atmospheric neutrinos Super Kamiokande (1998), the observations of solar neutrino oscillations, CHOOZE (1999), and the Sudbury Neutrino Observatory (SNO) (2001). The settlement of important outstanding neutrino oscillation issues was settled with reactor neutrinos, KamLAND (2003). More precise measurement of the oscillation parameters followed, culminating with the constraint of the small θ_{13}, providing upper and lower limits, T2K (2011), and the actual measurement of θ_{13}, Daya Bay (March 2012).

12.4 Aspects of neutrino masses and mixing

Within the Standard Model (SM), see, e.g., the companion book [Vergados (2017)], the neutrinos cannot acquire a mass like all the other fermions, because the right handed neutrino does not exist. If the right handed neutrino existed then the neutrino would acquire a mass through the Yukawa coupling, similarly to the up quarks:

$$\mathcal{L}_D = y_D(\bar{\nu}_L, \bar{e}_L)\begin{pmatrix} \phi^{*0} \\ -\phi^- \end{pmatrix}\nu_R + HC = y_D\left(\bar{\nu}_L\nu_R\phi^{*0} - \bar{e}\nu_R\phi^-\right) + HC. \tag{12.6}$$

In the unitary gauge we get the mass term

$$\mathcal{L}_D = y_D\bar{\nu}_L\nu_R\frac{\eta+\upsilon}{\sqrt{2}} + HC = \bar{\nu}_L\mathcal{M}_D\nu_R + \frac{\mathcal{M}_D}{\upsilon}\bar{\nu}_L\mathcal{M}_D\nu_R\eta + HC \tag{12.7}$$

where

$$\mathcal{M}_D = \frac{\upsilon}{\sqrt{2}}y_D. \tag{12.8}$$

The first term in Eq. (12.7) is called the Dirac mass term. The above relation can be generalized to a Dirac mass matrix by considering ν_L and ν_R as 3 generation vectors:

$$\nu_L = (\nu_{eL}, \nu_{\mu L}, \nu_{\tau L}),\ \nu_R = (\nu_{eR}, \nu_{\mu R}, \nu_{\tau R}).$$

This matrix can be non diagonal, if there exist lepton flavor violating interactions.

Anyway, if this is all there is, the above matrix can be diagonalized a la up quarks yielding the neutrino eigenstates and eigenmasses. In this case the mass eigenstates are of the Dirac type (the charge conjugate (antiparticle) is different from its parent state). This scenario is not accepted by most physicists, since it does not explain why the neutrinos are so much lighter than the other fermions. In fact in extensions of the standard model, like Grand Unified Theories, the neutrino mass ought to be as large as the up quark masses. In such theories as well as in string theories or theories in extra dimensions there also appear other types of mass as we will see immediately.

12.4.1 *The Majorana neutrino mass*

Since the neutrino is a neutral particle it is possible to consider mass terms, which do not involve the right handed neutrino, but the existing right handed antineutrino, i.e. of the type

$$\mathcal{L}_{mass} = m_{m_\nu} \bar{\nu}^c_R \nu_L .$$

These terms violate lepton number conservation, but such a conservation is the consequence of a global, not gauge symmetry and, thus, it is not a sacred law. In the current understanding of symmetries the above term cannot exist in this bare form but it must arise by a spontaneous symmetry breaking, i.e. out of a scalar acquiring a vacuum expectation value. In fact Weinberg [Weinberg (1980)] has shown that this can be achieved via two collaborating scalar isodoublets. The most economic way to achieve Majorana mass matrix is via the introduction of an isotriplet scalar $T^{I=1}_{I_3}$ particle, which couples to leptons with a Yukawa coupling. This isotriplet contains an electrically neutral component indicated as Δ^0, which can couple to neutrinos with a coupling f_{m_ν} as follows:

$$\mathcal{L}_{m_\nu} = f_{m_\nu} \bar{\nu}_L \mathcal{M}_{m_\nu} \nu^c_R \Delta^0 + HC. \tag{12.9}$$

If this acquires a vacuum expectation value $\langle \Delta^0 \rangle = \upsilon_T$, we get the Majorana mass matrix and the resulting eigenstates are Majorana neutrinos, i.e. the charge conjugate states coincides with the original, $\nu^C \propto \nu$:

$$\mathcal{L}_{m_\nu} = \bar{\nu}_L \mathcal{M}_{m_\nu} \nu^c_R + HC, \; \mathcal{M}_{m_\nu} = f_{m_\nu} \upsilon_T . \tag{12.10}$$

The above expression is the celebrated light neutrino Majorana mass matrix.

In the Weinberg model there is no elementary isotriplet, but the two isodoubles can couple to an effective isotriplet.

Again as in the case of the other fermion masses, such models cannot predict the numerical values of the entries of this matrix. One can arrange them to fit the data, if the vacuum expectation value is sufficiently small. This is all fine, except that neither the isotriplet has been found nor other effects attributed to it have yet been observed.

12.4.2 *The see-saw mechanism*

Once the right handed neutrinos have been introduced, one can construct another Majorana mass matrix of the form:

$$\mathcal{L}_{m_N} = \bar{\nu}_L^c \mathcal{M}_{m_N} \nu_R + HC, \quad \mathcal{M}_{m_N} = f_{m_N} \upsilon_S$$

where υ_S is the vacuum expectation value of an isosinglet scalar. Because this does not carry any SM quantum numbers, it is easy to introduce into the theory. In any case, this type of matrix in collaboration with the Dirac mass matrix we have discussed above, can yield an effective Majorana mass $(\mathcal{M}_{m_\nu})_{eff}$ term, producing a light neutrino, without the need of introducing an isotriplet or the Weinberg mechanism. The resulting 6×6 mass matrix is of the form:

$$\mathcal{L}_m = (\bar{\nu}_L \, \bar{\nu}_L^c) \begin{pmatrix} 0 & \mathcal{M}_D \\ \mathcal{M}_D^T & \mathcal{M}_{m_N} \end{pmatrix} \begin{pmatrix} \nu_R^c \\ \nu_R \end{pmatrix}. \tag{12.11}$$

One can assume that the Majorana mass involving the right-handed neutrino has eigenvalues, which are all much larger than the entries of the Dirac mass. Since the right handed neutrino has not been observed, it is natural to assume that it is too heavy. This way we obtain in perturbation theory an effective light Majorana mass of the form

$$(\mathcal{M}_{m_\nu})_{eff} = -\mathcal{M}_D^T \mathcal{M}_N^{-1} \mathcal{M}_D \tag{12.12}$$

similar in form with that obtained in the case of the isotriplet. In fact one can see that even if the Dirac mass entries are of the order of 100 MeV, the resulting neutrinos can be quite light, if the isosinglet neutrinos are extremely heavy. E.g.:

$$\frac{m_D^2}{M_N} = \frac{(100 \text{ MeV})^2}{10^{10} \text{ MeV}} = 10^{-6} \text{ MeV} = 1 \text{ eV},$$

which is of the order expected from neutrino oscillations data. This is the celebrated see saw mechanism of type I. The resulting eigenstates ν_j are necessarily Majorana particles, i.e. under charge conjugation $\nu_j^c = e^{-i\alpha_j}\nu_j$, $j = 1, 2, 3$, with α_j real parameters called Majorana phases.

The question of predicting the fermion masses remains open. Many efforts in this direction over the years have given some interesting models, but a real theory is missing.

The form of the unitary matrix U, which expresses the original weak eigenstates, the neutrino flavor states, in terms of the mass eigenstates is known as U^{PMNS} (Pontecorvo-Maki-Nakagawa-Sakata matrix). It is commonly parametrized in terms of three elementary rotations, one of which involves a phase δ, times a diagonal matrix of the Majorana phases, out of which only two are independent. It is expressed as follows:

$$U = \begin{pmatrix} 1 & 0 & 0 \\ 0 & c_{23} & s_{23} \\ 0 & -s_{23} & c_{23} \end{pmatrix} \begin{pmatrix} c_{13} & 0 & s_{13}e^{-i\delta} \\ 0 & 1 & 0 \\ -s_{13}e^{i\delta} & 0 & c_{13} \end{pmatrix} \begin{pmatrix} c_{12} & s_{12} & 0 \\ -s_{12} & c_{12} & 0 \\ 0 & 0 & 0 \end{pmatrix} \begin{pmatrix} 1 & 0 & 0 \\ & e^{i\alpha_{21}/2} & 0 \\ 0 & 0 & e^{i\alpha_{31}/2} \end{pmatrix}, \tag{12.13}$$

with

$$s_{ij} = \sin\theta_{ij}, \quad c_{ij} = \cos\theta_{ij}, \quad \alpha_{21} = \alpha_2 - \alpha_1, \quad \alpha_{31} = \alpha_3 - \alpha_1 \tag{12.14}$$

or

$$U = \begin{pmatrix} c_{12}c_{13} & c_{13}s_{12} & e^{i\delta}s_{13} \\ -c_{23}s_{12} - e^{-i\delta}c_{12}s_{13}s_{23} & c_{12}c_{23} - e^{-i\delta}s_{12}s_{13}s_{23} & c_{13}s_{23} \\ s_{12}s_{23} - e^{-i\delta}c_{12}c_{23}s_{13} & -e^{-i\delta}c_{23}s_{12}s_{13} - c_{12}s_{23} & c_{13}c_{23} \end{pmatrix}$$
$$\times \begin{pmatrix} 1 & 0 & 0 \\ & e^{i\alpha_{21}/2} & 0 \\ 0 & 0 & e^{i\alpha_{31}/2} \end{pmatrix}. \tag{12.15}$$

The Majorana phases do not appear in neutrino oscillations or in other lepton flavor changing processes. They appear only in lepton number violating processes, neutrinoless double beta decay etc.

12.5 Neutrino oscillations in the case of two generations

Let us consider two types of neutrinos ν_α and ν_β produced in weak interactions, say the electron and the muon neutrinos. Suppose, however, that these neutrinos are not eigenstates of the world Hamiltonian, in other words the Hamiltonian in flavor space is represented by a 2×2 non diagonal Hermitian matrix.

12.5.1 *Neutrino oscillations in the vacuum*

The eigenstates of the above Hamiltonian matrix are, say, ν_1 and ν_2, with masses m_1 and m_2 respectively. Then they can be expressed in terms of the original flavor states ν_α and ν_β i.e.

$$\nu_1 = \cos\theta\nu_\alpha + \sin\theta\nu_\beta, \quad \nu_2 = -\sin\theta\nu_\alpha + \cos\theta\nu_\beta$$

where the mixing angle θ depends on the interaction. We could as well have written:

$$\nu_\alpha = \cos\theta\nu_1 - \sin\theta\nu_2, \quad \nu_\beta = \sin\theta\nu_1 + \cos\theta\nu_2. \tag{12.16}$$

The above expressions hold say at $t = 0$.

Clearly these states are orthogonal, $\langle\nu_\alpha(0)|\nu_\beta(0)\rangle = 0$, but they are not stationary. The evolve according to the laws of quantum mechanics, e.g. at a later time:

$$\nu_\beta(t) = \sin\theta\nu_1 e^{-iE_1 t} + \cos\theta\nu_2 e^{-iE_2 t} \tag{12.17}$$

and

$$\langle\nu_\alpha(0)|\nu_\beta(t)\rangle = \cos\theta\sin\theta\left(e^{-iE_1 t} - e^{-iE_2 t}\right). \tag{12.18}$$

So these states are no longer orthogonal. Suppose we now define

$$\Delta_{12} = \frac{E_1 - E_2}{2}. \tag{12.19}$$

We get

$$\begin{aligned} P(\nu_\alpha(0) \to \nu_\beta(t)) &= |\langle\nu_\alpha(0)|\nu_\beta(t)\rangle|^2 = 2\cos\theta\sin\theta\,(1 - \cos(2\Delta_{12})) \\ &= \sin^2 2\theta \sin^2 \Delta_{12} t \end{aligned}$$

"appearance of ν_β" and

$$P(\nu_\alpha(0) \to \nu_\alpha(t)) = 1 - P(\nu_\alpha(0) \to \nu_\beta(t)) = 1 - \sin^2 2\theta \sin^2 \Delta_{12} t$$

"disappearance of ν_α".

Suppose that the beam is characterized by momentum p, then

$$\begin{aligned} \Delta_{12} &= \frac{1}{2}(E_1 - E_2) = \frac{1}{2}\left(\sqrt{p^2 + m_1^2} - \sqrt{p^2 + m_1^2}\right) \\ &\approx \frac{1}{2}\left(p\left(1 + \frac{m_1^2}{2p^2}\right) - p\left(1 + \frac{m_2^2}{2p^2}\right)\right) = \frac{m_1^2 - m_2^2}{4E}. \end{aligned}$$

We thus find:

$$\begin{aligned} P(\nu_\alpha(0) \to \nu_\beta(t)) &= \sin^2 2\theta \sin^2 \pi\frac{L}{\ell_{12}}, \\ P(\nu_\alpha(0) \to \nu_\alpha(t)) &= 1 - \sin^2 2\theta \sin^2 \pi\frac{L}{\ell_{12}}. \end{aligned} \tag{12.20}$$

With

$$\ell_{12} = \frac{4\pi E_\nu}{\Delta m_{12}^2}, \ \Delta m_{12}^2 = m_1^2 - m_2^2. \tag{12.21}$$

So a measurement at distance $L = ct$ will detect the appearance of a new neutrino with flavor β with amplitude $\sin^2 2\theta$ and an oscillation length ("period") ℓ_{12}. It will also detect a diminishing flux of neutrinos with flavor α (disappearance). The situation is exhibited in Fig. 12.1. Note that in the standard experiments one does not see the full waveform, but essentially one point if $\ell_{ij} \gg$ detector size.

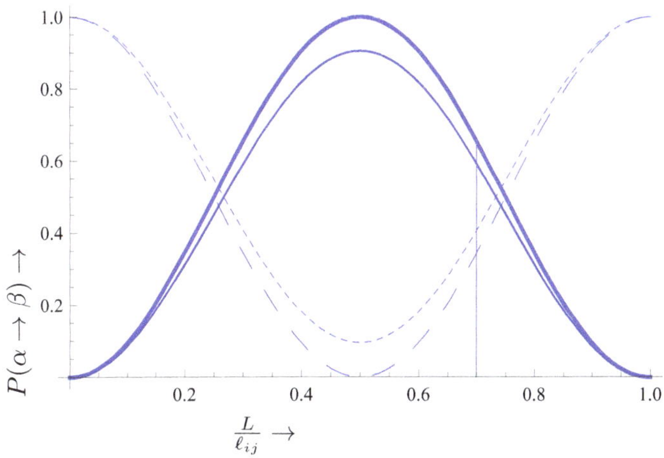

Fig. 12.1: Transition probability in the case of two flavors. The solid line describes the oscillation to one flavor, say β, from another one, say α. The thick line corresponds to mixing $\theta = \pi/4$, while the fine one to $\theta = \pi/5$. We will see that the first corresponds to atmospheric neutrinos, while the latter to solar neutrinos. With long (short) dashed line the neutrino disappearance is shown for $\theta = \pi/4$ ($\theta = \pi/5$) respectively. The intersection of the vertical line with the graphs indicates the values expected, if the detector is placed at $L = 0.7\ell_{ij}$. This value in the appearance case is, of course, above what is expected at $L = 0$, i.e. without neutrino oscillations.

The above treatment is known as neutrino oscillations in vacuum. Before proceeding further with such oscillations we will briefly consider neutrino oscillations in matter.

12.5.2 *Neutrino oscillations in matter*

The above quantum mechanical description can also be applied in the case of neutrino oscillations in matter, in the context of the WMS (Wolfenstein-Mikheyev-Smirnov) proposal, [Wolfenstein (1978)], [Mikheyev and Smirnov (1985)]. An in depth discussion of this matter is beyond the goals of the present book. The interested reader is referred to the literature, e.g. in a book devoted to the subject [Giunti and Kim (2007)].

For illustration purpose we will restrict our discussion in the case of only two generations. We start by recalling that in the case of vacuum oscillation, the time evolution of flavor eigenstates satisfies the Schrödinger equation:

$$i\frac{d}{dt}\begin{pmatrix}\nu_e\\ \nu_\mu\end{pmatrix} = H_v\begin{pmatrix}\nu_e\\ \nu_\mu\end{pmatrix}, \quad H_v = \frac{\Delta m_{21}^2}{E_\nu}\begin{pmatrix}-\cos 2\theta & \sin 2\theta\\ \sin 2\theta & \cos 2\theta\end{pmatrix}. \tag{12.22}$$

To see that this is equivalent with the procedure followed in the previous section we write the evolution equation for the mass eigenstates

$$i\frac{d}{dt}\begin{pmatrix}\nu_1\\ \nu_2\end{pmatrix} = \begin{pmatrix}E_1 & 0\\ 0 & E_2\end{pmatrix}\begin{pmatrix}\nu_1\\ \nu_2\end{pmatrix}. \tag{12.23}$$

Substituting in the above equation the mass eigenstates via the relationship:

$$\begin{pmatrix}\nu_1\\ \nu_2\end{pmatrix} = \begin{pmatrix}\cos\theta & \sin\theta\\ -\sin\theta & \cos\theta\end{pmatrix}\begin{pmatrix}\nu_e\\ \nu_\mu\end{pmatrix} \tag{12.24}$$

we obtain

$$\begin{pmatrix}\cos\theta & \sin\theta\\ -\sin\theta & \cos\theta\end{pmatrix} i\frac{d}{dt}\begin{pmatrix}\nu_e\\ \nu_\mu\end{pmatrix} = \begin{pmatrix}E_1 & 0\\ 0 & E_2\end{pmatrix}\begin{pmatrix}\cos\theta & \sin\theta\\ -\sin\theta & \cos\theta\end{pmatrix}\begin{pmatrix}\nu_e\\ \nu_\mu\end{pmatrix} \Rightarrow \tag{12.25}$$

$$i\frac{d}{dt}\begin{pmatrix}\nu_e\\ \nu_\mu\end{pmatrix} = \begin{pmatrix}\cos\theta & -\sin\theta\\ \sin\theta & \cos\theta\end{pmatrix}\begin{pmatrix}E_1 & 0\\ 0 & E_2\end{pmatrix}\begin{pmatrix}\cos\theta & \sin\theta\\ -\sin\theta & \cos\theta\end{pmatrix}\begin{pmatrix}\nu_e\\ \nu_\mu\end{pmatrix}. \tag{12.26}$$

The product of the above three matrices gives the matrix:

$$\begin{pmatrix}E_1 + E_2 + \frac{1}{2}\cos(2\theta)\,(E_1 - E_2) & \sin(2\theta)\,(E_1 - E_2)\,/2\\ \sin(2\theta)\,(E_1 - E_2)\,/2 & E_1 + E_2 - \frac{1}{2}\cos(2\theta)\,(E_1 - E_2)\end{pmatrix}.$$

Ignoring the irrelevant multiple $E_1 + E_2$ of the identity matrix and proceeding as in the previous section we find $E_1 - E_2 = \frac{m_1^2 - m_2^2}{2E_\nu} = -\frac{\Delta m_{21}^2}{2E_\nu}$ and the above matrix is reduced to the matrix entering Eq. (12.22).

We have judiciously chosen to write the evolution equation, Eq. (12.23), in flavor space, since it is in this basis that the neutrinos interact. When neutrinos travel through a dense medium (in the Sun or in the Earth, e.g.), their propagation can be significantly modified by the coherent forward scattering from particles they encounter along the way. As a result, the oscillation probability can be rather different than it is in vacuum. The flavor-changing mechanism in matter was named (MSW) after Mikheyev, Smirnov and Wolfenstein [Wolfenstein (1978)], [Mikheyev and Smirnov (1985)], who first pointed out that there is an interplay between flavor-non-changing neutrino-matter interactions and neutrino mass and mixing. The MSW effect stems from the fact that electron neutrinos (and antineutrinos) have different interactions with matter compared to other neutrino flavors. In particular, ν_e can have both charged current and neutral current elastic scattering with electrons, while ν_μ or ν_τ have only neutral current interactions with electrons, see, e.g., Fig. 7.13 of the companion book [Vergados (2017)]. This fact gives rise to an extra-potential [Giunti and Kim (2007)] of the form

$$V_e = \pm\sqrt{2}G_F n_e, \tag{12.27}$$

where n_e is the electron number density in matter, G_F is the Fermi constant, and the positive(negative) sign applies to electron neutrinos (antineutrinos).

Thus the matrix that regulates the neutrino propagation in matter is of the form:

$$H_m = \frac{\Delta m^2_{21}}{E_\nu}\begin{pmatrix} -\cos 2\theta & \sin 2\theta \\ \sin 2\theta & \cos 2\theta \end{pmatrix} + \begin{pmatrix} V_e & 0 \\ 0 & 0 \end{pmatrix}. \tag{12.28}$$

We find it convenient to sub tract $V_e/2$ times the identity matrix from this and rewrite it in the form:

$$H_m = \frac{\Delta m^2_{21}}{E_\nu}\begin{pmatrix} -\cos 2\theta + a & \sin 2\theta \\ \sin 2\theta & \cos 2\theta - a \end{pmatrix} \tag{12.29}$$

where

$$a = \frac{2\sqrt{2}G_F E n_e}{\Delta m^2_{21}}. \tag{12.30}$$

This matrix can be diagonalized with a unitary matrix with a mixing angle θ_m:

$$(1/2)\tan 2\theta_m = \frac{\sin 2\theta}{2(\cos 2\theta - a)} \Rightarrow \sin 2\theta_m = \frac{\sin 2\theta}{\sqrt{(\cos 2\theta - a)^2 + \sin^2 2\theta}} \tag{12.31}$$

while the eigenvalues are

$$\lambda = \pm\sqrt{a^2 - 2a\cos(2\theta) + 1}\frac{\Delta m^2_{21}}{4E_\nu} \tag{12.32}$$

which leads to:

$$\left(\Delta m_{21}^2\right)_m = \sqrt{a^2 - 2a\cos(2\theta) + 1}\Delta m_{21}^2. \tag{12.33}$$

The oscillation probability becomes:

$$P(\nu_e \to \nu_\mu) = \sin^2(2\theta_m)\sin^2\left(\frac{\left(\Delta m_{21}^2\right)_m L}{4E_\nu}\right). \tag{12.34}$$

The above expression looks quite simple, but, in fact, it is not. This is due to the fact that the parameter a, through the neutrino density, depends on the position in the sun. This makes the calculation complicated. Note, however, that

- The oscillation pattern of neutrinos is different from that of antineutrinos due to the fact that the corresponding parameter a being opposite, see Eqs. (12.31) and (12.33). This happens in this case without the need of CP violation required in vacuum oscillations, see Section 12.6 and the problem at the end of this chapter, Eq. (12.87).
- The condition of the resonance is:

$$a = \cos 2\theta \text{ (resonance condition)}. \tag{12.35}$$

  At the position of the resonance we find:

$$\sin^2(2\theta_m) = 1,\ \left(\Delta m_{21}^2\right)_m = \sin 2\theta \Delta m_{21}^2 \text{ (at the resonance)}. \tag{12.36}$$

  In other words the mixing parameter becomes maximal (unity) regardless of $\sin^2 2\theta$ in vacuum.
- If the angle θ is known to be in the interval $(-\pi/4, \pi/4)$ the resonance condition is satisfied only if $a > 0$. This, however, depends on the sign of Δm_{21}^2. Thus, in principle, neutrino oscillations can aid in settling the hierarchy problem (see Section 12.8). The reader should confirm the patterns exhibited in Fig. 12.11 at the end of this chapter.

One defines the length parameters

$$L_v = L_{12} = \frac{4\pi E_\nu}{\Delta m_{21}^2}, \text{ (oscillation length in vaumm)},$$

$$L_e = \frac{4\pi}{2\sqrt{2}G_F n_e} = ((\nu, e) \text{ interaction length}). \tag{12.37}$$

We thus find

$$\frac{L_v}{L_e} = \frac{2\sqrt{2}}{\Delta m_{21}^2} = a \Rightarrow L_v = L_e a. \tag{12.38}$$

In order for a neutrino to reach the resonance, Eq. (12.35), it is necessary:

$$L_v \geq L_e(\text{core})\cos 2\theta \Rightarrow (E_\nu)_{\text{min}} = \frac{\Delta m_{21}^2}{2\sqrt{2}G_F n_e(\text{core})}\cos 2\theta. \qquad (12.39)$$

The electron number density at the core of the sun is estimated to be 3×10^{31} m^{-3}. One can show that

$$\Delta m_{21}^2 = 8.0{\times}10^{-5}\,\text{eV}^2,\ \theta \approx \frac{\pi}{6}\ (\text{LMA-MSW solution}) \Rightarrow (E_\nu)_{\text{min}} = 1\,\text{MeV}. \qquad (12.40)$$

See problems in Section 12.11 at the end of the chapter, Eq. (12.86).

12.6 The three generation formalism

Let us assume that a neutrino flavor ν_α is produced at time $t = 0$. What is the probability to appear as flavor ν_β at a later time? We have

$$\nu_\alpha(t) = \sum_j U_{\alpha_j} e^{-iE_j t}|\nu_j\rangle, \quad \nu_\beta(0) = \sum_j U_{\beta_j}|\nu_j\rangle, \qquad (12.41)$$

with $\langle \nu_k|\nu_j\rangle = \delta_{ij}$. Thus we get:

$$P(\nu_\alpha \to \nu_\beta) = |\langle \nu_\beta(0)|\nu_\alpha(t)\rangle|^2, \quad \langle \nu_\beta(0)|\nu_\alpha(t)\rangle = \sum_j U_{\beta_j}^* U_{\alpha_j} e^{-iE_j t}, \qquad (12.42)$$

$$P(\nu_\alpha \to \nu_\beta) = |\langle \nu_\beta(0)|\nu_\alpha(t)\rangle|^2 = \sum_{j,k} U_{\beta_j}^* U_{\alpha_j} U_{\beta_k} U_{\alpha_k}^* e^{2i\Delta_{jk}L}, \qquad (12.43)$$

$$\Delta_{jk} = \frac{E_j - E_k}{2}, \quad L = ct. \qquad (12.44)$$

Noting that the states $\nu_\alpha(0)$ and $\nu_\beta(0)$ are orthogonal we get $\sum_{j,k} U_{\beta_j}^* U_{\alpha_j} U_{\beta_k} U_{\alpha_k}^* = \delta_{\alpha\beta}$, so that the above equation becomes:

$$P(\nu_\alpha \to \nu_\beta) = \delta_{\alpha\beta} - \sum_{j,k} U_{\beta_j}^* U_{\alpha_j} U_{\beta_k} U_{\alpha_k}^* (1 - e^{-2i\Delta_{jk}L}), \qquad (12.45)$$

which can be written as

$$P(\nu_\alpha \to \nu_\beta) = \delta_{\alpha\beta} - 2\text{Re}\left\{\sum_{j<k} U_{\beta_j}^* U_{\alpha_j} U_{\beta_k} U_{\alpha_k}^* (1 - e^{-2i\Delta_{jk}L})\right\}. \qquad (12.46)$$

For light neutrinos $E_j = \sqrt{p^2 + m_j^2} \approx p + m_j^2/2p$ with the momentum p independent of j. Thus

$$\Delta_{jk}L \approx \frac{m_j^2 - m_k^2}{4E_\nu}L = \pi\frac{L}{\ell_{jk}}, \quad \ell_{jk} = \frac{4\pi E_\nu}{m_j^2 - m_k^2} \qquad (12.47)$$

or

$$P(\nu_\alpha \to \nu_\beta) = \delta_{\alpha\beta} - 4\mathrm{Re}\left\{\sum_{j<k} U^*_{\beta_j} U_{\alpha_j} U_{\beta_k} U^*_{\alpha_k} \sin^2 \pi \frac{L}{\ell_{jk}}\right\}$$
$$-2\ \mathrm{Im}\left\{\sum_{j<k} U^*_{\beta_j} U_{\alpha_j} U_{\beta_k} U^*_{\alpha_k} \sin\left(2\pi \frac{L}{\ell_{jk}}\right)\right\}. \quad (12.48)$$

Note, however, that if the phase δ of the standard U_{PMNS} matrix vanishes we have:

$$\mathrm{Im}\left\{\sum_{j<k} U^*_{\beta_j} U_{\alpha_j} U_{\beta_k} U^*_{\alpha_k}\right\} = 0. \quad (12.49)$$

We should also mention that this term comes with opposite sign in the case of antineutrinos. Ignoring this term we get:

$$P(\nu_\alpha \to \nu_\beta) = \delta_{\alpha\beta} - 4\mathrm{Re}\left\{\sum_{j<k} U^*_{\beta_j} U_{\alpha_j} U_{\beta_k} U^*_{\alpha_k} \sin^2 \pi \frac{L}{\ell_{jk}}\right\}, \quad (12.50)$$

with ℓ_{jk} conveniently written as:

$$\ell_{jk} = 2.476\ \mathrm{km}\frac{E_\nu/1\ \mathrm{GeV}}{|\Delta m^2_{jk}|/1\ \mathrm{eV}^2} = 2.476\ \mathrm{m}\frac{E_\nu/1\ \mathrm{MeV}}{|\Delta m^2_{jk}|/1\ \mathrm{eV}^2}$$
$$= 2.476\ \mathrm{m}\frac{E_\nu/1\ \mathrm{keV}}{|\Delta m^2_{jk}|/10^{-3}\ \mathrm{eV}^2}. \quad (12.51)$$

In the special case of two generations we recover the formulas obtained above. Note that there exist 3 values of $|\Delta m^2_{jk}|$, with two of them independent. If these 3 values are different, one can isolate only one mode by a judicious choice of the source detector distance L. One can thus treat it as a two generation problem with an effective mixing angle θ_{eff}.

In the presence of three generations in the case of ν_e disappearance we get

$$P(\nu_e \to \nu_e) = 1 - 4\left[\cos^4\theta_{13}\cos^2\theta_{12}\sin^2\theta_{12}\sin^2\left(\pi\frac{L}{\ell_{21}}\right)\right.$$
$$\cos^2\theta_{13}\sin^2\theta_{12}\sin^2\theta_{13}\sin^2\left(\pi\frac{L}{\ell_{32}}\right)$$
$$\left.\cos^2\theta_{13}\cos^2\theta_{12}\sin^2\theta_{13}\sin^2\left(\pi\frac{L}{\ell_{31}}\right)\right]. \quad (12.52)$$

In the special case of $\ell_{31} \approx \ell_{32}$ and $\theta_{13} << 1$, we find

$$P(\nu_e \to \nu_e) = 1 - \left[\sin^2 2\theta_{12} \sin^2\left(\pi\frac{L}{\ell_{21}}\right) + \sin^2 2\theta_{13} \sin^2\left(\pi\frac{L}{\ell_{32}}\right)\right], \tag{12.53}$$

i.e. we have to deal with the superposition of two oscillations, one due to θ_{12} and the other due to θ_{13}.

For three generations we find that some useful relations which are going to be useful in the study of CP violation (see the problem in Section 12.11, Eq. (12.87)).

12.7 Neutrino oscillation experiments

Depending on the neutrino source we have the following possibilities:

- Solar neutrinos.
  They are neutrinos ν_e originating from the sun during the proton fusion. Since their energy is known their oscillation can be detected as neutrino disappearance provided that the original flux is known.
- Reactor neutrinos.
  These are primarily anti-neutrinos produced in β-decay. Their oscillation cannot be detected by transition to other flavors, since their energy is low.
- Accelerator neutrinos.
  In this case all neutrino flavors are produced. Transitions to all flavors can be detected.
- The atmospheric neutrinos.
  These are produced in the atmosphere from cosmic rays producing π^+ and π^-, which eventually produce neutrinos.

$$\pi^- \to \mu^- + \tilde{\nu}_\mu, \quad \mu^- \to e^- + \nu_\mu + \bar{\nu}_e, \tag{12.54}$$

$$\pi^+ \to \mu^+ + \nu_\mu, \quad \mu^+ \to e^+ + \tilde{\nu}_\mu + \nu_e. \tag{12.55}$$

It is seen from the last expression that the number of muon type to electron type neutrinos is 2:1. What is the ratio in the laboratory? Note that the ratio of the initial neutrino flux is adequate.

12.7.0.1 *The atmospheric neutrinos*

The following indication were chronologically observed:

(1) A difference in the muon neutrino to electron neutrino ratio. The measurement of the ratio:

$$R_{\mu,e} = \frac{N_\mu}{N_e}, \tag{12.56}$$

indicated a value deviating from the value 2. If one defines:

$$fR_{\mu,e} = \frac{R_{\mu,e}\ (\text{experiment})}{R_{\mu,e}\ (\text{no oscillation})}, \tag{12.57}$$

one expects in the absence of oscillations this to be unity, but it was found that [Fukuda *et al.* (1998a)]

$$fR_{\mu,e} = 0.63 \pm .03\ (\text{statistical}) \pm .05\ (\text{systematic}). \tag{12.58}$$

(2) A difference in the behavior of neutrinos, depending on whether they were coming from above (near zenith) or below (from the opposite side of the Earth), see Fig. 12.2. This could not be attributed to the magnetic field of the Earth, since the neutrino energy was high.

We define the asymmetry:

$$A = \frac{N\ (\text{above}) - N\ (\text{below})}{N\ (\text{below}) + N\ (\text{above})}. \tag{12.59}$$

Then, after a detailed analysis of the data [Fukuda *et al.* (1998b)], the obtained results are exhibited in Fig. 12.3. From the data of Fig. 12.3, one finds $A = 0.296 \pm 0.048$ (statistical) ± 0.01 (systematic). Also one finds:

$$\theta \approx \frac{\pi}{4}, \quad \Delta m^2 \approx 2.5 \times 10^{-3}\ (\text{eV/c}^2)^2. \tag{12.60}$$

Furthermore we deal with an oscillation of the type $\nu_\mu \to \nu_\tau$.

12.7.0.2 *Solar neutrinos*

The production mechanism and energies of the neutrinos originating from the sun have been discussed in Section 7.2.2. Also the difficulties in obtaining the neutrino fluxes have been outlined. For the readers convenience we present here again the spectra of neutrino fluxes on earth, see Fig. 12.4. In the following we will discuss the main experimental efforts concerned with neutrino detection.

The detection of solar neutrino oscillations is extremely difficult for the following reasons:

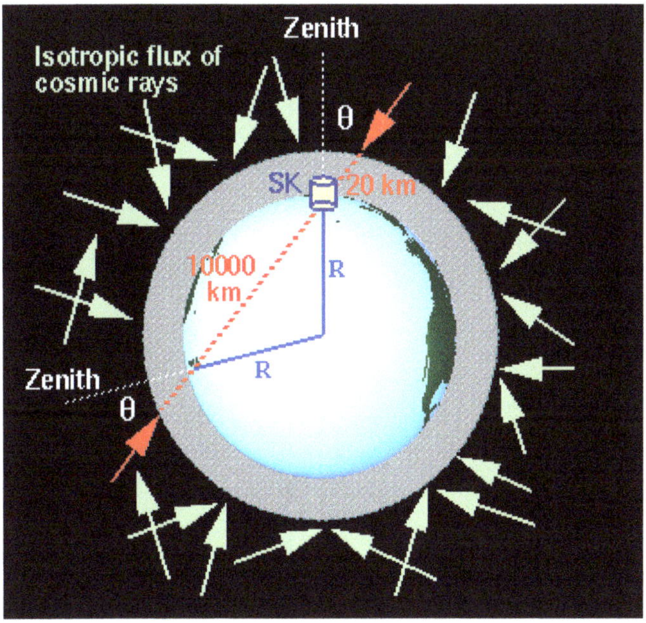

Fig. 12.2: There is no preferred direction for neutrinos from cosmic rays above 10 GeV, since the Earth's magnetic field is negligible.

- The energy of neutrinos with large flux is extremely low, i.e. below the energy threshold of the standard detectors. On the other hand the flux of the higher energy neutrinos is not adequate to give detectable events.
- The oscillation, manifested as a reduction of the neutrino flux in the detector, presupposes knowledge of the corresponding flux in the surface of the sun. This cannot be measured, but only estimated based on an assumed solar model. The existing solar models, even though they were considered otherwise reliable, could not exclude the possibility of an overestimate of the neutrino flux.
- The neutrinos traverse a long distance before reaching Earth. The oscillation could take place on their way to earth or inside the sun. The description of the oscillation inside the sun is quite complicated, due to the presence of electrons, whose distribution also depends on the solar model. This very interesting scenario, however, is not going to be discussed here in greater detailed than that already given in Section 12.5.2.

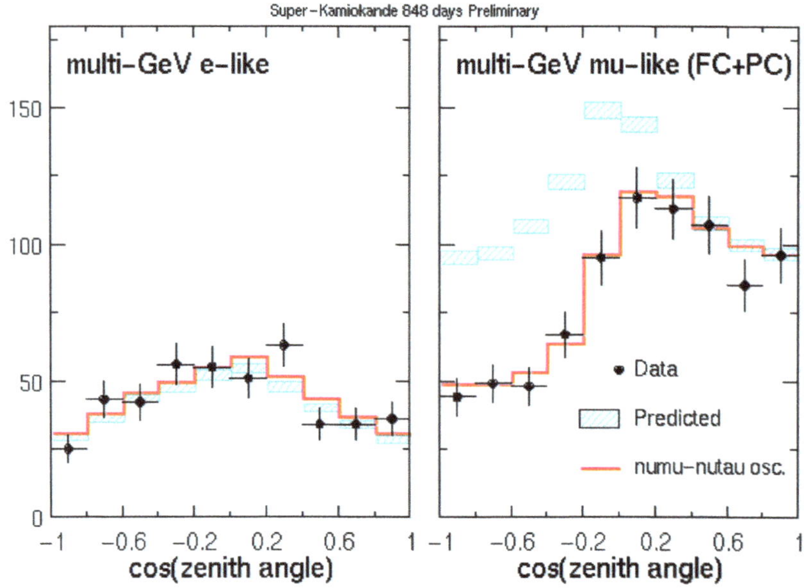

Fig. 12.3: The distribution of muon-like (on the left) and electron like events (on the right) exhibits a characteristic asymmetry. Note that $\cos\theta_{zenith} < 0$ corresponds to upward events.

For all these reasons the question of solar neutrino oscillations was finally resolved with the aid of reactor neutrino experiments. The most important stages are:

(1) Solar neutrinos

- Cl experiment, which began in 1968 known as Davies experiments:

$$\nu_e +^{37} C\ell \longrightarrow e^- +^{37} Ar, (E_{th} = 0.814 \text{ MeV}).$$

It is remarkable that Davies was able to extract a dosen of Ar atoms from tons of $CC\ell_4$.

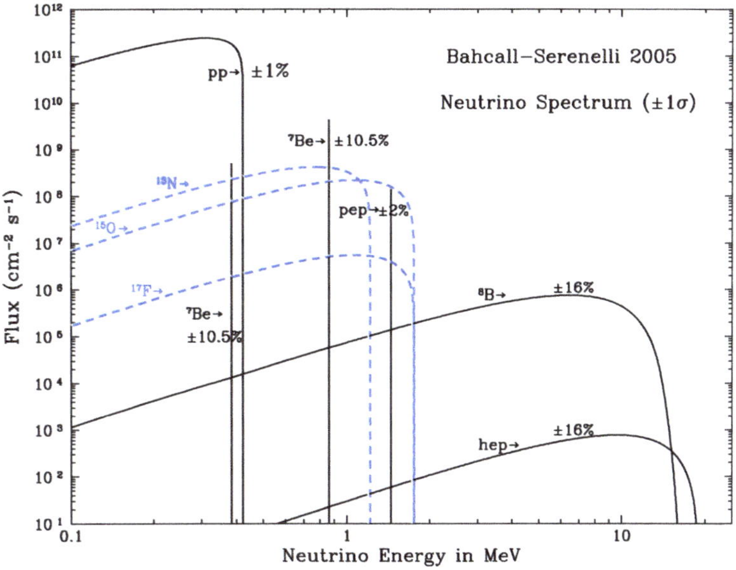

Fig. 12.4: The spectrum of solar neutrinos.

- The Gallium experiment, GNO, SAGE, GALLEX:

$$\nu_e + \ ^{71}Ga \longrightarrow e^- + \ ^{71}Ge, (E_{th} = 0.523 \text{ MeV}).$$

It involved 1000 tons D_2O, $E_\nu \geq 5$ MeV utilizing the reactions:

$$\nu_e + d \longrightarrow e^- + p + p \text{ (cc)},$$

$$\nu_\alpha + d \longrightarrow \nu_\alpha + p + n, \ (\alpha = e, \mu, \tau), \ \text{(nc)}$$

followed by: $n + p \longrightarrow d + \gamma$. Also the reaction:

$$\nu_\alpha + e^- \longrightarrow \nu_\alpha + e^-, \quad (\alpha = e, \mu, \tau) \text{ (ES)}.$$

In neutrino electron scattering (ES) all electrons contribute, but the electron neutrinos ν_e dominate. For the the boron neutrinos (see Fig. 12.4) it was found that

$$\Phi^{cc}_{SNO}(\nu_e) = (1.68^{+0.08}_{-0.09} \pm 0.015) \times 10^6 \text{ cm}^2 \text{ s}^{-1},$$

$$\Phi^{ES}_{SNO}(\nu_x) = (2.35 \pm 0.015 \pm 0.012) \times 10^6 \text{ cm}^2 \text{ s}^{-1},$$

$$\Phi^{nc}_{SNO}(\nu_x) = (5.2^{+0.7}_{-0.8}) \times 10^6 \text{ cm}^2 \text{ s}^{-1} \text{ neutrinos}.$$

Clearly, since the nc process is insensitive to neutrino oscillations, this gives us a measure of the initial flux, which was measured this way! This value was consistent with the predictions of the standard solar model (SSM) as well as the particle standard model (GSW):

$$\Phi^{nc}_{SSM} = 5.7 \times 10^6 \text{ cm}^2\text{s}^{-1} \text{ neutrinos.}$$

The cc process gives a measure of neutrino disappearance via:

$$\Phi^{cc}_{SNO} = \langle P_{ee}\rangle \Phi^{nc}_{SNO}, \quad \langle P_{ee}\rangle \approx 0.34^{0.07}_{-0.06}. \tag{12.61}$$

Furthermore

$$\Phi^{ES}_{SNO} = \left[\langle P_{ee}\rangle \Phi^{nc}_{SNO} + r_\sigma(1 - \langle P_{ee}\rangle)\right] \Phi^{nc}_{SNO}, \quad r_\sigma \approx 0.15, \tag{12.62}$$

where $r_\sigma$is the contribution of ν_μ and ν_τ.

(2) The reactor experiments, KamLAND.

$$\tilde{\nu}_e + p \longrightarrow e^+ + n, (E_{th} = 1.8 \text{ MeV}),$$

followed by: $n + p \longrightarrow d + \gamma$. Here, of course, anti-neutrinos are involved, but, as far as oscillations are concerned, their behavior is the same. These experiments [Araki *et al.* (2005)] not only removed the ambiguities of solar neutrinos, but they also improved the accuracy of the atmospheric neutrino measurements, yielding

$$\tan^2\theta \approx 0.45, \quad \Delta m^2 \approx (5.0 - 7.5) \times 10^{-5} \,(\text{eV}/\text{c}^2)^2,$$
$$\theta \approx \theta_{12}, \, \Delta m^2 = \Delta m^2_{12}. \tag{12.63}$$

Even though these experiments are ν_e disappearance experiments, a more detailed analysis reveals that it is an oscillation of the type $\nu_e \to \nu_\mu$.

(3) The NOSTOS experiment [Vergados *et al.* (2010)], [Vergados *et al.* (2012b)].

This employs monochromatic neutrinos resulting from electron capture, with energies around 10 keV which subsequently interact with electrons:

$$\nu_e + e^-_{\text{bound}} \longrightarrow \nu_e + e^-.$$

The evaluation of the oscillation probability is a bit tricky, since all neutrino flavors interact with electrons. We know that in the context of the standard model [Vergados (2017)] the ν_e interact both via the charged and the neutral current, while ν_μ and ν_τ do so only via the neutral current. Thus the total electron neutrino cross section in the presence of oscillations is given by:

$$\sigma_{\text{tot}} = P(\nu_e \to \nu_e)\sigma + \sum_{\alpha=\mu,\tau} P(\nu_e \to \nu_\alpha)\sigma'$$
$$= \sigma - (\sigma - \sigma') \sum_{\alpha=\mu,\tau} P(\nu_e \to \nu_\alpha), \tag{12.64}$$

where the last equation followed from unitarity. This equation can be rewritten as

$$\sigma_{\text{tot}} = \sigma \left\{1 - \left(1 - \frac{\sigma'}{\sigma}\right) \sum_{\alpha=\mu,\tau} P(\nu_e \to \nu_\alpha)\right\}$$
$$= \sigma \left\{1 - \left(1 - \frac{\sigma'}{\sigma}\right) \left[\sin^2 2\theta_{12} \sin^2\left(\pi \frac{L}{\ell_{21}}\right) + \sin^2 2\theta_{13} \sin^2\left(\pi \frac{L}{\ell_{32}}\right)\right]\right\}. \tag{12.65}$$

In the last step we have used Eq. (12.53). Thus the effective oscillation probability takes the form:

$$P(\nu_e \to \nu_e) = 1 - \chi(E_\nu) \left[\sin^2 2\theta_{12} \sin^2\left(\pi \frac{L}{\ell_{21}}\right) + \sin^2 2\theta_{13} \sin^2\left(\pi \frac{L}{\ell_{32}}\right)\right],$$
$$\chi(E_\nu) = 1 - \frac{\sigma'}{\sigma}. \tag{12.66}$$

The cross sections are functions of the neutrino energy. We see that, if $\sigma' = \sigma$, there is no oscillation, reminiscent of the absence of neutrino oscillations induced by the hadronic neutral current.

Since in such experiments the oscillation length ℓ_{32} can be a few meters, one can see the full oscillation or a major part of it (see Fig. 12.5), with a spherical TPC detector of radius about 5 meters, as opposed to the standard experiments, which can only see one point (indicated by a vertical bar in Fig. 12.1).

We note that in the reactor experiments one can also observe not only the strong oscillation due to the large angle θ_{12}, but also the weak one with the small oscillation length associated with the small angle θ_{13} (see Eq. (12.53)).

In summary with the discovery of neutrino oscillations quite a lot of information regarding the neutrino sector has become available (e.g., for a recent review see [Vergados *et al.* (2012a)]). More specifically we know:

- The mixing angles θ_{12}, θ_{23} and θ_{13}.

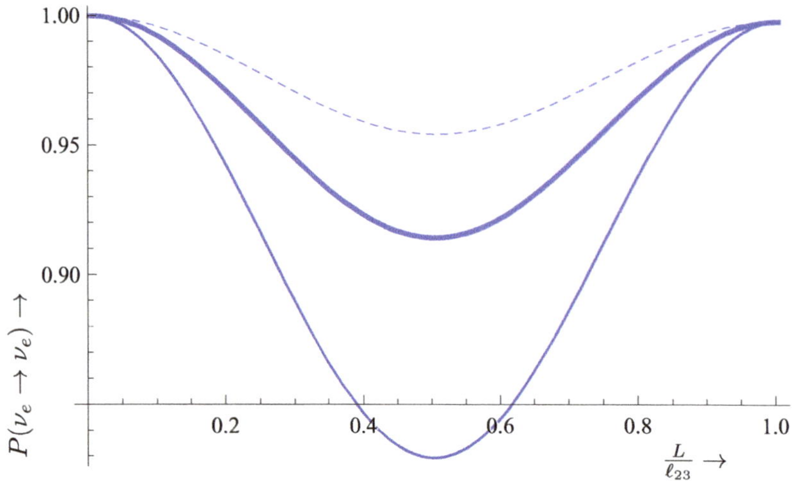

Fig. 12.5: The oscillation probability for the disappearance of electron neutrinos due to the small mixing angle θ_{13}. ℓ_{23} is the small oscillation length, which corresponds to the large mass squared difference Δm^2_{23}. The fine solid, thick solid and dashed lines correspond to $\sin^2 2\theta_{13} = 0.170, 0.085$ and 0.045. At the time the experiment was designed, $\sin^2 2\theta_{13}$ was not known. The values employed here are consistent what was found later (see text). The source is at the center of the spherical detector.

- The mass squared differences:
  $$\Delta^2_{\text{SUN}} = \Delta^2_{12} = m_2^2 - m_1^2 \quad \text{and } \Delta^2_{\text{ATM}} = |\Delta^2_{23}| = |m_3^2 - m_2^2|,$$
  entering the solar and atmospheric neutrino oscillation experiments. Note that we do not know the absolute scale of the neutrino mass and the sign of Δ^2_{23}.

Neutrino oscillation experiments cannot, of course, determine the absolute scale of the neutrino mass. The relevant neutrino oscillation parameters extracted are: the MINOS value $\Delta m^2_{\text{ATM}} = (2.43 \pm 0.13) \times 10^{-3}$ eV^2, the global fit value $\Delta m^2_{\text{SUN}} = (7.65^{+0.13}_{-0.20}) \times 10^{-5}$ eV^2, the solar-KamLAND value $\tan^2\theta_{12} = 0.452^{+0.035}_{-0.033}$ and the recent Daya Bay [An *et al.* (2012)] observation $\sin^2 2\theta_{13} = 0.092 \pm 0.016$ (stat) ± 0.005 (syst) with a significance of 5.2 standard deviations. We note that non-zero value of mixing angle θ_{13} was already observed also by the T2K ($0.04 < \sin^2 2\theta_{13} < 0.34$), the DOUBLE CHOOZ ($\sin^2 2\theta_{13} = 0.085 \pm 0.029$ (stat) ± 0.042 (syst) (68% CL)) and RENO ($\sin^2 2\theta_{13} = 0.103 \pm 0.013$ (stat) ± 0.011 (syst)) collaborations.

Table 12.1: Results of the global 3ν oscillation analysis, in terms of best-fit values and allowed 1, 2 and 3σ ranges for the 3ν mass-mixing parameters. We remind that Δm^2 is defined herein as $m_3^2 - (m_1^2 + m_2^2)/2$, with $+\Delta m^2$ for Normal Hierarchy (NH) and $-\Delta m^2$ for Inverted Hierarchy (IH). The notion of NH and IH is given below in the text.

Parameter	Best fit	1σ range	2σ range	3σ range
$\delta m^2/10^{-5}$ eV^2(NH-IH)	7.54	7.32 – 7.80	7.15 – 8.00	6.99 – 8.18
$\sin^2\theta_{12}/10^{-1}$ (NH-IH)	3.08	2.91 – 3.25	2.75 – 3.42	2.59 – 3.59
$\Delta m^2/10^{-3}$ eV^2 (NH)	2.43	2.37 – 2.49	2.30 – 2.55	2.23 – 2.61
$\Delta m^2/10^{-3}$ eV^2 (IH)	2.38	2.32 – 2.44	2.25 – 2.50	2.19 – 2.56
$\sin^2\theta_{13}/10^{-2}$ (NH)	2.34	2.15 – 2.54	1.95 – 2.74	1.76 – 2.95
$\sin^2\theta_{13}/10^{-2}$ (IH)	2.40	2.18 – 2.59	1.98 – 2.79	1.78 – 2.98
$\sin^2\theta_{23}/10^{-1}$ (NH)	4.37	4.14 – 4.70	3.93 – 5.52	3.74 – 6.26
$\sin^2\theta_{23}/10^{-1}$ (IH)	4.55	4.24 – 5.94	4.00 – 6.20	3.80 – 6.41

An analysis of the 3-generation data is given in Table 12.1 (see Capozzi *et al.* [Capozzi *et al.* (2014)]). For a more recent update see Esteban *et al.* [Esteban *et al.* (2019)]. Finally, thanks to the combination of the latest accelerator and reactor neutrino data, we have obtained an enhanced sensitivity to the CP violation phase. The preferred values for δ are around 1.5π for both mass hierarchies [Forero *et al.* (2014)]. On the other hand, values close to 0.5π are disfavored at $1.8\sigma(2.5\sigma)$ for normal (inverted) mass ordering.

At this point we would like to exhibit the essential features of the K-M (Kobayashi-Maskawa) charged mixing matrix for hadrons, which is almost diagonal, and the PNMS (Pontecorvo-Maki-Nakagawa-Sakata) matrix for leptons, which is characterized by large off diagonal elements.

$$\text{K-M} \approx \begin{pmatrix} 0.975 & 0.221 & 0.003 \\ 0.221 & 0.975 & 0.040 \\ 0.009 & 0.039 & 0.999 \end{pmatrix}, \text{PNMS} \approx \begin{pmatrix} 0.82 & 0.55 & 0.15 \\ -0.49 & 0.52 & 0.70 \\ 0.30 & -0.65 & 0.70 \end{pmatrix}. \quad (12.67)$$

Theoretically the matrix PNMS is not very well understood. It involves physics beyond the standard model. Horizontal discreet symmetries predict, to leading order, the tri-bimaxial mixing matrix:

$$\text{PMNS} \approx \begin{pmatrix} -2/\sqrt{6} & 1/\sqrt{3} & 0 \\ 1/\sqrt{6} & 1/\sqrt{3} & 1/\sqrt{2} \\ 1/\sqrt{6} & 1/\sqrt{3} & -1/\sqrt{2} \end{pmatrix}. \quad (12.68)$$

12.8 The hierarchy problem

Based on the above we find that neutrino oscillations cannot determine the absolute scale of the neutrino mass. They cannot determine the sign of the mass squared differences. Of the three mass squared differences only two are independent since we have a relation between them have, e.g.

$$\Delta_{13}^2 = \Delta_{12}^2 + \Delta_{23}^2$$

Δ_{12}^2 can be chosen to be positive. The sign of Δ_{23}^2 can be determined by other experiments, e.g. by neutrino oscillations in vacuum, see Section 12.5.2 or by measuring the difference in the appearance oscillation probability between neutrinos and antineutrinos. In the latter case we note that only the last term in Eq. (12.48) survives, which involves the imaginary part. The dependence of the sinusodial part of the oscillation length involves the mass squared differences linearly. For the moment this type of information is not available. So the following scenarios emerge:

- Normal Spectrum (NS), $m_1 < m_2 < m_3$:

$$\Delta m^2_{\text{SUN}} = m_2^2 - m_1^2 \;,\; \Delta m^2_{\text{ATM}} = m_3^2 - m_1^2,$$

$$m_0 = m_1,\; m_2 = \sqrt{\Delta m^2_{\text{SUN}} + m_0^2}\;,\; m_3 = \sqrt{\Delta m^2_{\text{ATM}} + m_0^2}.$$

- Inverted Spectrum (IS), $m_3 < m_1 < m_2$:

$$\Delta m^2_{\text{SUN}} = m_2^2 - m_1^2 \;,\; \Delta m^2_{\text{ATM}} = m_2^2 - m_3^2,$$

$$m_0 = m_3,\; m_2 = \sqrt{\Delta m^2_{\text{ATM}} + m_0^2},$$

$$m_1 = \sqrt{\Delta m^2_{\text{ATM}} - \Delta m^2_{\text{SUN}} + m_0^2}.$$

The flavor content and the order of the eigenstates is exhibited in Fig. 12.6. We note in the normal hierarchy (normal order) the lowest eigenstate is mostly electron neutrino, following the hierarchy of the charged lepton partners. On the other hand in the invented hierarchy (inverted order) scenario, the lowest eigenstate contains very little electron neutrino.

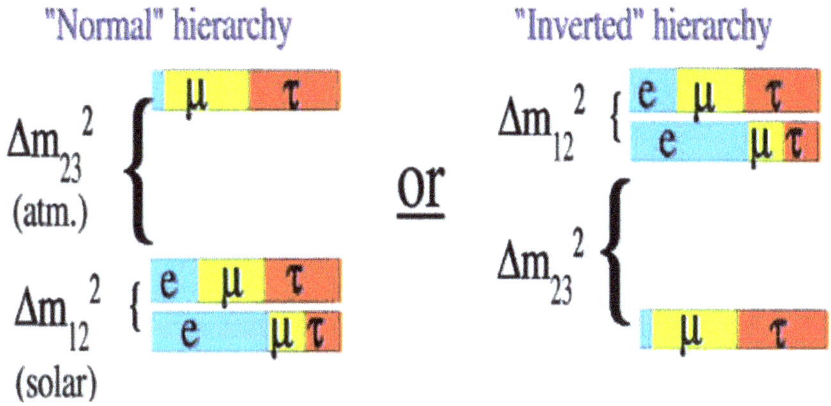

Fig. 12.6: The ordering and the flavor content of the neutrino mass eigenstates. The absolute scale, m_{ν_1} in the normal hierarchy and m_{ν_3} in the inverted one, is not known.

12.9 The absolute scale of the neutrino mass

For determination of an absolute scale of the neutrino mass the relevant neutrino oscillation parameters are summarized in the review [Vergados *et al.* (2012a)] with references to the original sources given there. These are: the MINOS value $\Delta m^2_{\text{ATM}} = (2.43 \pm 0.13) \times 10^{-3}$ eV^2, the global fit value $\Delta m^2_{\text{SUN}} = (7.65^{+0.13}_{-0.20}) \times 10^{-5}$ eV^2, the solar-KamLAND value $\tan^2\theta_{12} = 0.452^{+0.035}_{-0.033}$ and the recent Daya Bay observation $\sin^2 2\theta_{13} = 0.092 \pm 0.016$ (stat) ± 0.005 (syst) with a significance of 5.2 standard deviations. We note that non-zero value of mixing angle θ_{13} was already observed also by the T2K ($0.04 < \sin^2 2\theta_{13} < 0.34$). The absolute scale m_0 of neutrino mass can in principle be determined by the following observations:

- Neutrinoless double beta decay.
  There exist nuclei, which are absolutely stable against ordinary weak decays, but they are unstable against double beta decay (see Fig. 12.7). This can occur with the emission of two neutrinos $2\nu\beta\beta$ decay, which is not exotic, but a very slow second order weak interaction, and has, in fact, been observed in all nuclear systems allowing it, with life times in the range $10^{19} - 10^{24}$ y (the longest time ever measured). If, however, there exist lepton number violating interactions and the neutrinos happen to be Majorana particles

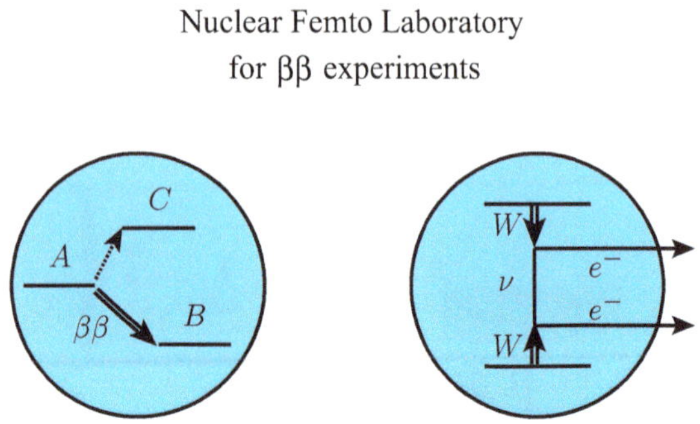

Fig. 12.7: Schematic diagrams of $\beta\beta$ decays in nuclear femto (10^{-15} m) laboratories, where single β-decay is not allowed but Double Beta Decay (DBD) may occur. Left-hand side: DBD decay scheme. Right-hand side: Neutrinoless DBD, $0\nu\beta\beta$ decay, with a virtual Majorana ν exchanged between two nucleons.

with an appreciable mass, it can also proceed without emission of neutrinos, $0\nu\beta\beta$ decay as shown in Fig. 12.8. It is somewhat favored kinematically against the standard $2\nu\beta\beta$ decay. It is suppressed by the smallness of the effective Majorana neutrino mass [Vergados *et al.* (2012a)].

The effective light neutrino mass $\langle m_\nu \rangle$ extracted in such experiments is given as follows [Vergados *et al.* (2012a)]:

$$\begin{aligned}\langle m_\nu \rangle &= \sum_k^3 (U_{ek})^2 \, m_k \\ &= c_{12}^2 c_{13}^2 m_1 + c_{13}^2 s_{12}^2 e^{i\alpha_{21}} m_2 + s_{13}^2 e^{i\alpha_{31}} m_3. \quad (12.69)\end{aligned}$$

The relative phases, $\alpha_{31} = 2\,(\alpha_3 - \alpha_1 + \delta)$, $\alpha_{21} = 2\,(\alpha_2 - \alpha_1)$, result from both the Majorana phases α_k, $k = 1, 2, 3$ of the neutrino and the CP violating phase of the matrix U_{PMNS}.

There exists some evidence that there exist neutrinos which do not participate in weak interactions, e.g. right handed neutrinos, which are called sterile neutrinos. These neutrinos, needed to explain the reactor neutrino anomaly as well as the short-baseline neutrino oscillation anomalies [Gariazzo *et al.* (2016)], may be light,

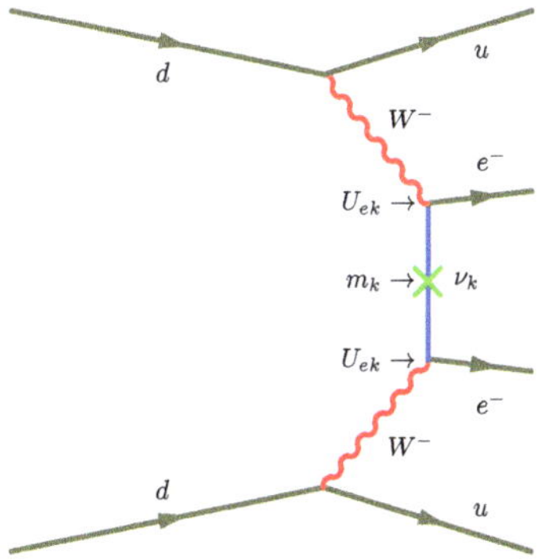

Fig. 12.8: A Feynman diagram leading to neutrinoless double beta decay at the quark level. The chiralites are such that only the mass term in the propagator $\approx \left(m_k + \gamma_\lambda p_\nu^\lambda\right)/p_\nu^2$ contributes, weighted by the square of the amplitude for neutrino production, i.e. U_{ek}^2, with the Majorana phases absorbed in it.

i.e. of the order of few eV to keV in mass, and may couple to the ordinary electron neutrino with relatively large amplitude. Thus they can contribute to the above light neutrino effective mass appearing in neutrinoless double beta decay, i.e.

$$\begin{aligned}\langle m_\nu \rangle &= \sum_k^4 (U_{ek})^2\, m_k \\ &= C_{14}^2\left(c_{12}^2 c_{13}^2 m_1 + c_{13}^2 s_{12}^2 e^{i\alpha_{21}} m_2 + s_{13}^2 e^{i\alpha_{31}} m_3\right) \\ &\quad + m_4 s_{14}^2 e^{i\alpha_{41}}. \end{aligned} \tag{12.70}$$

Since the mass m_4 is much heavier than the other three masses, even with a coupling of the order of $s_{14}^2 = 10^{-2} - 10^{-3}$ it will make a large contribution to the above matrix element. We are not, however, going to include the sterile neutrino contribution in our analysis.

- The neutrino mass extracted from ordinary beta decay, see, e.g., Section 13, is given by:

$$\begin{aligned}\langle m_\nu \rangle_{\text{decay}} &= \sqrt{\sum_k^3 |U_{ek}|^2 m_k^2} \\ &= \sqrt{c_{12}^2 c_{13}^2 m_1^2 + c_{13}^2 s_{12}^2 m_2^2 + s_{13}^2 m_3^2}, \end{aligned} \tag{12.71}$$

assuming, of course, that the three neutrino states cannot be resolved.

For a review of such experiments (KATRIN, MARE, ECHO, Mainz and Troitsk etc.) see [Drexlin *et al.* (2013)]. The most promising is the KATRIN triton decay experiment [Bohn *et al.* (1999)].

- From astrophysical and cosmological observations (see, e.g., the recent summary [Vergados *et al.* (2012a)]).

$$m_\nu = \sum_{k}^{3} m_k \leq m_{\text{astro}}. \tag{12.72}$$

The current limit m_{astro} depends on the type of observation. Thus CMB (Cosmic Microwave Background)primordial gives 1.3 eV, CMB+distance 0.58 eV, galaxy distribution and and lensing of galaxies 0.6 eV. On the other hand the largest photometric red shift survey yields 0.28 eV, see the review [Vergados *et al.* (2012a)] and references to the original work there in. For purposes of illustration we will take a world average of $m_{\text{astro}} = 0.71$ eV.

The above mass combinations can be written as follows:

(1) Normal Hierarchy (NH), $m_1 \ll m_2 \ll m_3$.
In this case we set $m_0 = m_1$ and

$$\Delta m^2_{\text{SUN}} = m_2^2 - m_1^2 \,, \quad \Delta m^2_{\text{ATM}} = m_3^2 - m_1^2.$$

Thus:

- Triton decay.

$$\langle m_\nu \rangle_{\text{decay}} = \sqrt{c_{12}^2 c_{13}^2 m_0^2 + s_{12}^2 c_{13}^2 \left(\Delta m^2_{\text{SUN}} + m_0^2\right) + s_{13}^2 (\Delta m^2_{\text{ATM}} + m_0^2)}. \tag{12.73}$$

- Cosmological bound:

$$m_\nu = m_0 + \sqrt{\Delta m^2_{\text{SUN}} + m_0^2} + \sqrt{(\Delta m^2_{\text{ATM}} + m_0^2)}. \tag{12.74}$$

- $0\nu\beta\beta$ decay:

$$\langle m_\nu \rangle = c_{12}^2 c_{13}^2 m_0 + s_{12}^2 c_{13}^2 e^{i\alpha_{21}} \sqrt{\Delta m^2_{\text{SUN}} + m_0^2} + s_{13}^2 e^{i\alpha_{31}} \sqrt{(\Delta m^2_{\text{ATM}} + m_0^2)}, \tag{12.75}$$

Note the presence, in this case, of the Majorana phases as we have mentioned earlier.

By assuming NH, i.e., that m_0 is negligibly small ($m_1 \ll \sqrt{\Delta m^2_{\mathrm{SUN}}}$, $m_2 \simeq \sqrt{\Delta m^2_{\mathrm{SUN}}}$, and $m_3 \simeq \sqrt{\Delta m^2_{\mathrm{ATM}}}$), we obtain

$$\begin{aligned} |\langle m_\nu \rangle| &\simeq |c^2_{13} s^2_{12} \sqrt{\Delta m^2_{\mathrm{SUN}}} + s^2_{13} \sqrt{\Delta m^2_{\mathrm{ATM}}} e^{-2i\alpha_2}| \\ &\leq 4 \cdot 10^{-3} \text{ eV}. \end{aligned} \tag{12.76}$$

(2) Inverted Hierarchy (IH), $m_3 \ll m_1 < m_2$:
In this case we set $m_0 = m_3$ and

$$\Delta m^2_{\mathrm{SUN}} = m^2_2 - m^2_1 \ , \ \Delta m^2_{\mathrm{ATM}} = m^2_2 - m^2_3.$$

- Triton decay.

$$\begin{aligned} \langle m_\nu \rangle_{\mathrm{decay}} = \big(& s^2_{13} m^2_0 + s^2_{12} c^2_{13} (\Delta m^2_{\mathrm{ATM}} + m^2_0) \\ &+ c^2_{12} c^2_{13} (\Delta m^2_{\mathrm{ATM}} - \Delta m^2_{\mathrm{SUN}} + m^2_0)\big)^{1/2}. \end{aligned} \tag{12.77}$$

- Cosmological bound:

$$m_\nu = m_0 + \sqrt{(\Delta m^2_{\mathrm{ATM}} + m^2_0)} + \sqrt{(\Delta m^2_{\mathrm{ATM}} - \Delta m^2_{\mathrm{SUN}} + m^2_0)}. \tag{12.78}$$

- $0\nu\beta\beta$ decay:

$$\begin{aligned} \langle m_\nu \rangle = &s^2_{13} e^{i\alpha_{31}} m_0 + e^{i\alpha_{21}} s^2_{12} c^2_{13} \sqrt{(\Delta m^2_{\mathrm{ATM}} + m^2_0)} \\ &+ c^2_{12} c^2_{13} \sqrt{(\Delta m^2_{\mathrm{ATM}} - \Delta m^2_{\mathrm{SUN}} + m^2_0)}. \end{aligned} \tag{12.79}$$

Since in IH scenario m_0 is negligibly small ($m_1 \simeq m_2 \simeq \sqrt{\Delta m^2_{\mathrm{ATM}}}$ and $m_0 \ll \sqrt{\Delta m^2_{\mathrm{ATM}}}$), we find

$$|\langle m_\nu \rangle| \simeq \sqrt{\Delta m^2_{\mathrm{ATM}}} c^2_{13} (1 - \sin^2 2\theta_{12} \sin^2 \alpha_{12})^{\frac{1}{2}}. \tag{12.80}$$

The phase difference α_{12} is the only unknown parameter in the expression for $|\langle m_\nu \rangle|$. From (12.80) we obtain the following inequality [Vergados *et al.* (2012a)]

$$1.5 \cdot 10^{-2} \text{ eV} \leq |\langle m_\nu \rangle| \leq 5.0 \cdot 10^{-2} \text{ eV}. \tag{12.81}$$

(3) Quasi-degenerate (QD) spectrum, $m_0 = m_1 \simeq m_2 \simeq m_3$. Then

- Triton decay and Cosmology:

$$\langle m_\nu \rangle_{\mathrm{decay}} = m_0, \quad m_\nu = 3m_0. \tag{12.82}$$

- $0\nu\beta\beta$ decay:

The effective Majorana mass is relatively large in this case and for both types of the neutrino mass spectrum is given by the expression

$$m_0 |1 - 2c^2_{13} c^2_{12}| \leq |\langle m_\nu \rangle| \leq m_0. \tag{12.83}$$

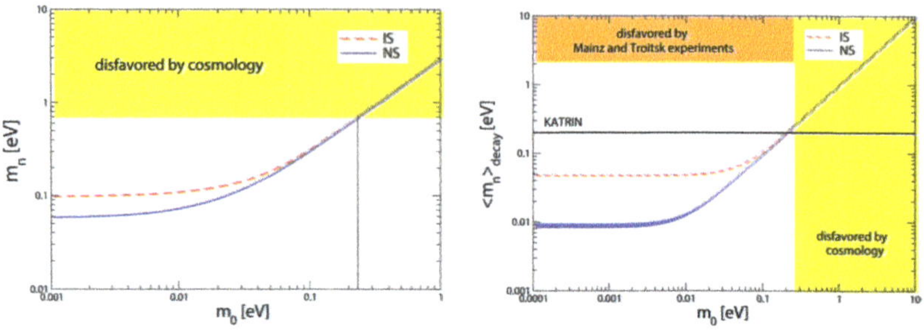

Fig. 12.9: The neutrino mass limits in eV as a function of mass of the lowest eigenstate m_0 also in eV, extracted from cosmology (left panel) triton decay (right panel). It is assumed that $\Delta m^2_{\rm ATM} = (2.43 \pm 0.13) \times 10^{-3}$ eV^2, $\Delta m^2_{\rm SUN} = (7.65^{+0.13}_{-0.20}) \times 10^{-5}$ eV^2, $\tan^2\theta_{12} = 0.452^{+0.035}_{-0.033}$ and $\sin^2 2\theta_{13} = 0.092 \pm 0.016$, see the review [Vergados *et al.* (2012a)]. From the current upper limit of 2.2 eV of the Mainz and Troitsk experiments we deduce a lowest neutrino mass of 2.2 eV both for the NS and IS. From the astrophysical limit value of 0.71 eV the corresponding neutrino mass extracted is about 0.23 eV for the NS and IS.

The above results are exhibited in Fig. 12.9 for the tritium β-decay and cosmological limits as a function of the lowest neutrino mass.[2]

The case of the $0\nu\beta\beta$-decay, both for the NS and the IS scenarios, is exhibited in Fig. 12.10. Shown is the allowed range values of $|\langle m_\nu \rangle|$ as a function of the lowest mass eigenstate m_0. For the values of the parameter $\sin^2 2\theta_{13}$ the new Double Chooz data are used [Vergados *et al.* (2012a)]. The IH allowed region for $|\langle m_\nu \rangle|$ is presented by the region between two parallel lines in the upper part of Fig. 12.10. The NH allowed region for $|\langle m_\nu \rangle| \approx$ few meV is compatible with m_0 smaller than 10 meV. The quasi-degenerate spectrum can be determined, if m_0 is known from future β-decay experiments KATRIN [Bohn *et al.* (1999)] and MARE [Drexlin *et al.* (2013)] or from cosmological observations. The lowest value for the sum of the neutrino masses, which can be reached in future cosmological measurements, is about (0.05–0.1) eV. The corresponding values of m_0 are in the region

[2]While the manuscript was on preparation a neutrino mass measurement was reported by the Karlsruhe Tritium Neutrino experiment KATRIN [Aker *et al.* (2019)]. A fit of the integrated electron spectrum over a narrow interval around the kinematic end point at 18.57 keV gives an effective neutrino mass square value $\langle m^2_\nu \rangle = -1.0^{0.9}_{-1.1}$ eV^2. From this an upper limit of 1.1 eV (90% confidence level) on the absolute mass scale of neutrinos has been obtained. This value coincides with the KATRIN sensitivity. This is cutting in half the existing limit derived from similar experiments.

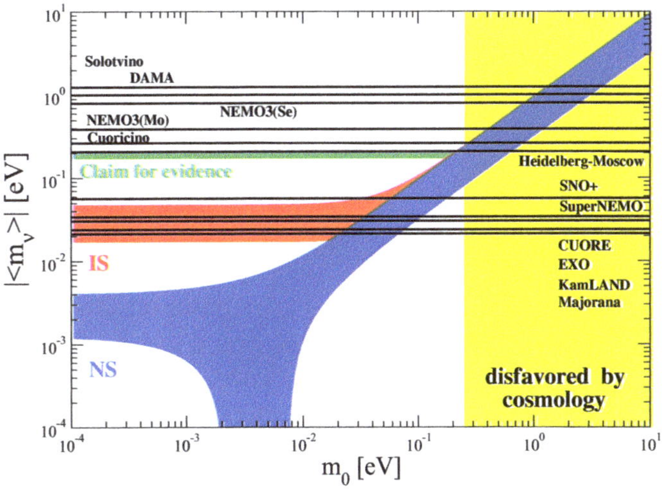

Fig. 12.10: We show the allowed range of values for $|\langle m_\nu \rangle|$ as a function of the lowest mass eigenstate m_0 using the three standard neutrinos for the cases of normal (NS, $m_0 = m_1$) and inverted (IS, $m_0 = m_3$) spectrum of neutrino masses. Also shown are the current experimental limits and the expected future results [Vergados *et al.* (2012a)] (QRPA NMEs with CD-Bonn short-range correlations and $g_A^{eff} = 1.25$ are assumed). Note that in the inverted hierarchy there is a lower bound, which means that in such a scenario the $0\nu\beta\beta$-decay should definitely be observed, if the experiments reach the required level. The same set of neutrino oscillation parameters as in Fig. 12.9 is considered. For an updated version of this figure see, e.g., [Hagedorn *et al.* (2018)].

where the IS and the NS predictions for $|\langle m_\nu \rangle|$ differ significantly from each other.

From the most precise experiments on the search for $0\nu\beta\beta$-decay by using the available nuclear matrix elements [Vergados *et al.* (2012a)], the following stringent bounds were inferred:

$$\begin{aligned} |\langle m_\nu \rangle| &< (0.20 - 0.32)\ eV \quad (^{76}\mathrm{Ge}), \\ &< (0.33 - 0.46)\ eV \quad (^{130}\mathrm{Te}), \\ &< (0.17 - 0.30)\ eV \quad (^{136}\mathrm{Xe}). \end{aligned} \tag{12.84}$$

In future experiments CUORE, EXO, MAJORANA, SuperNEMO, SNO+, KamLAND-Zen and others, see [Vergados *et al.* (2012a)] for a review, a sensitivity of

$$|\langle m_\nu \rangle| \simeq \text{a few } 10^{-2} \text{ eV} \tag{12.85}$$

is expected to be reached, which is the region of the IH of neutrino masses. In the case of the normal mass hierarchy $|\langle m_\nu \rangle|$ is too small in order to be probed in the $0\nu\beta\beta$-decay experiments of the next generation.

12.10 Neutrinos as probes

We have seen the very interesting physics that are associated with neutrinos. Furthermore in recent years neutrinos are beginning to be thought of as probes for studying other systems, in particular the interior of large objects, like the interior of stars in their last stages (supernova explosions), the interior of the Earth (geoneutrinos) and the deep sky (neutrino telescopes). The essential reason is that they can travel undistorted large distances with almost the speed of light. They are not affected by magnetic fields or other objects on their way and they, thus, carry undistorted information about their source. The remote control and/or surveillance of reactors is another useful application. In particular, these possibilities can be fully exploited, if low threshold detectors involving the neutral current interaction are employed, since these are not affected by neutrino oscillations. One of the real challenges will be to detect the neutrino background radiation, which may provide a picture of the Universe before the microwave black body radiation. Detailed study of such objects is, however, beyond our goals.

12.11 Problems

(1) Show that the minimum neutrino energy solution to reach the resonance region, Section 12.5.2, is:

$(E_\nu)_{\text{min}} = 1$ MeV, using

$$\Delta m^2_{21} = 8.0 \times 10^{-5} \text{ eV}^2,\ \theta \approx \frac{\pi}{6} \text{ (LMA-MSW solution).} \tag{12.86}$$

Hint: Assume that the neutrino number density in the core of the sun is 3×10^{31} m^{-3} and $G_F m_p^2 = 1.0 \times 10^{-5}$.

(2) Using the data of the previous problem determine and plot the oscillation pattern assuming assuming $a = 1/2$, $a = 1/4$, $a = 0$, $a = -1/4$, see Section 12.5.2. Confirm that you get the patterns shown in Fig. 12.11.

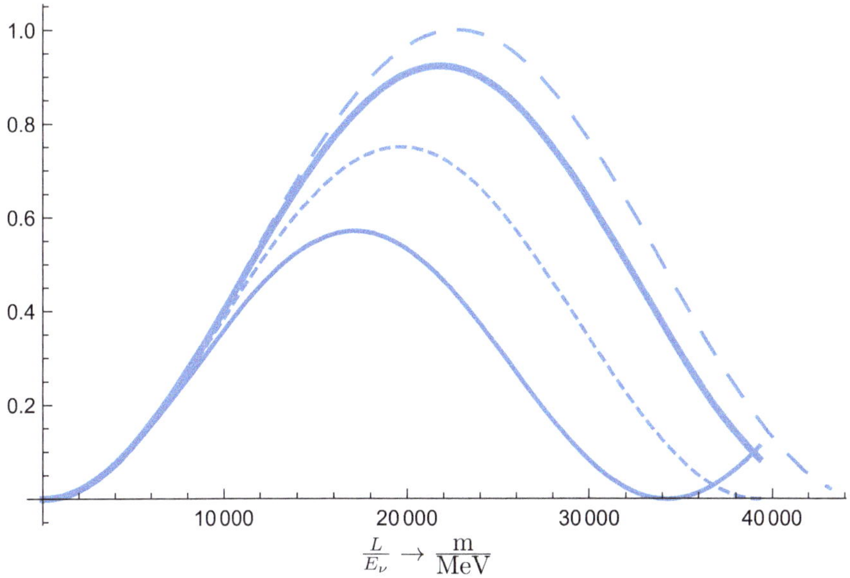

Fig. 12.11: We show the oscillation patterns in the case of $a = 1/4$ (thick solid line), $a = 0$ (dotted line), $a = -1/4$ (solid line) and the resonance $a = 1/2$ (dashed line).

(3) Using the three generation formula of neutrino oscillations Eq. (12.48), and assuming that the phase δ of antineutrinos is opposite to that of neutrinos, find expressions for the asymmetry parameter $r_{\alpha,\beta}$ defined by:

$$r_{\alpha,\beta} = \frac{P(\nu_\alpha \rightarrow \nu_\beta) - P(\tilde{\nu}_\alpha \rightarrow \tilde{\nu}_\beta)}{P(\nu_\alpha \rightarrow \nu_\beta) + P(\tilde{\nu}_\alpha \rightarrow \tilde{\nu}_\beta)}. \tag{12.87}$$

Chapter 13

Weak interactions — Beta decays-Neutrino interactions

13.1 History of beta decay

Beta decay has a long history. It first appeared in 1933, when it was found that the neutron was unstable and decayed into a proton, an electron and a neutrino. The observed life time was very long and Fermi understood that the interaction responsible for the decay must be very weak. He named it weak and the name stuck. It was also realized that the weak interaction can manifests itself only when a neutrino, the only particle which does not feel the strong and electromagnetic interaction, is involved.[1] The second remarkable fact was that the weak interaction was of extremely short range. So in 1934 Fermi proposed that it was a **current-current contact** interaction of the form (in a modern notation)

$$H_W = G_F \left((J^+)^\lambda (j^-)_\lambda + (J^-)^\lambda (j^+)_\lambda \right) \tag{13.1}$$

with

$$\begin{aligned} (J^+)^\lambda &= \bar{p}\gamma^\lambda n \leftrightarrow (n \to p),\ (j^-)_\lambda \\ &= \bar{e}\gamma_\lambda \nu_e \leftrightarrow \nu_e \to e^- \text{ or } (e^-, \tilde{\nu}_e) \text{ production} \end{aligned} \tag{13.2}$$

where γ_λ are the familiar gamma matrices of the Dirac theory, discussed in detail in Section 16.2.3, while p, n, e and ν represent the the proton, neutron, electron and neutrino spinors[2] fields respectively, see Appendix C, Chapter 16. Also

$$\begin{aligned} (J^-)^\lambda &= \bar{n}\gamma^\lambda p \leftrightarrow (p \to n),\ (j^+)_\lambda \\ &= \bar{\nu}\gamma_\lambda e \leftrightarrow e^- \to \nu_e \text{ or } (e^+, \nu_e) \text{ production.} \end{aligned} \tag{13.3}$$

The process $p \to n + e + \tilde{\nu}_e$ cannot take place in free space.

[1]The neutrino can be either a real particle or virtual state like in neutrinoless double beta decay.

[2]At that time, of course, no distinction was made between neutrino and antineutrino.

In this context nuclear processes studied were the following

$$A(N,Z) \rightarrow A(N-1,Z+1)+e^-+\bar{\nu}_e \text{ or } A(N,Z) \rightarrow A(N+1,Z-1)+e^++\nu_e \tag{13.4}$$

the first is called negaton decay or simply nuclear decay and the second positon decay. Another process caused by the weak interaction is electron capture, in which a bound electron of the atom is absorbed by the nucleus

$$e_b^- + A(N,Z) \rightarrow A(N,Z-1) + \nu_e. \tag{13.5}$$

This is an interesting process and it will be discussed below.[3]

Processes involving neutrino-nucleus scattering

$$\nu_e + A(N,Z) \rightarrow A(N-1,Z+1)+e^-,\ \tilde{\nu}_e + A(N,Z) \rightarrow A(N+1,Z-1)+e^+ \tag{13.8}$$

were not considered at that time.

But then came the news that shocked the word: The fall of parity conservation. It was proposed by Lee and Yang [Lee and Yang (1956)] and experimentally found by Wu and co-workers [Wu *et al.* (1957)]. Furthermore the neutrino is very prejudiced: always left handed! This fact was found experimentally by Goldhaber and co-workers [Goldhaber *et al.* (1958)]. As a result, after efforts by a lot of people, including Lee, Yang, Feynman etc., the contact Fermi interaction was improved and became Vector-Axial (V-A) theory.

$$H_W = \frac{G_F}{\sqrt{2}}\left((J^+)^\lambda (j^-)_\lambda + (J^-)^\lambda (j^+)_\lambda\right) \tag{13.9}$$

with

$$(J^+)^\lambda = \bar{p}\gamma^\lambda (g_V - g_A\gamma_5)n,\ (j^-)_\lambda = \bar{e}\gamma_\lambda(1-\gamma_5)\nu_e \tag{13.10}$$

where γ_5 is the familiar from the Dirac theory gamma-5 matrix, see Appendix C, Chapter 16.

This was fine except that it violates unitarity and leads to divergence contributions to the cross section at high energies. After many attempts

[3]We are not going to consider here other very interesting decays, second order weak processes like:

$$A(N,Z) \rightarrow A(N-2,Z+2) + 2e^- + 2\bar{\nu}_e, 2\nu\beta\beta \text{ decay} \tag{13.6}$$

or the lepton violating

$$A(N,Z) \rightarrow A(N-2,Z+2) + 2e^-, 0\nu\beta\beta \text{ decay}. \tag{13.7}$$

The last process is very interesting, since it provides the best way for the determination of the scale of neutrino mass (see Chapter 12). The interested reader is referred to the literature, e.g. [deShalit and Feshbach (1974)], [Vergados *et al.* (2012a)].

a solution to this problem came with the invention of the Standard Model (SM) of particle physics or GSW [Glashow (1961)], [Weinberg (1967)], [Salam (1968)]. This model is discussed in detail in many books on particle theory, e.g. in the companion book [Vergados (2017)]. So the interaction for hadrons at the quark level becomes

$$\frac{g}{2\sqrt{2}}\left(\bar{u}\gamma^{\lambda}(1-\gamma_5)d\right)W_{\lambda}^{-},\quad \frac{g}{2\sqrt{2}}\left(\bar{d}\gamma^{\lambda}(1-\gamma_5)u\right)W_{\lambda}^{+} \tag{13.11}$$

or

$$\frac{g}{2\sqrt{2}}\left(\bar{q}\gamma^{\lambda}(1-\gamma_5)\tau_{\pm}q\right)W_{\lambda}^{\mp}, \tag{13.12}$$

where τ_+, τ_- are the isospin raising and lowering operators i.e. $\tau_+|d\rangle=|u\rangle$, $\tau_-|u\rangle=|d\rangle$.

For leptons it is of the form

$$\frac{g}{2\sqrt{2}}\left(\bar{\nu}_e\gamma^{\lambda}(1-\gamma_5)e\right)W_{\lambda}^{-},\quad \frac{g}{2\sqrt{2}}\left(\bar{e}\gamma^{\lambda}(1-\gamma_2)\nu_e\right)W_{\lambda}^{+}. \tag{13.13}$$

Where g is a coupling constant $W_{\lambda}^{\pm}$ are two charged heavy bosons with helicity λ.

It is easy to show that at the nucleon level the interaction takes the form

$$\frac{g}{2\sqrt{2}}\left(\bar{N}\gamma^{\lambda}(g_V-g_A\gamma_5)\tau_{\pm}N\right)W_{\lambda}^{\pm}, \tag{13.14}$$

where now we use the nuclear convention for isospin,[4] namely

$$\left|I=\frac{1}{2},I_3=\frac{1}{2}\right\rangle=n,\ \left|I=\frac{1}{2},I_3=-\frac{1}{2}\right\rangle=p,\ I_+|p\rangle=|n\rangle,\ I_-|n\rangle=|p\rangle. \tag{13.15}$$

Since the mediator between the currents massive the propagator becomes

$$H_W=\left(\frac{g}{2\sqrt{2}}\right)^2\frac{1}{\mathbf{q}^2+m_W^2}\left((J^+)^{\lambda}(j^-)_{\lambda}+(J^-)^{\lambda}(j^+)_{\lambda}\right). \tag{13.16}$$

For very small momentum transfer this equation is reduced to Eq. (13.9) with

$$\frac{g^2}{m_W^2}=4\sqrt{2}G_F. \tag{13.17}$$

We thus recover the old current-current contact interaction.

At this point we should mention that the standard model predicts the existence of a neutral gauge boson Z^0. This, of course, leads to a new interaction at the nucleon level of the form

$$H_W=\frac{G_F}{2\sqrt{2}}(J^Z)^{\lambda}(j^Z)_{\lambda} \tag{13.18}$$

[4]In nuclear physics the isospin indicated by T rather I, but we are not going to use this notation here.

where

$$(j^Z)_\lambda = \bar{\nu}\gamma_\lambda(1-\gamma_5)\nu,\ (J^{Z)})^\lambda = \bar{N}\gamma^\lambda(1-\gamma_5)\tau_3 N. \tag{13.19}$$

These currents became known as neutral currents. Their existence was confirmed experimentally in 1973, in a neutrino experiment in the Gargamelle bubble chamber at CERN [Hajert *et al.* (1973)]. This was a triumph for the SM. The $W^\pm$ and Z^0 were discovered in 1983 at CERN [Arnison *et al.* (1983a)], [Arnison *et al.* (1983b)], [Arnison *et al.* (1983c)]. The accepted values of their mass are

$$m_W = 80.425(38)\ \text{GeV/c}^2, \quad m_{Z^0} = 91.1876(21)\ \text{GeV/c}^2.$$

The hadronic current J^λ gives rise to the nuclear Matrix element $M_{i\to f}$ as will be discussed below.

Nuclear beta decay is discussed in a number of books, to mention, e.g., [Konopinski (1966)], [Roy and Nigam (1967)], [deShalit and Feshbach (1974)], [Cottingham and Greenwood (1986)] among others.

13.2 Kinematics of beta decay

It can be shown, using standard techniques, as given, e.g., in the companion book [Vergados (2017)], and similar techniques as those of the nuclear physics books [Roy and Nigam (1967)], [deShalit and Feshbach (1974)], that for a non polarized nucleus the invariant amplitude takes the form:

$$\begin{aligned}|ME_{i\to f}|^2 = |ME_{i\to f}|^2_F \, &\frac{2(E_e E_\nu + \mathbf{p}_e.\mathbf{p}_\nu)}{m_e m_\nu} \\ + |ME_{i\to f}|^2_{GT} \, &\frac{2(E_e E_\nu - \mathbf{p}_e.\mathbf{p}_\nu)}{m_e m_\nu}\end{aligned} \tag{13.20}$$

E_e, E_ν are the total electron and neutrino energies respectively with $\mathbf{p}_e$ and $\mathbf{p}_\nu$ the corresponding momenta of the electron and the neutrino. The two nuclear matrix elements denoted by F (Fermi) and GT (Gamow-Teller) will be discussed in the next section.

The width for beta decay can be cast in the form:[5]

$$\begin{aligned}d\Gamma = |ME_{i\to f}|^2 \frac{G_F^2}{2} \frac{m_e m_\nu}{E_e E_\nu} \frac{2(E_e E_\nu \pm \mathbf{p}_e.\mathbf{p}_\nu)}{m_e m_\nu} \frac{d^3\mathbf{p}_e}{(2\pi)^3} \frac{d^3\mathbf{p}_\nu}{(2\pi)^3} \frac{d^3\mathbf{p}_A}{(2\pi)^3} \\ (2\pi)\delta\left(\Delta - E_e - E_\nu\right)(2\pi)^3\delta\left(\mathbf{p}_A + \mathbf{p}_e + \mathbf{p}_\nu\right)\end{aligned} \tag{13.21}$$

[5]We find it convenient to use here the natural system of units, see Section 9.1.

($\pm$ is understood to mean that, strictly speaking, we must put a plus for F and a minus for GT, but it does not affect the total width). Δ is the available energy[6] and $\mathbf{p}_A$ the momentum of the recoiling nucleus.

We assumed that the initial nucleus was at rest and the the recoiling energy of the outgoing nucleus is negligible. The conservation of energy and momenta is expressed by the relevant δ functions. Integrating over the momenta eliminates the dependence on the recoiling nucleus

$$d\Gamma = \left|ME_{i\to f}\right|^2 G_F^2 \frac{E_e E_\nu \pm \mathbf{p}_e.\mathbf{p}_\nu}{E_e E_\nu} \frac{1}{(2\pi)^5} d^3\mathbf{p}_e d^3\mathbf{p}_\nu \delta\left(\Delta - E_e - E_\nu\right). \tag{13.22}$$

Integrating over the neutrino momentum with the aid of the energy δ function we get

$$d\Gamma = \left|ME_{i\to f}\right|^2 \frac{G_F^2}{2} \frac{2}{(2\pi)^5}(4\pi)d^3\mathbf{p}_e(\Delta - E_e)^2. \tag{13.23}$$

Integrating over the direction of the electron momenta we get

$$d\Gamma = \left|ME_{i\to f}\right|^2 G_F^2 \frac{1}{(2\pi)^5}(4\pi)^2\sqrt{E_e^2 - m_e^2}E_e dE_e(\Delta - E_e)^2. \tag{13.24}$$

This expression in normally expressed in standard units with the energy given in units of m_e as follows:

$$d\Gamma = \left|ME_{i\to f}\right|^2 \frac{G_F^2 m_e^4}{2\pi^3} \frac{c^2}{\hbar^6} m_e c^2 \sqrt{\epsilon^2 - 1}\epsilon(\epsilon_0 - \epsilon)^2 d\epsilon \tag{13.25}$$

with $\epsilon_0 = \Delta/m_e$ and $\epsilon = E_e/m_e$.

The total width is

$$\Gamma = \left|ME_{i\to f}\right|^2 \frac{G_F^2 m_e^4}{2\pi^3} \frac{c^2}{\hbar^6} m_e c^2 f(\epsilon_0),$$

$$f(\epsilon_0) = \int_1^{\epsilon_0} \rho(\epsilon_0)d\epsilon,\ \rho(\epsilon_0) = \sqrt{\epsilon^2 - 1}\epsilon(\epsilon_0 - \epsilon)^2. \tag{13.26}$$

$$G_F m_e^2 = N\left(G_F m_p^2\right) \frac{m_e^2}{m_p^2} \frac{\hbar^3}{c},$$

see Section 9.1, where $N\left(G_F m_p^2\right)$ is the numerical value of the quantity which is $\approx 1.05 \times 10^{-5}$. We thus obtain

$$\Gamma = K\left|ME_{i\to f}\right|^2 \frac{1}{2\pi^3} m_e c^2 f(\epsilon_0),\ K \approx 9.3 \times 10^{-24} \tag{13.27}$$

the life time is

$$\tau = \frac{1}{K\left|ME_{i\to f}\right|^2 f(\epsilon_0)}(2\pi^3)\frac{\hbar}{m_e c^2}. \tag{13.28}$$

[6]The available energy Δ is normally given as the energy difference between the initial and final nuclear ground states. For transitions to excited states one should use $\Delta - E_x$ with E_x the excitation energy.

The half-life is given by

$$\tau_{\frac{1}{2}} = \ln(2)\frac{1}{K\left|ME_{i\to f}\right|^2 f(\epsilon_0)}(2\pi^3)\frac{\hbar}{m_e c^2}. \tag{13.29}$$

From this we get the celebrated expression ft value

$$ft \equiv f(\epsilon_0)t_{\frac{1}{2}} = \ln(2)\frac{1}{K\left|ME_{i\to f}\right|^2}(2\pi^3)\frac{\hbar}{m_e c^2}, \; \frac{\hbar}{m_e c^2} \approx 1.285\times 10^{-21}\text{ s}. \tag{13.30}$$

This expression is very useful. It allows one to extract the nuclear matrix element from the measurement of the half life and the spectral function f. In doing so, one must be careful. In Ref. [deShalit and Feshbach (1974)], e.g., one finds

$$t_{\frac{1}{2}} = \ln(2)\frac{1}{\left|ME_{i\to f}\right|^2}\frac{2\pi^3}{\Gamma^2}\frac{\hbar}{m_e c^2}$$

with a different definition of Γ.

At this point we should mention that the above derivation assumes that the outgoing charged lepton, e^- or e^+, is described by a plane wave. This is not true, since the Coulomb field causes a distortion given by the Fermi function $F(Z,\epsilon,R)$. Thus

$$\rho(\epsilon_0,Z,R)=\sqrt{\epsilon^2-1}\epsilon(\epsilon_0-\epsilon)^2F(Z,\epsilon,R)d\epsilon,\; f(\epsilon_0,Z,R)=\int_1^{\epsilon_0}\rho(\epsilon,Z,R)d\epsilon. \tag{13.31}$$

Furthermore we have neglected the small neutrino mass. If that is taken into account one obtains

$$\begin{aligned}\rho(\epsilon,Z,\nu) &= \sqrt{\epsilon^2-1}\sqrt{(\epsilon_0-\epsilon)^2-\nu^2}\epsilon(\epsilon_0-\epsilon)F(Z,\epsilon),\\ f(\epsilon_0,Z,\nu) &= \int_1^{\epsilon_0-\nu}\rho(\epsilon,Z,\nu)d\epsilon, \nu = \frac{\sqrt{\langle m_\nu\rangle^2}}{m_e}.\end{aligned} \tag{13.32}$$

The precise meaning of $\langle m_\nu\rangle^2$ may become clear in the discussion of triton decay in Section 12.9. If needed, both of these corrections are understood in the ft value given in the expression (13.30).

The Fermi function at the origin [deShalit and Feshbach (1974)]

$$F(Z,\epsilon,0) = \left|\frac{\eta}{1-e^{-2\pi\eta}}\right|, \quad \eta = \pm\frac{\alpha Z\epsilon}{\sqrt{\epsilon^2-1}}. \tag{13.33}$$

The above simple approximation is not adequate, if relativistic corrections must be taken into account. In this case one adopts a Fermi function, which is evaluated at some appropriate distance from the origin. The appropriate expression now becomes [Landau and Lifshitz (1977)], [deShalit and Feshbach (1974)]:

$$F(Z,\epsilon,R) = (2kR)^{2(s-!)}\frac{2(1+s)}{s^2+\eta^2}e^{\pi\eta}\left|\frac{\Gamma(s+1+i\eta)}{\Gamma(2s+1)}\right|^2, \quad s=\sqrt{1-(\alpha Z)^2}.$$

It can be shown [Landau and Lifshitz (1977)] that the most appropriate distance to use is given by:

$$R \approx \frac{1}{2}\left(\eta^2 - s(s-1)\right)\frac{1}{m_e\alpha Z}.$$

Thus

$$k = \alpha Z m_e \Rightarrow 2kR = \eta^2 - s(s-1).$$

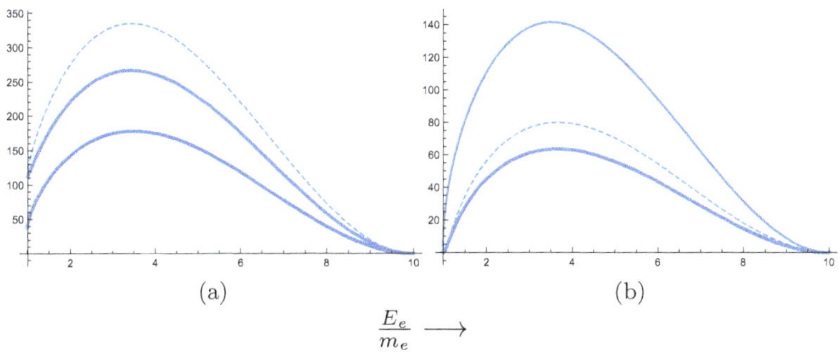

Fig. 13.1: The distribution ρ, spectrum of $e^{\pm}$ in the case of beta decay, as a function of the energy in units of the electron mass. (a) in the case of electron emission and (b) in the case of positron emission. The solid line corresponds to zero charge (no Fermi function correction), while the thick solid line corresponds to the Fermi function with $R=0$ and $Z=30$. The dotted line corresponds to $R \neq 0$, as discussed in the text, also with $Z=30$.

In Fig. 13.1 we exhibit the spectrum ρ for three cases, no Coulomb correction, with Coulomb correction with $R=0$ and with Coulomb correction with $R \neq 0$: (a) electron and (b) positron emission. A typical value of $Z=30$ has been employed. In Fig. 13.2 we exhibit the same quantity using a Fermi function with $R \neq 0$ for various charges.

It is interesting to consider the expected spectra of triton

$$^3_1\text{H} \rightarrow\ ^3_2\text{He} + e^{-1} + \bar{\nu}_e \qquad (13.34)$$

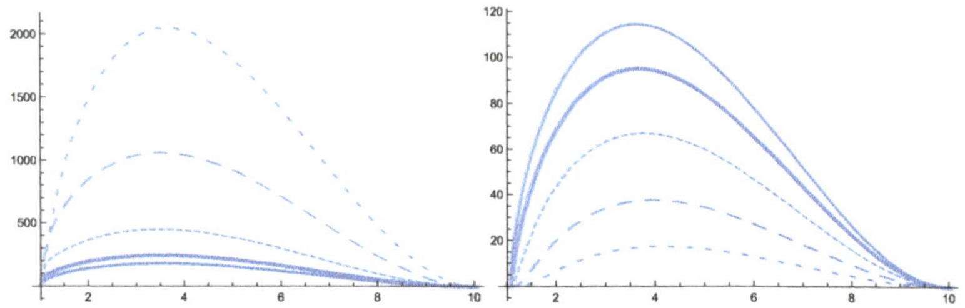

(a) (b)

$\frac{E_e}{m_e} \longrightarrow$

Fig. 13.2: The distribution ρ, spectrum of $e^{\pm}$ in the case of beta decay, as a function of the energy in units of the electron mass. (a) in the case of electron emission and (b) in the case of positron emission. In both cases we used a Fermi function with $R \neq 0$, as discussed in the text. The solid, thick solid, dotted, dashed and dot-dashed curves correspond to $Z = 20$, 40, 70 and 100.

with an end point electron kinetic energy of 18.591 keV. Because of its small end point energy, this decay is considered to be the best place to extract the absolute scale of neutrino mass, see Section 12.9. This kinetic energy corresponds to $\epsilon_0 = 1.03638$. In this case the expected spectra are exhibited in Fig. 13.3(a) for massless neutrino and in Fig. 13.3(b), if the neutrino has a non zero mass, say of 1 eV. Both spectra are exhibited in Fig. 13.3(c) near the end point. One notices the suppression in the expected rate if the neutrino has a mass.

13.3 The beta decay nuclear matrix element

Let us now consider the hadronic current. Its time component is $\bar{p}\gamma_0(g_V - g_A\gamma_5)n$. For non relativistic nucleus the γ_5 term makes a negligible contribution. Thus

$$\begin{aligned} g_V\bar{p}\gamma_0 n &= g_V(\phi_p, 0)\begin{pmatrix} 1 & 0 \\ 0 & -1 \end{pmatrix}\begin{pmatrix} 1 & 0 \\ 0 & -1 \end{pmatrix}\begin{pmatrix} \phi_n \\ 0 \end{pmatrix} \\ &= g_V(\phi_p, 0)\begin{pmatrix} 1 & 0 \\ 0 & 1 \end{pmatrix}\begin{pmatrix} \phi_n \\ 0 \end{pmatrix} = g_v\phi_p\phi_n \rightarrow \psi_N\tau_-\psi_N. \end{aligned}$$

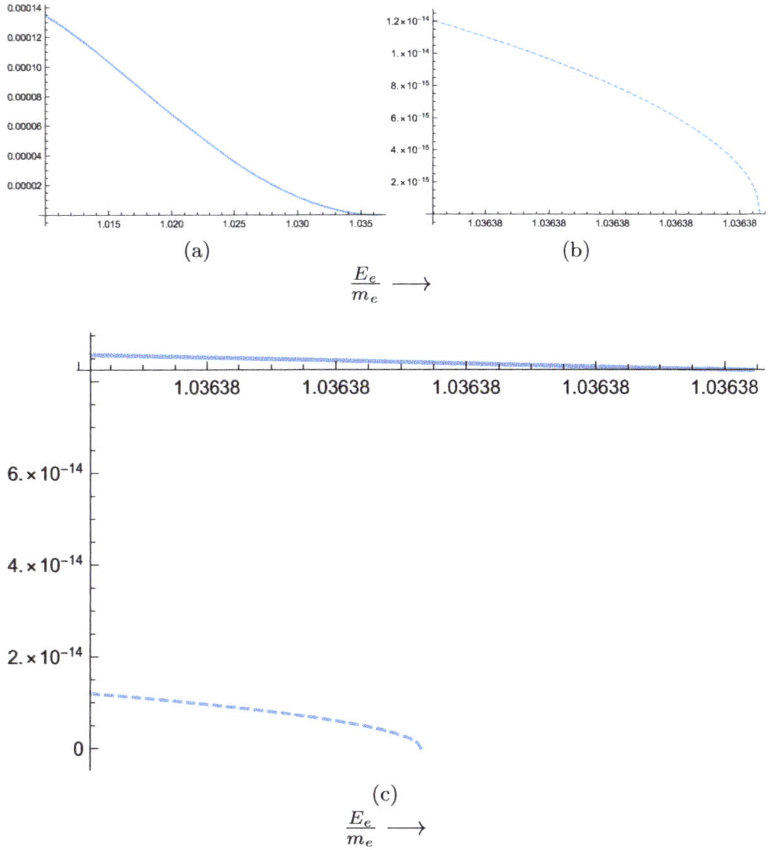

Fig. 13.3: (a) The expected distribution ρ, spectrum of e^- in the case of the triton beta decay, assuming zero neutrino mass. (b) The expected spectrum assuming a neutrino with average mass of 1 eV. One notices the dip of the spectrum near the end point. (c) The previous 2 spectra in the region near the endpoint. One sees the suppression of the spectrum in the case of a massive neutrino. All spectra are plotted as a function of the energy in units of the electron mass.

So the transition operator at the nucleon level becomes $g_v\tau_\pm$. In the case of many nucleons

$$g_V\Omega_F, \Omega_F = \sum_n \tau_\pm(n) = T_\pm.$$

In the case of the Axial current for non relativistic nucleons we get

$$g_A\bar{p}\gamma_0 n = g_A(\phi_p, 0)\begin{pmatrix} 1 & 0 \\ 0 & -1 \end{pmatrix}\begin{pmatrix} 0 & \sigma_k \\ -\sigma_k & 0 \end{pmatrix}\begin{pmatrix} \phi_n \\ 0 \end{pmatrix} = g_A(\phi_p, 0)\begin{pmatrix} \sigma_k & 0 \\ 0 & \sigma_k \end{pmatrix}\begin{pmatrix} \phi_n \\ 0 \end{pmatrix}$$
$$= g_A\phi_p\sigma_k\phi_n \rightarrow \psi_N\tau_-\sigma_k\psi_N.$$

In the case of many nucleons[7]

$$g_A\Omega_{GT}, \Omega_{GT} = \sum_n \tau_\pm(n)\boldsymbol{\sigma}(n).$$

Let us now evaluate the nuclear matrix elements. We will consider a transition characterized by good angular quantum numbers, i.e. from a state $|J_iM_i\rangle$ to a state $|J_fM_f\rangle$. The nuclear matrix element is defined by

$$|ME_{i\rightarrow f}|_F^2 = g_V^2\frac{1}{2J_i+1}\sum_{M_i,M_f}|\langle J_f, M_f|\Omega_F|J_iM_i\rangle|^2,$$
$$|ME_{i\rightarrow f}|_{GT}^2 = g_A^2\frac{1}{2J_i+1}\sum_{M_i,M_f}|\langle J_f, M_f|\Omega_{GT}|J_iM_i\rangle|^2. \quad (13.35)$$

In the above expression we followed the standard procedure by averaging over the initial magnetic sub-states and summing over all the final ones.

Sometimes the notation

$$|M_F|^2 = \frac{|ME_{i\rightarrow f}|_F^2}{g_V^2}, \; |M_{GT}|^2 = \frac{|ME_{i\rightarrow f}|_{GT}^2}{g_A^2} \quad (13.36)$$

is employed.

Let us begin with Fermi transition. The operator is spin independent and, as a result, the angular momentum cannot change, $J_f = J_i$, $M_f = M_i$ and the matrix element is independent of the of M_i and M_f. As a result the summation over the sub-states yields $(2J_i + 1)$. Thus

$$|ME_{i\rightarrow f}|_F^2 = g_v^2\left|\langle J_f|\sum_n \tau_\pm(n)|J_i\rangle\right|^2. \quad (13.37)$$

The nuclear states have a good isospin projection I_3 and, usually, good isospin I. In such a case we get:

$$|ME_{i\rightarrow f}|_F^2 = g_V^2\left((I \pm I_3)(I \mp +I_3 + 1)\right), \text{ for } e^\mp \quad (13.38)$$

[7]Strictly speaking in transforming the operator into the coordinate space we should have written $\Omega_F = \sum_n \tau_\pm(n)e^{\mathbf{q}\cdot\mathbf{r}(n)}$, $\Omega_{GT} = \sum_n \tau_\pm(n)\boldsymbol{\sigma}(n)e^{\mathbf{q}\cdot\mathbf{r}(n)}$ where $\mathbf{q}$ is the momentum transfer to the nucleus. Then the nuclear matrix elements will then be functions of q^2, i.e. $\left|M_F(q^2)\right|^2$, $\left|M_{GT}(q^2)\right|^2$. This will affect the momentum integrals appearing in the kinematics. In the case of the allowed β decay, however, the momentum transfer to the nucleus is small and such refinements are not needed.

where I_3 characterizes the initial state. In the case of the neutron decay the isospin matrix element is unity. Thus

$$|ME_{n\to p}|^2_F = g_V^2.$$

In the case of Gamow-Teller transition

$$|ME_{i\to f}|^2_{GT} = g_A^2 \frac{1}{2J_i+1}$$
$$\sum_{M_i,M_f} \langle J_f, M_f|\sum_n \tau_\pm(n)\boldsymbol{\sigma}(n)|J_iM_i\rangle . \langle J_i, M_i|\sum_n \tau_\pm(n)\boldsymbol{\sigma}(n)|J_fM_f\rangle^*. \tag{13.39}$$

In the above scalar product is understood as

$$\boldsymbol{\sigma}.\boldsymbol{\sigma} = 4\sum_{q=-1,0,1}(-1)^q T^1_q T^1_{-q},\ T^1_1 = -\frac{1}{\sqrt{2}}s_+, T^1_{-1} = \frac{1}{\sqrt{2}}s_-,\ T^1_0 = s_0.$$

Before proceeding with the calculation, it is worth examining the simple example of neutron decay. The relevant matrix elements, in the notation $|j,m\rangle$, are

$$\left\langle \frac{1}{2}\,\frac{1}{2}\right| 2T^1_1 \left|\frac{1}{2}\,-\frac{1}{2}\right\rangle = -\sqrt{2}\left\langle \frac{1}{2}\,\frac{1}{2}\right| s_+ \left|\frac{1}{2}\,-\frac{1}{2}\right\rangle = -\sqrt{2},$$
$$\left\langle \frac{1}{2}\,\frac{1}{2}\right| 2T^1_{-1} \left|\frac{1}{2}\,-\frac{1}{2}\right\rangle = \sqrt{2}\left\langle \frac{1}{2}\,-\frac{1}{2}\right| s_- \left|\frac{1}{2}\,\frac{1}{2}\right\rangle = \sqrt{2}$$
$$\left\langle \frac{1}{2}\pm\frac{1}{2}\right| 2T^1_0 \left|\frac{1}{2}\pm\frac{1}{2}\right\rangle = \sqrt{2}\left\langle \frac{1}{2}\pm\frac{1}{2}\right| 2s_0 \left|\frac{1}{2}\pm\frac{1}{2}\right\rangle = \pm 1 \tag{13.40}$$

the isospin matrix element is unity as before.

$$|ME_{n\to 5}|^2_{GT} = g_A^2\frac{1}{2}(2+2+1+1) = 3g_A^2. \tag{13.41}$$

This brute force is hard to generalize. But we suspect the matrix elements involving the σ operator with the various angular momentum projections are related via the Clebsch-Gordan (C-G) coefficients of the rotation group. In fact

$$\left\langle \frac{1}{2}m_f\right| 2T^1_q \left|\frac{1}{2}\,m_i\right\rangle = \frac{1}{\sqrt{2}}\left\langle \frac{1}{2}m_i, 1q\right|\left.\frac{1}{2},\,m_f\right\rangle\left\langle \frac{1}{2}\right\|\sigma\left\|\frac{1}{2}\right\rangle. \tag{13.42}$$

From tables of C-G or Mathematica we find:

$$\left\langle \frac{1}{2}-\frac{1}{2},1,1\right|\left.\frac{1}{2}\frac{1}{2}\right\rangle = -\frac{\sqrt{2}}{\sqrt{3}}, \left\langle \frac{1}{2}\frac{1}{2},1,-1\right|\left.\frac{1}{2}\,-\frac{1}{2}\right\rangle = \frac{\sqrt{2}}{\sqrt{3}}$$

$$\left\langle \frac{1}{2}\pm\frac{1}{2},1,0\right|\left.\frac{1}{2}\pm\frac{1}{2}\right\rangle = \pm\frac{1}{\sqrt{3}}.$$

Multiplying these coefficients with the other two factors appearing in Eq. (13.42) and comparing with the corresponding factors in Eq. (13.40) we find that they are consistent provided that $\langle \frac{1}{2}||\sigma||\frac{1}{2}\rangle = \sqrt{6}$.

We now generalize this result for a general tensor operator T^k_q:

$$\langle J_f M_f|T^k_q|J_i M_i\rangle = \frac{1}{\sqrt{2J_f+1}}\langle J_i M_i, kq|J_f, M_f\rangle\langle J_f||T^k||J_i\rangle \tag{13.43}$$

where $\langle J_f||T^k||J_i\rangle$ is the reduced ME of the operator, see, e.g., [deShalit and Feshbach (1974)], [Hecht (2000)] and Section 14.4. It depends on the nature of the operator, but it is independent of the M's. For some examples and applications, see Section 14.4 of the Appendix A.

Returning now to Eq. (13.39) we have:

$$\begin{aligned}|ME_{i\to f}|^2_{GT} = g_A^2\frac{1}{2J_i+1} \\ \sum_{M_i,M_f,q}(-1)^q\langle J_f, M_f|\sum_n \tau_\pm(n)\boldsymbol{\sigma}_q(n)|J_i M_i\rangle. \\ \langle J_i, M_i|\sum_n \tau_\pm(n)\boldsymbol{\sigma}_{-q}(n)|J_f M_f\rangle^*\end{aligned} \tag{13.44}$$

or

$$\begin{aligned}|ME_{i\to f}|^2_{GT} = g_A^2\frac{1}{2J_i+1} \\ \sum_{M_i,M_f,q}\langle J_f, M_f|\sum_n \tau_\pm(n)\boldsymbol{\sigma}_q(n)|J_i M_i\rangle\langle J_f, M_f|\sum_n \tau_\pm(n)\boldsymbol{\sigma}_q(n)|J_f M_f\rangle\end{aligned} \tag{13.45}$$

$$\begin{aligned}|ME_{i\to f}|^2_{GT} = g_A^2\frac{1}{2J_i+1} \\ \frac{1}{2J_f+1}\langle J_f||T^1||J_i\rangle^2\sum_{M_i,M_f,q}\langle J_i M_i 1q|J_f, M_f\rangle^2\end{aligned} \tag{13.46}$$

but from the unitarity of the C-G coefficients we get

$$\sum_{M_f}\left(\sum_{M_i,q}\langle J_i M_i 1q|J_f, M_f\rangle^2\right) = \sum_{M_f} = 2j_f + 1.$$

Thus finally

$$|ME_{i\to f}|^2_{GT} = g_A^2\frac{1}{2J_i+1}\langle J_f||T^1||J_i\rangle^2. \tag{13.47}$$

Table 13.1: The Gamow-Teller ME $|M_{GT}|^2 = \frac{|ME_{J_i \to J_f}|^2_{GT}}{g_A^2}$ for single particle states $n\ell J$. They are independent of n. For $\ell = 0$ only the first entry should be considered.

	$J_i = \ell + \frac{1}{2}$	$J_i = \ell - \frac{1}{2}$
$J_f = \ell + \frac{1}{2}$	$\frac{J_f+1}{J_f}$	$\frac{2J_f+1}{J_f}$
$J_f = \ell - \frac{1}{2}$	$\frac{2J_f+1}{J_f+1}$	$\frac{J_f}{J_f+1}$

The evaluation of the many body reduced matrix element can be obtained in the context of the shell model. This general case is beyond the scope of this book. The following selection rules hold:

$$\Delta J = |J_f - J_i| \leq 1,\ J_i = 0, J_f = 0 \text{ forbidden, no parity change}$$
$$\Delta I = |I_f - I_i| \leq 1,\ I_i = 0, I_f = 0 \text{ forbidden, } \Delta I_3 = \pm 1 \text{ for } e\mp. \quad (13.48)$$

For the Fermi transitions the isospin selection rules hold, but $J_i = J_f$.

In the special case of the neutron decay we recover the above result:

$$|ME_{n\to p}|^2_{GT} = g_A^2 \frac{1}{2}(\sqrt{6})^2 = 3g_A^2.$$

In the case the nucleus can be described by a single shell model states $n\ell J_i$, $J_i = \ell \pm 1/2$ and $n\ell J_f$, $J_f = \ell \pm 1/2$ the reduced matrix element can be evaluated, see, e.g., [deShalit and Feshbach (1974)], [Hecht (2000)] and Section 14.4. Using such reduced ME one finds the G-T matrix element given in Table 13.1. They are independent of n.

13.4 Applications

We will begin by evaluating the nuclear matrix elements for the triton decay, ${}^3_1\text{H} \to {}^3_2\text{He} + e^- + \tilde{\nu}_e$. For the Fermi transition we clearly have $|M_F|^2 = 1$ and so we need evaluate $|M_{GT}|^2$.

In the simple shell model description the A $= 3$ system is composed of three $0s1/2$ nucleons. In the isospin basis we write:

$${}^3_1\text{H} = |(0s1/2)^3 I = 1/2, I_3 = 1/2, s = 1/2\rangle,$$

$${}^3_1\text{He} = |(0s1/2)^3 I = 1/2, I_3 = -1/2, s = 1/2\rangle.$$

In evaluation the matrix element it is convenient to express it in terms of a product of a two nucleon and one nucleon wave function

$${}^3_1\text{H} = |(0s1/2)^3 I = 1/2, I_3 = 1/2; s = 1/2\rangle$$

$$= \frac{1}{\sqrt{2}}([0s1/2)^2(1;0)] \times 0s1/2]1/2, 1/2; 1/2 + [0s1/2)^2(0;1)]$$
$$\times\, 0s1/2]1/2, 1/2; 1/2)$$

where (0;1) and (1;0) are the only $(I; J)$ values allowed for the two nucleon system allowed by antisymmetry. The expression for ${}^3_1\mathrm{He}$ is analogous. Isospin conservation requires that the reduced ME to be of the form

$$\langle {}^3_2\mathrm{He}||\Omega_{GT}||{}^3_1\mathrm{H}\rangle = 3\frac{1}{2}$$

$$([0s1/2)^2(1;0) \times 0s1/2]1/2, -1/2; 1/2||\Omega_{GT}||[0s1/2)^2(1;0)$$
$$\times\, 0s1/2]1/2, 1/2; 1/2+$$
$$[0s1/2)^2(0;1) \times 0s1/2]1/2, -1/2; 1/2||\Omega_{GT}||[0s1/2)^2(0;1)$$
$$\times\, 0s1/2]1/2, 1/2; 1/2),$$

with the factor of 3 coming from the fact that 3 particles can interact. But

$$[0s1/2)^2(1;0) \times 0s1/2]1/2, -1/2; 1/2||\Omega_{GT}||[0s1/2)^2(1;0)$$
$$\times\, 0s1/2]1/2, 1/2; 1/2$$
$$= \langle 1, 0, 1/2, 1/2|1/2, 1/2\rangle\langle 1, 0, 1/2, -1/2|1/2, -1/2\rangle$$
$$\langle 1/2, -1/2|\tau_+|1/2, 1/2\rangle R_0.$$

The first factors are isospin C-G coefficients $\frac{1}{\sqrt{3}}\frac{-1}{\sqrt{3}} = -\frac{1}{3}$

$$[0s1/2)^2(0;10) \times 0s1/2]1/2, -1/2; 1/2|\Omega_{GT}|[0s1/2)^2(0;1)$$
$$\times\, 0s1/2]1/2, 1/2; 1/2$$
$$= \langle 0, 0, 1/2, 1/2|1/2, 1/2\rangle\langle 0, 0, 1/2, -1/2|1/2, -1/2\rangle$$
$$\langle 1/2, -1/2|\tau_+|1/2, 1/2\rangle R_1$$

the isospin C-G coefficients now yield 1. R_0 and R_1 are the spin reduced matrix elements.

$$R_0 = \langle[0 \times 1/2]1/2||\boldsymbol{\sigma}||[0 \times 1/2]1/2\rangle = \langle 1/2||\boldsymbol{\sigma}||1/2\rangle$$

$$R_1 = \langle[1 \times 1/2]1/2||\boldsymbol{\sigma}||[1 \times 1/2]1/2\rangle = -\frac{1}{3}\langle 1/2||\boldsymbol{\sigma}||1/2\rangle.$$

Finally

$$\langle {}^3_2\mathrm{He}|\Omega_{GT}|{}^3_1\mathrm{H}\rangle = 3\frac{1}{2}\left(\frac{-1}{3} + \frac{-1}{3}\right)\langle 1/2||\boldsymbol{\sigma}||1/2\rangle = -\langle 1/2||\boldsymbol{\sigma}||1/2\rangle$$

and

$$|M_{GT}\left({}^3_1\mathrm{H} \rightarrow {}^3_2\,\mathrm{He} + e^- + \tilde{\nu}_e\right)|^2 = \frac{1}{2}\left(-\langle 1/2||\boldsymbol{\sigma}||1/2\rangle\right)^2 = 3, \qquad (13.49)$$

i.e. the matrix elements are the same as in neutron decay.

For a nuclear physicist it is not surprising that the reduced ME for the A = 3 system is the opposite of that for a 0s1/2 nucleon. It can be viewed as 0s1/2 hole on the closed os shell. This can be used for other nuclei with 1 particle less than the the closed shell. As we mentioned the single particle approach can be used for one particle outside a closed shell, see Fig. 2.22. This way one can easily evaluate with the aid of Table 13.1 the G-T matrix elements for the following transitions:

- ${}^{11}_{6}\mathrm{C} \rightarrow {}^{11}_{5}\mathrm{B}$, (one hole, ${}^{12}_{6}\mathrm{C}$ assumed as closed shell).
- ${}^{13}_{7}\mathrm{N} \rightarrow {}^{13}_{6}\mathrm{C}$, (one particle, ${}^{12}_{6}\mathrm{C}$ assumed as closed shell).
- ${}^{15}_{8}\mathrm{O} \rightarrow {}^{15}_{7}\mathrm{N}$, (one hole, ${}^{16}_{8}\mathrm{O}$ assumed as closed shell).
- ${}^{17}_{9}\mathrm{F} \rightarrow {}^{17}_{8}\mathrm{O}$, (one particle, ${}^{16}_{8}\mathrm{O}$ assumed as closed shell) etc.

The evaluation of the Fermi and Gamow-Teller matrix elements is left as an exercise. The goodness of the approximation depends on the goodness of the closed shell description. At some point this approach dos not work, since the protons and neutrons occupy single particle levels, which do not have the same n and/or ℓ. Then the beta decays are not of the allowed Fermi and Gamow-Teller type discussed here, with the selection rules given above. Such decays are are of a different type and hey are called **forbidden beta decays**. The reader interested on such topics is referred to the literature, e.g. [Konopinski (1966)], [deShalit and Feshbach (1974)].

Anyway using the fact that:

$$|ME_{n\rightarrow p}|^2 = f_V^2 + 3f_A^2 \approx 5.613$$

from Eq. (13.30) we obtain the $ft = 1062$ s. One can check the value of the matrix element obtained above in the one nucleon approximation by comparing the corresponding ft values with those of Table 13.2. One must first check, however, whether the processes indicated in this table are beta decays or electron captures.

13.5 Electron capture

The relevant process is indicated by

$$e_b^- + A(N,Z) \rightarrow A(N,Z-1) + \nu_e, \tag{13.50}$$

where the absorbed electron is in a bound atomic orbit. Electron capture is the primary decay mode for isotopes with a relative superabundance of protons in the nucleus, but with insufficient energy difference between the isotope and its prospective daughter (the isobar with one less positive

Table 13.2: The ft values for various weak transitions in the case of mirror nuclei taken from Konopinski, [Konopinski (1966)].

parent	daughter	single particle transition	ft[s]
${}^{3}_{1}\mathrm{H}$	${}^{3}_{2}\mathrm{He}$	$0s_{1/2} \to 0s_{1/2}$	1092
${}^{7}_{4}\mathrm{Be}$	${}^{7}_{3}\mathrm{Li}$	$0p_{3/2} \to 0p_{3/2}$	1730
${}^{11}_{6}\mathrm{C}$	${}^{11}_{5}\mathrm{B}$	$0p_{3/2} \to 0p_{3/2}$	1730
${}^{13}_{7}\mathrm{N}$	${}^{13}_{6}\mathrm{C}$	$0p_{1/2} \to 0p_{1/2}$	3876
${}^{15}_{8}\mathrm{O}$	${}^{15}_{7}\mathrm{N}$	$0p_{1/2} \to 0p_{1/2}$	3876
${}^{17}_{9}\mathrm{F}$	${}^{17}_{8}0$	$0d_{5/2} \to 0d_{5/2}$	1915
${}^{19}_{10}\mathrm{Ne}$	${}^{19}_{9}\mathrm{F}$	$1s_{1/2} \to 1s_{1/2}$	1092
${}^{25}_{13}\mathrm{Al}$	${}^{25}_{12}\mathrm{Mg}$	$0d_{5/2} \to 0d_{5/2}$	1915
${}^{27}_{14}\mathrm{Si}$	${}^{27}_{13}\mathrm{Al}$	$0d_{5/2} \to 0d_{5/2}$	1915
${}^{29}_{15}\mathrm{P}$	${}^{29}_{14}\mathrm{Si}$	$1s_{1/2} \to 1s_{1/2}$	1092
${}^{31}_{16}\mathrm{S}$	${}^{31}_{15}\mathrm{P}$	$1s_{3/2} \to 1s_{1/2}$	1092
${}^{33}_{17}\mathrm{Cl}$	${}^{33}_{16}\mathrm{S}$	$0d_{3/2} \to 0d_{3/2}$	3090
${}^{35}_{18}\mathrm{Ar}$	${}^{35}_{17}\mathrm{Cl}$	$0d_{3/2} \to 0d_{3/2}$	3090
${}^{39}_{20}\mathrm{Ca}$	${}^{39}_{19}\mathrm{K}$	$0d_{3/2} \to 0d_{3/2}$	3090
${}^{41}_{20}\mathrm{Ca}$	${}^{41}_{19}\mathrm{K}$	$0f_{7/2} \to 0f_{7/2}$	8476

charge) for the nuclide to decay by emitting a positron. Positon decay, left process in Eq. (13.4), dominates, however, if the radioactive isotopes involved have sufficient energy.

It is an old process considered soon after Fermi's theory of beta decay. The theory of electron capture was first discussed by Gian-Carlo Wick in 1934 and further developed by Hideki Yukawa and others. K-electron capture was first observed by Luis Alvarez in vanadium-48 [Alvarez (1937)].

Some light nuclei weak transitions, which can only proceed via electron capture, are:

$$e_b^- + {}^{7}_{4}\mathrm{Be} \to {}^{7}_{3}\mathrm{Li}$$

$$e_b^- + {}^{26}_{13}\mathrm{Al} \to {}^{26}_{12}\mathrm{Mg}$$

$$e_b^- + {}^{37}_{18}\mathrm{Ar} \to {}^{37}_{17}\mathrm{Cl}$$

$$e_b^- + {}^{40}_{19}\mathrm{K} \to {}^{40}_{18}\mathrm{Ar}$$

$$e_b^- + {}^{59}_{28}\mathrm{Ni} \to {}^{59}_{27}\mathrm{Co}.$$

The transition rate can be obtained in a way analogous to Eq. (13.21) except that the nuclear matrix element involves the bound electron wave function and the kinematics is different, in other words:

$$d\Gamma = |ME_{i\to f}\phi_b|^2 \frac{G_F^2}{2}\frac{m_e m_\nu}{E_e E_\nu}\frac{2(E_e E_\nu \pm \mathbf{p}_e.\mathbf{p}_\nu)}{m_e m_\nu}\frac{d^3\mathbf{p}_\nu}{(2\pi)^3}\frac{d^3\mathbf{p}_A}{(2\pi)^3}$$
$$\times (2\pi)\delta\left(\Delta - \epsilon_b - E_\nu\right)(2\pi)^3\delta\left(\mathbf{p}_A + \mathbf{p}_\nu\right). \quad (13.51)$$

Where ϵ_b the binding energy of the absorbed electron and ϕ_b its wave function. Integrating over the nuclear recoil momentum withe the aid of the momentum conserving δ function and over the neutrino momentum taking note of the energy conserving δ function we obtain:

$$\Gamma = |ME_{i\to f}\phi_b|^2 G_F^2 \frac{4\pi}{(2\pi)^2}(\Delta - \epsilon_b)^2. \quad (13.52)$$

To a good approximation it is sufficient to replace the bound electron matrix element by its wave function at the origin and thus

$$|ME_{i\to f}\phi_b|^2 = |ME_{i\to f}|^2 \phi_b^2(0). \quad (13.53)$$

Where the nuclear matrix element can be evaluated as in positon decay. We thus can write (in natural units)

$$\Gamma = \kappa(Z)\,|ME_{i\to f}|^2 \left(G_F^2 m_e^4\right)\frac{1}{2\pi^3}m_e f(\epsilon_0),$$
$$f(\epsilon_0) = \left(\epsilon_0 - \frac{\epsilon_b}{m_e}\right)^2, \quad \kappa(Z) = 2\pi^2\frac{\phi_b^2(0)}{m_e^3}. \quad (13.54)$$

Thus all the above formulas hold except that

$$|ME_{i\to f}|^2 \to \kappa(Z)\,|ME_{i\to f}|^2.$$

Thus, e.g.,

$$f(\epsilon_0)t_{\frac{1}{2}} = \ln(2)\frac{1}{K\kappa(Z)\,|ME_{i\to f}|^2}(2\pi^3)\frac{\hbar}{m_e c^2}, \quad \frac{\hbar}{m_e c^2} \approx 1.285\times 10^{-21}\ \text{s}. \quad (13.55)$$

The $N = 1(n_r = 0, \ell = 0)$ hydrogenic wave function is given by:

$$\phi_b(r) = \frac{a^{3/2}}{\sqrt{2}}e^{-ar}, \ a = 2\alpha Z\frac{m_e c^2}{\hbar c} \quad (13.56)$$

$$\kappa(Z) = 2\pi^2\frac{1}{2}(2\alpha Z)^3 = 8\pi^2(\alpha Z)^3.$$

For $(\alpha Z) < 0.233$ one finds $\kappa < 1$.

$\kappa(Z)$ can also be evaluated for other electron orbits. If necessary, especially for high Z, distortion factors can be incorporated taking into account the finite nucleus effects but we will elaborate here on such matters, see, e.g., [Konopinski (1966)].

Let us now consider the decay of $^{41}_{20}$Ca. In this case the available energy is only 0.44 ± 0.02 MeV. So positron emission is energetically forbidden. For the allowed electron capture the nuclear transition is a single particle one, $0f_{7/2} \to 0f_{7/2}$, with $\kappa(20) = 0.246$. The nuclear matrix element can be obtained from Table 13.1, but the ft value in Table 13.2 must be obtained using this value of κ.

13.6 Neutrino nucleus scattering

In this case a flux of neutrinos, measured in particles per unit area per unit time, is falling on a nuclear target. The process in question is:

$$\nu_e + A(N,Z) \to A(N-1,Z+1) + e^- \text{ or } \tilde{\nu}_e + A(N,Z) \to A(N+1,Z-1) + e^+ \tag{13.57}$$

involving the charged current and

$$\nu + A(N,Z) \to A(N,Z) + \nu \text{ or } \tilde{\nu}A(N,Z) \to A(N,Z)\tilde{\nu}, \text{ neutral current} \tag{13.58}$$

(with ν any neutrino flavor). If the neutrino is of sufficient energy, the final nucleus can be in an excited state. The daughter nucleus can also break up, but we will not concern ourselves here with this last possibility.

The relevant quantity to evaluate is the cross section for the process, see Section 9.6 or, e.g., the companion book [Vergados (2017)].

13.7 The charged current contribution

The process is similar to beta decay, except that now we are talking about a cross section rather than the rate and we have two body rather than three body kinematics. The differential cross section is given by:

$$d\sigma = |ME_{i \to f}|^2 \frac{1}{\upsilon} \frac{G_F^2}{2} \frac{m_e m_\nu}{E_e E_\nu} \frac{2(E_e E_\nu \pm \mathbf{p}_e . \mathbf{p}_\nu)}{m_e m_\nu} \frac{d^3\mathbf{p}_\nu}{(2\pi)^3} \frac{d^3\mathbf{p}_A}{(2\pi)^3}$$
$$(2\pi)\delta\left(E - \Delta - E_x - E_\nu\right)(2\pi)^3\delta\left(\mathbf{p}_A + \mathbf{p}_\nu\right). \tag{13.59}$$

Where Δ is the difference in the energy of the nuclear ground states, $(E_f)_{gs} - (E_i)_{gs}$, E is the oncoming neutrino energy, E_x and E_ν the excitation energy of the nucleus and the energy of the outgoing neutrino energy.

We neglected the nuclear recoil energy. Proceeding as above with the integration over the nuclear momentum and the outgoing neutrino momentum we find:

$$\sigma(E) = |ME_{i\to f}|^2 \frac{1}{v} G_F^2 \frac{4\pi}{(2\pi)^2} (-\Delta - E_x + E)^2 . \tag{13.60}$$

Since the velocity of the neutrino is c, in standard units we obtain:

$$\sigma(E) = |ME_{i\to f}|^2 \frac{1}{\pi} K \left(\frac{\hbar c}{m_e c^2}\right)^2 \frac{(-\Delta + E - E_x)^2}{m_e^2 c^4}. \tag{13.61}$$

As expected the cross section is given in units of length^2. One finds

$$\sigma(E) = |ME_{i\to f}|^2 \Lambda \frac{(-\Delta + E - E_x)^2}{m_e^2 c^4} \tag{13.62}$$

with

$$\Lambda = \frac{1}{\pi} K \left(\frac{\hbar c}{m_e c^2}\right)^2 = K \frac{1}{\pi} \left(\frac{197 \times 10^{-13} \text{ MeV-cm}}{0.511 \text{ MeV}}\right)^2 = 4.38 \times 10^{-45} \text{ cm}^2$$
$$= 4.38 \times 10^{-21} \text{ b} = 4.38 \times 10^{-9} \text{ pb}$$

$1 \text{ b} = 10^{-24} \text{ cm}^2$.

The number of events per unit time is given by

$$\frac{dn(E)}{dt} = \Phi \sigma(E) \tag{13.63}$$

where Φ is the neutrino flux. I.e. the number of neutrinos per unit area per unit time.

The above hold for a monochromatic neutrino beam. The neutrino sources, however, are not monochromatic but they are characterized by a neutrino spectrum $df(E_\nu)/dE_\nu$ such that the total number of neutrinos emitted is $\int f(E_\nu) dE_\nu$. Then the neutrino flux depends on the neutrino energy E_ν. Thus the number of events per unit time are given by

$$\frac{dn(E_\nu)}{dE_\nu dt} = \Phi(E_\nu) \sigma(E_\nu). \tag{13.64}$$

If the flux in normally expressed in units $\text{cm}^{-2}\text{s}^{-1}$, see, e.g., the spectrum of boron neutrinos in Fig. 12.4. Thus the rate per second per unit energy becomes

$$\frac{dn(E_\nu)}{dE_\nu dt} = |ME_{i\to f}|^2 \frac{(-\Delta - E_x + E_\nu)^2}{m_e^2 c^4} \Phi(E_\nu) \times (4.38 \times 10^{-45} \text{ s}^{-1}). \tag{13.65}$$

The corresponding rate per year is

$$\frac{dn(E_\nu)}{dE_\nu dt} = |ME_{i\to f}|^2 \frac{(-\Delta - E_x + E)^2}{m_e^2 c^4} \Phi(E_\nu) \times (1.37 \times 10^{-37}\ \mathrm{y}^{-1}). \quad (13.66)$$

This is, of course, per nucleus. Where in the target one may have N_A nuclei so the expected event rate becomes:

$$\frac{dN(dE_\nu)}{dE_\nu dt} = N_A \frac{dn(E_\nu)}{dE_\nu dt}. \quad (13.67)$$

One expresses N_A per kg of target. So $N_A \approx 6 \times 10^{26}/A$, where A is the nuclear mass number. Thus

$$\frac{dN(E_\nu)}{dE_\nu dt} = \frac{|ME_{i\to f}|^2}{A} \frac{(-\Delta - E_x + E_\nu)^2}{m_e^2 c^4} \Phi(E_\nu) \times (0.83 \times 10^{-10}\ \text{kg-y}^{-1}). \quad (13.68)$$

One can obtain the total rate by integrating over the spectrum

$$\frac{dN}{dt} = \frac{|ME_{i\to f}|^2}{A} f_t(E_\nu - E_x + \Delta) \times (0.83 \times 10^{-10}\ \text{kg-y}^{-1}), \quad (13.69)$$

with

$$f_t(E_x + \Delta) = \int_{E_x+\Delta}^{(E_\nu))_{max}} \frac{(E_\nu - E_x - \Delta)^2}{m_e^2 c^4} \Phi(E_\nu) dE_\nu. \quad (13.70)$$

Where $(E_\nu))_{max}$ is the maximum neutrino energy.

As an application we will consider $f_t(x)$ obtained:

a) Using antineutrinos with the flux of reactor antineutrinos shown in Fig. 13.4. We used the neutrino spectra of [Davis *et al.* (1979)] and [Hayes and Vogel (2016)].

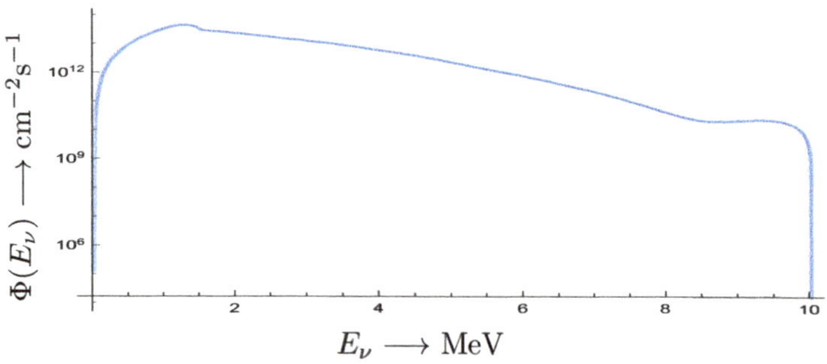

Fig. 13.4: The reactor neutrino in units of $\mathrm{cm}^{-2}\mathrm{s}^{-1}$ as a function of the neutrino energy flux in MeV. It is obtained at a distance of 10 m from a reactor emitting 10^{23} neutrinos per second.

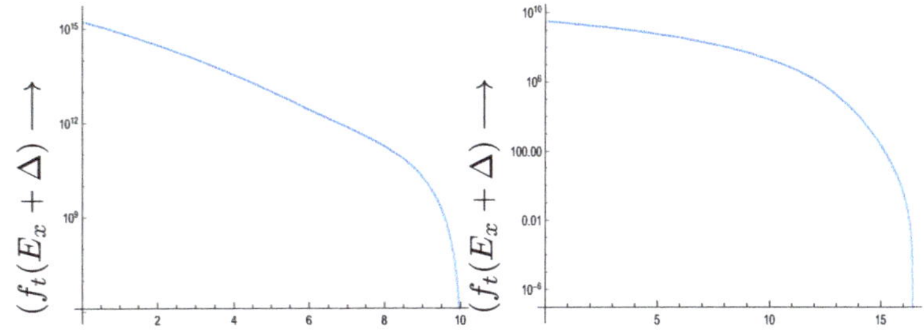

(a) (b)

$(E_x + \Delta) \longrightarrow$ MeV

Fig. 13.5: The function $f_t(E_x + \Delta)$ as a function of $E_x + \Delta$ obtained: (a) with the flux of reactor antineutrinos and (b) with the flux of solar boron neutrinos.

(b) Using neutrinos with the solar boron neutrino flux exhibited in Fig. 12.4. Then one finds the functions $f_t(E_x + \Delta)$ exhibited in Fig. 13.5.

To obtain the total event rate for boron solar neutrinos one must multiply $(0.83 \times 10^{-10}$ kg-y$^{-1})f_t(E_x + \Delta)$, with $f_t(E_x + \Delta)$ read from Fig. 13.5 at the appropriate value of $E_x + \Delta$, with $\frac{|ME_{i \to f}|^2}{A}$ for the specific nucleus.

13.8 The neutral current contribution

The coupling is smaller by a factor of 2, see Eq. (13.18). Now the Fermi transition operator $\Omega_f \approx \sum_n \frac{1+\tau_3(n)}{2}$, see the companion book [Vergados (2017)]. This means that only the neutrons contribute. On the other hand the G-T matrix element is similar to that of the charge current, $\Omega_{GT} = \sum_n \tau_3(n)\boldsymbol{\sigma}(n)$. Thus

$$\sigma(E)_{NC} = |ME_{i \to f}|^2 \, \frac{\Lambda}{4} \frac{(E_\nu - E_x)^2}{m_e^2 c^4}. \tag{13.71}$$

With this expression of the cross section Eq. (13.64) becomes

$$\left. \frac{dn(E_\nu)}{dE_\nu dt} \right|_{NC} = \Phi(E_\nu)\sigma(E_\nu)_{NC} \tag{13.72}$$

where now $\Delta = 0$. With these modifications one proceeds as in the case of charged current discussed above.

Of particular interest is the elastic neutrino-nucleus scattering. Then the quantity $f_t(0)$ attains the maximum of $f_t(x)$. Furthermore in the case of the Fermi interaction all neutrons in the nucleus contribute coherently and and we obtain

$$M_F = N^2 \Rightarrow \tag{13.73}$$

$$\frac{dN}{dt} = \frac{1}{4}\frac{N^2}{A} f_t(0) \times \left(0.83 \times 10^{-10}\ (\text{kg-y})^{-1}\right). \tag{13.74}$$

The Fermi interaction dominates and one gets a very large event rate especially[8] for neutron rich nuclei, that is:

$$\frac{dN}{dt} = 3.5 \times 10^4 \frac{N^2}{A}, \text{ per kg-y} \tag{13.75}$$

for reactor antineutrinos, see Fig. 13.5. For boron neutrinos the effect is much smaller

$$\frac{dN}{dt} = 0.071 \frac{N^2}{A}, \text{ per kg-y.} \tag{13.76}$$

The smaller value is due to the fact that now the neutrino flux is much lower.

The neutron coherence is an advantage of neutral current detectors, in particular spherical TPC detectors discussed in Section 11.4.6.2.

The only disadvantage from an experimental point of view is that there is no good experimental signature to look for in such process, since the final nucleus cannot decay. So one can only measure the energy of the recoiling nucleus, which is not easy for low counting rates. Great progress has been made in recent years in this direction, motivated in large part by direct dark matter searches, see, e.g., the companion book [Vergados (2017)]. After collisions with other particles in the target, the energy of the recoiling nucleus can be converted into light, heat and ionization. Thus it can be detected by scintillation, ionization and bolometric detectors discussed in Chapter 11, Section 11.4. A problem which remains is the fact that, since the counting rates expected in dark matter searches are quite low, their signature cannot be easily distinguished from that of unavoidable backgrounds.

The rate, of course, for coherent neutrino nucleus scattering is expected to be much larger with a source of more energetic neutrinos, like e.g. in the case of the detection of supernova neutrinos [Giomataris and Vergados (2006b)] or from exotic neutron spallation sources [Vergados *et al.* (2009)]. So such experiments can only benefit from the experiences gained in direct dark matter searches.

[8]The rate in the case of heavy nuclei is going to be reduced, if the nuclear form factor, Section 2.1.2, is employed, since $|M_F(q^2)|^2 = N^2 \left(F(q^2)\right)^2$.

13.9 Suppression of allowed beta transitions

There exist nuclear systems, which can undergo both beta and inverse beta decays. In some cases these processes suffer from suppression of the rates from those expected in typical allowed Fermi and Gamow-Teller. One such example is the $A = 14$ system exhibited in Fig. 13.6.

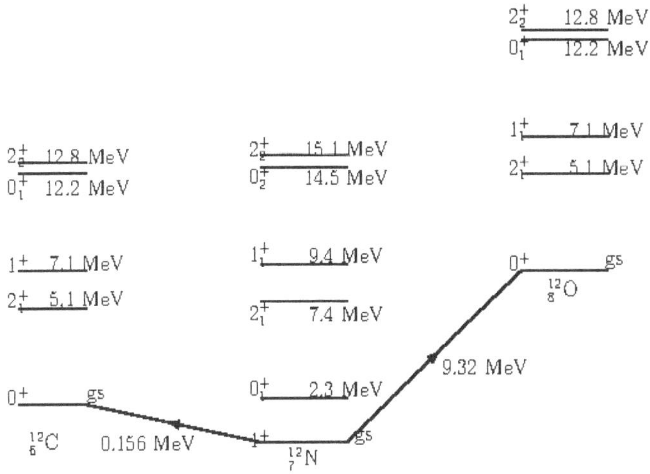

Fig. 13.6: The theoretically obtained energy levels of the isotriplet $A = 14$ system. The isotriplet states of $^{14}_{7}$ have been shifted by the coulomb energy.

The ground state of the $^{14}_{7}\mathrm{N}$ is 1^+, it has angular momentum 1 and parity $+$, and isospin 0. The ground states of $^{14}_{6}\mathrm{C}$ and $^{14}_{8}\mathrm{O}$ are 0^+ and isospin 1. The decays

$$^{14}_{6}\mathrm{C} \rightarrow \, ^{14}_{7}\mathrm{N} + \tilde{\nu}_e + e^-, \, ^{14}_{8}\mathrm{O} \rightarrow \, ^{14}_{7}\mathrm{N} + \nu_e + e^+$$

look like bona fide allowed β^- and β^+ G-T decays respectively, since the Fermi type decays are forbidden due to angular momentum election rules. The ft values however, are very long, namely $\log ft^- = 9.04$ and $\log ft^+ = 7.27$ respectively. Could they be forbidden decays?

Before we answer that let us look and the nuclear wave functions. With respect to the $^{16}_{8}\mathrm{O}$ closed shells they are expected to be 2 p-shell holes. The $^{14}_{7}\mathrm{N}1^+$ has isospin 0, i.e. its wave function is antisymmetric in isospin space. So

- both space and spin parts must be symmetric, $L = 0, 2;\ s = 1$
- both space and spin parts must be anti symmetric, $L = 1;\ s = 0$.

The wave 0^+ wave functions of ${}^{14}_{6}C$ and ${}^{14}_{8}O$ have isospin 1, i.e. its wave function is symmetric in isospin space. So for the total wave function to be completely antisymmetric

- the space must be symmetric and spin part must be antisymmetric, $L = 0;\ s = 0$
- the space must be antisymmetric and spin part must be symmetric, $L = 1;\ s = 1$.

Thus

$$\psi\left({}^{14}_{7}\text{N}1^+\right) = \left|\alpha\left[L=0, s=1\right]^{J=1} + \beta\left[L=2, s=1\right]^{J=1} + \gamma\left[L=1, s=0\right]^{J=1}\right\rangle \tag{13.77}$$

with $\alpha^2 + \beta^2 + \gamma^2 = 1$. The other two wave functions are of the form:

$$\psi\left({}^{14}_{6}\text{C}^+\right)\ \text{gs or}\ \psi\left({}^{14}_{8}\text{O}^+\right) gs = \left|\xi\left[L=0, s=0\right]^{J=0} + \eta\left[L=1, s=1\right]^{J=0}\right\rangle. \tag{13.78}$$

The corresponding first excited 0^+_1 state is

$$0^+_1 = -\eta\left|\left[L=0, s=0\right]^{J=0} + \xi\left[L=1, s=1\right]^{J=0}\right\rangle$$

with $\xi^2 + \eta^2 = 1$.

Furthermore

$$2^+_1 = \xi_2\left|\left[L=2, s=0\right]^{J=2} + \eta_2\left[L=1, s=1\right]^{J=2}\right\rangle$$

$$2^+_2 = -\eta_2\left|\left[L=2, s=0\right]^{J=2} + \xi_2\left[L=1, s=1\right]^{J=2}\right\rangle$$

with $\xi_2^2 + \eta_2^2 = 1$.

We thus see that the G-T transition will be greatly suppressed, if β is close to one. Shell model calculations show that this indeed the case. In fact one finds [Baer *et al.* (1975)] that

$$\alpha = 0.1636,\ \beta = 0.9564, \gamma = 0.3410,$$

$$\xi = 0.7980,\ \eta = 0.6027,\ \xi_2 = 0.9068,\ \eta_2 = -0.4216.$$

This trend of large β, i.e. $L = 2$ component, is confirmed by other authors [Genz *et al.* (1991)] and references therein.

Using the above wave function we obtain $M_{GT} = -0.1135$. We thus find $\log(ft_{1/2} = 5.3$, which is a bit too far from experiment. The simple model is not adequate. The more elaborate calculations in [Genz *et al.* (1991)] yield for β^- $\log ft^- = 9.48$, which is fairly close to the observed one. In spite of the fact these authors claim that this agreement does not originate from a fit, but it is the result of a calculation with realistic wave functions, their arguments are not convincing. They also obtain $\log ft^+ = 7.27$ for β^+, which in amazing agreement with experiment. Such an agreement with experiment does not constitute a proof that there was no fine tuning involved. We will, thus, use the value of $M_{GT} = -0.1135$ obtained theoretically, in full knowledge that there uncertainties in it.

Let us now consider the neutrino scattering

$$\nu_e + {}^{14}_{7}\mathrm{N}\,(1^+) \rightarrow {}^{14}_{8}\mathrm{O}\,(J^+) + e^- .$$

This process has a high threshold, 9.32 MeV for gs transitions. Even excited states can, however, be populated using solar Boron neutrinos, which have a maximum energy of about 17 MeV.

The second possibility is

$$\nu_\alpha(\tilde{\nu}_\alpha) + {}^{14}_{7}\mathrm{N}(1^+) \rightarrow {}^{14}_{7}\mathrm{N}(J^+) + \nu_\alpha(\tilde{\nu}_\alpha).$$

This can be populated either using neutrinos of any flavor, e.g. boron neutrinos, or antineutrinos of any flavor, e.g. reactor antineutrinos.

Finally we will consider the process

$$\tilde{\nu}_e + {}^{14}_{7}\mathrm{N}(1^+) \rightarrow {}^{14}_{6}\mathrm{C}(J^+) + e^+$$

using, e.g., reactor antineutrinos. Of particular interest is the ground state to ground state transition

$$\tilde{\nu}_e + {}^{14}_{7}\mathrm{N}(1^+) \rightarrow {}^{14}_{6}\mathrm{C}(0^+) + e^+ .$$

This possibility generated a lot of excitement, because of the low threshold of $0.156 + 0.511 = 0.667$ MeV involved. Current neutrino detectors are based on the inverse beta decay:

$$\tilde{\nu}_e + p \rightarrow n + e^+ .$$

This process, however, has the disadvantage that the antineutrino energy must be at least 1.806 MeV. This threshold comes from the fact that the neutron is heavier than the proton and the positron must be created. Thus neutrino spectra, e.g. in reactors, can only be measured above 1.8 MeV. The low threshold for the A = 14 system opens a new window, but, unfortunately, the nuclear ME involved is very small $\frac{M_{GT}}{\sqrt{3}} = -0.0655$ and uncertain. Anyway we exhibit the spectrum obtained from each target in the case of reactor antineutrinos in Fig. 13.7.

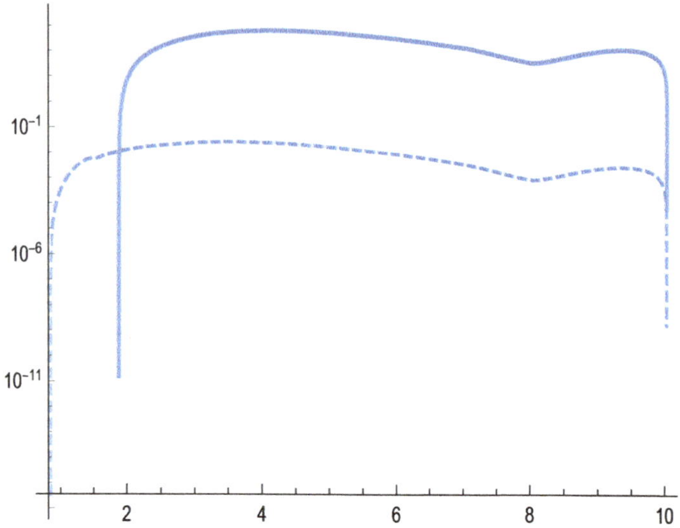

Fig. 13.7: The spectra obtained in the case of scattering of reactor antineutrinos. (a) In the case ${}^{14}_{7}$N to the ${}^{14}_{6}$C gs (dotted line) and (b) in inverse beta decay (continuous line).

We see that the effect of threshold is visible in the case of low energy neutrinos. It maybe more important in the case of reactor neutrinos resulting from some fission fragments, see e.g. [Hayes and Vogel (2016)]. In any case the ${}^{14}_{7}$N to the ${}^{14}_{6}$C gs may be used to explore the low energy regime for which experimental information cannot be obtained from the inverse beta decay.

In any case let us now continue with neutrino scattering ${}^{14}_{7}$N leading to ${}^{14}_{6}$C. The final states of interest are the ground state $|J = 0,\ I = 1\rangle$ and the excited states $|2^+_1\rangle$, $|1^+_1\rangle$, $|0^+_1\rangle$ and $|2_2\rangle$ with isospin $I = 0$ at energies $5. + 0.156 = 5.156$, $7.1 + 0.156 = 7.256$, $12.2 + 0.156 = 12.356$ and $12.8 + 0.156 = 18.956$ respectively. Using the wave functions of Ref. [Baer *et al.* (1975)] we obtain the nuclear matrix elements -0.0665, -1.0793, 0.3422, 0.0972 and 0.0853 respectively.

We will also consider transitions to the excited states of ${}^{14}_{7}$N with the same quantum numbers as those of ${}^{14}_{6}$C, which belong to the same isospin multiplet. The excitation energies are slightly different [Baer *et al.* (1975)], namely 2.3, 7.4, 9.4, 14.5 and 15.1 due to Coulomb interaction.

The obtained neutrino-nitrogen scattering results for in the case of reactor antineutrinos are summarized in Table 13.3.

Table 13.3: Results on the $\tilde{\nu}_e - {}^{14}_{6}$N scattering involving reactor antineutrinos, both due to the charged and the neutral weak currents.

	state	b+E_x(MeV)	$M^2_{GT}/3$	$f_t(b+E_x)$	$\frac{d\Gamma}{dt}$(kg-y) $^{-1}$
	0^+	0.156	0.4422×10^{-2}	1.5×10^{15}	61
	2_1^+	5.256	1.1513	7.1×10^{12}	75,0
${}^{14}_{6}$	1_1^+	7.156	0.1171	5.1×10^{11}	0.55
	0_1^+	12.356	0.0085	0	0
	2_2^+	12.956	0.693×10^{-3}	0	0
	0^+	2.3	0.004422	2.23938×10^{14}	9.02
	2_1^+	7.4	1.1513	4.21149×10^{11}	4.42
${}^{14}_{7}$N	1_1^+	9.4	0.1171	2.54174×10^{9}	0.0027132
	0_1^+	14.5	0.009459	0	0
	2_2^+	15.1	0.007372	0	0

We should mention that the suppression of the transition to the excited 0^+ state of ${}^{14}_{7}$N is confirmed by its weak M1 transition to the ground state. On the other hand the M1 transition of the 2_1 excited state is strong.

The largest by far rate is due to the coherent antineutrino to ground state scattering, see Eq. (13.75). Thus for A = 14 and N = 7 we find for reactor antineutrinos

$$\frac{dN}{dt} = \frac{7}{8} f_t(0) \times 0.83 \times 10^{-10} = 1.2 \times 10^5 \text{ events per kg-y} \tag{13.79}$$

while for B neutrinos

$$\frac{dN}{dt} = 0.245 \text{ events per kg-y.} \tag{13.80}$$

The neutrino-nitrogen scattering data for solar B neutrinos are summarized in Table 13.4. The suppression of the rates in the case ${}^{14}_{7}\mathrm{N} \rightarrow {}^{14}_{8}\mathrm{O}$ is understood due to the large threshold of 9.3 MeV.

Table 13.4: Results on the $\nu_e - {}^{14}_{6}\text{N}$ scattering involving boron solar neutrinos, both due to the charged and the neutral weak currents.

	state	b+E_x(MeV)	$M_{GT}^2/3$	$f_t(b+E_x)$	$\frac{d\Gamma}{dt}$(kg-y) $^{-1}$
${}^{14}_{8}\text{O}$	0^+	9.3	0.4422×10^{-2}	3.4×10^9	1.53×10^{-6}
	2_1^+	14.4	1.1513	1.9×10^4	2.05×10^{-8}
	1_1^+	16.4	0.1171	7.3×10^{-4}	7.8×10^{-15}
	0_1^+	21.5	0.009459	0	0
	2_2^+	22.1	0.007272	0	0
${}^{14}_{7}\text{N}$	0^+	2.3	0.4442×10^{-2}	1.8×10^9	7.08×10^{-5}
	2_1^+	7.08	1.1513	1.6×10^8	0.0017
	1_1^+	9.4	0.1171	3.5×10^7	3.70×10^{-5}
	0_1^+	14.5	0.009459	1.3×10^3	1.14×10^{-10}
	2_2^+	15.1	0.007272	90	6.00×10^{-12}

Chapter 14

Appendix A: Aspects of the shell model wave functions — Reduced matrix elements

In this chapter we are going to discuss the motion of a particle in a spherically symmetric nuclear potential. We will develop the tools for evaluating nuclear matrix elements in the context of the shell model, for more details on such matters see [Brown (2020)]. We will thus define the reduced matrix elements of spherical tensors and calculate them for some simple operators encountered in nuclear physics. Furthermore we will learn to express the two nucleon harmonic oscillator wave function in relative and center of mass coordinates.

14.1 Review of solving the Schrödinger equation

The Schrödinger equation for a particle of mass m in a spin-independent central potential, $U_0(r)$, $r = |\mathbf{r}|$, is

$$H_0|\alpha\rangle = (T + U|\alpha\rangle = \epsilon_\alpha|\alpha\rangle, \tag{14.1}$$

where T is the kinetic energy operator given in coordinate space by

$$T = -\frac{\hbar^2}{2m}\nabla^2 = -\frac{\hbar^2}{2m}\left(\frac{1}{r^2}\frac{d}{dr}r^2\frac{d}{dr} - \frac{1}{r^2}\boldsymbol{\ell}^2\right)$$

the ket $|\alpha\rangle$ is the single-particle wave function, and ϵ_α is the eigenvalue. $\boldsymbol{\ell}$ is the orbital angular momentum operator in units of $\hbar$. The coordinate space solution ϕ_α of this equation is written as:

$$\left[-\frac{\hbar^2}{2m}\left(\frac{1}{r^2}\frac{d}{dr}r^2\frac{d}{dr} - \frac{1}{r^2}\boldsymbol{\ell}^2\right) + U(r)\right]\phi_\alpha(\mathbf{r}) = \epsilon_\alpha\phi_\alpha(\mathbf{r}). \tag{14.2}$$

The solution, which is finite and single valued over the surface of a sphere of radius r can be chosen to be the form

$$\phi_\alpha = \psi_{n_r,\ell}(r)Y^\ell_m(\hat{r}),\ Y^\ell_m(\hat{r}) = Y^\ell_m(\theta,\phi)$$

i.e. the α stands for the set of quantum numbers n_r, ℓ, m. In the above expression ψ_{n_r} is the radial solution, with n_r indicating the number of times the radial function crosses the r axes, i.e. taking values $0, 1, 2, \cdots$ (sometimes, however, a different notation $1, 2, \cdots$ is adopted). $Y^{\ell}_{m}(\theta, \phi)$ are the spherical harmonics, with $\ell = 0, 1, 2, \cdots$ and m an integer, $|m| \leq \ell$. The spherical harmonics are eigensolutions of the operators $\boldsymbol{\ell}^2$ and $\boldsymbol{\ell}_z$, i.e.:

$$\boldsymbol{\ell}^2 Y^{\ell}_{m}(\theta, \phi) = \ell(\ell+1) Y^{\ell}_{m}(\theta, \phi),\ \boldsymbol{\ell}_z Y^{\ell}_{m}(\theta, \phi) = m Y^{\ell}_{m}(\theta, \phi). \tag{14.3}$$

They form an orhonormal set:

$$\int_0^{\pi} \sin\theta d\theta \int_0^{2\pi} d\phi \left(Y^{\ell'}_{m'}(\theta, \phi)\right)^* Y^{\ell}_{m}(\theta, \phi) = \delta_{\ell\ell'} \delta_{mm'}. \tag{14.4}$$

The radial wave function $\psi_{n_r,\ell}(r)$ is a solution of the one-dimensional radial equation:

$$\left[-\frac{\hbar^2}{2m}\left(\frac{1}{r^2}\frac{d}{dr}r^2\frac{d}{dr} - \frac{1}{r^2}\ell(\ell+1)\right) + U(r)\right]\psi_{n_r,l}(r) = \epsilon_{n_r,\ell}\psi_{n_r,l}(r). \tag{14.5}$$

Setting

$$\psi_{n_r,l}(r) = \frac{1}{r} R_{n_r,l}(r) \tag{14.6}$$

the previous equation can be written as:

$$-\frac{\hbar^2}{2m}\left(R''_{n_r,\ell}(r) - \frac{\ell(\ell+1)}{r^2} R_{n_r,\ell}(r)\right) + U(r) R_{n_r,\ell}(r) = \epsilon_{n_r,\ell} R_{n_r,\ell}(r) \tag{14.7}$$

where the potential was assumed to be spherically symmetric and, hence, the radial solution and the eigenvalues are independent of the quantum number m. $R''(r)$ indicates the second derivative of $R(r)$.
Remarks:

- The eigenvalues are independent of m. This is due to the assumed spherical symmetry of the problem.
- Equation (14.7) can be solved analytically only in a few cases. These include the Coulomb potential $U(r) \propto 1/r$ and the harmonic oscillator potential $U(r) \propto r^2$.
- In these two cases the symmetry of the problem is higher than spherical symmetry. Thus the eigenvalues are independent of the orbital angular momentum. They depend on a quantum number called **major quantum number**. In this case we say that the eigensolutions corresponding to the same major quantum number, but different ℓ, are degenerate.

- There exist, however, efficient numerical methods for obtaining the eigensolutions numerically for the cases of practical interest to nuclear physics.
- The radial functions can be chosen to be real. They can also be chosen to be non negative at the origin. Some people choose them to be non-negative at infinity.
- Only the solutions with zero orbital angular momentum, $\ell = 0$, are non zero at the origin.

14.2 The harmonic oscillator potential

In this case the potential takes the form:

$$U = \frac{1}{2}kr^2. \tag{14.8}$$

In this case (14.5) takes the form:

$$\left[-\frac{\hbar^2}{2m}\left(\frac{1}{r^2}\frac{d}{dr}r^2\frac{d}{dr} - \frac{\ell(\ell+1)}{r^2}\right) + \frac{1}{2}kr^2\right]\psi(r) = E\psi(r) \tag{14.9}$$

with E being the energy. Before proceeding further we like to transform the last equation to one involving dimensionless variables. Introduce:

$$r = b\xi,\ \omega = \sqrt{\frac{k}{m}},\ b = \sqrt{\frac{\hbar}{m\omega}},\ \psi(r) = \psi(b\xi) = u(\xi).$$

Then Eq. (14.9) becomes

$$\left[\frac{1}{\xi^2}\frac{d}{d\xi}\xi^2\frac{d}{d\xi}u(\xi) - \left(\frac{\ell(\ell+1)}{\xi^2} + \xi^2\right) + 2\epsilon\right]u(u) = 0 \tag{14.10}$$

with $\epsilon = E/\hbar\omega$. We now notice that for small ξ, ϵ and ξ^2 terms become negligible and the angular momentum term becomes dominant. In this case $u(\xi) \propto \xi^\ell$. On the other for large ξ the potential becomes dominant and

$$u(\xi) \propto e^{-\xi^2/2}(\text{ large } \xi).$$

We thus write

$$u(\xi) = \xi^\ell e^{-\xi^2/2}\upsilon(\xi)$$

with $\upsilon(\xi)$ obeying the equation:

$$\xi\upsilon'' + 2(\ell + 1 - \xi^2)\upsilon' + 2\xi\left(2\epsilon - 2\ell - 3\right)\upsilon = 0.$$

Unfortunately this equation does not coincide with any of the classical equation of mathematical physics. Consider however the transformation:

$$\xi = \sqrt{\rho},\ \upsilon(\xi) = \upsilon(\sqrt{\rho}) = L(\rho).$$

Then one finds that $L(\rho)$ satisfies the equation:

$$\rho L'' + (\nu + 1 - \rho)L' + \alpha L = 0, \; \nu = \ell + \frac{1}{2}, \; \alpha = \frac{1}{2}(\epsilon - \nu - 1). \tag{14.11}$$

This is the adjoined Laguerre equation, which has been extensively studied in mathematical physics. Taking everything into account we can write

$$u(\xi) = \xi^{\ell} e^{-\xi^2/2} L_{\alpha}^{\ell+1/2}(\xi^2),$$

$L_{\alpha}^{\ell+1/2}(x)$ the Laguerre functions. It turns out that this function is square integrable in the domain $[0, \infty)$ only if $\alpha = n_r =$ non negative integer. Then the last two equations of (14.11) yield:

$$\epsilon = 2n_r + \ell + 3/2 \rightarrow E_N = \left(N + \frac{3}{2}\right)\hbar\omega, \; N = 2n_r + \ell = 0, 1, 2, \ldots, \tag{14.12}$$

i.e. the energy is quantized and N is the principal quantum number. For the lowest N values we find:

$$N = 0 \Rightarrow (n_r = 0, \ell = 0), \; N = 1 \Rightarrow (n_r = 0, \ell = 1), \; N = 2 \Rightarrow$$
$$(n_r = 0, \ell = 2), (n_r = 1, \ell = 0), N = 3 \Rightarrow (n_r = 0, \ell = 3), \; (n_r = 1, \ell = 1). \tag{14.13}$$

The wave function becomes:

$$u(\xi) = \tilde{C}_{n_r,\ell} \xi^{\ell} e^{-\xi^2/2} L_{n_r}^{\ell+1/2}(\xi^2), \tag{14.14}$$

with $\tilde{C}_{n_r,\ell}$ a normalization constant, which can be chosen to be:

$$\tilde{C}_{n_r,\ell} = \left[\frac{2^{n_r+\ell+2} n_r!}{\sqrt{\pi}\,(2(n_r + \ell) + 1)!!}\right]^{1/2}.$$

In the usual units:

$$\psi(r) = C_{n_r,\ell} \left(\frac{r}{b}\right)^{\ell} e^{-r^2/(2b^2)} L_{n_r}^{\ell+1/2}\left(\frac{r^2}{b^2}\right), \quad b = \sqrt{\frac{\hbar}{m\omega}} \tag{14.15}$$

with

$$C_{n_r,\ell} = \left[\frac{2^{n_r+\ell+2} n_r!}{\sqrt{\pi}\,(2(n_r + \ell) + 1)!!}\right]^{1/2} \left(\sqrt{\frac{m\omega}{\hbar}}\right)^{3/2}$$

b is the size parameter of the harmonic oscillator. Some H.O radial functions are exhibited in Figs. 14.1 and 14.2. The reader should be able to identify these functions from their nodes. One finds

$$b = \frac{\hbar c}{\sqrt{mc^2 \hbar\omega}} = 41.5 \text{ fm} \left(\frac{\hbar\omega}{1 \text{ MeV}}\right)^{-1/2}$$

b is related to the root mean square radius. In fact using the Virial theorem we have:

$$\langle\psi|U|\psi\rangle = \langle\psi|T|\psi\rangle = \frac{1}{2}\langle\psi|H|\psi\rangle = \frac{1}{2}\left(N+\frac{3}{2}\right)\hbar\omega$$

but

$$U = \frac{1}{2}kr^2 = \frac{1}{2}m\omega^2 r^2 \Rightarrow \frac{1}{2}m\omega^2\langle|r^2|\rangle = \frac{1}{2}\left(N+\frac{3}{2}\right)\hbar\omega.$$

Thus

$$\langle|r^2|\rangle = \left(N+\frac{3}{2}\right)\frac{\hbar}{m\omega} \Rightarrow \langle|r^2|\rangle = \left(N+\frac{3}{2}\right)b^2.$$

From Eq. (14.13) we find that the number of states with the same major harmonic oscillator quantum number, i.e. the number of degenerate states, is

$$g(N) = \frac{1}{2}(N+1)(N+2).$$

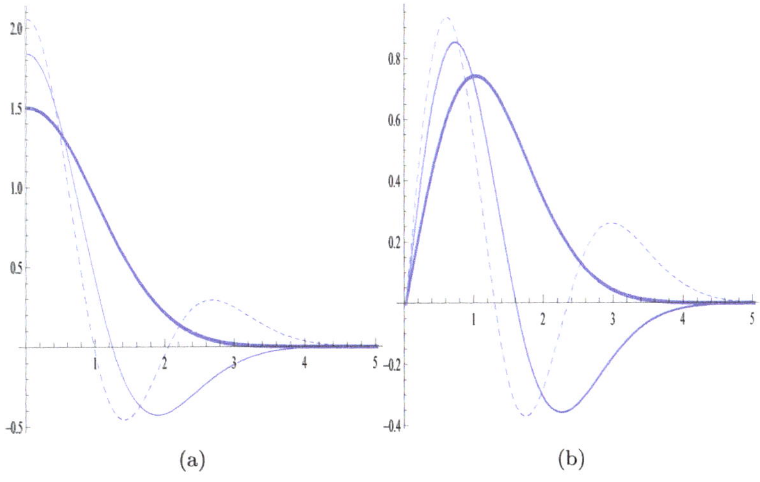

Fig. 14.1: Radial harmonic oscillator wf plotted as a function of r/b for (a) $0s$, $1s$ and $2s$ and (b) $0p$, $1p$ and $2p$.

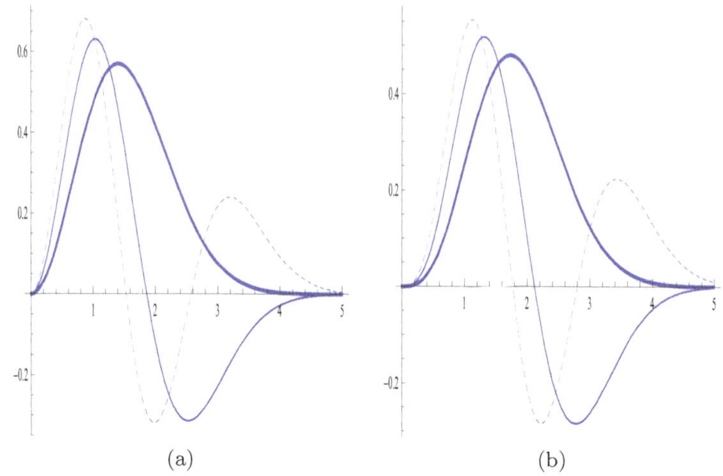

Fig. 14.2: Radial harmonic oscillator wf plotted as a function of r/b for (a) $0d$, $1d$ and $2d$ and (b) $0f$, $1f$ and $2f$.

14.3 The spin orbit coupling

The nucleons have spin $s = 1/2$. The spin wave function is indicated by the spin projection m_s on the quantization axis.

$$\chi^s_{m_s}, \; m_s = \pm 1/2.$$

So a set of single particle basis states is given by:

$$\psi_{n_r,\ell}(r) Y^\ell_{m_\ell}(\hat{r}) \chi^s_{m_s}.$$

Alternatively we could obtain a more convenient basis by coupling the orbital and spin to a total angular momentum j

$$|n_r, \ell, j, m\rangle \equiv \psi_{n_r,\ell}(r) \left[Y^\ell \otimes \chi^s \right]^j_m, \; j = \ell + \frac{1}{2}, \left| \ell - \frac{1}{2} \right|.$$

The two sets of states are related by a unitary transformation:

$$|n_r, \ell, j, m\rangle = \sum_{m_\ell, m_s} \left\langle \ell m_\ell, \frac{1}{2} m_s | jm \right\rangle \psi_{n_r,\ell}(r) Y^\ell_{m_\ell}(\hat{r}) \chi^s_{m_s}.$$

The quantity $\langle \ell m_\ell, \frac{1}{2} m_s | jm \rangle$ is called Clebsch-Gordan (C-G) coefficient and vanishes unless $m_\ell + m_s = m$. They are presented as a matrix with the columns labeled by j for a given m and the rows by m_s (for a given m, $m_\ell = m - m_s$), see Table 14.1. For a general discussion of C-G coefficients

Table 14.1: The C-G coefficients $\langle \ell m_\ell, \frac{1}{2} m_s | jm \rangle$. $H(\ell)$ is the Heaviside function, i.e. it unity for $\ell > 0$ and zero otherwise.

	$\lvert j = \ell - 1/2, m\rangle$	$\lvert j = \ell + 1/2, m\rangle$
$m_\ell = m + 1/2, m_s = -1/2$	$\sqrt{\frac{(1+2m+2\ell}{2(1+2\ell)}} H(\ell)$	$\sqrt{\frac{(1-2m+2\ell}{2(1+2\ell)}}$
$m_\ell = m - 1/2, m_s = 1/2$	$-\sqrt{\frac{(1-2m+2\ell}{2(1+2\ell)}} H(\ell)$	$\sqrt{\frac{(1+2m+2\ell}{2(1+2\ell)}}$

involving in angular momentum coupling see any book on angular momentum, a Group Theory book, e.g., [Vergados (2016)], or more advanced nuclear physics textbook, e.g. [deShalit and Feshbach (1974)].

If there exists a spin dependent interaction the above-mentioned degeneracy could be partly removed. Suppose, e.g. that we consider a central spin orbit interaction:

$$U_{\ell s} = U_1(r) 2\boldsymbol{\ell}.\mathbf{s}, \ \mathbf{s} = \text{the spin operator}. \tag{14.16}$$

Then

$$a_\ell = \langle jm | 2\boldsymbol{\ell}.\mathbf{s} | jm \rangle = \langle jm | \mathbf{j}^2 - \boldsymbol{\ell}^2 - \mathbf{s}^2 | jm \rangle = j(j+1) - \ell(\ell+1) - \frac{3}{4}. \tag{14.17}$$

One finds that for $\ell > 0$

$$a_\ell = \begin{cases} \ell, \ j = \ell + 1/2 \\ -\ell - 1, \ j = \ell - 1/2 \end{cases}. \tag{14.18}$$

In other words, since experimentally we find that, unlike atomic physics, the partner with $j = \ell + 1/2$ comes lower in energy, the potential $U_1(r)$ must be negative. The energy difference between the two partners is proportional to $2\ell + 1$).

The radial equation to be solved now in the presence of the two potentials is:

$$-\frac{\hbar^2}{2m}\left(R''_{n_r,\ell}(r) - \frac{\ell(\ell+1)}{r^2} R_{n_r,\ell}(r)\right) + \left(U(r) + a_\ell U_1(r)\right) R_{n_r,\ell}(r)$$
$$= \epsilon_{n_r,\ell} R_{n_r,\ell}(r). \tag{14.19}$$

It is clear that in this case the eigenvalues depend on ℓ even if both $U(r)$ and $U_1(r)$ are central. The reader is encouraged to try to find the eigenvalues in the special case that both potentials are harmonic.

14.4 Evaluation of one-body matrix elements

In this section we will consider the evaluation of 1-body matrix elements involving operators, which are assumed to have proper transformations under rotations, i.e. they be spherical tensors T^k_q, of rank k, a non negative integer, and projection q, $-k \le q \le k$, see e.g. [Hecht (2000)]. Examples are, the angular momentum itself with rank $k = 1$, the spherical harmonic Y^ℓ_m with rank ℓ etc. For any such operator we need to evaluate the matrix element $\langle j'm'|T^k_q|j,m\rangle$. It turns out that this ME can be factored as a product, a part, which is a Clebsch-Gordan (C-G) coefficient, and a part that is independent of m, m', q. This is a statement of the well known Wigner-Eckart theorem:

$$\langle j'm'|T^k_q|j,m\rangle = \frac{1}{\sqrt{2j'+1}}\langle (j,m),(k,q)|(j',m')\rangle\langle j'||T^k||j\rangle. \tag{14.20}$$

Where the double barred coefficient is called **reduced matrix element** and it depends on the nature of the operator. Examples:
i) T^k_q the identity operator 1.

$$\langle j'm'|1|j,m\rangle = \frac{1}{\sqrt{2j'+1}}\langle (j,m),(0,0)|(j'.m')\rangle\langle j'||1||j\rangle \Rightarrow$$

$$\delta_{j,j'}\delta_{m,m'} = \frac{1}{\sqrt{2j'+1}}\langle (j,m),(0,0)|(j'.m')\rangle\langle j'||1||j\rangle \Rightarrow$$

$$\langle j'||1||j\rangle \Rightarrow= \sqrt{2j'+1}\delta_{j,j'} \tag{14.21}$$

(the C-G is unity).
ii) The operator is that of the angular momentum itself, J^1_q

$$\langle j'm'|J^1_q|j,m\rangle = \frac{1}{\sqrt{2j'+1}}\langle (j,m),(1,q)|(j'.m')\rangle\langle j'||\mathbf{J}||j\rangle.$$

This must hold for any q. Selecting $q = 0$ we find

$$\langle j'm'|J_0|j,m\rangle = \frac{1}{\sqrt{2j'+1}}\langle (j,m),(1,0)|(j'.m')\rangle\langle j'||\mathbf{J}||j\rangle$$

$$\langle j'||\mathbf{J}||j\rangle = \sqrt{2j'+1}\frac{\langle j'm|J_0|j,m\rangle}{\langle (j,m),(1,0)|(j',m)\rangle} = \sqrt{2j'+1}\frac{\delta_{j,j'}\delta_{m,m'}m}{m/\sqrt{j(j+1)}} \Rightarrow$$

$$\langle j'||\mathbf{J}||j\rangle = \delta_{j,j'}\sqrt{j(j+1)(2j+1)}. \tag{14.22}$$

This, of course, holds for orbital angular momentum and spin operators

$$\langle \ell||\mathbf{L}||\ell\rangle = \sqrt{\ell(\ell+1)(2\ell+1)},$$

$$\langle 1/2||\mathbf{S}||1/2\rangle = \sqrt{\frac{3}{2}}, \quad \langle 1/2||\boldsymbol{\sigma}||1/2\rangle = 2\langle 1/2||\mathbf{S}||1/2\rangle = \sqrt{6}.$$

iii) The operator is a spherical harmonic.
Now the crucial relation [Hecht (2000)] is the reduction of two spherical harmonics of the same argument into one:

$$Y_m^\ell(\hat{r})Y_q^L(\hat{r}) = \sum_{\mathcal{L}} \langle(\ell,m),(L,q)|\mathcal{L},m)\rangle\sqrt{\frac{(2\ell+1)(2L+1)}{(2\mathcal{L}+1)4\pi}}$$
$$\langle(\ell,0),(L,0)|(\mathcal{L},0)Y^{\mathcal{L}}_{m+q}(\hat{r})\rangle. \qquad (14.23)$$

Hence

$$\langle\ell' m'|Y_q^L|(\ell,m)\rangle = \sum_{\mathcal{L}} \langle(\ell,m),(L,q)|\mathcal{L},m)\rangle\sqrt{\frac{(2\ell+1)(2L+1)}{2(\mathcal{L}+1)4\pi}}$$
$$\langle(\ell,0),(L,0)|(\mathcal{L},0)\rangle \int \left(Y_m^{\ell'}\right)^* Y^{\mathcal{L}}_{m+q} d\Omega. \qquad (14.24)$$

The last integral is $\delta_{\ell',\mathcal{L}}\delta_{m',m+q}$. On the other hand by definition

$$\langle\ell',m'|Y_q^L|(\ell,m)\rangle = \frac{1}{\sqrt{2\ell'+1}}\langle(\ell,m),(L,q)|(\ell',m)\rangle\langle\ell'||Y^L||\ell\rangle. \qquad (14.25)$$

Combining the last equations and remembering the condition $\delta_{\ell',\mathcal{L}}$ we find

$$\langle\ell'||Y^L||\ell\rangle = \sqrt{\frac{(2\ell+1)(2L+1)}{4\pi}}\langle(\ell,0),(L,0)|(\ell',0)\rangle. \qquad (14.26)$$

iii) Now will find the reduced ME of the spin and the orbital angular momentum operators, $\Omega^1_q = \mathbf{s}, \boldsymbol{\ell}$, in the the coupled system $\ell(1/2)jm\rangle$. The reduced ME is determined by the relationship:

$$\langle\ell(1/2)j'||\Omega||\ell(1/2)j\rangle = C_{j',j}\langle j,m|\Omega_0|j,m\rangle,\ m=\min(j,j'),$$
$$C_{j',j} = \frac{\sqrt{(2j'+1)}}{\langle(j',j'),(1,0)|(j,j)\rangle},\ j',j=\ell+1/2,\ell-1/2. \qquad (14.27)$$

The first part of the expression is a geometric factor involving the angular momentum quantum numbers, while the second is the matrix element involving the the maximum allowed value of m. The geometric factor can easily be computed using Mathematica to obtain the C-G coefficients. The result is:

$$C_{j,j} = \sqrt{\frac{j+1}{j}},\ C_{j,j-1} = -C_{j-1,j} = \sqrt{j(2j+1)}. \qquad (14.28)$$

Regarding the evaluation of the ME we consider the following possibilities:
a) $j=j'=\ell+1/2$

$$\langle(\ell(1/2)\ell+1/2,\ell+1/2|\Omega_0|(\ell(1/2)\ell+1/2,\ell+1/2\rangle$$

$$= \langle(\ell\ell)(1/2,1/2)|\Omega_0|(\ell\ell)(1/2,1/2)\rangle = \begin{cases} 1/2 & \text{for } \Omega = \mathbf{s} \\ \ell & \text{for } \Omega = \boldsymbol{\ell} \end{cases}.$$

b) $j = j' = \ell - 1/2$. Now

$$(\ell 1/2)\ell{-}1/2, \ell{-}1/2 = K_{1/2}^{(\ell-1/2)}(\ell,\ell{-}1)(1/2,1/2) + K_{-1/2}^{\ell-1/2}(\ell.\ell{+}1)(1/2,-1/2)$$

with the C-G coefficients given

$$K_{1/2}^{\ell-1/2} = \langle(\ell,\ell-1),(1/2,1/2)|(\ell-1/2,\ell-1/2)\rangle = -\frac{1}{\sqrt{2\ell+1}},$$

$$K_{-1/2}^{\ell-1/2} = \langle(\ell,\ell),(1/2,-1/2)|(\ell-1/2,\ell-1/2)\rangle = \sqrt{\frac{2\ell}{2\ell+1}}.$$

For the spin operator

$$\langle(\ell 1/2)\ell-1/2,\ell-1/2|\Omega_0|(\ell 1/2)\ell-1/2,\ell-1/2\rangle$$

$$= \left(K_{1/2}^{\ell-1/2}\right)^2 \frac{1}{2} + \left(K_{-1/2}^{\ell-1/2}\right)^2 \left(-\frac{1}{2}\right) = -\frac{j}{2(j+1)}.$$

For the orbital angular momentum operator

$$\langle(\ell 1/2)\ell-1/2,\ell-1/2|\Omega_0|(\ell 1/2)\ell-1/2,\ell-1/2\rangle$$

$$= \left(K_{1/2}^{\ell-1/2}\right)^2 (\ell-1)\ell + \left(K_{-1/2}^{\ell-1/2}\right)^2 \ell = \ell - \frac{1}{2\ell+1}.$$

c) $j' = \ell - 1/2, j = \ell + 1/2$. Now we need extra C-G coefficients, which are orthogonal to the previous:

$$K_{-1/2}^{\ell+1/2} = -K_{1/2}^{\ell-1/2},\; K_{1/2}^{\ell+1/2} = K_{-1/2}^{\ell-1/2}.$$

Then the spin ME is

$$\langle(\ell 1/2)\ell-1/2,\ell-1/2|\Omega_0|(\ell 1/2)\ell+1/2,\ell-1/2\rangle$$

$$= K_{-1/2}^{\ell-1/2}K_{-1/2}^{\ell+1/2}(-)\frac{1}{2} + K_{1/2}^{\ell-1/2}K_{1/2}^{\ell+1/2}\frac{1}{2} = -\frac{\sqrt{2j-1}}{2j}.$$

The orbital matrix element is

$$\langle(\ell 1/2)\ell-1/2,\ell-1/2|\Omega_0|(\ell 1/2)\ell+1/2,\ell-1/2\rangle$$

$$= K_{-1/2}^{\ell-1/2}K_{-1/2}^{\ell+1/2}\ell + K_{1/2}^{\ell-1/2}K_{1/2}^{\ell+1/2}(\ell-1) = \frac{\sqrt{2\ell}}{2\ell+1}.$$

d) $j' = \ell + 1/2, j = \ell - 1/2$. The matrix element is the same as in c).

We summarize our results in Table 14.2.

Table 14.2: The reduced ME of $\langle(\ell 1/2)j'||\Omega|(\ell 1/2)j\rangle$, $\Omega = \boldsymbol{\sigma}, \boldsymbol{\ell}$.

	$\langle(\ell 1/2)j'\|\|\boldsymbol{\sigma}\|\|(\ell 1/2)j\rangle$	
	$\ell+1/2$	$\ell-1/2$
$\ell+1/2$	$\frac{1}{2}\sqrt{\frac{2\ell+3}{2\ell+1}}$	$\sqrt{\frac{2\ell(\ell+1)}{2\ell+1}}$
$\ell-1/2$	$-\sqrt{\frac{2\ell(\ell+1)}{2\ell+1}}$	$-\frac{1}{2}\sqrt{\frac{2\ell+1}{2\ell-1}}$
	$\langle(\ell 1/2)j'\|\|\boldsymbol{\ell}\|\|(\ell 1/2)j\rangle$	
	$\ell+1/2$	$\ell-1/2$
$\ell+1/2$	$\ell\sqrt{\frac{2\ell+3}{2\ell+1}}$	$\sqrt{\frac{\ell(\ell+1)}{2\ell+1}}$
$\ell-1/2$	$-\sqrt{\frac{\ell(\ell+1)}{2\ell+1}}$	$(\ell-\frac{1}{2\ell+1})\sqrt{\frac{2\ell+1}{2\ell-1}}$

iv) Now will find the reduced ME of the electric dipole operator in the the coupled system $(\ell 1/2)jm\rangle$.

This can be expressed [Hecht (2000)] as follows:

$$\langle(\ell' 1/2)j'||\mathbf{D}||(\ell 1/2)j\rangle = (-1)^{\ell-\ell'+j'-j}\langle\ell'||\mathbf{D}||\ell\rangle U(1/2,\ell,j',1,j,\ell')$$

where $U(1/2,\ell,j',1,j,\ell')$ is the unitary Racah function, which can be computed by Mathematica codes. The reduced ME $\langle\ell'||\mathbf{D}||\ell\rangle$ can be obtained using the result of Eq. (14.26), by noting that $\mathbf{D_0} = r\sqrt{\frac{4\pi}{3}}Y_0^1(\hat{r})$, i.e.

$$\langle\ell'||\mathbf{D}||\ell\rangle = \sqrt{2\ell+1}\langle(\ell,0),(1,0)|(\ell'0)\langle n_r'\ell'|r|n_r\ell\rangle.$$

We thus have the constraints

$$\ell' = |\ell \pm 1|, \quad |j'-j| \le 1, \quad N' = |N \pm 1|.$$

We thus find

$$\langle\ell'||\mathbf{D}||\ell\rangle = \langle n_r'\ell'|r|n_r\ell\rangle \times \begin{cases} -\sqrt{\ell}, & \ell' = \ell-1 \\ \sqrt{\ell+1}, & \ell' = \ell+1 \end{cases}.$$

The radial integral can easily be evaluated using HO wave functions.

The above expressions impose constraints on the allowed values of the Racah function. In fact the non-vanishing values of the Racah function are given in Table 14.3.

14.5 Problems

(1) In many cases we have an operator, which depends on the relative coordinate of two particles $|\mathbf{r}_1 - \mathbf{r}_2|$. Suppose we want to evaluate

Table 14.3: The allowed values of $U(1/2,\ell,j',1,j,\ell')$ relevant for the calculation of electric dipole transitions for any given value of ℓ.

j'	j	ℓ'	U
$\ell-3/2$	$\ell-1/2$	$\ell-1$	1
$\ell-1/2$	$\ell-1/2$	$\ell-1$	$-\frac{1}{\sqrt{\ell(2\ell+1)}}$
$\ell-1/2$	$\ell+1/2$	$\ell-1$	$\frac{\sqrt{2\ell^2+\ell-1}}{\sqrt{\ell(2\ell+1)}}$
$\ell+1/2$	$\ell-1/2$	$\ell+1$	$\frac{\sqrt{\ell(2\ell+3))}}{\sqrt{(\ell+1)(2\ell+1)}}$
$\ell+1/2$	$\ell+1/2$	$\ell+1$	$\frac{1}{\sqrt{(\ell+1)(2\ell+1)}}$
$\ell+3/2$	$\ell+1/2$	$\ell+1$	1

the matrix element of such an operator sandwiched between two particle wave functions. To accomplish this one may expand the operator in terms of the coordinates $\mathbf{r}_1$ and $\mathbf{r}_2$, known as Legendre expansion. This is trivial in he case of the Gaussian interaction and relatively easy in the case of the Coulombic potential, but it is not easy in the general case. In the case of the harmonic oscillator wave functions, however, such an expansion is not needed. One has the option of transforming the two particle wave function in terms of products of relative and CM harmonic oscillator wave functions. These are known as Brody-Moshinsky brackets, [Brody and Moshinsky (1960)], [deShalit and Feshbach (1974)]. These are tabulated but in the simple cases of the p-shell it is instructive to derive them. Thus, consider

$$\mathbf{r}_1 = \frac{1}{\sqrt{2}}(\mathbf{R}+\mathbf{r}), \quad \mathbf{r}_2 = \frac{1}{\sqrt{2}}(\mathbf{R}-\mathbf{r})$$

show that:

$$\psi_{0s}(\mathbf{r}_1)\psi_{0s}(\mathbf{r}_2) = \psi_{0s}(\mathbf{r})\psi_{0s}(\mathbf{R}).$$

Hint: This is a consequence of the fact that the transformation of coordinates is orthogonal and, as a result,

$$\mathbf{r}_1^2 + \mathbf{r}_2^2 = \mathbf{r}^2 + \mathbf{R}^2,$$

yielding the exponent of the HO of the right form.

(2) Preceding as above show that

$$[\psi_{0p}(\mathbf{r}_1) \times \psi_{0p}(\mathbf{r}_2)]^{L=0} = \frac{1}{\sqrt{2}}(\psi_{0s}(\mathbf{r})\psi_{1s}(\mathbf{R}) - \psi_{0s}(\mathbf{R})\psi_{1s}(\mathbf{r})). \tag{14.29}$$

Hint: The exponent being of the correct form we notice the the rest of the wave function is proportional to

$$fp = \frac{3}{4\pi}\frac{[\mathbf{r}_1 \times \mathbf{r}_2]^{L=0}}{b^2} = -\frac{3}{4\pi}\frac{1}{\sqrt{3}}\frac{\mathbf{r}_1.\mathbf{r}_2}{b^2}$$

the first factor comes from the spherical harmonic part. Thus

$$fp = -\frac{\sqrt{3}}{4\pi}\frac{1}{2}(R^2 - r^2) = -\frac{\sqrt{3}}{4\pi}\frac{1}{2}(R^2 - 3/2 - (r^2 - 3/2)).$$

This has the right structure for a $1s0s$ type HO function. The factor $(1/4\pi)$ is absorbed in the angular part. Taking into account the normalization of he radial part we find

$$\psi = \frac{c_{0p}c_{0p}}{c_{0s}c_{1s}}fp = \frac{c_{0p}}{c_{0s}}fp = \frac{1}{\sqrt{2}}(\psi_{0s}(r)\psi_{0s}(R) - \psi_{0s}(R)\psi_{0s}(r)).$$

(3) Show further that

$$[\psi_{0p}(\mathbf{r}_1) \times \psi_{0p}(\mathbf{r}_2)]^{L=2} = \frac{1}{\sqrt{2}}(\psi_{0s}(\mathbf{r})\psi_{0d}(\mathbf{R}) - \psi_{0s}(\mathbf{R})\psi_{0d}(\mathbf{r})). \tag{14.30}$$

Hint: We can proceed as in the previous case, but it is best to use the reduction of spherical[1] harmonics. Thus

$$fp = \sqrt{9/5}]\sqrt{2/3}\frac{1}{2}(R^2Y^0(\hat{r})Y^2(\hat{R}) - r^2Y^2(\hat{r})Y^0(\hat{R}))$$

taking into account the normalizations we find

$$[\psi_{0p}(\mathbf{r}_1) \times \psi_{0p}(\mathbf{r}_2)]^{L=2} = \frac{c_{0p}c_{0p}}{c_{0s}c_{0d}}fp$$

$$= \frac{1}{\sqrt{2}}(\psi_{0s}(\mathbf{r})\psi_{0d}(\mathbf{R}) - \psi_{0s}(\mathbf{R})\psi_{0d}(\mathbf{r})).$$

(4) Show further that

$$[\psi_{0p}(\mathbf{r}_1) \times \psi_{0p}(\mathbf{r}_2)]^{L=1} = [\psi_{0p}(\mathbf{r}) \times \psi_{0p}(\mathbf{R})]^{L=1}. \tag{14.32}$$

[1]Using Eq. (14.23), it is easily proved in the angular momentum theory that

$$\left[Y^{\ell_1}(\hat{r}) \times Y^{\ell_2}(\hat{r})\right]^L = \sqrt{\frac{(2\ell_1+1)+(2\ell_2+1)}{4\pi(2L+1)}}Y^L(\hat{r})\langle(\ell_1,0),(\ell_2,0)|L,0\rangle, \tag{14.31}$$

the last factor being a Clebsch-Gordan coefficient. Here $\langle(1,0),(1,0)|2,0\rangle = \sqrt{\frac{2}{3}}$, $\langle(1,0),(1,0)|0,0\rangle = -\frac{1}{\sqrt{3}}$.

Chapter 15

Appendix B: Construction of wave functions of given symmetry: SU(2) and SU(3)-Young tableaux

15.1 The angular momentum-SU(2) symmetry

In the case of angular momentum the construction of states of a given symmetry, symmetric or antisymmetric under the exchange of two particles, is quite simple. All one needs to know is the rules of addition of angular momentum states, see e.g. [Vergados (2016)], Chapter 2, or nuclear physics books like [deShalit and Feshbach (1974)]. We summarize the essential facts as follows:

i) an angular momentum state is described by two integers or half integers denoted by $|j,m\rangle$, $m=j,j-1,j-2,\cdots,j$.

ii) Combining two angular momenta, e.g. $|j_1,m_1\rangle$ and $|j_2,m_2\rangle$ we can obtain states of angular momentum j, $|(j_1,j_2)jm\rangle$ with $m=m_1+m_2$.

iii) The transformation between the two is given in terms of Clebsch-Gordan (C-G)[1] coefficients $\langle j_1,m_1;j_2,m_2|j,m\rangle$, which can be chosen to be real. In other words

$$|(j_1,j_2)j,m\rangle=\sum_{m_1}\langle j_1,m_1;j_2,m-m_1|j,m\rangle|j_1,m_1\rangle|j_2,m-m_1\rangle,$$

$$|j_1,m_1\rangle;|j_2,m_2\rangle=\sum_{j}\langle j_1,m_1;j_2,m_2|j,m_1+m_2\rangle|(j_1,j_2)j,m_1+m_2\rangle.$$

iv) The C-G coefficients can now be computed with convenient codes found, e.g., in Mathematica.

v) The C-G coefficients have a number of symmetries.[2] Among them the most useful are.

$$\langle j_1,m_1;j_2,m_2|j,m\rangle=(-1)^{j_1+j_2-j}\langle j_2,m_2;j_1,m_1|j,m\rangle, \tag{15.1}$$

[1]For clarity sometimes we may write $\langle(j_1,m_1)(j_2,m_2)|(j,m)\rangle$.

[2]Those not familiar with C-G coefficients may find it useful to look at the first problem at the end of this chapter.

$$\langle j_1, m_1; j_2, m_2 | j, m \rangle = (-1)^{j_1+j_2-j} \langle j_1, -m_1; j_2, -m_2 | j, -m \rangle.$$

Let us now consider two particles, labeled 1 and 2, with angular momenta $|j_1, m_1\rangle_1$ and $|j_2, m_2\rangle_2$ respectively. In the special case the two particles are identical particles $j_2 = j_1$ with j_1 being half integral, i.e. $j_1 = (1/2)(2n+1)$, $n = 0, 1, 2, \cdots$. The state with total angular momentum j is written as

$$\psi = ((j_1)_1 (j_1)_2) \, j, m\rangle = \sum_{m_1} \langle j_1, m_1; j_1, m - m_1 | j, m \rangle |j_1, m_1\rangle_1 |j_1, m - m_1\rangle_2.$$

Then applying the exchange operator P_{12} we obtain

$$P_{12}\psi = \sum_{m_1} \langle j_1, m_1; j_1, m - m_1 | j, m \rangle |j_1, m_1\rangle_2 |j_1, m - m_1\rangle_1$$

$$= (-1)^{2j_1 - j} \sum_{m_1} \langle j_1, m - m_1; j_1, m_1; | j, m \rangle |j_1, m_1\rangle_2 |j_1, m - m_1\rangle_1$$

$$= (-1)^{2j_1 - j} \sum_{m_1} \langle j_1, m - m_1; j_1, m_1; | j, m \rangle |j_1, m - m_1\rangle_1 |j_1, m_1\rangle_2 = (-1)^{2j_1 - j} \psi.$$

This implies that, if $j = 0, 2, \ldots, 2j_1 - 1$ the state is antisymmetric, while if $j = 1, 2, \ldots, 2j_1$ the state is symmetric. Applying this in the case of isospin $I_1 = 1/2$ we find that the two particle isospin wave function with total isospin I is symmetric if $I = 1$ and antisymmetric if $I = 0$. Thus the product wave function containing both angular momentum j and isospin I is antisymmetric as follows:

$$I = 1 \text{ requires } j = 0, 2, \cdots 2j_1 - 1, \; I = 0 \text{ requires } j = 1, 3, \cdots 2j_1.$$

While for the symmetric case:

$$I = 0 \text{ requires } j = 0, 2, \cdots 2j_1 - 1, \; I = 1 \text{ requires } j = 1, 3, \cdots 2j_1.$$

Clearly no constraint is required if $j_1 \neq j_2$.

On the other hand if $j_1 = j_2 = \ell$, $j = L$ with ℓ, L = non negative integers, we find that the state with angular momentum L even is symmetric under the exchange of the 1 and 2 and antisymmetric, if it is odd. In this case we have the $L - S$ coupling scheme $(\ell \otimes \ell)L$, S, J, where the particle indices have been omitted with the understanding that the first is 1 and the second is 2. Thus to get an antisymmetric wave function for identical two particles the allowed I, L, S combinations are:

$$I = 1 \text{ requires } L = \text{even } S = 0 \text{ or } L = \text{odd } S = 1$$

$$I = 0 \text{ requires } L = \text{odd } S = 0 \text{ or } L = \text{even } S = 1$$

with the angular momentum coupling condition $|L - S| \leq J \leq J + L$ satisfied. In other words the condition is $L + S + I =$ odd integer.

In shell model calculations one may choose to work in the proton-neutron representation. Then the above rules apply for the proton and neutron components of the wave function separately thinking of them as having isospin 1.

For more identical particles the situation is not so simple. One may use elementary methods, see problems at the end of this chapter and in particular Eq. (15.42).

For actual calculations, however. one needs more sophisticated techniques as we will see below, Section 15.5.

15.2 Youmg tableaux

The construction of wave functions of a given symmetry is a little involved, see specialized books on group theory, e.g. [Chen *et al.* (2002)]. For our purposes we will outline method of the Young tableaux (diagram). For applications of the Young tableaux to physics we refer the reader to the literature [Vergados (2016)].

The Young tableaux consists of r boxes arranged in at most r rows and each row containing at most r boxes. The length of each row not being longer than the previous one.

As an example we will consider the $U(3)$. In this the allowed representations are classified by Young tableaux containing at most three rows with f_1 boxes in the first row, f_2 boxes in the second row and f_3 in the last row, such that $f_1 \geq f_2 \geq f_3$, see, e.g., Ref. [Vergados (2016)]. A typical Young tableaux $[f_1, f_2, f_3]$ is:

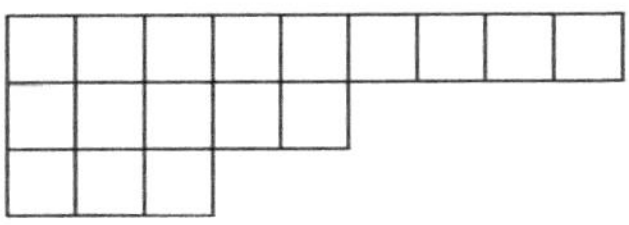

This is so because the representations of $U(3)$ can be labeled by the representations of the symmetric group S_n $n = f_1 + f_2 + f_3$. Some essential facts about the tableaux:

- A single row pattern represents a symmetric representation.
- A single column pattern represents an antisymmetric representation.

- In all other cases wa have mixed symmetry.
- When restricted to $SU(3)$ columns of 3 boxes can be eliminated.

15.3 Kronecker products $[f] \otimes [f']$

Suppose we are given two representations $[f]$ and $f']$. The product $[f])\otimes[f']$ is, in general, a reducible representation, but it can be expressed as a linear combination of irreducible tensors.

$$[f] \otimes [f') = \sum_{[f'']} \alpha_{[f],[f'],f'']}[f''].$$

The question of evaluating the coefficients $\alpha_{[f],[f'],[f'']}$ is complicated and it will be addressed later. At this point we would like to see how one can find the possible Young tableaux $[f'']$ corresponding to the "product" $[f] \otimes [f']$. We will assume that the first tableaux corresponds to the symmetry group S_r and the second to the symmetry group $S_{r'}$, while the product to that of $S_{r+r'}$. We will not prove the relevant theorems but we will only state the general rules:

(1) We select the most complicated symmetry $[f]$, usually containing the largest number of boxes, and write the corresponding Young tableaux.
(2) In the Young tableaux corresponding to the symmetry $[f']$ we put some numbers, e.g. the number 1 in the first row, the number 2 in the second row etc. We add the boxes of $[f']$ one by one on $[f]$ in all possible ways to yield an acceptable Young Tableaux but with the following restrictions:
(3) the number of boxes in $[f'']$ must be equal to the sum of the number of boxes of $[f]$ and $[f']$.
(4) The boxes with the same number cannot belong to the same column.
(5) In the resulting Young Tableaux when reading the elements of a row from right to left every number 2 must be preceded by at least one number 1, every number three must be preceded by at least one number 2 etc.

The product obtained this way will be indicated as "external product". If any of the Young tableaux in the product contains one box, there is no need to put numbers.

We will apply them in some simple cases of practical interest and, in particular, we will consider the reduction in the special case of $U(3)$. This characterizes the flavor symmetry of quarks u, d, s) or their color symmetry (g, r, b). Then the number of rows of the Young tableaux cannot exceed 3. After performing the reduction, if we restrict ourselves to $SU(3)$, columns with length longer two can be deleted. So two non negative integers specify the representation, e.g. (λ, μ) with $\lambda = f_1 - f_2$ and $\mu = f_2 - f_3$.

Let us begin with the very simple cases:

i) Consider the reduction

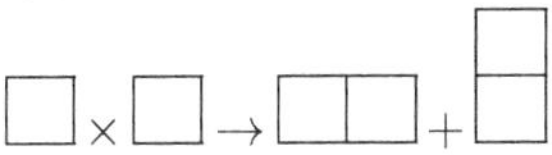

So we take the product of the fundamental representation [1,0,0] or $(\lambda, \mu) = (1, 0)$ with itself. On the right-hand side the first is the symmetric representation and the second is the antisymmetric one. We write

$$\begin{array}{c} (1,0) \times (1,0) \rightarrow (2,0) + (0,1) \\ \underline{3} \;\times\; \underline{3} \;\rightarrow\; \underline{6} + \underline{3} \end{array}$$

We know that their dimensions are 6 and 3 and we check that $3 \times 3 = 6 + 3$.

ii) Consider the reduction

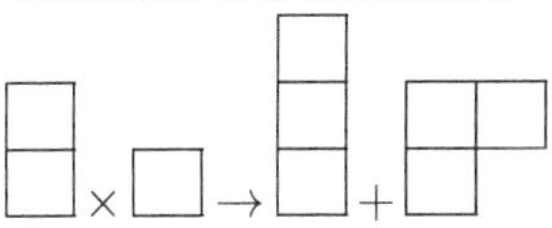

i.e. we take the product of the antisymmetric representation [1,1] with highest weight $(\lambda, \mu) = (0, 1)$ found above with the fundamental. On the right hand side the first is the antisymmetric representation [1,1,1] of $U(3)$, which, restricted to $SU(3)$, becomes $(\lambda, \mu) = (0, 0)$ and the second is the mixed symmetry with highest weight $(\lambda, \mu) = (1, 1)$. We know that the dimension of (0,0) is 1. Furthermore the (01) has the same quantum number with the antiquark, i.e. the same isospin I and opposite additive quantum numbers I_3 and Y. The rule of dimensions gives: $(1, 1) = 3 \times 3 - 1 = 8$. We thus write

$$\begin{array}{c} (0,1) \times (1,0) \rightarrow (0,0) + (1,1) \\ \underline{3}^* \;\times\; \underline{3} \;\rightarrow\; \underline{1} + \underline{8} \end{array}$$

Clearly these last rules apply in the case of the coupling of an antiquark to a quark to produce a meson.

iii) Consider the reduction

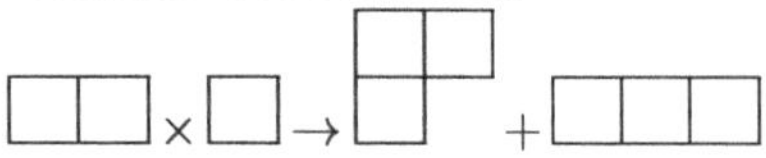

i.e. we take the product of the symmetric representation with highest weight (2,0) with the fundamental representation with highest weight (1,0). On the right-hand side the first is the mixed symmetry representation with highest weight (1,1), which we obtained above, and the second is the symmetric representation with highest weight (3,0). We know that the dimension of (1,1) is 8. So the dimension of the symmetric one is $(3, 0) = 6 \times 3 - 8 = 10$. We thus write

$$\begin{array}{c} (2,0) \times (1,0) \to (1,1) + (3,0) \\ \underline{6} \;\; \times \;\; \underline{3} \;\; \to \;\; \underline{8} + \underline{10} \end{array}$$

iv) Consider the reduction

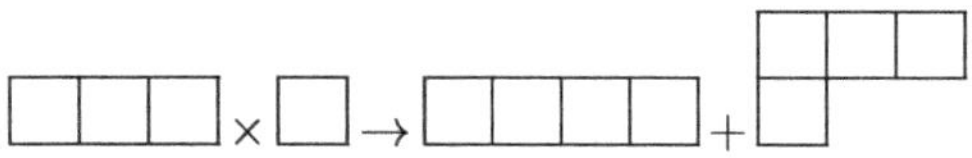

The [4], and [3,1] have highest weight (4,0) and (2,1). Thus if we know that the dimension of [4] is 15 we can obtain the dimensions all the rest.

$$\begin{array}{c} (3,0) \times (1,0) \to (4,0) + (2,1) \\ \underline{10} \;\; \times \;\; \underline{3} \;\; \to \;\; \underline{15} + \underline{15} \end{array}$$

(there exist two representations with dimension 15).

v) Consider the reduction:

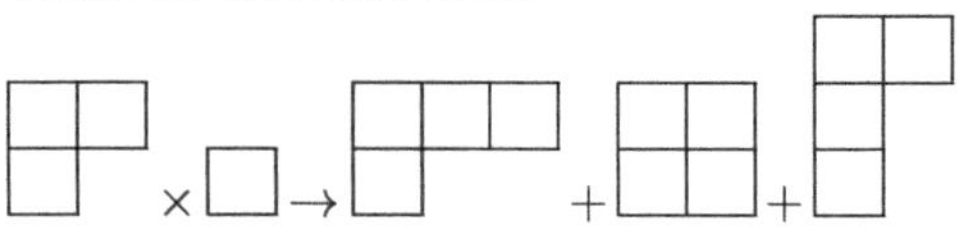

The last diagram restricted to SU(3) becomes . Furthermore the [2,2] has highest weight (0,2). We know that the dimension of (2,1) is 15 and obtain dim(0,2) = 6. We thus write

$$\begin{array}{c} (1,1) \times (1,0) \to (2,1) + (0,2) + (1,0) \\ \underline{8} \;\; \times \;\; \underline{3} \;\; \to \;\; \underline{15} + \underline{6} + \underline{3} \end{array}$$

In the next text we do not restrict ourselves to the $SU(3)$ symmetry and again we consider some simple cases.

i) One of the Young tableaux has only one box. Let us consider the two simple cases 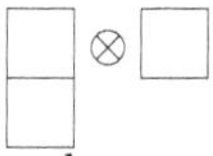 and 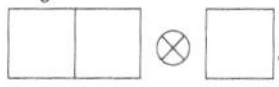. Since the second tableaux in both cases has only one box there is no need to put any numbers in the second box. Thus:

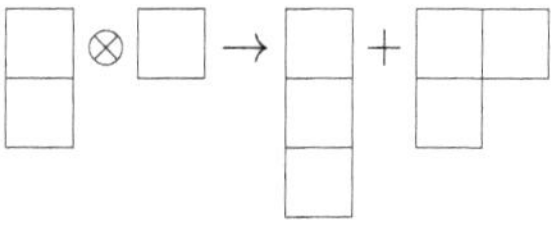

(15.2)

From this reduction we can find the dimension of [2,1] in the case of n objects. All we need to do is to equate the dimensions of the left and right of the above equation:

$$N([1^2, n]) \times N([1, n]) = N([[1, 1, 1], n]) + N([2, 1], n) \Rightarrow$$

$$\frac{1}{2}n(n-1)n = \frac{1}{6}n(n-1)(n-2) + N([2, 1], n) \Rightarrow$$

$$N([2, 1], n) = \frac{1}{3}n(n+1)(n-1) \Rightarrow$$

$$N([2, 1]3) = 8 \Rightarrow [2, 1] \leftrightarrow \underline{8}.$$

Similarly we find

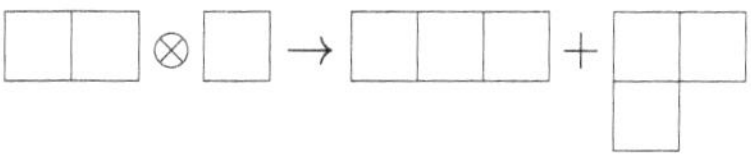

(15.3)

The dimensional condition yields:

$$N([2], n]) \times N([1, n]) = N([3], n]) + N([2, 1], n) \Rightarrow$$
$$\frac{1}{2}n(n+1)n = \frac{1}{3}n(n+1)(n-1) + N([3], n) \Rightarrow$$
$$N([3], n) = \frac{1}{6}n(n+1)(n+2) \Rightarrow N([3], 3) = 10.$$

The reductions

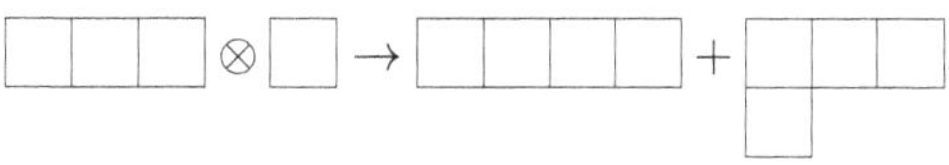

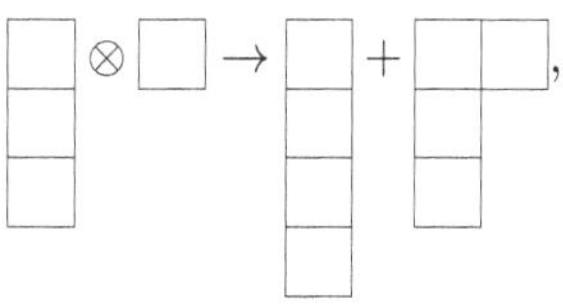

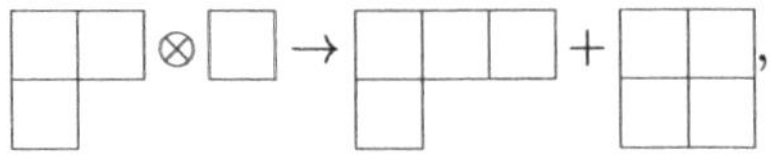

can be obtained equally simply. Also with the rule of dimensions we can find the dimensions of $N([3,1],n)$, $N([2,1,1],n)$ and $N([2,2],n)$ from $N([2,1],n)$ found above and the known dimensions of $N([4],n)$ and $N([1,1,1,1],n)$.

ii) Let us consider the case 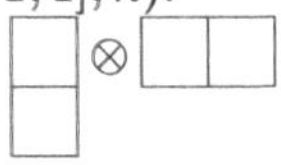. We write it as 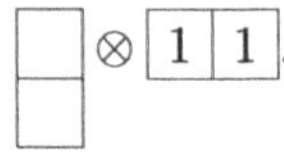.

Then

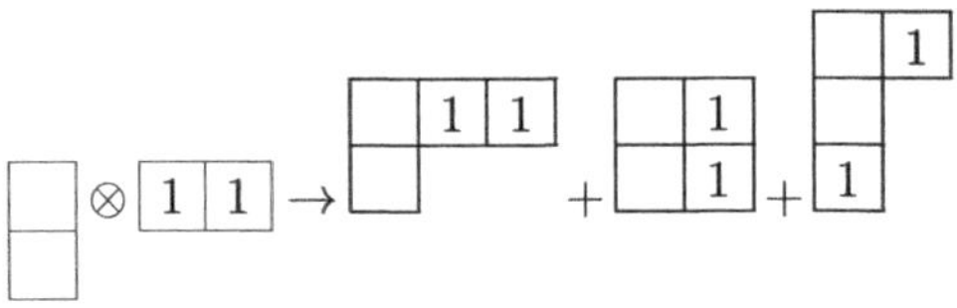

The middle diagram is not consistent with above rules and it is omitted. Thus as a result:

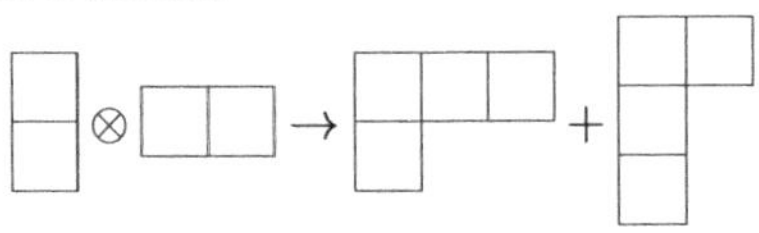

(15.4)

All representations that appear in the product space are single.

iii) Let us now consider a more complicated case 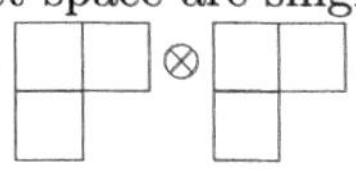. We write it as 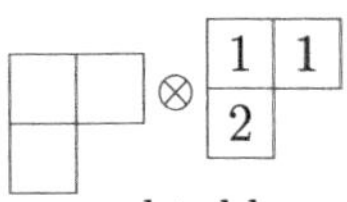. Then we proceed successively for each row of the second tableaux.

As a first step we have:

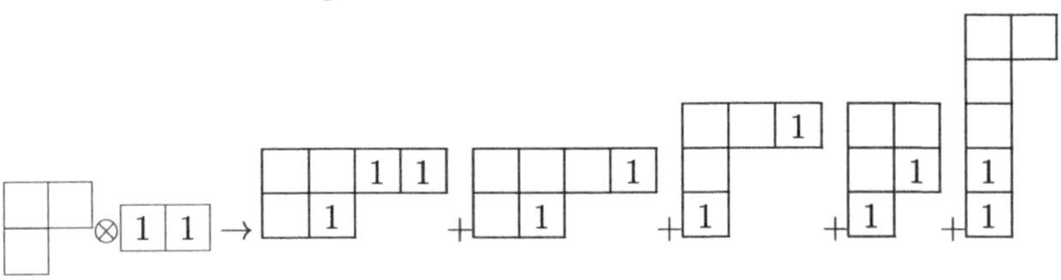

The last Young tableaux dos not satisfy the above rules and is omitted. Thus the correct expression is:

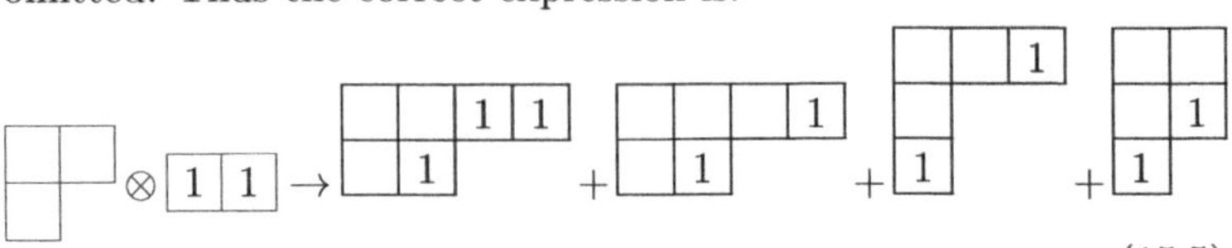

(15.5)

We now continue with the second step:

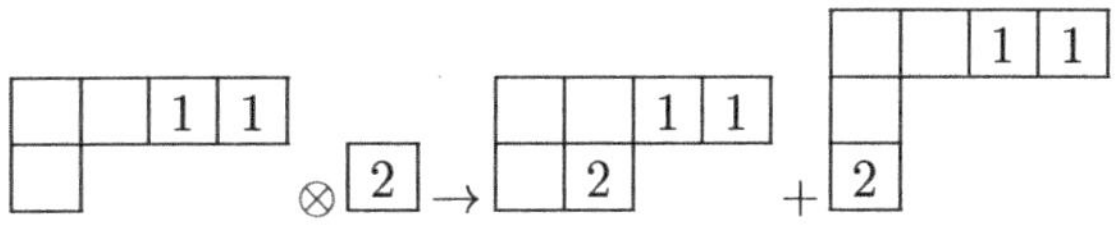

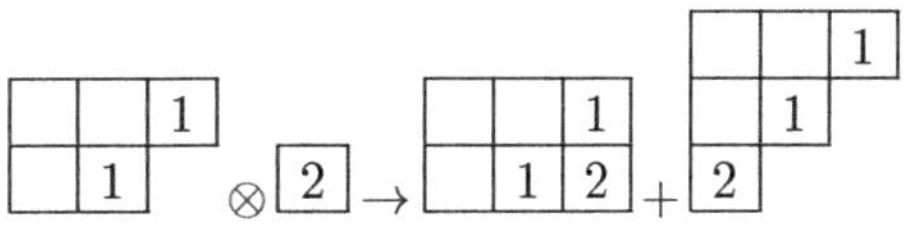

The diagram

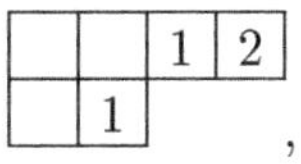

does not satisfy the above rules and has been omitted.

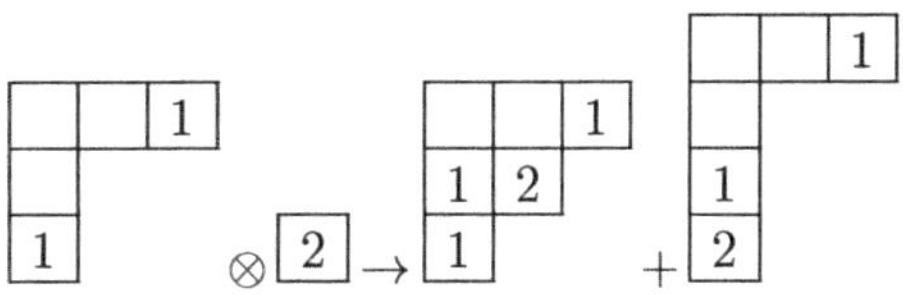

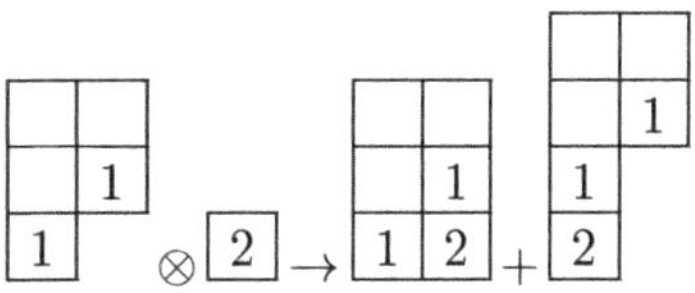

As a result:

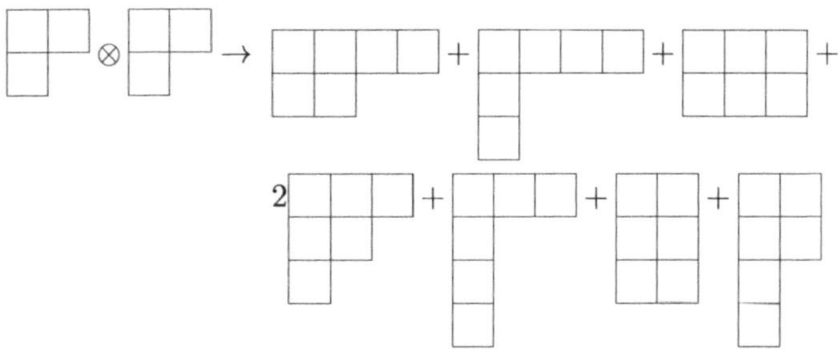

(15.6)

Note that the above rules yield the multiplicities, if any. E.g. the [3,2,1] obtained in two different ways:

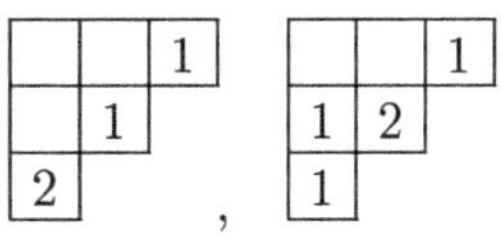

15.4 Construction of functions of a given S_n symmetry — Young operators

In each box we put one of r different indices, e.g. 1, 2, $\cdots$ n. The Young tableaux is called standard, if the indices cannot decrease in any row, e.g. from left to right, and must increase in any column, e.g. from up to down. Sometimes this arrangement will be denoted by

$$[f_1, f_2, \ldots, f_r]$$

where r is the number of rows, f_1 is the number of boxes in the first row, f_2 in the second row etc. c is the number of columns. For $n = 2$ objects we have two possibilities:

$$\begin{array}{|c|c|}\hline 1 & 2 \\ \hline\end{array}\,, \quad \begin{array}{|c|}\hline 1 \\ \hline 2 \\ \hline\end{array}$$

corresponding to [2] $(r = 1, c = 0)$ and $\begin{bmatrix}1\\1\end{bmatrix} \equiv [1^2]$. $(r = 0, c = 1)$ both are standard.

For $n = 3$ objects the standard tableaux are:

$$\begin{array}{|c|c|c|}\hline 1 & 2 & 3 \\ \hline\end{array}\,, \quad \begin{array}{|c|c|}\hline 1 & 2 \\ \hline 3 \\ \cline{1-1}\end{array}\,, \quad \begin{array}{|c|c|}\hline 1 & 3 \\ \hline 2 \\ \cline{1-1}\end{array}\,, \quad \begin{array}{|c|}\hline 1 \\ \hline 2 \\ \hline 3 \\ \hline\end{array} \qquad (15.7)$$

We see that the [2,1] of the symmetric group is 2-dimensional. All others are 1-dimensional.

The k numbers in a given row indicate that we should consider all possible permutations of S_k,$k = 1, 2, \ldots, r$, involving the numbers present in that row. Then sum them up with a + sign regardless of whether they are even or odd. The resulting operator is symmetric. The ℓ numbers in a given column indicate that we should consider all possible permutations of S_ℓ, $\ell = 1, 2, \ldots, c$. Then construct an operator summing them up with the sign + or − dependent on whether they are even or odd permutations.

Let us begin by considering a problem in which two particles can occupy two levels, say α and β.

i) Consider the symmetric case [2], $(r = 1,\ c = 0)$.

Suppose that the 2 particles occupy both levels. The occupation is designated by assigning to each level a number indicating the particle occupying it. Only the permutations associated with the row need be considered and they are the identity E and the permutation $p_{12} = (12)$. The corresponding Young operator $T([2]) = E + (12)$ is. If the particles occupy different levels have the symmetric wave function can be obtained by applying the operator in any product of the two functions, e.g. particle 1 occupies the level α and 2 the level β i.e. $\alpha(1)\beta(2)$. Thus

$$\psi^{[2]}_{\alpha,\beta} = \frac{1}{\sqrt{2}}T\left([2]\right)|\alpha(1)\beta(2)\rangle = \frac{1}{\sqrt{2}}(\alpha(1)\beta(2) + \alpha(2)\beta(1)). \qquad (15.8)$$

The factor $\frac{1}{\sqrt{2}}$ was introduced to make $\psi_{\alpha,\beta}$ normalized, once the single particle states have been normalized.

If the two particles can occupy the same level, the solution is trivial and one does not need the power of young tableaux. Clearly the symmetric combinations are

$$\psi^{[2]}_{\alpha,\alpha} = \alpha(1)\alpha(2),\ \psi^{[2]}_{\beta,\beta} = \beta(1)\beta(2).$$

ii) Consider the anti-symmetric case [1,1], ($r = 0$, $c = 1$).
Only the permutations associated with the column need be considered and they are the identity E and the permutation $p_{12} = (12)$ as before, but now the corresponding Young operator is $T\left([1^2]\right) = E - (12)$.

Clearly in this case the two particles must occupy different levels (Pauli principle).

$$\psi^{[1^2]}_{\alpha,\beta} = \frac{1}{\sqrt{2}}T\left([1^2]\right)|\alpha(1)\beta(2)\rangle = \frac{1}{\sqrt{2}}(\alpha(1)\beta(2) - \alpha(2)\beta(1)). \qquad (15.9)$$

Let us now continue by considering three particles which can occupy three levels, say α, β and γ.

i) Consider the symmetric case [3], ($r = 1$, $c = 0$).
Then $T\left([3]\right) = E + (12) + (13) + (23) + (123) + (132)$ and

$$\begin{aligned}\psi^{[3]}_{\alpha,\beta,\gamma} &= \frac{1}{\sqrt{6}}T\left([6]\right)|\alpha(1)\beta(2)\gamma(3)\rangle \\ &= \frac{1}{\sqrt{6}}(\alpha(1)\beta(2)\gamma(3) + \alpha(2)\beta(1)\gamma(3) + \alpha(3)\beta(2)\gamma(1) \\ &\quad +\alpha(1)\beta(3)\gamma(2) + \alpha(2)\beta(3)\gamma[1] + \alpha(3)\beta(1)\gamma[2]).\end{aligned} \qquad (15.10)$$

If two particles can occupy the same level, say α and the other is put in level γ one finds

$$\begin{aligned}\psi^{[3]}_{\alpha,\gamma} &= \frac{1}{\sqrt{3}}T\left([6]\right)|\alpha(1)\alpha(2)\gamma(3)\rangle \\ &= \frac{1}{\sqrt{3}}(\alpha(1)\alpha(2)\gamma(3) + \alpha(2)\alpha(3)\gamma(1) + \alpha(1)\alpha(3)\gamma(2)).\end{aligned} \qquad (15.11)$$

The other two possibilities can be found similarly.

ii) Consider the completely anti-symmetric case $[1^3]$, $(r = 0,\ c = 1)$. Then $T\left([1^3]\right) = E - (12) - (13) - (23) + (123) + (132)$ and

$$\begin{aligned}\psi^{[1^3]}_{\alpha,\beta,\gamma} &= \frac{1}{\sqrt{6}} T\left([1^3]\right) |\alpha(1)\beta(2)\gamma(3)\rangle \\ &= \frac{1}{\sqrt{6}}(\alpha(1)\beta(2)\gamma(3) - \alpha(2)\beta(1)\gamma(3) - \alpha(3)\beta(2)\gamma(1) \\ &\quad -\alpha(1)\beta(3)\gamma(2) + \alpha(2)\beta(3)\gamma[1] + \alpha(3)\beta(1)\gamma[2]).\end{aligned} \tag{15.12}$$

iii) Consider the mixed symmetry case $[2, 1]$, $(r = 1,\ c = 1)$. We distinguish two possibilities:

a) The numbers 1,2 are in the row while in the column we have 1,3. In the row we associate the operator $T\left([2]\right)_S = E + (12)$ while in the column we have the operator $T\left([1^2]\right)_A = E - (13)$. Thus

$$T\left([2,1]_1\right) = T\left([1^2]\right)_A T\left([2]\right)_S = (E-(13))(E+(12)) = E+(12)-(13)-(132)$$

$$\begin{aligned}\psi^{[21]_1}_{\alpha,\beta,\gamma} &= \frac{1}{2} T\left([21]_1\right) |\alpha(1)\beta(2)\gamma(3)\rangle \\ &= \frac{1}{2}(\alpha(1)\beta(2)\gamma(3) + \alpha(2)\beta(1)\gamma(3) - \alpha(3)\beta(2)\gamma(1) - \alpha(3)\beta(1)\gamma(2)).\end{aligned} \tag{15.13}$$

In the other possibility the 2 and 3 are interchanged. Thus we find:

$$T\left([2,1]_2\right) = (E - (12))(E + (13)) = E - (12) + (13) - (123)$$

$$\begin{aligned}\psi^{[21]_2}_{\alpha,\beta,\gamma} &= \frac{1}{2} T\left([21]_2\right) |\alpha(1)\beta(2)\gamma(3)\rangle \\ &= \frac{1}{2}(\alpha(1)\beta(2)\gamma(3) - \alpha(2)\beta(1)\gamma(3) + \alpha(3)\beta(2)\gamma(1) - \alpha(2)\beta(3)\gamma(1)).\end{aligned} \tag{15.14}$$

15.5 Angular momentum decomposition of many particle configurations with single particle spin j

In nuclear physics one must construct antisymmetric wave functions of r particles, each particle composed of angular momentum j and isospin $1/2$. The isospin part of the wave function is represented by a Young tableaux $\tilde{f}$ with at most two rows. Columns of two rows do not contribute to the isospin (the corresponding pair has isospin zero). To obtain a totally antisymmetric wave function, the angular momentum part must have an appropriate symmetry with Young tableaux f, obtained from that of isospin by exchanging rows and columns, i.e. it must have at most two columns. Thus, e.g., one finds

$$r = 3 \leftrightarrow \tilde{f} = [3], [2, 1] \leftrightarrow I = 3/2.1/2 \leftrightarrow [f] = [1^3], [2, 1]$$

$$r = 4 \leftrightarrow \tilde{f} = [4], [3, 1], [2, 2] \leftrightarrow I = 2,\ 1,\ 0 \leftrightarrow [f] = [1^4], [2, 1^2], [2, 2].$$

In general

$$r = \text{odd}] \leftrightarrow \tilde{f} = [r], [r-1, 1] \cdots [r^{[(r-1)/2]}, 1] \leftrightarrow I = r/2.(r-1)/2 \cdots 1/2 \leftrightarrow$$

$$[f] = [1^r], [2, 1^{r-1}], [2^2, 1^{r-2}], \cdots [2^{[(r-1)/2]}, 1]$$

$$r = \text{even}] \leftrightarrow \tilde{f} = [r], [r-1, 1] \cdots [r^{[(r)/2]}] \leftrightarrow I = r/2.(r-1)/2 \cdots 1, 0 \leftrightarrow$$

$$[f] = [1^r], [2, 1^{r-1}], [2^2, 1^{r-2}], \cdots [2^{[(r)/2]}].$$

The isospin part with one row is completely symmetric and is associated with completely antisymmetric angular momentum wave function, which can be decomposed in J with elementary methods, see Section 15.1. For the other components the decomposition is not easy but it can be obtained by noting that it corresponds to the reduction of the group $SU(2j+1)$, transforming the $(2j+1)$ components of a given j to each other, to the angular momentum group $SU(2)$ or $SO(3)$, using group theoretical techniques, see [Vergados (2016)]. For the reader interested in shell model calculations we have included here the decomposition of some of the common needed configurations, see Tables 15.1, 15.2 and 15.3, found in the above reference.

The reader should check the results of this table with those obtained in Section 15.1.

15.6 The SU(3) group

15.6.1 *The flavor SU(3) group*

We recall that this group has 8 generators. These can be chosen to be the Gell-Mann matrices λ_i discussed, e.g., in Ref. [Vergados (2016)] (see also Section 15.7). We find it more convenient to use a basis starting with the operators:

$$A_{xy} = a_x^+ a_y,\ x, y = u, d, s$$

Table 15.1: The reduction $SU(2j+1) \supset SO(3)$ for $j=\frac{3}{2}$ and $r \leq 4$.

r	[f]	J	dim[f]
0	[0]	0	1
1	[1]	$\frac{3}{2}$	4
2	[2]	1,3	10
	[11]	0,2	6
3	[21]	$\frac{1}{2}, \frac{3}{2}, \frac{5}{2}, \frac{7}{2}$	20
4	[22]	0, $(2)^2$, 4	20
	[211]	1, 2, 3	15

Table 15.2: The reduction $SU(2j+1) \supset SO(3)$ for $j=5/2$ and $r \leq 6$.

r	$[f]$	J	dim $[f]$
0	[0]	0	1
1	[1]	5/2	6
2	[2]	1, 3, 5	21
	[11]	0, 2, 4	15
3	[21]	$1/2, 3/2, (5/2)^2, (7/2)^2, 9/2, 11/2, 13/2$	70
	[111]	3/2, 5/2, 9/2	20
4	[22]	$(0)^2, (2)^2, 3, (4)^2, 5, (6)^2, 8$	105
	[211]	$(1)^2, (2)^2, (3)^3, (4)^2, (5)^2, 6, 7$	105
5	[221]	$(1/2)^2, (3/2)^3, (5/2)^4, (7/2)^4, (9/2)^4, (11/2)^3, (13/2)^2,$ 15/2, 17/2	210
	[2111]	$1/2, (3/2)^2, (5/2)^2, (7/2)^2, (9/2)^2, 11/2, 13/2$	84
6	[222]	$(1)^3, 2, (3)^5, (4)^2, (5)^3, (6)^2, (7)^2, 9$	175
	[2211]	$(0)^2, 1, (2)^5, (3)^3, (4)^5, (5)^2, (6)^3, 7, 8$	189
	[21111]	1, 2, 3, 4, 5	35

where the quark creation and annihilation operators satisfy:

$$\left[a_i, a_j^+\right] = \delta_{ij},\ \left[a_i^+, a_j^+\right] = 0,\ [a_i, a_j] = 0. \tag{15.15}$$

Table 15.3: The reduction $SU(2j+1) \supset SO(3)$ for $j = 7/2$ and $r \leq 4$.

r	$[f]$	J	dim $[f]$
0	[0]	0	1
1	[1]	7/2	8
2	[2]	1, 3, 5, 7	36
	[11]	0, 2, 4, 6	28
3	[21]	$1/2,\ 3/2,\ (5/2)^2,\ (7/2)^3,\ (9/2)^2,\ (11/2)^2,\ (13/2)^2,\ 15/2,\ 17/2,\ 19/2$	168
	[111]	3/2, 5/2, 7/2, 9/2, 11/2, 15/2	56
4	[22]	$0^3,\ 2^4,\ 3^2,\ 4^5,\ 5^2,\ 6^5,\ 7^2,\ 8^3,\ 9,\ 10^2,\ 12$	336
	[211]	$1^3,\ 2^3,\ 3^5,\ 4^4,\ 5^5,\ 6^4,\ 7^4,\ 8^2,\ 9^2,\ 10,\ 11$	378
	[1111]	$0,\ 2^2,\ 4^2,\ 5,\ 6,\ 8$	70

Then a convenient basis for $SU(3)$, i.e. the subset of traceless generators, is

$$A_{xy} = a_x^+ a_y,\ x \neq y,\ x, y = u, d, s, H_1 = \frac{1}{2}(A_{uu} - A_{dd}),$$
$$H_2 = \frac{1}{2\sqrt{3}}(A_{uu} + A_{dd} - 2A_{ss}). \tag{15.16}$$

For $U(3)$ we have the extra generator

$$B = \frac{1}{3}\left(A_{uu} + A_{dd} + A_{ss}\right). \tag{15.17}$$

Sometimes we may use the operators:

$$I_+ = A_{ud},\ I_- = A_{du},\ I_3 = H_1,\ Y = \frac{\sqrt{2}}{3}H_2,\ I^2 = \frac{1}{2}\left(I_+I_- + I_-I_+\right) + H_3^2.$$

Thus defining:

$$|1\rangle = a_u^+|0\rangle,\ |2\rangle = a_d^+|0\rangle,\ |3\rangle = a_s^+|0\rangle,$$

we obtain

$$I_+|1\rangle = 0, I_+|2\rangle = |1\rangle, I_+|3\rangle = 0,$$

$$I_-|1\rangle = |2\rangle, I_-|2\rangle = 0, I_-|3\rangle = 0,$$

$$A_{ds}|1\rangle = 0, A_{ds}|2\rangle = 0, A_{ds}|3\rangle = |2\rangle,$$

$$H_1|1\rangle = \frac{1}{2}|1\rangle, H_1|2\rangle = -\frac{1}{2}|2\rangle, H_1|3\rangle = 0,$$

$$H_2|1\rangle = \frac{1}{2\sqrt{3}}|1\rangle, H_2|2\rangle = \frac{1}{2\sqrt{3}}|2\rangle, H_2|3\rangle = -\frac{1}{\sqrt{3}}|3\rangle.$$

These lead to the matrices, i.e. the representation:

$$I_+ \leftrightarrow E_{ud} = \begin{pmatrix} 0\,1\,0 \\ 0\,0\,0 \\ 0\,0\,0 \end{pmatrix}, I_- \leftrightarrow E_{du} = \begin{pmatrix} 0\,0\,0 \\ 1\,0\,0 \\ 0\,0\,0 \end{pmatrix}$$

$$A_{ds} \leftrightarrow E_{ds} = \begin{pmatrix} 0\,0\,0 \\ 0\,0\,1 \\ 0\,0\,0 \end{pmatrix}, A_{sd} \leftrightarrow Esd = \begin{pmatrix} 0\,0\,0 \\ 0\,0\,0 \\ 0\,1\,0 \end{pmatrix}$$

$$A_{us} \leftrightarrow E_{us} = \begin{pmatrix} 0\,0\,1 \\ 0\,0\,0 \\ 0\,0\,0 \end{pmatrix}, A_{su} \leftrightarrow Esu = \begin{pmatrix} 0\,0\,0 \\ 0\,0\,0 \\ 1\,0\,0 \end{pmatrix}$$

$$H_1 \leftrightarrow h_1 = \frac{1}{2}\begin{pmatrix} 1 & 0 & 0 \\ 0 & -1 & 0 \\ 0 & 0 & 0 \end{pmatrix}, H_2 \leftrightarrow h_2 = \frac{1}{2\sqrt{3}}\begin{pmatrix} 1 & 0 & 0 \\ 0 & 1 & 0 \\ 0 & 0 & -2 \end{pmatrix}. \quad (15.18)$$

The above operators form an algebra $\mathcal{L}$ in the sense that the commutator of any two of them can be expressed as linear combination of these operators. Symbolically:

$$\begin{aligned}[X_i, X_j] &= X_iX_j - X_jX_i] \\ &= \sum_k C^k_{i,j}X_k,\ X_i \in \mathcal{L}, X_j \in \mathcal{L}, X_k \in \mathcal{L},\ i,j = 1,2,\ldots,n. \quad (15.19)\end{aligned}$$

Where $C^k_{i,j}$ are the structure constants of the algebra in the chosen basis. n is the number of elements of the algebra, in our case $n = 8$. The elements $I_\pm$, $I_3 = H_1$ form a closed set

$$[I_3, I_+] = I_+,\ [I_3, I_-] = -I_-,\ [I_+, I_-] = 2I_3. \quad (15.20)$$

These are familiar from the angular momentum theory. So this is a subalgebra, which can be identified with the (strong) isotopic spin $SU_I(2)$. We also note that these three operators commute with the operator H_2. So we write

$$SU_I(2) \otimes U(1) \subset SU(3). \quad (15.21)$$

The elements H_1 and H_2 commute, $[H_1, H_2] = 0$ and they form an Abelian sub-algebra with elements $\ell = 2$. From the form of the above matrices it is straightforward to compute all commutators. We will follow a more elegant mathematical procedure. The commutators can be expressed in terms of six vectors, defined in the ℓ dimensions, which we call roots of the algebra and uniquely characterize it. These are:

$$\alpha^{(1)} = (1,0),\ \alpha^{(2)} = \left(\frac{1}{2}, -\frac{\sqrt{3}}{2}\right),\ \alpha^{(3)} = \left(\frac{1}{2}, \frac{\sqrt{3}}{2}\right) \tag{15.22}$$

and their opposites

$$\alpha^{(-1)} = (-1,0),\ \alpha^{(-2)} = \left(-\frac{1}{2}, \frac{\sqrt{3}}{2}\right),\ \alpha^{(-3)} = \left(\frac{1}{2}, -\frac{\sqrt{3}}{2}\right). \tag{15.23}$$

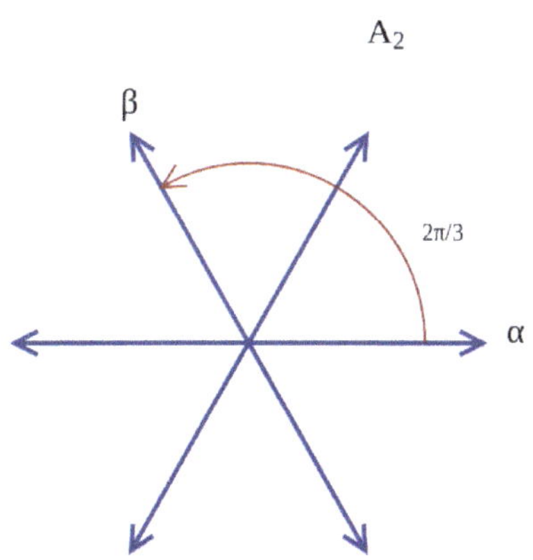

Fig. 15.1: The root system for the algebra of SU(3) (A_2).

These roots are represented in Fig. 15.1. The first set, whose first non zero components are positive are called positive roots. Some roots can be expressed in terms of a subset of the roots, e.g.

$$\alpha^{(1)} = \alpha^{(2)} + \alpha^{(3)},\ \alpha^{(1)} + \alpha^{(-2)} = \alpha^{(3)},\ \alpha^{(1)} + \alpha^{(-3)} = \alpha^{(2)}. \tag{15.24}$$

Analogous relations hold for the opposite of the above roots.

Let us for notation convenience define:

$$Q_1 = \frac{1}{\sqrt{3}} E_{ud},\ Q_{-1} = \frac{1}{\sqrt{3}} E_{du},\ Q_2 = \frac{1}{\sqrt{3}} E_{sd},\ Q_{-2} = \frac{1}{\sqrt{3}} E_{ds},$$

$$Q_3 = \frac{1}{\sqrt{3}} E_{us},\ Q_{-3} = \frac{1}{\sqrt{3}} E_{su}.$$

Then one finds

$$[h_i, Q_j] = \alpha_j^{(j)} Q_j,$$
$$[Q_i, Q_{-i}] = \sum_k \tilde{\alpha}_k^{(i)} h_k,$$
$$[Q_i, Q_j] = \tilde{C}_{i,j}^k Q_k, \tag{15.25}$$

with

$$\tilde{C}_{i,j}^k = \begin{cases} C_{i,j}^k, \text{ if } \alpha^{(k)} = \alpha^{(i)} + \alpha^{(j)} \\ 0, \quad \text{otherwise} \end{cases}. \tag{15.26}$$

The quantities $\tilde{\alpha}^{(i)}$ are contragredient to the roots with the metric specified by

$$g_{i,j} = \sum_{k=1,2,3} \alpha_i^{(k)} \alpha_j^{(k)} \Rightarrow g = \begin{pmatrix} \frac{3}{2} & 0 \\ 0 & \frac{3}{2} \end{pmatrix}. \tag{15.27}$$

The inverse matrix is

$$g^{-1} = \begin{pmatrix} \frac{2}{3} & 0 \\ 0 & \frac{2}{3} \end{pmatrix} \tag{15.28}$$

$$\tilde{\alpha}^{(i)} = g^{-1}\alpha^{(i)} \Rightarrow \tilde{\alpha}^{(1)} = \left(\frac{2}{3}, 0\right), \tilde{\alpha}^{(2)} = \left(\frac{1}{3}, -\frac{1}{\sqrt{3}}\right), \tilde{\alpha}^{(3)} = \left(\frac{1}{3}, \frac{1}{\sqrt{3}}\right). \tag{15.29}$$

Finally we come to the evaluation of the structure constants $\tilde{C}_{i,j}^k$. It is clear that the anti-symmetry holds, i.e. $\tilde{C}_{j,i}^k = -\tilde{C}_{i,j}^k$. Furthermore we need evaluate only those terms dictated by Eq. (15.24). By explicit calculation of the commutators of the relevant matrices we obtain

$$\tilde{C}_{2,3}^1 = \frac{1}{\sqrt{3}}, \tilde{C}_{1,-2}^3 = \frac{1}{\sqrt{3}}, \tilde{C}_{1,-3}^2 = -\frac{1}{\sqrt{3}}. \tag{15.30}$$

Those involving the opposite roots are obtained as follows:

$$Q_{-i}Q_{-j} - Q_{-j}Q_{-i} = (Q_j Q_i - Q_i Q_j)^T$$

where T stands for the transpose of the matrix. Thus

$$Q_{-i}Q_{-j} - Q_{-j}Q_{-i} = \tilde{C}_{j,i}^k (Q_k)^T = \tilde{C}_{j,i}^k Q_{-k}.$$

Therefore

$$[Q_2, Q_3] = \frac{1}{\sqrt{3}} Q_1, \; [Q_1, Q_{-2}] = \frac{1}{\sqrt{3}} Q_3, \; [Q_1, Q_{-3}] = -\frac{1}{\sqrt{3}} Q_2. \tag{15.31}$$

From those we obtain:

$$[Q_{-2}, Q_{-3}] = -\frac{1}{\sqrt{3}}Q_{-1},\ [Q_{-1}, Q_2] = -\frac{1}{\sqrt{3}}Q_{-3},\ [Q_{-1}, Q_3] = \frac{1}{\sqrt{3}}Q_{-2}. \tag{15.32}$$

The algebra of matrix representation is isomorphic with the algebra of the operators as given by Eq. (15.16). The commutators for the operators can be obtained from the structure constants given above. The notation becomes a little tricky, but we give the simple rules, after the right correspondence in notation:

- The structure constants of the first equation of (15.25) remain the same.
- The structure constants of the second equation of (15.25) must be multiplied by 3.
- The structure constants of the third equation of (15.25) must be multiplied by $\sqrt{3}$.

One must compare this with the structure constants in the Gell-Mann basis discussed in Section 15.7.

15.6.2 *Construction of states of a given symmetry*

In Section 15.6.1 we show that there appear two types of representation of the flavor group, namely:

- the **fundamental representation** specified by $\underline{3}, (I, I_3), Y$ as follows:
$$u = \underline{3}\left(\frac{1}{2}, \frac{1}{2}\right), \frac{1}{3},\ d = \underline{3}\left(\frac{1}{2}, -\frac{1}{2}\right), \frac{1}{3},\ s = \underline{3}(0,0), -\frac{2}{3},\ Y = B - S, \tag{15.33}$$
- the **regular representation** specified by the matrices given by Eq. (15.18).

From the fundamental representation one can construct representations of larger dimensions. In particular representations, which cannot be reduced to smaller dimensions by similarity transformation called irreducible. In other words one cannot find a transformation matrix S such that for any $X \in \mathcal{L}$

$$\Gamma(X) \Rightarrow S^{-1}\Gamma(X)S = \begin{pmatrix} \Gamma_1(X) & 0 \\ 0 & \Gamma_2(X) \end{pmatrix},\ X \in \mathcal{L}, \tag{15.34}$$

where $\Gamma(X)$, $\Gamma_1(X)$ and $\Gamma_2(X)$ have dimensions $n \times n$, $n_1 \times n_1$ and $(n - n_1) \times (n - n_1)$ respectively, with $n_1 < n$.

To this end we will find very useful to consider the group S_n of all permutations of n objects. We will find that the irreducible representations are given by a set of integers $[f_1, f_2, f_3]$ subject to the condition $f_1 \geq f_2 \geq f_3$. When restricted to $SU(3)$ two indices suffice, namely

$$(\lambda, \mu) \text{ with } \lambda = f_1 - f_2, \mu = f_2 - f_3. \tag{15.35}$$

The conjugate of (λ, μ) is (μ, λ). For $\lambda\mu$ the representation is self conjugate (real).

The states in a given representation are specified the isospin quantum numbers I, I_3 and the eigenvalue of Y. Some people use instead of Y the eigenvalue of:

$$S = Y + B,\ B = \text{baryon number}.$$

The determination of these quantum numbers contained in a given representation (λ, μ) is beyond the scope of this book. It can be shown, see e.g. [Vergados (2016)] that the following formula holds:

$$Y = -\frac{1}{3}\left(2\lambda + \mu - 3(p+q)\right), I = \frac{\mu}{2} + \frac{1}{2}(p-q),\ I_3 = I - r$$
$$0 \leq p \leq \lambda,\ 0 \leq q \leq \mu,\ 0 \leq r \leq 2I. \tag{15.36}$$

Let us begin with the fundamental representation of $U(3)$. It is clear that $f_1 = 1$, $f_2 = f_3 = 0$ i.e. $\lambda = 1$ and $\mu = 0$. From the above formula we get $q = 0$ and $p = 0, 1$. For $p = 0$ we get $I = 0$, $Y = -2/3$ to be identified with the s =quark, while for $q = 1$ we get $I = 1/2$, $Y = 1/3$ leading to $I = 1/2$, $I_3 = 1/2$, $Y = 1/3$ to be identified with the u quark and $I = 1/2$, $I_3 = -1/2$, $Y = 1/3$ to be identified with the d quark.

15.6.3 *The color SU(3) group*

We proceed as in the previous section except that now the identification of the quantum numbers is different. The basic representation now $(\lambda\mu) = (10)$ contains three states defined by their color $(10)_c$, $c = r$, g and b. The two particle state q^2 contains a color symmetric six-dimensional representation with $(\lambda\mu) = (20) = \underline{6}$ and a color antisymmetric 3-dimensional representation with $(\lambda\mu) = (01) = \underline{3}^*$.

The three quark states q^3 come in many possibilities. A symmetric state with $(\lambda\mu) = (30) = \underline{10}$, an antisymmetric state with $(\lambda\mu) = (00) = \underline{1}$ and two 8-dimensional representations with $(\lambda\mu) = (11) = \underline{8}$. Out of these only the singlet $\underline{1}$ contains a colorless component and only this one can be realized in hadrons. The same is true for the $q\bar{q}$, which contains a color singlet and a color octet. Thus the color singlet combinations of interest in hadrons are

$$|q^3\rangle = \frac{1}{\sqrt{6}}\left((q_r q_g - q_g q_r) + (q_g q_b - q_b q_g) + (q_b q_r - q_r q_b)\right) \qquad (15.37)$$

$$|q\bar{q}\rangle = \frac{1}{\sqrt{3}}\left(q_r \bar{q}_r + q_g \bar{q}_b + q_b \bar{q}_b\right). \qquad (15.38)$$

We also encounter $3^2 - 1 = 8$ colored gauge bosons, which can transmit color. These are six non diagonal $G^1 \leftrightarrow \tilde{t}^1_{rg}$, $G^2 \leftrightarrow \tilde{t}^2_{rb}$, $G^4 \leftrightarrow \tilde{t}^4_{gb}$ and their adjoined ones $G^5 \leftrightarrow \left(\tilde{t}^1_{rg})\right)^+ = \tilde{t}^5_{gr}$, $G^6 \leftrightarrow \left(\tilde{t}^2_{rg})\right)^+ = \tilde{t}^6_{br}$, $G^7 \leftrightarrow \left(\tilde{t}^4_{rg})\right)^+ = \tilde{t}^7_{bg}$ as well as two diagonal

$$G^3 \leftrightarrow \tilde{t}^3 = \frac{1}{\sqrt{2}}(rr - gg),\; G^8 \leftrightarrow \tilde{t}^8 = \frac{1}{\sqrt{6}}(rr + gg - 2bb).$$

Only the non zero entries of the matrices $\tilde{t}^a_{\alpha\beta}$ have been explicitly written.

These correspond to the matrices A and H_1 and H_2 of the previous section but with the diagonal ones normalized differently

$$\mathrm{tr}\left(\left(\tilde{t}^a\right)^+ \tilde{t}^b\right) = \delta_{ab}.$$

Ignoring the overall strength (color charge, $g_r = g_s = 4\pi\alpha_s$) and omitting the gluon propagator, the interaction between similar colors is

$$\langle rr|V|rr\rangle = \langle gg|V|gg\rangle = \left(\frac{1}{6} + \frac{1}{2}\right) = \frac{2}{3}.$$

The first contribution comes from the exchange of the gluon G^8 and the second one comes from G^3 (see Fig. 15.2(b)). Similarly from the exchange of the gluon G^8 (see Fig. 15.2(c)) we get

$$\langle bb|V|bb\rangle = \frac{1}{6}2^2 = \frac{2}{3}.$$

In a similar fashion we get

$$\langle gr|V|gr\rangle = \langle rg|V|rg\rangle = \left(\frac{1}{6} - \frac{1}{2}\right) = -\frac{1}{3} \text{ (see Fig. 15.2(b))}$$

$$\langle rb|V|rb\rangle = \langle br|V|br\rangle = \frac{1}{6}(-2) = -\frac{1}{3} \text{ (see Fig. 15.2(c))}.$$

Furthermore

$$\langle \alpha\beta|V|\beta\alpha\rangle = 1,\; \alpha \neq \beta \text{ (see Fig. 15.2(a))}.$$

In this last case the off diagonal gluon is exchanged, $G^{1,2,4,5,6,7}$. All other matrix elements are zero.

The same results hold in the case of two antiquarks. Let us suppose that the two quarks are $q_\alpha(1)q_\beta(2)$, where α and β are color indices, taking values r, g, b, and (1) (2) label the particles. For simplicity of notation we will drop the particle index, with the understanding that particle one will be first and particle second, by writing $q_\alpha(1)q_\beta(2) \Leftrightarrow \alpha\beta = rr, rg,$ etc. The Feynman diagrams leading to the interaction between two quarks are exhibited in Fig. 15.2.

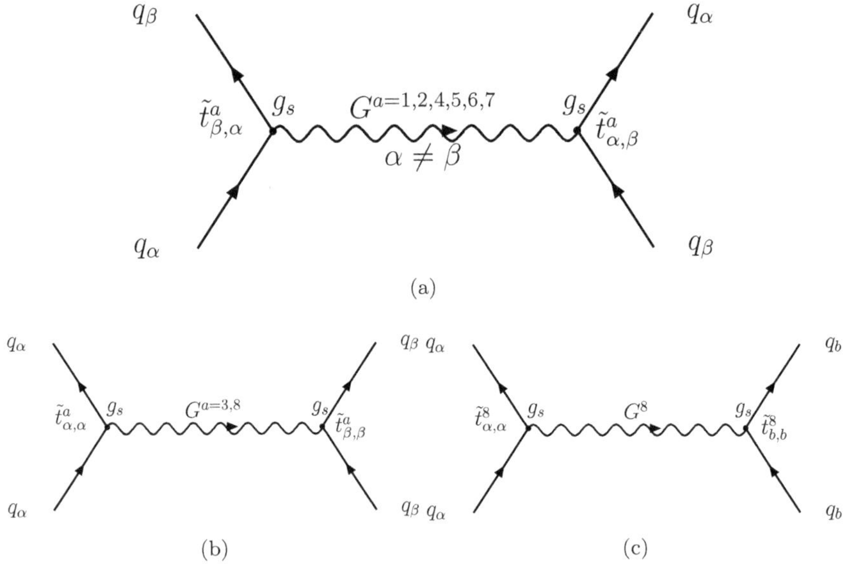

Fig. 15.2: The strong interaction mediated by gluons, with the Lorentz index suppressed. Shown are the color changing gluon interaction ($\tilde{t}^a_{\alpha\beta}$, $\alpha \neq \beta$, which destroys a quark β and converts to a quark α) (a), the color conserving interaction involving quarks other than b ($\tilde{t}^{3,8}_{\alpha\beta}$, α, $\beta = r, g$) (b) and the color conserving interaction in which at least one of the quarks is blue ($\tilde{t}^{3,8}_{\alpha\alpha}$, $\alpha = r, g, b$) (c). Note that all operators are normalized to unity $\mathrm{tr}\big((\tilde{t}^a)^+ \tilde{t}^b\big) = \delta_{a,b}$.

The above product of two quarks, namely,

$$rr, gg, bb, rg, rb, gb, gr, br, bg,$$

constitutes a basis in color space, $\underline{3} \otimes \underline{3}$, which yields a reducible nine dimensional representation of $SU(3)$. The interaction between the first 3 is diagonal. Diagonalizing the interaction between the remaining six states one finds three eigenstates with eigenvalue $-\frac{4}{3}$ of the form

$$\Leftrightarrow \tilde{b} = \frac{1}{\sqrt{2}}\left(rg - gr\right), \tilde{g} = \frac{1}{\sqrt{2}}\left(rb - br\right), \tilde{r} = \frac{1}{\sqrt{2}}\left(gb - bg\right)$$

and three eigenstates with eigenvalue,

$$\frac{2}{3} \Leftrightarrow \frac{1}{\sqrt{2}}\left(rg + gr\right), \ \frac{1}{\sqrt{2}}\left(rb + br\right), \ \frac{1}{\sqrt{2}}\left(gb + bg\right),$$

(see Eq. (15.43) in the problem section at the end of this chapter). Thus the six symmetric in color index states with eigenvalue 2/3 are associated with the irreducible representation $(\lambda\mu) = (2,0) = \underline{6}$ and the three antisymmetric ones associated with the irreducible representation $(\lambda\mu) = (0,1) = \underline{3}^*$.

The above results can be more generally be given in terms of the eigenvalues of invariant Casimir operator [Vergados (2016)], which in the case of $SU(3)$ is given by

$$C(\lambda, \mu) = \frac{2}{3}\left(\lambda^2 + \mu^2 + \lambda\mu + 3(\lambda + \mu)\right). \tag{15.39}$$

In the present case, however, we are interested in the invariant two body interaction for between the irreducible representations of the two particle system, sextet ant antitriplet, namely[3]

$$\boldsymbol{\lambda}_1.\boldsymbol{\lambda}_2 = \frac{1}{2}\left(C(\lambda, \mu) - 2C(1,0)\right)$$

where λ and μ specify the irreducible representation of $SU(3)$. Thus:

$$\text{triplet:}(\lambda, \mu) = (1,0) \rightarrow C(\lambda, \mu) = \frac{8}{3}$$

$$\text{anti-triplet:}(\lambda, \mu) = (0,1) \rightarrow C(\lambda, \mu) = \frac{8}{3} \rightarrow \boldsymbol{\lambda}_1.\boldsymbol{\lambda}_2 = -\frac{4}{3}$$

$$\text{sextet:}(\lambda, \mu) = (2,0) \rightarrow C(\lambda, \mu) = \frac{20}{3} \rightarrow \boldsymbol{\lambda}_1.\boldsymbol{\lambda}_2 = \frac{2}{3}.$$

[3]Recall the familiar case of $U(2)$

$$\boldsymbol{\sigma}_1.\boldsymbol{\sigma}_2 = 4\mathbf{s}_1.\mathbf{s}_2 = 2\left((\mathbf{s}_1 + \mathbf{s}_2)^2 - \mathbf{s}_1^2 - \mathbf{s}_2^2\right) = 2\mathbf{S}^2 - 2\mathbf{s}_1^2 - 2\mathbf{s}_2^2, \ \mathbf{S} = \mathbf{s}_1 + \mathbf{s}_2,$$

with eigenvalue

$$\langle S|\boldsymbol{\sigma}_1.\boldsymbol{\sigma}_2|S\rangle = 2S(S+1) - 3$$

where S is the spin of the two particle state.

Let us now examine the interaction between quarks and antiquarks. The antiquark viewed as the antisymmetric combination of two quarks given above,[4] but we prefer to use the Casimir operator.

We have seen that:

$$\underline{3} \otimes \underline{3}^* = \underline{1} + \underline{8}.$$

As a result we can write the quark antiquark interaction as:

$$\boldsymbol{\lambda}_1.\boldsymbol{\lambda}_2 = \frac{1}{2}\left(C(\lambda,\mu) - C(1,0) - C(0,1)\right)$$

$$\text{singlet:}(\lambda,\mu) = (0,0) \rightarrow C(\lambda,\mu) = 1 \rightarrow \boldsymbol{\lambda}_1.\boldsymbol{\lambda}_2 = -\frac{8}{3}$$

$$\text{octet:}(\lambda,\mu) = (1,1) \rightarrow C(\lambda,\mu) = \frac{18}{3} \rightarrow \boldsymbol{\lambda}_1.\boldsymbol{\lambda}_2 = \frac{1}{3}.$$

Before concluding this section we should mention that in the context of the one gluon exchange potential between the above allowed color combinations is proportional to $\boldsymbol{\lambda}_1.\boldsymbol{\lambda}_2\alpha_s$.

[4]As an exercise one can proceed as above, i.e. one can evaluate the matrix element of the gluon between the relevant three quark states, noting that one of the quarks of the antisymmetric combination is a spectator (not interacting). Thus we find the non vanishing independent ME involving different colors are:

$$\langle r\bar{b}|V|r\bar{b}\rangle = \frac{1}{3},\ \langle r\bar{g}|V|r\bar{g}\rangle = \frac{1}{3},\ \langle g\bar{b}|V|g\bar{b}\rangle = \frac{1}{3}.$$

For the similar color combination we get:

$$\langle r\bar{r}|V|r\bar{r}\rangle = \langle g\bar{g}|V|g\bar{g}\rangle = \langle b\bar{b}|V|b\bar{b}\rangle = -\frac{2}{3},$$

$$\langle r\bar{r}|V|g\bar{g}\rangle = \langle r\bar{r}|V|b\bar{b}\rangle = \langle g\bar{g}|V|b\bar{b}\rangle = -1.$$

Note the change in sign compared to the $q-q$ interaction. This usually is attributed to the "opposite color charge of anti-quarks", but note the notational difference in the last equation. We have seen, however, that it comes as a transformation from the picture of anti-quarks as an antisymmetric combination of quarks.

The product of a quark and anti-quark transforms under $SU(3)$ like $\underline{3} \otimes \underline{3}^*$, which is reducible. One can easily see that the above above six combinations $q_\alpha \bar{q}_\beta$, $\alpha \neq \beta$ transform as the six members of the octet. Diagonalizing the above matrix in the case of similar color indices we find the eigensolutions:

$$\epsilon_0 = -\frac{8}{3} \leftrightarrow r_0 = \frac{1}{\sqrt{3}}(|r\bar{r}\rangle + |g\bar{g}\rangle + b\bar{b}\rangle),$$

$$\epsilon_1 = \frac{1}{3} \leftrightarrow r_8^1 = \frac{1}{\sqrt{3}}(|r\bar{r}\rangle - |g\bar{g}\rangle),$$

$$\epsilon_2 = \frac{1}{3} \leftrightarrow r_8^2 = \frac{1}{\sqrt{6}}(|r\bar{r}\rangle + |g\bar{g}\rangle - 2b\bar{b}\rangle).$$

The first transforms like the singlet, while the other two are the additional members of the octet.

15.7 The SU(3) Gell-Mann matrices

Consider the special case of $n = 3$ Hermitian matrix α, associated with $U(3)$. Then

$$\alpha = \text{Hermitian} \Rightarrow U = e^{i\alpha} = \text{unitary}.$$

If the matrix α is traceless, it depends on $n^2 - 1 = 8$ parameters and is a associated with the unimodular $SU(3)$.

$$\alpha = \text{traceless} \rightarrow \det U = e^{itr(\alpha)} = e^0 = 1.$$

Proceeding as above we select a basis set of traceless 3×3 Hermitian matrices:

$$\lambda_1 = \begin{pmatrix} 0 & 1 & 0 \\ 1 & 0 & 0 \\ 0 & 0 & 0 \end{pmatrix}, \lambda_2 = \begin{pmatrix} 0 & -i & 0 \\ i & 0 & 0 \\ 0 & 0 & 0 \end{pmatrix}, \lambda_3 = \begin{pmatrix} 1 & 0 & 0 \\ 0 & -1 & 0 \\ 0 & 0 & 0 \end{pmatrix} \lambda_4 = \begin{pmatrix} 0 & 0 & 1 \\ 0 & 0 & 0 \\ 1 & 0 & 0 \end{pmatrix},$$
$$\lambda_5 = \begin{pmatrix} 0 & 0 & -i \\ 0 & 0 & 0 \\ i & 0 & 0 \end{pmatrix}, \lambda_6 = \begin{pmatrix} 0 & 0 & 0 \\ 0 & 0 & 1 \\ 0 & 1 & 0 \end{pmatrix}, \lambda_7 = \begin{pmatrix} 0 & 0 & 0 \\ 0 & 0 & -i \\ 0 & i & 0 \end{pmatrix}, \lambda_8 = \frac{1}{\sqrt{3}} \begin{pmatrix} 1 & 0 & 0 \\ 0 & 1 & 0 \\ 0 & 0 & -2 \end{pmatrix}. \quad (15.40)$$

With this selection:

$$tr(\lambda_i \lambda_j) = 2\delta_{ij}$$

$$[\lambda_\alpha, \lambda_\beta] = 2i f_{\alpha\beta\gamma} \lambda_\gamma.$$

Summation over repeated indices is understood. Sometimes we will use $T_k = \frac{1}{2}\lambda_k$. Then

$$T_k = \frac{1}{2}\lambda_k \Rightarrow [T_\alpha, T_\beta] = i f_{\alpha\beta\gamma} T_\gamma.$$

It is adequate to consider $\alpha \leq \beta$. Then the non zero values of the parameters $f_{\alpha\beta\gamma}$ are given in Table 15.4. Note that in the special case of $\alpha = 4$, $\beta = 5$ two terms appear in the commutator with the indicated structure constants. Similarly for $\alpha = 7$, $\beta = 8$.

To each generator λ_k, there corresponds a group element, $U_k = e^{i\theta\lambda_k}$.

15.8 Problems

(1) Without tables of C-G coefficients

- Discuss the symmetry of two particles with spin 1/2 each and total spin S.

Table 15.4: The SU(3) structure functions.

μ	ν	ρ	$f_{\mu\nu\rho}$
1	2	3	1
1	3	2	-1
1	4	7	$\frac{1}{2}$
1	5	6	$-\frac{1}{2}$
1	6	5	$\frac{1}{2}$
1	7	4	$-\frac{1}{2}$
2	3	1	1
2	4	6	$\frac{1}{2}$
2	5	7	$\frac{1}{2}$
2	6	4	$-\frac{1}{2}$
2	7	5	$\frac{1}{2}$
3	4	5	$\frac{1}{2}$
3	5	4	$-\frac{1}{2}$
3	6	7	$-\frac{1}{2}$

;

μ	ν	ρ	$f_{\mu\nu\rho}$
3	7	6	$\frac{1}{2}$
4	5	3	$\frac{1}{2}$
		8	$\frac{\sqrt{3}}{2}$
4	6	2	$\frac{1}{2}$
4	7	1	$\frac{1}{2}$
4	8	5	$-\frac{\sqrt{3}}{2}$
5	6	1	$-\frac{1}{2}$
5	7	2	$\frac{1}{2}$
5	8	4	$\frac{\sqrt{3}}{2}$
6	7	3	$-\frac{1}{2}$
		8	$\frac{\sqrt{3}}{2}$
6	8	7	$-\frac{\sqrt{3}}{2}$
7	8	6	$\frac{\sqrt{3}}{2}$

- Repeat the previous in the case of isospin 1/2 paricles.
- Repeat the above procedure in the case of two particles possessing both spin and isospin as above.

Hints:
The standard angular momentum operators are the sum of the operators for each particle, i.e.:

$$J_+ = (J_+)_1 + (J_+)_2,\ J_- = (J_-)_1 + (J_-)_2,\ J_0 - = (J_0)_1 + (J_0)_2.$$

In the special case of $j_1 = j_2 = 1/2$, which covers the spin or isospin we have the states

$$m = 1 \leftrightarrow |1,1\rangle = |1/2,1/2\rangle_1 |1/2,1/2\rangle_2,$$

$$m = -1 \leftrightarrow |1,1\rangle = |1/2,-1/2\rangle_1 |1/2,-1/2\rangle_2.$$

These are states of good angular momentum $j = 1$, $m = 1$ and $j = 1$, $m = -1$, since the first is killed by J_+, while the second is killed by J_-. The C-G coefficients are unity.

In the case of $m = 0$ states we can have two possibilities:

$$m = 0 \leftrightarrow |0,0\rangle$$
$$= \frac{1}{\sqrt{2}} \left(|1/2,1/2\rangle_1 |1/2,-1/2\rangle_2 - |1/2,-1/2\rangle_1 |1/2,1/2\rangle_2 \right)$$

$$m = 0 \leftrightarrow |1,0\rangle$$
$$= \frac{1}{\sqrt{2}}\left(|1/2,1/2\rangle_1|1/2,-1/2\rangle_2 + |1/2,-1/2\rangle_1|1/2,1/2\rangle_2\right).$$

The first of these is indeed a $j = 0$ state since it is killed by either J_+ or J_-. The second is a $j = 1$ state since it gives $|1,1\rangle$ after acting with J_+ and $|1,-1\rangle$ after acting with J_-. Thus the C-G coefficients are:

$$\langle 1/2,1/2;1/2,-1/2|1,0\rangle = \langle 1/2,-1/2;1/2,1/2|1,0\rangle = \frac{1}{\sqrt{2}}$$

$$\langle 1/2,1/2;1/2,-1/2|0,0\rangle = -\langle 1/2,-1/2;1/2,1/2|1,0\rangle = \frac{1}{\sqrt{2}}.$$

Proceeding in an analogous fashion one can construct the C-G

$$\langle j_1, m_1; 1/2, m_2|j,m\rangle,\ j = j_1 + 1/2 \text{ and } j = |j_1 - 1/2|.$$

In any case one can see that the state with spin $j = 0$ is antisymmertric with respect with the exchange of two particles, while that of spin 1 is symmetric. If we consider also the quarks $q = (u,d)$ or nucleons $N = (p,n)$ which are characterized by isospin $I = 1/2$ the two particle system with $I = 1$ is symmetric, while the $I = 0$ is antisymmetric. Thus, if we want to have both spin and isospin degrees of freedom, the combined wave function can be written as a product of a spin and isospin part. Thus to satisfy the Pauli principle the state must be completely antisymmetric. This requires that it must be a product of a $j = 0$ and an $I = 1$ or a $j = 1$ and an $I = 0$.

(2) One way to find the allowed angular momenta for antisymmetric configurations involving n particles with angular momentum j each is to use the m scheme as follows:

i) Find the possible $m_1, m_2, \cdots, m_n$ combinations allowed by the Pauli principle. It is adequate to consider $M = \sum m_1 + m_2 + \cdots + m_n \geq 0$. The others are understood to exist.

ii) From this set select the highest M. Then the highest angular momentum allowed is $J = M$. Next strike out the allowed values of $M = J, J-1, \cdots$ which go with it.

iii) From the remaining you find the maximum M and this is your next J. Proceed as in ii).

iv) Continue till the set is exhausted. Thus you have obtained all the allowed values of J that are allowed in the given antisymmetric configuration.

This rule also applies in the case one is working in the proton-neutron representation, in which case one encounters antisymmetric proton and neutron configurations.

Note: This procedure is not needed in the m-scheme, but then all $m_1, m_2, \ldots, m_n$ combinations for all values of M must be considered in the shell model basis. This, of course, leads to a huge basis for the Hamiltonian matrix.

(3) As an application of the previous rule consider the case $n = 3$ and examine the following cases:

- $j = 3/2$. Then show that:
  The allowed combinations are corresponding to $M = 3/2, 1/2$. These are adequate for the $J = 3/2$.

  We could have guessed this since three particles with $j = 3/2$ correspond to a single hole in this j-shell.
- $j = 5/2$. Then show that the allowed combinations are

$$(m_1, m_2, m_3) = (5/2, 3/2, 1/2),$$

$$(5/2, 3/2, -1/2), (5/2, 3/2, -3/2), (5/2, 3/2, -5/2)$$

  plus those found above for $j = 3/2$. Thus we obtain the allowed possibilities are $M = 7/2$, $5/2$, $2 \times 3/2$, $2 \times 1/2$. Then $J = 7/2$ is the highest allowed. The values $M = 7/2$, $5/2$, $3/2$, $1/2$ go with it. The remaining two $M = 3/2$, $1/2$ are sufficient for $J = 3/2$. So $J = 7/2$, $J = 3/2$ are the allowed values.
- $j = 7/2$. Then show that:
  The allowed combinations are

$$(m_1, m_2, m_3) = (7/2, 5/2, 3/2),$$

$$(7/2, 5/2, 1/2), (7/2, 5/2, -1/2), (7/2, 5/2, -3/2),$$

$$(7/2, 5/2, -5/2), (7/2, 5/2, -7/2)$$

  plus those found in the previous case. Thus we obtain the allowed possibilities are $M = 15/2$, $13/2$, $11/2$, $2\times 9/2$, $2\times 7/2$, $2\times 5/2$, $3\times 3/2$, $3\times 1/2$. Then the value $J = 15/2$ is the highest allowed. The values $M = 15/2$, $13/2$, $11/2$, $9/2$, $7/2$, $5/2$, $3/2$, $1/2$ go with it. Among the remaining the maximum two is $M = 9/2$ which takes with it the needed m values. Striking them out we are left with $3/2$, $1/2$, which are sufficient for a $J = 3/2$. So $J = 15/2$, $9/2$, $3/2$ are the allowed values.

- $j = 9/2$. Then proceeding as in previous case show that the allowed M values are

$$M = 21/2, 19/2, 17/2, 2\times 15/2, 2\times 13, 2\times 11/2, 3\times 9/2, 3\times 7/2, \\ 3\times 5/2, 4\times 3/2, 4\times 1/2. \tag{15.41}$$

Then show that

$$J = 21/2, 15/2, 9/2, 5/2, 3/2 \tag{15.42}$$

are the allowed values.

(4) Show that the non-diagonal matrix V and the matrix of eigenvectors S, column vectors, discussed in Section 15.6.3 are:

$$V = \left(\begin{array}{c|cccccc} & rg & rb & gb & gr & br & bg \\ \hline rg & -\frac{1}{3} & 0 & 0 & 1 & 0 & 0 \\ rb & 0 & -\frac{1}{3} & 0 & 0 & 1 & 0 \\ gb & 0 & 0 & -\frac{1}{3} & 0 & 0 & 1 \\ gr & 1 & 0 & 0 & -\frac{1}{3} & 0 & 0 \\ br & 0 & 1 & 0 & 0 & -\frac{1}{3} & 0 \\ bg & 0 & 0 & 1 & 0 & 0 & -\frac{1}{3} \end{array}\right),$$

$$S = \begin{pmatrix} \frac{1}{\sqrt{2}} & 0 & 0 & \frac{1}{\sqrt{2}} & 0 & 0 \\ 0 & \frac{1}{\sqrt{2}} & 0 & 0 & \frac{1}{\sqrt{2}} & 0 \\ 0 & 0 & \frac{1}{\sqrt{2}} & 0 & 0 & \frac{1}{\sqrt{2}} \\ -\frac{1}{\sqrt{2}} & 0 & 0 & \frac{1}{\sqrt{2}} & 0 & 0 \\ 0 & -\frac{1}{\sqrt{2}} & 0 & 0 & \frac{1}{\sqrt{2}} & 0 \\ 0 & 0 & -\frac{1}{\sqrt{2}} & 0 & 0 & \frac{1}{\sqrt{2}} \end{pmatrix} \tag{15.43}$$

$S^+VS =$ diagonal matrix$(-4/3, -4/3, -4/3, 2/3, 2/3, 2/3)$.

Chapter 16

Appendix C: Elements of the Dirac theory

16.1 Preliminaries; the Klein-Gordon equation

Recall from ordinary quantum mechanics for a free particle

$$E = \frac{p^2}{2m},$$
$$E \to \hbar i \frac{\partial}{\partial t}, \quad \mathbf{p} \to -i\hbar i \boldsymbol{\nabla},$$
$$i\frac{\partial \psi}{\partial t} = -\frac{\nabla^2}{2m}\psi.$$

In the presence of a potential:

$$H = \frac{p^2}{2m} + V \Rightarrow H = -\hbar^2 \frac{\nabla^2}{2m} + V.$$

This leads to the Schrödinger equation

$$\Rightarrow i\hbar \frac{\partial \psi}{\partial t} = H\psi. \tag{16.1}$$

Then one defines the probability and current density

$$\rho = \psi^* \psi \geq 0,$$
$$\mathbf{J}_\rho = \frac{\hbar i}{2m}(\psi \nabla \psi^* - \psi^* \nabla \psi), \tag{16.2}$$

with the conservation law:

$$\frac{\partial \rho}{\partial t} + \boldsymbol{\nabla} \cdot \mathbf{J} = 0. \tag{16.3}$$

In going to the relativistic case, we found it convenient to use the natural system of units ($\hbar = 1$, $c = 1$), see Chapter 9. Thus the energy momentum relation becomes:

$$p_\mu p^\mu = m^2 \Rightarrow E^2 = \mathbf{p}^2 + m^2$$

or setting $x^0 = ct$:

$$p_\mu = i\frac{\partial}{\partial x^\mu} \equiv i\partial_\mu, \quad p^2 = -\partial_\mu\partial^\mu, \quad \partial_\mu\partial^\mu = \frac{\partial^2}{c^2\partial t^2} - \nabla^2.$$

Thus for a scalar particle:

$$\left(\frac{\partial^2}{\partial t^2} - \nabla^2\right)\phi + m^2\phi^2 = 0 \text{ (Klein-Gordon equation).} \qquad (16.4)$$

From the Klein-Gordon equation we get:

$$\phi^*\left(\frac{\partial^2\phi}{\partial t^2} - \nabla^2\phi + m^2\phi\right) = 0,$$
$$\phi\left(\frac{\partial^2\phi^*}{\partial t^2} - \nabla^2\phi^* + m^2\phi^*\right) = 0.$$

Subtracting the two we get:

$$\left(\phi^*\frac{\partial^2\phi}{\partial t^2} - \phi\frac{\partial^2\phi^*}{\partial t^2}\right) - (\phi^*\nabla^2\phi - \phi\nabla^2\phi^*) = 0 \Rightarrow$$

$$\frac{\partial}{\partial t}\left(\phi^*\frac{\partial\phi}{\partial t} - \phi\frac{\partial\phi^*}{\partial t}\right) = \nabla.\left(\phi^*\boldsymbol{\nabla}\phi - \phi\boldsymbol{\nabla}\phi^*\right).$$

Defining:

$$\rho = \frac{i}{2}\left(\phi^*\frac{\partial\phi}{c\partial t} - \phi\frac{\partial\phi^*}{c\partial t}\right), \quad \mathbf{J} = \frac{i}{2}\left(\phi^*\boldsymbol{\nabla}\phi - \phi\boldsymbol{\nabla}\phi^*\right),$$

we get the conservation law:

$$\frac{\partial\rho}{c\partial t} + \boldsymbol{\nabla}\cdot\mathbf{J} = 0. \qquad (16.5)$$

We note that the space component for the current has the proper form. The time component, however, is unusual. It contains a time derivative. This happens, because unlike the Schrödinger equation, which is first order in the time derivative, the Klein-Gordon equation is of second order. This, of course, presents a problem, since ρ is not always positive, and cannot be interpreted as a probability.

The problem persists even in the case of a plane wave:

$$\phi = Ne^{-i(Et-\mathbf{p}\cdot\mathbf{x})}, \; E = \pm\sqrt{p^2+m^2}$$

(for the normalization N see problem 1 at the end of this chapter). Then in the case of a free particle we have:

$$\rho = |N|^2E, \quad E = \pm\sqrt{p^2+m^2}$$

i.e. ρ can become negative for negative energy. The presence of negative energies lead to the abandonment of the Klein-Gordon equation, quite prematurely as we will see later.

We can write the above conservation equation in covariant form:

$$\partial_\mu j^\mu = 0,$$

with

$$j_\mu = \frac{i}{2}\left(\phi^* \partial_\mu \phi - i\phi\partial_\mu \phi^*\right) = (\rho, J),\ j^\mu = (\rho, -J).$$

We thus find in the case of a free particle:

$$J^\mu = |N|^2 (E, \mathbf{p}) = |N|^2 p_\mu.$$

16.2 The Dirac equation

Motivated by the non relativistic case, which is first order in time, Dirac attempted to set up an equation of the type:

$$\begin{aligned} &i\frac{\partial \psi(x,t)}{\partial t} = H\psi(x), \\ &H = -i\left(\alpha_1 \frac{\partial}{\partial x^1} + \alpha_2 \frac{\partial}{\partial x^2} + \alpha_3 \frac{\partial}{\partial x^3}\right) + \beta m \end{aligned} \tag{16.6}$$

with α_i, $i = 1, 2, 3$, β to be determined. In operator form:

$$\begin{aligned} i\frac{\partial}{\partial t} &= -i\boldsymbol{\alpha}\cdot\boldsymbol{\nabla} + \beta m, \\ -\frac{\partial^2}{\partial t^2} &= \left(-i\boldsymbol{\alpha}\cdot\boldsymbol{\nabla} + \beta m\right)\left(-i\boldsymbol{\alpha}\cdot\boldsymbol{\nabla} + \beta m\right), \end{aligned}$$

that is

$$\begin{aligned} -\frac{\partial^2}{\partial t^2}\psi = -\alpha_i^2 \frac{\partial^2 \psi}{\partial x_i^2} - \sum_{i<j}\left(\alpha_i\alpha_i + \alpha_j\alpha_i\right)\frac{\partial^2 \psi}{\partial x_i \partial x_i} - im\left(\alpha_i\beta + \beta\alpha_i\right)\frac{\partial \psi}{\partial x_i} \\ +\beta^2 m^2 \psi. \end{aligned} \tag{16.7}$$

This should be brought to the form:

$$-\frac{\partial^2}{\partial t^2} = -\nabla^2 + m^2.$$

So we demand:

$$\begin{aligned} &\alpha_1^2 = \alpha_2^2 = \alpha_3^2 = \beta^2 = 1, \\ &\alpha_i\alpha_j + \alpha_j\alpha_i = 0, \quad i \neq j, \\ &\alpha_i\beta + \beta\alpha_i = 0. \end{aligned} \tag{16.8}$$

These conditions cannot be satisfied by ordinary numbers. Dirac realized that it is possible to have them satisfied if α_i, β are matrices. He then proceeded to construct these matrices starting from the Pauli matrices σ_i, which satisfy

$$\sigma_i\sigma_j + \sigma_j\sigma_i = 2I\delta_{ij}.$$

His choice was:

$$\alpha_i = \begin{pmatrix} 0 & \sigma_i \\ \sigma_i & 0 \end{pmatrix}, \quad \beta = \begin{pmatrix} 1 & 0 \\ 0 & -1 \end{pmatrix}, \tag{16.9}$$

i.e.

$$\alpha_1 = \begin{pmatrix} 0 & \sigma_1 \\ \sigma_1 & 0 \end{pmatrix} = \begin{pmatrix} 0&0&0&1 \\ 0&0&1&0 \\ 0&1&0&0 \\ 1&0&0&0 \end{pmatrix}, \quad \alpha_2 = \begin{pmatrix} 0&0&0&-i \\ 0&0&i&0 \\ 0&-i&0&0 \\ i&0&0&0 \end{pmatrix},$$

$$\alpha_3 = \begin{pmatrix} 0&0&1&0 \\ 0&0&0&-1 \\ 1&0&0&0 \\ 0&-1&0&0 \end{pmatrix}, \quad \beta = \begin{pmatrix} 1&0&0&0 \\ 0&1&0&0 \\ 0&0&-1&0 \\ 0&0&0&-1 \end{pmatrix}, \tag{16.10}$$

$$\{\alpha_i, \alpha_i\} = 2\delta_{ij}I, \ \{\alpha, \beta\} = 0, \ \beta^2 = 1.$$

In practice it is more convenient to use a different set of matrices

$$\gamma^\mu = (\gamma_0, \gamma_i), \quad \gamma_0 = \beta, \ \gamma_i = \beta\alpha_i$$

or more explicitly the Dirac matrices:

$$\gamma^0 = \begin{pmatrix} 1 & 0 \\ 0 & -1 \end{pmatrix}, \gamma^k = \begin{pmatrix} 0 & \sigma_k \\ -\sigma_k & 0 \end{pmatrix}. \tag{16.11}$$

These satisfy:

$$\{\gamma_\mu, \gamma_\nu\} = (\gamma_\mu\gamma_\nu + \gamma_\nu\gamma_\mu) = 2g_{\mu\nu}. \tag{16.12}$$

One then can show:

$$(\gamma_0)^2 = 1, \ (\gamma_i)^2 = -1, \ (\gamma^0)^\dagger = \gamma^0, \ (\gamma^k)^\dagger = -\gamma^k.$$

Then the Dirac equation reads:

$$i\beta\frac{\partial\psi}{\partial t} = (-i\beta\boldsymbol{\alpha}\cdot\boldsymbol{\nabla} + m)\psi = 0 \Rightarrow$$

$$i\gamma^0\frac{\partial\psi}{\partial t} = (-i\boldsymbol{\gamma}\cdot\boldsymbol{\nabla} + m)\psi \text{ or}$$

$$i\gamma^0\frac{\partial\psi}{\partial t} + i\boldsymbol{\gamma}\cdot\boldsymbol{\nabla}\psi - m\psi = 0,$$

i.e. more compactly in covariant form

$$(i\gamma^\mu \partial_\mu - m)\psi = 0 \text{ (The Dirac equation)}, \tag{16.13}$$

with

$$\partial_\mu = \frac{\partial}{\partial x^\mu} = \left(\frac{\partial}{\partial x^0}, \frac{\partial}{\partial x^1}, \frac{\partial}{\partial x^2}, \frac{\partial}{\partial x^3}\right) = \left(\frac{\partial}{\partial x_0}, -\frac{\partial}{\partial x_1}, -\frac{\partial}{\partial x_2}, -\frac{\partial}{\partial x_3}\right).$$

Taking the Hermitian conjugate of (16.13) we get:

$$-i\partial_\mu \psi^+ (\gamma^\mu)^+ - m\psi^+ = 0$$

or

$$-i\partial_\mu \psi^+ (\gamma^\mu)^+ \gamma_0 - m\psi^+ \gamma_0 = 0.$$

Defining now $\bar{\psi} = \psi^+ \gamma_0$ and noting that $\gamma_0 \gamma_\mu^+ \gamma_0 = \gamma_\mu$ we get:

$$\bar{\psi}(i\gamma^\mu \partial_\mu + m) = 0 \text{ (conjugate Dirac equation)}, \tag{16.14}$$

with the understanding that ∂_μ acts on the left. Equation (16.14) is called the conjugate of Eq. (16.13).

Multiplying (16.13) from the left with $\bar{\psi}$ and (16.14) from the right with ψ and adding the two we find:

$$\bar{\psi}\gamma^\mu(\partial_\mu)\psi + (\partial_\mu \bar{\psi})\gamma^\mu \psi = 0.$$

This can be written as:

$$\partial_\mu \left(\bar{\psi}\gamma^\mu \psi\right) = 0$$

or finally

$$\partial_\mu j^\mu = 0, \quad j^\mu = \bar{\psi}\gamma^\mu \psi. \tag{16.15}$$

This defines the (probability) current which is conserved. In particular we find:

$$j^0 = \bar{\psi}\gamma^0 \psi = \psi^+ \psi = |\psi_1|^2 + |\psi_2|^2 + |\psi_3|^2 + |\psi_4|^2 \geq 0, \tag{16.16}$$

which is positive definite. Thus the Dirac equation, being of first order in time, avoids the problem of negative probability, a problem that plagued the Klein-Gordon equation.

16.2.1 *The spinors u and v*

For a free particle, $\psi = ue^{-i(Et-\mathbf{p}\cdot\mathbf{x})}$, $i\partial_\mu = (i)(-ip_\mu) = p_\mu \Rightarrow$

$$(i\gamma^\mu\partial_\mu - m)\psi = 0 \Leftrightarrow (p_\mu\gamma^\mu - m)u = 0 \tag{16.17}$$

and $\bar{\psi}= \bar{u}e^{i(Et-\mathbf{p}\cdot\mathbf{x})}$, $i\partial_\mu = (i)ip_\mu = -p_\mu \Rightarrow$

$$\bar{\psi}(i\gamma^\mu\partial_\mu + m) = 0 \Leftrightarrow \bar{u}(\gamma^\mu p_\mu - m) = 0. \tag{16.18}$$

Note that in momentum space both expressions contain $-m$.

Multiplying Eq. (16.17) from the left with $\bar{\psi}\gamma^\nu$ and Eq. (16.18) from the right with $\gamma^\nu\psi$ and adding the two we get:

$$\bar{u}\left(\gamma^\mu\gamma^\nu + \gamma^\nu\gamma^\mu\right)p_\mu u = 2m\bar{u}\gamma^\nu u.$$

Then using the fundamental relation, Eq. (16.12), we get:

$$\bar{u}\gamma^\nu u = \bar{u}u\frac{p^\nu}{m}. \tag{16.19}$$

Returning to the plane wave solution we write:

$$\psi \sim e^{ipx}u \text{ (outgoing)}, \psi \sim e^{-ipx}u \text{ (incoming)}, px = Et - \mathbf{x}\cdot\mathbf{p}. \tag{16.20}$$

In this context we choose:

$$\text{positive energy: } \psi^{(+)} = e^{-ipx}u, \text{ negative energy: } \psi^{(-)} = e^{ipx}\upsilon$$

we find respectively

$$(\not{p} - m)u = 0, \quad (\not{p} + m)\upsilon = 0 \tag{16.21}$$

(note the sign difference). The quantity $\not{p}$ defined by:

$$\not{p} = p_\mu\gamma^\mu = E\gamma^0 - \mathbf{p}_k\gamma^k = \begin{pmatrix} E & -\mathbf{p}.\boldsymbol{\sigma} \\ \mathbf{p}.\boldsymbol{\sigma} & -E \end{pmatrix}. \tag{16.22}$$

We will consider each case separately.

- The positive energy solution:

$$(\not{p} - m)u = 0,$$

$$u = \begin{pmatrix} \phi \\ \chi \end{pmatrix}, \ \phi, \chi \text{ two component spinors:}$$

$$\begin{pmatrix} 1 \\ 0 \end{pmatrix} \Leftrightarrow S_z = \frac{1}{2}, \begin{pmatrix} 0 \\ 1 \end{pmatrix} \Leftrightarrow S_z = -\frac{1}{2}. \tag{16.23}$$

Then noting that

$$\not{p} = \begin{pmatrix} E & -\mathbf{p}\cdot\boldsymbol{\sigma} \\ \mathbf{p}\cdot\boldsymbol{\sigma} & -E \end{pmatrix},$$

we find

$$\begin{pmatrix} E-m & -\mathbf{p}\cdot\boldsymbol{\sigma} \\ \mathbf{p}\cdot\boldsymbol{\sigma} & -E-m \end{pmatrix}\begin{pmatrix} \phi \\ \chi \end{pmatrix} = 0. \tag{16.24}$$

That is

$$(E-m)\phi - \mathbf{p}\cdot\boldsymbol{\sigma})\chi = 0,$$

$$(E+m)\chi = \mathbf{p}\cdot\boldsymbol{\sigma}\phi \Rightarrow \tag{16.25}$$

$$\chi = \frac{\mathbf{p}\cdot\boldsymbol{\sigma}}{E+m}\phi, \text{ or } u = N\begin{pmatrix} \phi \\ \frac{\mathbf{p}\cdot\boldsymbol{\sigma}}{E+m}\phi \end{pmatrix}. \tag{16.26}$$

Since the two component spinors are normalized, $|\phi|^2 = 1$, the normalization condition $\bar{u}u = 1$ yields:

$$|N|^2\left(1 - \frac{(\mathbf{p}\cdot\boldsymbol{\sigma})(\mathbf{p}\cdot\boldsymbol{\sigma})}{(E+m)^2}\right) = 1, \quad |N|^2\left(1 - \frac{p^2}{(E+m)^2}\right)$$

$$= 1 \Rightarrow |N|^2\left(1 - \frac{E-m}{E+m}\right) = 1$$

or

$$|N|^2\frac{2m}{E+m} \Rightarrow N = \sqrt{\frac{E+m}{2m}}.$$

Thus

$$u = \sqrt{\frac{E+m}{2m}}\begin{pmatrix} \phi \\ \frac{\mathbf{p}\cdot\boldsymbol{\sigma}}{E+m}\phi \end{pmatrix}. \tag{16.27}$$

We note that, with this normalization:

$$u^\dagger u = \frac{E+m}{2m}\left[1 + \frac{E^2-m^2}{(E+m)^2}\right] = \frac{E+m}{2m}\left(\frac{E+m+E-m}{E+m}\right) = \frac{E}{m}.$$

Some authors use the normalization:

$$\bar{u}u = 2m \Rightarrow u = \sqrt{(E+m)}\begin{pmatrix} \phi \\ \frac{\mathbf{p}\cdot\boldsymbol{\sigma}}{E+m}\phi \end{pmatrix},$$

in which case:

$$u^\dagger u = (E+m)\frac{2E}{E+m} = 2E.$$

- Negative energy solution:

$$(\not{p}+m)v=0,$$

$$\begin{pmatrix} E+m & -\mathbf{p}\cdot\boldsymbol{\sigma} \\ \mathbf{p}\cdot\boldsymbol{\sigma} & -E+m \end{pmatrix}\begin{pmatrix}\phi\\ \chi\end{pmatrix}=0$$

$$(E+m)\phi=(\mathbf{p}\cdot\boldsymbol{\sigma})\chi, \quad (\mathbf{p}\cdot\boldsymbol{\sigma})\phi=E-m)\chi.$$

Solving again for the coefficient of $E+m$, in this case ϕ, we get

$$\phi=\frac{\mathbf{p}\cdot\boldsymbol{\sigma}}{E+m}\chi \Rightarrow v=N\begin{pmatrix}\frac{\mathbf{p}\cdot\boldsymbol{\sigma}}{E+m}\chi\\ \chi\end{pmatrix}. \tag{16.28}$$

If we choose again:

$$N=\sqrt{\frac{E+m}{2m}} \Rightarrow v=\sqrt{\frac{E+m}{2m}}\begin{pmatrix}\frac{\mathbf{p}\cdot\boldsymbol{\sigma}}{E+m}\chi\\ \chi\end{pmatrix}.$$

We can, of course, rewrite it as:

$$v=\sqrt{\frac{E+m}{2m}}\begin{pmatrix}\frac{\mathbf{p}\cdot\boldsymbol{\sigma}}{E+m}\phi\\ \phi\end{pmatrix}. \tag{16.29}$$

Then

$$\begin{aligned}\bar{v}v&=\frac{E+m}{2m}\left[\frac{p^2}{(E+m)^2}-1\right]=\frac{E+m}{2m}\left[\frac{(E+m)(E-m)}{(E+m)^2}-1\right]\\ &=\frac{E+m}{2m}\left[\frac{E-m}{E+m}-1\right]=\frac{E+m}{2m}\frac{-2m}{E+m}=-1\end{aligned}$$

i.e.

$$v^\dagger v=E/m.$$

Sometimes the choice $N=\sqrt{E+m}$ is made. Then

$$v^\dagger v=(E+m)\left[\frac{(E+m)(E-m)}{(E+m)^2}+1\right]=(E+m)\frac{2E}{E+m}=2E.$$

Summarizing for plane wave solutions, by choosing the momentum as the quantization axis, we have:

$$u_1=\sqrt{\frac{E+m}{2m}}\begin{pmatrix}1\\0\\ \frac{p}{E+m}\\0\end{pmatrix}, \quad v_1=\sqrt{\frac{E+m}{2m}}\begin{pmatrix}\frac{p}{E+m}\\0\\1\\0\end{pmatrix},$$

$$u_2=\sqrt{\frac{E+m}{2m}}\begin{pmatrix}0\\1\\0\\-\frac{p}{E+m}\end{pmatrix}, \quad v_2=\sqrt{\frac{E+m}{2m}}\begin{pmatrix}0\\-\frac{p}{E+m}\\0\\1\end{pmatrix}. \tag{16.30}$$

In the static limit:

$$u_1 = \begin{pmatrix} 1 \\ 0 \\ 0 \\ 0 \end{pmatrix}, u_2 = \begin{pmatrix} 0 \\ 1 \\ 0 \\ 0 \end{pmatrix}, \upsilon_1 = \begin{pmatrix} 0 \\ 0 \\ 1 \\ 0 \end{pmatrix}, \upsilon_2 = \begin{pmatrix} 0 \\ 0 \\ 0 \\ 1 \end{pmatrix}. \tag{16.31}$$

In this presentation we can see that one cannot throw away the negative energy solutions, since, then, one does not have a complete set.

With the above ingredients, one can summarize the results as follows:

$$\begin{gathered} \bar{u}(\mathbf{p}, s)u(\mathbf{p}, s) = 1 \ , \ \bar{\upsilon}(\mathbf{p}, s)\upsilon(\mathbf{p}, s) = -1 \\ \bar{u}(\mathbf{p}, s)\gamma^\mu u(\mathbf{p}, s) = \frac{1}{m}p^\mu \ , \ \bar{\upsilon}(\mathbf{p}, s)\gamma^\mu \upsilon(\mathbf{p}, s) = -\frac{1}{m}p^\mu \\ \sum_{s=\pm 1/2} u(\mathbf{p}, s)\bar{u}(\mathbf{p}, s) = \frac{1}{2m}(\not{p} + m) \ , \ \upsilon(\mathbf{p}, s)\bar{\upsilon}(\mathbf{p}, s) = \frac{1}{2m}(\not{p} - m), \\ \not{p} = p_\mu \gamma^\mu. \end{gathered} \tag{16.32}$$

Note: To prove the second of the above equations you observe that the right hand side must be proportional to p^μ, the only available 4-vector. You write it as Λp^μ. Then you multiply the resulting equation with p_μ and you make a Lorentz contraction. After that, using the Dirac equation, you find $\Lambda = 1/m$ and $\Lambda = -1/m$ for the spinors u and υ respectively.

The third equation is, of course, a matrix equation.

Returning back to the free Dirac equation and considering fermion spinor

$$u = \begin{pmatrix} \phi \\ \chi \end{pmatrix},$$

we get:

$$\begin{aligned} &(p_\mu \gamma^\mu - m)\begin{pmatrix} \phi \\ \chi \end{pmatrix} = 0, \quad p_\mu \gamma^\mu = E_0 \gamma^0 - \mathbf{p} \cdot \boldsymbol{\gamma}, \\ &\Rightarrow \begin{pmatrix} E_0 - m & -\mathbf{p} \cdot \boldsymbol{\sigma} \\ \mathbf{p} \cdot \boldsymbol{\sigma} & -E - m \end{pmatrix} \begin{pmatrix} \phi \\ \chi \end{pmatrix} = 0 \end{aligned} \tag{16.33}$$

or

$$\begin{aligned} &(E - m)\phi = (\mathbf{p} \cdot \boldsymbol{\sigma})\chi, \\ &(\mathbf{p} \cdot \boldsymbol{\sigma})\phi = (E + m)\chi \Rightarrow \\ &(E - m)\phi = \frac{\mathbf{p} \cdot \boldsymbol{\sigma}}{E + m}\mathbf{p} \cdot \boldsymbol{\sigma}\phi \Rightarrow \\ &((E - m) - \frac{p^2}{E + m})\phi = 0. \end{aligned} \tag{16.34}$$

This is an eigenvalue problem, leading to the solution:

$$E - m = \frac{p^2}{E+m} \Rightarrow E^2 - m^2 = p^2 \Rightarrow E = \pm\sqrt{m^2 + p^2}.$$

In other words the energy can be both positive and negative.

16.2.2 *Projection operators*

The notion of projection operators is useful in computation. A projection operator projects a vector defined into a space into a subspace. Consider, e.g., a complete set of states $\mathcal{N} = \{[|i\rangle, i = 1, 2, \ldots, N\}$ and a complete set of states $\mathcal{N}' = \{|\alpha\rangle, \alpha = 1, 2, \ldots, N'\}$, $\mathcal{N}' \subset \mathcal{N}$. Then a projection operator P is defined by:

$$P|\psi\rangle = \sum_{\alpha=1}^{N'} c_\alpha |\alpha\rangle, \; c_\alpha = \text{constants}, \; \psi \in \mathcal{N}.$$

A special case is if:

$$P|\psi\rangle = |\psi\rangle.$$

Now we notice the following:

- $(\not p + m)/2m$ is a projection operator for the spinor u.
  Indeed using the Dirac equation, we get:

$$\frac{\not p + m}{2m} u = \frac{\not p - m}{2m} u + \frac{2m}{2m} u = u. \tag{16.35}$$

- $(- \not p + m)/2m$ is a projection operator for the spinor v.
  Again

$$\frac{- \not p + m}{2m} v = \frac{- \not p - m}{2m} v + \frac{2m}{2m} v = v. \tag{16.36}$$

16.2.3 *Various representations of the Dirac matrices*

There exist many equivalent representations of the Dirac matrices. The most common are:

- The above introduced Dirac-Pauli representation γ^μ_{DP}:

$$\gamma^0 = \begin{pmatrix} 1 & 0 \\ 0 & -1 \end{pmatrix}, \quad \boldsymbol{\gamma} = \begin{pmatrix} 0 & \boldsymbol{\sigma} \\ -\boldsymbol{\sigma} & 0 \end{pmatrix}, \quad \gamma^5 = \begin{pmatrix} 0 & 1 \\ 1 & 0 \end{pmatrix}.$$

- The Weyl representation γ_W^μ, to be discussed below in connection with the helicity:

$$\gamma^5 = \begin{pmatrix} -1 & 0 \\ 0 & 1 \end{pmatrix}, \quad \boldsymbol{\gamma} = \begin{pmatrix} 0 & \boldsymbol{\sigma} \\ -\boldsymbol{\sigma} & 0 \end{pmatrix}, \quad \gamma^0 = \begin{pmatrix} 0 & 1 \\ 1 & 0 \end{pmatrix}.$$

This is accomplished by a unitary matrix U, which diagonalizes the Dirac matrix γ_5, e.g.

$$U = \frac{1}{\sqrt{2}} \begin{pmatrix} 1 & 1 \\ -1 & 1 \end{pmatrix}, \quad U^\dagger = \frac{1}{\sqrt{2}} \begin{pmatrix} 1 & -1 \\ 1 & 1 \end{pmatrix},$$

$$U^\dagger U = \frac{1}{2} \begin{pmatrix} 1 & 1 \\ -1 & 1 \end{pmatrix} \begin{pmatrix} 1 & -1 \\ 1 & 1 \end{pmatrix} = \frac{1}{2} \begin{pmatrix} 2 & 0 \\ 0 & 2 \end{pmatrix} = \mathbf{1}. \tag{16.37}$$

Indeed this is equivalent to the D-P representation, since:

$$\boxed{\gamma_{DP}^\mu = U^\dagger \gamma_W^\mu U.} \tag{16.38}$$

- The Majorana representation γ_M^μ

$$\gamma_M^0 = \begin{pmatrix} 0 & \sigma_2 \\ \sigma_2 & 0 \end{pmatrix}, \gamma_M^1 = \begin{pmatrix} i\sigma_3 & 0 \\ 0 & i\sigma_3 \end{pmatrix},$$

$$\gamma_M^2 = \begin{pmatrix} 0 & -\sigma_2 \\ \sigma_2 & 0 \end{pmatrix}, \gamma_M^3 = \begin{pmatrix} -i\sigma_1 & 0 \\ 0 & -i\sigma_1 \end{pmatrix}, \gamma_M^5 = \begin{pmatrix} \sigma_2 & 0 \\ 0 & -\sigma_2 \end{pmatrix}. \tag{16.39}$$

This is equivalent to the Dirac-Pauli representation since:

$$\gamma_{DP}^\mu = U^\dagger \gamma_M^\mu U,$$

$$U = \frac{1}{\sqrt{2}} \gamma_{DP}^0 \left(1 + \gamma_{DP}^2\right).$$

16.3 Interpretation of negative energy solutions

What is the interpretation of the negative energy solutions? Dirac gave the following interpretation:

(1) Fermions obey the Pauli principle.
(2) In the ground state of the system all negative states are fully occupied.
(3) Given enough energy $> 2mc^2$, a negative energy particle, e.g. an electron, can be excited. Leaving behind an electron hole $\Leftrightarrow$ positron. Thus an electron positron pair is created out of the vacuum.

The above interpretation is fine with fermions, which obey the Pauli Principle and the Dirac equation.

In general consider the phase:

$$e^{-i(-E)(-t)} = e^{-iEt}.$$

Noting that $Et < 0$, either if $E < 0$, $t > 0$ or $E > 0$, $t < 0$, we may make the correspondence:

A particle with $E < 0$, $t < 0 \Leftrightarrow$ antiparticle with $E > 0$, $t > 0$.

Thus suppose the time t runs upward, $t \uparrow$, we have:

e^+ momentum p$\downarrow$ $(E < 0) \Leftrightarrow e^-$ momentum p$\uparrow$ $(E > 0)$.

It can be generalized to any kind of particles if one interprets the negative energy states as associated with antiparticles. The negative energy particles can be thought as traveling backwards in time, a picture first proposed by Stueckelberg and Feynman. The situation is exhibited in Figs. 16.1 and 16.2. Note that in Fig. 16.2 all actual particles move forward in time, the intermediate ones, called "virtual", can move both ways.

Given enough energy particle-antiparticle pairs can be crated out of the vacuum. So the negative value of the probability density encountered in the Klein-Gordon equation is simply a manifestation of the fact that in relativity the number of particles plus antiparticles is not conserved. One can produce particles and antiparticles out of the vacuum. The only restriction is that conservation laws have to be obeyed (energy, momentum, angular momentum, all kinds of "charges", including, of course, the electric charge). Conservation of charge implies that the number of charged particles must be the same with the number of their anti antiparticles. This restriction, of course, does not apply if the particles are absolutely neutral.

The Klein-Gordon equation was prematurely abandoned. It gave us the benefit of the invention of the Dirac theory. It is valid for scalar particles in the same way as the Dirac theory governs the behavior of spin 1/2 particles.

16.4 Transformation of spinors under space inversion

The space inversion is defined

$$P : \mathbf{x} \to -\mathbf{x},\, t \to t.$$

Under such a transformation:

$$\begin{aligned}&\mathbf{p} \to -\mathbf{p}, \mathbf{L} \to \mathbf{L}\,(\mathbf{L} = \mathbf{x} \times \mathbf{p}), \mathrm{s} \to \mathrm{s}\,(\mathrm{s} = \text{spin operator}),\\ &\mathbf{E} \to -\mathbf{E}, \mathbf{B} \to \mathbf{B} \text{ etc.}\end{aligned} \tag{16.40}$$

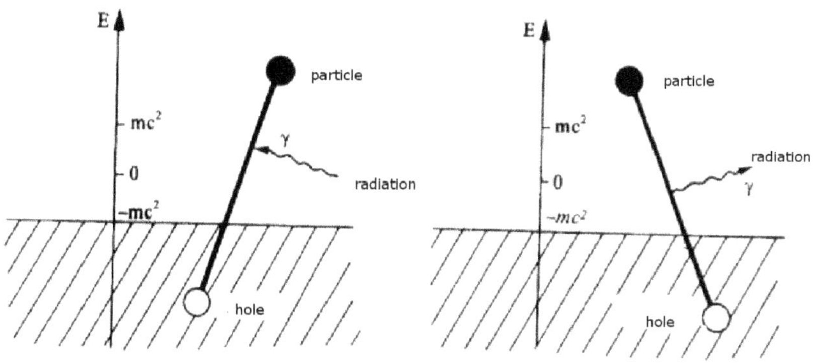

Fig. 16.1: (a) Pair production. A particle of negative energy absorbs radiation and it is transformed into a state of positive energy. (b) Pair annihilation. A particle with positive energy falls in a hole, an entity with negative energy, producing radiation with positive energy.

In the case of the Dirac spinors this operator is generalized to be the product of two operators, P followed by multiplication with γ_5, i.e.

$$\mathcal{P} = \gamma_0 P. \tag{16.41}$$

Then

$$\psi(\mathbf{x}, t) \rightarrow \psi'(\mathbf{x}, t) = \mathcal{P}\psi(\mathbf{x}, t).$$

In the special case of a free particle:

$$\psi(\mathbf{x}, t) = e^{i(Et - \mathbf{p}.\mathbf{x})}\sqrt{\frac{E+m}{2m}}\begin{pmatrix} \phi \\ \frac{\mathbf{p}\cdot\boldsymbol{\sigma}}{E+m}\phi \end{pmatrix}. \tag{16.42}$$

We find

$$\psi'(\mathbf{x}, t) = e^{i(Et + \mathbf{p}.\mathbf{x})}\sqrt{\frac{E+m}{2m}}\begin{pmatrix} \phi \\ \frac{-\mathbf{p}\cdot\boldsymbol{\sigma}}{E+m}\phi \end{pmatrix}, \tag{16.43}$$

as expected. Also

$$u(\mathbf{p}, s) = \sqrt{\frac{E+m}{2m}}\begin{pmatrix} \phi \\ \frac{\mathbf{p}\cdot\boldsymbol{\sigma}}{E+m}\phi \end{pmatrix} \rightarrow u'(\mathbf{p}', s) = \gamma_0 u(\mathbf{p}, s)$$
$$= \sqrt{\frac{E+m}{2m}}\begin{pmatrix} \phi \\ \frac{-\mathbf{p}\cdot\boldsymbol{\sigma}}{E+m}\phi \end{pmatrix}. \tag{16.44}$$

With this symmetry we associate a multiplicative quantum number π, called parity, which takes values ± 1. Parity is conserved in electromagnetic and strong interactions, but, to the surprise of everyone, it was found to be violated in weak interactions (see [Vergados (2017)] Section 12.1.1).

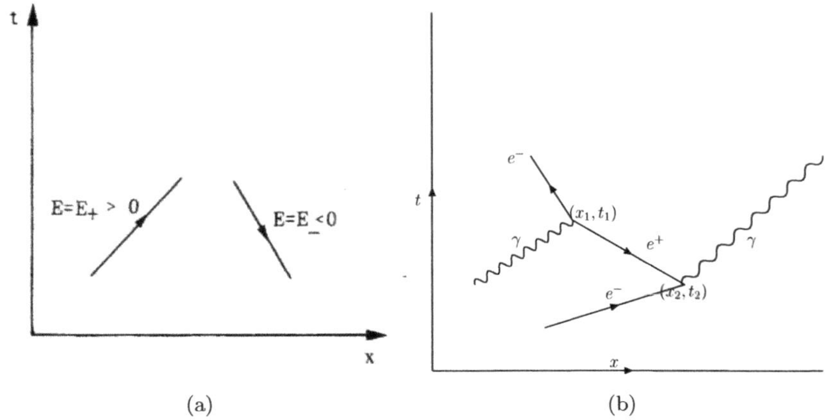

Fig. 16.2: The Stueckelberg-Feynman picture of negative energy interpreted as motion backwards in time. (a) A particle with positive energy + moves as usual. The antiparticle with negative energy − can be thought of as a particle with negative energy moving backwards in time. Both of them are moving to the right, x increases. (b) At the space-time point $(x_1,\, t_1)$ one sees a photon producing an electron positron pair. The electron is moving forward in time, while the anti-particle (positron) moves backward in time. It the space time point $(x_2,\, t_2)$ this positron meets another electron and they annihilate into a photon. In this process the positron is virtual.

16.5 The notion of helicity

(1) Ordinary quantum mechanics.
The helicity operator is defined by

$$h = \boldsymbol{\sigma} \cdot \hat{p}.$$

It has two eigenvectors:

$$\psi_1 = \frac{1}{2}(1 - \boldsymbol{\sigma} \cdot \hat{p})\phi, \quad \psi_2 = \frac{1}{2}(1 + \boldsymbol{\sigma} \cdot \hat{p})\phi.$$

Indeed

$$h\psi_1 = \frac{1}{2}(\boldsymbol{\sigma} \cdot \hat{p} - \boldsymbol{\sigma} \cdot \hat{p}\boldsymbol{\sigma} \cdot \hat{p})\psi = -\frac{1}{2}(1 - \boldsymbol{\sigma} \cdot \hat{p})\phi = -\psi_1,$$
$$h\psi_2 = \frac{1}{2}(1 + \boldsymbol{\sigma} \cdot \hat{p})\phi = \psi_2. \tag{16.45}$$

Thus, taking the direction of momentum as the quantization axis, we see that ψ_2 corresponds to the spin parallel to the momentum and ψ_1 to the case the spin is in the opposite to the momentum direction.

(2) Massless particles in the Dirac theory.

In the context of the special theory of relativity, the helicity has a meaning only for massless particles. For massive particles it is not an invariant quantity. A particle's helicity can be positive in one frame, but it may become negative in another frame, since the particle's momentum may become opposite.

In particle physics we deal with chiral fields defined by:

$$\psi_L = \frac{1}{2}(1-\gamma_5)\psi \text{ (left)}, \; \psi_R = \frac{1}{2}(1+\gamma_5)\psi \text{ (right)}, \; \gamma_5 = \begin{pmatrix} 0 & 1 \\ 1 & 0 \end{pmatrix}. \tag{16.46}$$

Note that since $\gamma_5^2 = 1$ it is trivial to shoe

$$\gamma_5\psi_L = -\psi_L, \quad \gamma_5\psi_R = \psi_L R$$

that is to say the chiral fields are eigenfields of γ_5. One also defines the helicity operator:

$$\Sigma = \begin{pmatrix} \boldsymbol{\sigma}\cdot\hat{p} & 0 \\ 0 & \boldsymbol{\sigma}\cdot\hat{p} \end{pmatrix}. \tag{16.47}$$

We consider the following cases:

α) Positive energy:

$$\psi = \begin{pmatrix} 1 \\ \boldsymbol{\sigma}\cdot\hat{p} \end{pmatrix}\phi, \tag{16.48}$$

$$\begin{aligned}\psi_L &= \frac{1}{2}(1-\gamma_5)\psi = \frac{1}{2}(1-\gamma_5)\begin{pmatrix} 1 \\ \boldsymbol{\sigma}\cdot\hat{p} \end{pmatrix}\phi \\ &= \frac{1}{2}\begin{pmatrix} 1 & -1 \\ -1 & 1 \end{pmatrix}\begin{pmatrix} 1 \\ \boldsymbol{\sigma}\cdot\hat{p} \end{pmatrix}\phi = \begin{pmatrix} \frac{1}{2}(1-\boldsymbol{\sigma}\cdot\hat{p})\phi \\ -\frac{1}{2}(1-\boldsymbol{\sigma}\cdot\hat{p})\phi \end{pmatrix}.\end{aligned}$$

Similarly

$$\psi_R = \frac{1}{2}(1+\gamma_5)\psi = \begin{pmatrix} \frac{1}{2}(1+\boldsymbol{\sigma}\cdot\hat{p})\phi \\ \frac{1}{2}(1+\boldsymbol{\sigma}\cdot\hat{p})\phi \end{pmatrix}. \tag{16.49}$$

Then it is trivial to show that

$$\Sigma\psi_L = -\psi_L, \quad \Sigma\psi_R = \psi_R.$$

For positive energy solutions the $+(-)$ helicity coincides with the right (left) chirality.

β) Negative energy:

$$\psi = \begin{pmatrix} -\boldsymbol{\sigma}\cdot\hat{p} \\ 1 \end{pmatrix}\phi, \tag{16.50}$$

$$\psi_L = \frac{1}{2}(1-\gamma_5)\psi = \frac{1}{2}(1-\gamma_5)\begin{pmatrix} -\boldsymbol{\sigma}\cdot\hat{p} \\ 1 \end{pmatrix}\phi$$
$$= \frac{1}{2}\begin{pmatrix} 1 & -1 \\ -1 & 1 \end{pmatrix}\begin{pmatrix} -\boldsymbol{\sigma}\cdot\hat{p} \\ 1 \end{pmatrix}\phi = \begin{pmatrix} -\frac{1}{2}(1+\boldsymbol{\sigma}\cdot\hat{p}) \\ \frac{1}{2}(1+\boldsymbol{\sigma}\cdot\hat{p}) \end{pmatrix}\phi,$$

$$\psi_R = \frac{1}{2}(1+\gamma_5)\psi = \frac{1}{2}(1+\gamma_5)\begin{pmatrix} -\boldsymbol{\sigma}\cdot\hat{p} \\ 1 \end{pmatrix}\phi$$
$$= \frac{1}{2}\begin{pmatrix} 1 & 1 \\ 1 & 1 \end{pmatrix}\begin{pmatrix} -\boldsymbol{\sigma}\cdot\hat{p} \\ 1 \end{pmatrix}\phi = \begin{pmatrix} \frac{1}{2}(1-\boldsymbol{\sigma}\cdot\hat{p})\phi \\ \frac{1}{2}(1-\boldsymbol{\sigma}\cdot\hat{p})\phi \end{pmatrix}, \tag{16.51}$$

$$\Sigma\psi_L = \psi_L, \quad \Sigma\psi_R = -\psi_R.$$

For negative energy solutions the $-(+)$ helicity coincides with the right (left) chirality.

We stress that the chiral fields are not eigenfields of the helicity operator, if they have non zero mass.

(3) The Weyl representation of spinors.
We choose the Weyl representation of the Dirac matrices.

$$\boldsymbol{\gamma} = \begin{pmatrix} 0 & \boldsymbol{\sigma} \\ -\boldsymbol{\sigma} & 0 \end{pmatrix}, \quad \gamma^0 = \begin{pmatrix} 0 & 1 \\ 1 & 0 \end{pmatrix}, \quad \gamma^5 = \begin{pmatrix} -1 & 0 \\ 0 & 1 \end{pmatrix}.$$

Then

$$\psi = \begin{pmatrix} \phi \\ \chi \end{pmatrix},$$
$$\psi_L = \begin{pmatrix} 1 & 0 \\ 0 & 0 \end{pmatrix}\begin{pmatrix} \phi \\ \chi \end{pmatrix} = \begin{pmatrix} \phi \\ 0 \end{pmatrix},$$
$$\psi_R = \begin{pmatrix} 0 & 0 \\ 0 & 1 \end{pmatrix}\begin{pmatrix} \phi \\ \chi \end{pmatrix} = \begin{pmatrix} 0 \\ \chi \end{pmatrix},$$
$$\Leftrightarrow \phi = \psi_L, \chi = \psi_R. \tag{16.52}$$

The Dirac equation becomes:

$$\begin{pmatrix} 0 & E-\mathbf{p}\cdot\boldsymbol{\sigma} \\ E+\mathbf{p}\cdot\boldsymbol{\sigma} & 0 \end{pmatrix}\begin{pmatrix} \phi \\ \chi \end{pmatrix} = m\begin{pmatrix} \phi \\ \chi \end{pmatrix} \rightarrow$$

$$(E - \mathbf{p}\cdot\boldsymbol{\sigma})\chi = m\phi, \quad (E + \mathbf{p}\cdot\boldsymbol{\sigma})\phi = m\chi \rightarrow$$

$$(E - \mathbf{p}\cdot\boldsymbol{\sigma})\psi_R = m\psi_L, \quad (E + \mathbf{p}\cdot\boldsymbol{\sigma})\psi_L = m\psi_R \text{ (Weyl equations).} \tag{16.53}$$

For massless particles ($m = 0$) the Weyl equations become:

$$(E - \mathbf{p}\cdot\boldsymbol{\sigma})\psi_R = 0, \; (E + \mathbf{p}\cdot\boldsymbol{\sigma})\psi_L = 0. \tag{16.54}$$

Thus:

- $E > 0$

$$\begin{aligned}(1 - \boldsymbol{\sigma}\cdot\hat{p})\psi_R = 0 \Rightarrow \boldsymbol{\sigma}\cdot\hat{p}\psi_R = \psi_R,\\ (1 + \boldsymbol{\sigma}\cdot\hat{p})\psi_L = 0 \Rightarrow \boldsymbol{\sigma}\cdot\hat{p}\psi_L = -\psi_L,\end{aligned} \tag{16.55}$$

$$\psi_R \leftrightarrow \text{ positive helicity,} \quad \psi_L \leftrightarrow \text{ negative helicity.}$$

- $E < 0$

$$\begin{aligned}(1 + \boldsymbol{\sigma}\cdot\hat{p})\psi_R = 0 \Rightarrow \boldsymbol{\sigma}\cdot\hat{p}\psi_R = -\psi_R,\\ (1 - \boldsymbol{\sigma}\cdot\hat{p})\psi_L = 0 \Rightarrow \boldsymbol{\sigma}\cdot\hat{p}\psi_L = \psi_L,\end{aligned} \tag{16.56}$$

$$\psi_R \leftrightarrow \text{ negative helicity,} \quad \psi_L \leftrightarrow \text{ positive helicity.}$$

The only particles believed to be massless are the photon and the neutrino. While the photon is known to exist in both helicities, the neutrino has been found to only left handed (helicity -1) (see [Vergados (2017)] Section 12.2)).

16.6 Charge conjugation for 4-spinors

Charge conjugation is an operation which changes the sign of the electric charge of any particle. The charge conjugation operator will be denoted by C. Under C

$$q \rightarrow -q, \; \mathbf{J} \rightarrow -\mathbf{J}, \; \mathbf{E} \rightarrow -\mathbf{E}, \; \mathbf{B} \rightarrow -\mathbf{B} \tag{16.57}$$

(see [Vergados (2017)] Section 12.3). The charge conjugation in Dirac theory is in addition defined by:

$$\psi \rightarrow \psi^c = C(\bar{\psi})^T = C\gamma_0\psi^*, \text{ (under charge conjugation transformation).}$$

Then ψ^c satisfies the Dirac equation

$$(i\partial\gamma^\mu - m)\,\psi^c = 0, \tag{16.58}$$

provided that C is non singular and satisfies the condition:

$$C^{-1}\gamma_\mu C = -\left(\gamma_\mu\right)^T. \tag{16.59}$$

Indeed then

$$C^{-1}\psi^c = \bar{\psi}^T \to \left(C^{-1}\psi^c\right)^T = \bar{\psi} \to \bar{\psi} = \left(\psi^c\right)^T\left(C^{-1}\right)^T.$$

Thus the Dirac equation:

$$\bar{\psi}\,(i\partial_\mu\gamma^\mu + m) = 0$$

implies

$$\left(\psi^c\right)^T\left(C^{-1}\right)^T\left(i\partial_\mu\gamma^\mu + m\right) = 0 \Rightarrow \left(i\partial_\mu\left(\gamma^\mu\right)^T + m\right)C^{-1}\psi^c$$

$$\Rightarrow \left(i\partial_\mu C\left(\gamma^\mu\right)^T C^{-1} + m\right)\psi^c.$$

That is ψ^c satisfies the Dirac equation, (16.58), provided that C is non singular and it can be chosen to meet the condition of Eq. (16.59). In fact such solutions exist and one may impose the additional conditions:

$$C^{-1} = -C = C^+ = C^T.$$

A possible choice:

$$C = i\gamma_2\gamma_0.$$

One can show that under the charge conjugation transformation:

$$\bar{\psi} \to \left(\bar{\psi}\right)^c = C\gamma_0\left(\bar{\psi}\right)^* = C\gamma_0\left(\psi^+\gamma_0\right)^* = C\gamma_0\gamma_0\psi^T = C\psi^T.$$

Another useful relation is

$$\bar{\psi^c} = \left(\psi^c\right)^+\gamma_0 = \left(C\gamma_0\psi^*\right)^+\gamma_0 = \psi^T\gamma_0 C^+\gamma 0 = -\psi^T\gamma_0 C\gamma 0 = \psi^T C,$$

since $\gamma_0 C\gamma 0 = \gamma_0(i\gamma_2\gamma_0)\gamma_0 = (i\gamma_0\gamma_2) = -(i\gamma_2\gamma_0) = -C$.

With this formalism Dirac showed that if a fermion exists, its charge conjugate also should exist, which was identified as its antiparticle. This proved to be one of the triumphs of physics.

Chiral states transform analogously:

$$\psi_{L,R} \to \psi^c_{L,R} = P_{L,R}\psi^c, \quad P_L = \frac{1}{2}(1-\gamma_5), \quad P_R = \frac{1}{2}(1+\gamma_5).$$

Note, however, that:

$$\left(\psi_R\right)^c = C\left(\bar{\psi}_R\right)^T = C\left(\bar{\psi}\frac{(1-\gamma_5)}{2}\right)^T = C\frac{(1-\gamma_5)}{2}\left(\bar{\psi}\right)^T$$

$$= \frac{(1-\gamma_5)}{2} C \left(\bar{\psi}\right)^T = \left(\psi^c\right)_L = \psi^c_L.$$

Similarly:

$$\left(\psi_L\right)^c = \left(\psi^c\right)_R = \psi^c_R.$$

We have only two independent chiral combinations. Usually these are taken to be:

$$(\psi_L, \psi_R) \text{ or } (\psi_L, (\psi_L)^c).$$

Charge conjugation is found to s a good symmetry in the case of the electromagnetic and strong interactions, but it is violated in weak interactions (see [Vergados (2017)]).

By combining the charge conjugation (C) and space inversion symmetry $\mathcal{P}$ we obtain a new symmetry CP. This symmetry has also been found to be violated in weak interactions (see [Vergados (2017)]). For many years the neutral Kaon system was the only one, which exhibited such violation, but in recent years it has been seen in few other systems as well.

16.7 Types of currents

We consider the following forms:

- Scalar form:

$$\bar{u}u \Leftrightarrow \text{ transforms like a scalar; one component.} \tag{16.60}$$

- 4-vector form:

$$\bar{u}\gamma_\mu u \Leftrightarrow \text{ transforms like a vector; 4 components.} \tag{16.61}$$

- Tensor form:

$$\bar{u}\frac{\sigma_{\alpha\beta}}{\sqrt{2}}u, \; \sigma_{\alpha\beta} = \frac{1}{2}\left(\gamma_\alpha\gamma_\beta - \gamma_\beta\gamma_\alpha\right), \tag{16.62}$$

  i.e. it transforms like a tensor (6 components).
- Pseudo-scalar[1] form:

$$\bar{u}\gamma_5 u \Leftrightarrow \text{ transforms like a pseudoscalar; one component.} \tag{16.63}$$

[1]Show that the matrix γ_5 given by Eq. (16.46) satisfies the equations

$$\gamma_5 = i\gamma^0\gamma^1\gamma^2\gamma^3, \quad [\gamma_5, \gamma^\mu] = 0, \; \mu = 0, 1, 2, 3.$$

- Pseudo-vector form:

$$\bar{u}\gamma_\mu\gamma_5 u \Leftrightarrow \text{ transforms like a pseudovector; 4 components.} \tag{16.64}$$

We thus have a total of 16 components. The matrices

$$O_i = \{1,\, \gamma_\mu,\, \sigma_{\alpha\beta},\, \gamma_5,\, \gamma_\mu\gamma_5\}$$

form a complete set in the space formed by any combination of products of γ matrices.

16.7.1 *Problems*

(1) In the case of the Klein-Gordon the normalized free particle solutions are label by their momentum and are indicated by

$$\phi_{\mathbf{p}}^{(+)}(x) = \frac{1}{(2\pi)^{3/2}}\frac{1}{\sqrt{2\omega_p}}e^{-i(\omega_p t - \mathbf{p}\cdot\mathbf{x})},$$

$$\phi_{\mathbf{p}}^{(-)}(x) = \frac{1}{(2\pi)^{3/2}}\frac{1}{\sqrt{2\omega_p}}e^{i(\omega_p t + \mathbf{p}\cdot\mathbf{x})}$$

with $\omega_p = \sqrt{p^2+m^2}$. The scalar product is defined by

$$\langle\psi_1|\psi_2\rangle = i\int d^3\mathbf{x}\,(\psi_1^*(\mathbf{x})\partial_0\psi_2(\mathbf{x}) - \psi_2(\mathbf{x})\partial_0\psi_1^*(\mathbf{x}))\,.$$

Show that:

$$\langle\phi_{\mathbf{p}}^{(+)}(x)|\phi_{\mathbf{p}}^{(+)}(x)\rangle = 1,\ \langle\phi_{\mathbf{p}}^{(-)}(x)|\phi_{\mathbf{p}}^{(-)}(x)\rangle = -1,$$
$$\langle\phi_{\mathbf{p}}^{(+)}(x)|\phi_{\mathbf{p}}^{(-)}(x)\rangle = 0.$$

(2) Consider the the Dirac equation

$$(i\partial_\mu\gamma^\mu - m)\psi = 0 \tag{16.65}$$

- show that it is invariant under a set global Abelian transformations

$$S = e^{ig\theta}, \quad \theta = \text{constant real function} \tag{16.66}$$

but it is not invariant under a set of local (gauge) transformations $\theta = \theta(\mathbf{x}, x_0)$.

- Modify it suitably so that it is invariant under gauge transformations with $g = e$. Do you recognize the result?

(3) Repeat the previous procedure in the case of Klein-Gordon equation.

$$(\partial_\mu\partial^\mu + m^2)\phi = 0. \tag{16.67}$$

(4) Imagine that in the world there exit three spin 1/2 fermions with the same mass obeying the above Dirac equation with ψ representing three 4 component spinors, i.e. the column $\psi = (\psi_1, \psi_2, \psi_3)^T$. Repeat the analysis of problem 1, if the group of transformations is the familiar group $SU(3)$. Do you notice anything unusual?

(5) Repeat the analysis of the previous problem if instead of spin 1/2 particles one has three scalar particles $\phi = (\phi_1, \phi_2, \phi_3)^T$.

(6) Using the Dirac equation show that the current

$$J_\mu = \bar{\psi}\gamma_\mu\gamma_5\psi \tag{16.68}$$

is not conserved so long as the fermions ψ have a non zero mass.

(7) A chiral global transformation on the fermion fields is defined as

$$\psi_L \to \psi' L = e^{i\alpha_L \gamma_5}\psi_L, \psi_R \to \psi' R = e^{i\alpha_L \gamma_5}\psi_R,$$

$$\psi_L = \frac{1}{2}(1-\gamma_5)\psi,\ \psi_R = \frac{1}{2}(1+\gamma_5)\psi.$$

Noting that the the eigenvalues of γ_5 are ± 1 with ψ_L and ψ_R show that:

$$\psi'_L = e^{-i\alpha_L}\psi_L,\ \psi'_R = e^{i\alpha_L}\psi_R.$$

Consider now the Lagrangian

$$\mathcal{L} = Y_f \bar{\psi}_L \psi_R \phi + \text{HC}, Y_f \text{ Yukawa coupling constant}$$

where HC stands for Hermitian Conjugate and ϕ is a scalar field transforming as $\phi \to \phi' = e^{-i\alpha}\phi$. Show that the above Lagrangian is invariant under global chiral transformations provided that the corresponding charges obey $\alpha = \alpha_R + \alpha_L$.

(8) For the Dirac matrices:

i) Show that the members of the set:

$$O = \{1, \gamma_\mu,\ \sigma_{\mu\nu},\ \gamma_5,\ \gamma_\mu\gamma_5\}$$

are linearly independent and constitute a basis in the 16 dimensional space of the spinors.

It may help to show first that for any two members of the set $\text{tr}(O_i O_j) = N_i \delta_{i,j},\ i,\ j = 1, \cdots 5$ so that a proper scalar product can be defined.

ii) Using the above scalar product express each of the expressions

$$\sigma_{\mu\nu}\gamma_5,\ \gamma_\lambda\gamma_\mu\gamma_\nu \text{ and } \gamma_\lambda\gamma_\mu\gamma_\nu\gamma_5$$

in terms of the above set.

(9) The electron magnetic moment. Suppose that the interaction of a photon with an electron is given by:

$$\mathcal{L} = -eJ_\lambda A^\lambda,\ J_\lambda = \bar{u}\gamma_\lambda u \tag{16.69}$$

where e is the unit of charge and A^λ describes the photon field (vector potential).

- Concentrate on the space part of this interaction $\lambda = k \neq 0$. Show that:

$$\mathcal{L} = e\bar{u}(\mathbf{p}_1', s')\boldsymbol{\gamma}.\mathbf{A}u(\mathbf{p}_1, s). \tag{16.70}$$

Thus

$$\mathcal{L} = e\frac{E+m_e}{2m_e}\left(\phi^+\left(1, \frac{\boldsymbol{\sigma}.\mathbf{p}_1'}{E+m_e}\right)\begin{pmatrix} 0 & \boldsymbol{\sigma}.\mathbf{A} \\ \boldsymbol{\sigma}.\mathbf{A} & 0 \end{pmatrix}\begin{pmatrix} \phi \\ \phi\frac{\boldsymbol{\sigma}.\mathbf{p}_1}{E+m_e} \end{pmatrix}\right). \tag{16.71}$$

- Show further that in the space of two component spinors one can write:

$$\mathcal{L} = \phi^+\Omega\phi,\ \Omega = e\left(\frac{\boldsymbol{\sigma}.\mathbf{p}_1'}{2m_e}\boldsymbol{\sigma}.\mathbf{A} + \boldsymbol{\sigma}.\mathbf{A}\frac{\boldsymbol{\sigma}.\mathbf{p}_1}{2m_e}\right). \tag{16.72}$$

- Using the fact that the Pauli matrices obey the rule:

$$\boldsymbol{\sigma}_\alpha\boldsymbol{\sigma}_\beta = \delta_{\alpha\beta} + i\epsilon_{\alpha\beta\gamma}\boldsymbol{\sigma}_\gamma$$

with $\epsilon_{\alpha\beta\gamma}$ the completely antisymmetric symbol. show that

$$(\boldsymbol{\sigma}.\mathbf{B})(\boldsymbol{\sigma}.\mathbf{C}) = \mathbf{B}.\mathbf{C} + i\boldsymbol{\sigma}.(\mathbf{B}\times\mathbf{C}).$$

As a result

$$\Omega = \frac{e}{2m_e}\left(\mathbf{P}.\mathbf{A} - i\boldsymbol{\sigma}.(\mathbf{q}\times\mathbf{A})\right),\ \mathbf{P} = \mathbf{p}_1 + \mathbf{p}_1',\ \mathbf{q} = \mathbf{p}_1 - \mathbf{p}_1' \tag{16.73}$$

where $\mathbf{q}$ is the momentum transfer to the photon.

- Noting that

$$\mathbf{q}\times\mathbf{A} = \frac{\hbar}{i}\boldsymbol{\nabla}\times\mathbf{A} = \frac{\hbar}{i}\mathbf{B} \tag{16.74}$$

where $\mathbf{B}$ is the magnetic field, show further that

$$\Omega = \frac{e}{2m_e}\left(\mathbf{P}.\mathbf{A} - \hbar\boldsymbol{\sigma}.\mathbf{B}\right) = \frac{e}{2m_e}\mathbf{P}.\mathbf{A} + \boldsymbol{\mu}_e.\mathbf{B} \tag{16.75}$$

the second term yields the interaction of the electron spin with the magnetic field, i.e.

$$\boldsymbol{\mu}_e = -\frac{e\hbar}{2m_e}\boldsymbol{\sigma} = -\mu_B\boldsymbol{\sigma} = -g_s\mu_B\mathbf{s} \tag{16.76}$$

where $\mathbf{s}$ is the spin of the electron and

$$g_s = 2,\ \mu_B = \frac{e\hbar}{2m_e} = \text{ Bohr's magneton}, e > 0 \tag{16.77}$$

$\boldsymbol{\mu}_e$ is the magnetic moment of the electron.

- Similarly for other elementary fermions, e.g. for the muon:

$$\boldsymbol{\mu}_\mu = \frac{m_e}{m_\mu}\boldsymbol{\mu}_e \tag{16.78}$$

  i.e. the muon magnetic moment is about 210 times smaller than that of the electron.

So in the Dirac theory g_s takes twice the classical value, in agreement with experiment. In practice, of course, one should consider other higher order (loop) corrections, yielding the $g_s - 2$ corrections.

Bibliography

Aad, G. *et al.* (2012). *Phys. Lett. B* **710**, p. 1.
Aaij, R. *et al.* (2019). *Phys. Rev. Lett.* **122**, p. 222001, arXiv:1904.03947 [hep-exp].
Abdon, P. *et al.* (2007). *New J. Phys.* **9**, p. 170, arXiv:physics/0702190 (physics.ins-det).
Abe, K. *et al.* (2002). *Phys. Rev. Lett.* **89**, p. 200802, see also *Phys. Rev. Lett.* 88 (2002) 181803.
Ahn, D. S. *et al.* (2019). *Phys. Rev. Lett.* **123**, p. 212501.
Aker, M. *et al.* (2019). *Phys. Rev. Lett.* **123**, p. 221802.
Alexopoulos, T. *et al.* (2011). *Nucl. Istr. Meth. A* **640**, p. 110.
Alvarez, L. W. (1937). *Phys. Rev.* **52**, p. 134.
An, F. J. *et al.* (2012). *Phys. Rev. Lett.* **108**, p. 171803, arXiv:1203.1669[hep-ex].
Antonyan, S. T. and Biermann, P. (2001). Test of primary model predictions by EAS size spectra, *proceedings of 27th ICRC (Hamburg)* **NA**, p. 1, arXiv:astro-ph/0106091.
Araki, T. *et al.* (2005). *Phys. Rev. Lett.* **94**, p. 081801.
Arnison, G. *et al.* (1983a). *Phys. Lett. B.* **122**, p. 103.
Arnison, G. *et al.* (1983b). *Phys. Lett. B.* **129**, p. 273.
Arnison, G. *et al.* (1983c). *Phys. Lett. B.* **216**, p. 398.
Audi, G., Wapstra, A., and Thibault, C. (2003). *Nuc. Phys. A* **729**, p. 337.
Baade, W. and Zwicky, F. (1934). *Proc. Nat. Ac. Sci. Amer.* **20**, p. 85.
Baer, H. W. *et al.* (1975). *Phys. Rev. C* **12**, p. 921.
Bahcall, J. N., Serenelli, A. M., and Basu, S. (2005). *Astrophysical Journal* **681**, p. 185.
Bai, J. *et al.* (2003). *Phys. Rev. Lett.* **91**, p. 022001, arXiv:hep-ex/0303006.
Benedikt, M. and Zimmermann, F. (2018). *CERN Courier* **58**, p. 1, retrieved 4 July 2.
Berezinskii, V. S. *et al.* (1990). *Astrophysics of Cosmic Rays* (North-Holland (Amsterdam)).
Boguta, J. and Bodmer, A. (1977). *Nuc. Phys.* **292 A**, p. 413.
Bohn, J. *et al.* (1999). *Nucl. Instrum. Meth.* **421 A**, p. 356.
Bohr, N. (1936). *Nature* **137**, p. 344.
Brody, T. and Moshinsky, M. (1960). *Tables of Transformation Brackets* (Insituto De Physica (Mexico)).

Brown, B. A. (2020). *Lecture Notes in Nuclear Structure Physics* (Michigan State University).
Campbell, M. *et al.* (2005). *Nucl. Instrum. Meth. A* **540**, p. 295, arXiv: physics/0409048(hep-ex).
Capozzi, F., Fogli, G., Lisi, E., Marrone, A., Montanino, D., and Palazzo, A. (2014). *Phys. Rev. D* **89**, p. 093018, arXiv:1312.2878 [hep-ph].
Charpak *et al.* (1968). *Nuc. Instr. Meth. A* **62**, p. 262.
Chatrchyan, S. *et al.* (2012). *Phys. Lett. B* **716**, p. 30.
Chen, J. Q., Ping, J., and Wang, F. (2002). *Group Representation Theory for Physicists* (World Scientific (2nd edition)).
Clarfield, G. H. and Wiecek, W. M. (194). *Nuclear America: Military and Civilian Nuclear Power in the United States 1940–1980* (Harper and Row).
Colas, P., Giomataris, I., and Lepeltier, V. (2004). *Nuclear Instruments and Methods in Physics Research A* **535**, p. 226, arXiv:physics/0412057 (physics.ins-det).
Cottingham, W. N. and Greenwood, D. A. (1986). *An introduction to nuclear physics* (Cambridge University Press).
Dalitz, R. H. (1953). *Phil. Mag.* **44**, p. 1068.
Dalrymple, G. B. (1994). *The age of the earth* (Stanford University Press), ISBN 9780804723312.
Davis, B. R., Vogel, P., Mann, F.M. and Schenter, R. (1979). *Phys. Rev. C* **19** **(6)**, p. 2259.
Derre, J., Giomataris, I., Perroud, J. P., Rebourgeard, P., and Charpak, G. (1999). *Nucl. Istr. Meth. A* **449**, p. 554.
Derre, J., Giomataris, I., Perroud, J. P., Rebourgeard, P., and Charpak, G. (2000). *Nucl. Istr. Meth. A* **449**, p. 314.
deShalit, A. and Feshbach, H. (1974). *Theoretical Nuclear Physics*, Vol. I (John Wiley and sons (New York)).
deShalit, A. and Talmi, I. (1963). *Nuclear Shell Theory* (Academic Press).
Donev, J. *et al.* (2017). *Energy Education* **Breeder reactor**, p. 1, https://energyeducation.ca/encyclopedia/Breeder_reactor. http://hyperphysics.phy-astr.gsu.edu/hbase/nucene/fasbre.html.
Drexlin, G., Hannen, V., Mertens, S., and Weinheimer, C. (2013). *AHEP* **2013**, p. 2013, artcle ID 293986.
Dyar, M. D., Agresti, D. G., Schaefer, M. W., Grant, C. A., and Sklute, E. C. (2006). *Ann. Rev. Earth. Planet. Sci.* **34**, p. 83, downloaded from arjournals.annualreviews.org.
Eden, B. (1999). *Encyclopaedia Britannica (Electronic Resources Review)*, p. 9.
Englert, F. and Brout, R. (1964). *Phys. Rev. Lett.* **13**, p. 321.
Erwin, A., March, R., Walker, W., and West, E. (1961). *Phys. Rev. Lett.* **6**, p. 628.
Esteban, I., Gonzalez-Garcia, M., Hernandez-Cabezudo, A., Maltoni, M., and Schwetz, T. (2019). *JHEP* **01**, p. 106, arXiv:1811.05487 [hep-ph].
Fabjan, C. W. (1987). *Experimental Techniques in High-Energy Physics - ed. T. Ferbel* (AddisonWesley (Menlo Park)).
Forero, D. V., Tortola, M., and Valle, J. W. F. (2014). *Phys. Rev. D* **90**, p. 093006, arXiv:1405.7540 [hep-ph].

Frauenfelder, H. and Henley, E. M. (1974). *Subatomic Physics* (Prentice Hall).
Fukuda, Y. *et al.* (1998a). *Phys. Rev. Lett.* **81**, p. 1562, super-Kamiokande Collaboration; (hep-ex/9807003).
Fukuda, Y. *et al.* (1998b). *Phys. Lett. B* **436**, p. 33, hep-ex/9805006.
Gailard, M. *et al.* (1975). *Rev. Mod. Phys.* **47**, p. 277.
Gaisser, T. K. (1990). *Cosmic Rays and Particle Physics* (Cambridge Univ. Press).
Gambhir, Y. K., Ring, P., and Thimet, A. (1990). *Ann. Phys. (N.Y.)* **198**, p. 132.
Gariazzo, S. *et al.* (2016). *J. Phys. G* **43**, p. 033001, see also C. Giunti and T. Lasserre (arXiv:1901.08330 [hep-ph]).
Gell-Mann, M. (1953). *Phys. Rev.* **92**, p. 833.
Gell-Mann, M. (1964). *Phys. Lett.* **8**, p. 214.
Genz, H., Kiihner, G., Richter, A., and Behrens, H. (1991). *Z. Phys. A* **341**, p. 9.
Ginzburg, V. L. and Syrovatskii, S. I. (1964). *The origin of cosmic rays* (Pergamon Press (Oxford)).
Giomataris, I. (1998). *Nucl. Instr. Meth. A* **419**, p. 239.
Giomataris, I., Irastorza, I., Savvidis, I., Andriamonje, S., Aune, S., Chapelier, M., Charvin, P., Colas, P., Derre, J., Ferrer, E., Gros, M., Navick, X., Salin, P., and Vergados, J. D. (2008) **3**, p. 09007, arXiv:0807.2802 (physics.ins-de).
Giomataris, I. *et al.* (2006). *Nucl. Instr. Meth. A* **560**, p. 405.
Giomataris, Y., Ang, J. P., Roberts, P. R., and Charpak, G. (1996). *Nuc. Instr. Meth. A* **376**, p. 29.
Giomataris, Y. and Vergados, J. (2004). *Nucl. Instr. Meth. A* **530**, p. 330.
Giomataris, Y. and Vergados, J. (2006a). *Phys. Lett. B* **634**, p. 23, hep-ex/0503029.
Giomataris, Y. and Vergados, J. (2006b). *Phys. Lett. B* **634**, p. 23, arXiv:hep-ex/0503029 (astro-ph), arXiv:1603.03966 (hep-ph).
Giunti, C. and Kim, C. (2007). *Fundamentals of Neutrino Physics and Astrophysics* (Oxford University Press).
Glashow, S. L. (1961). *Nuc. Phys.* **22 (4)**, p. 579.
Glashow, S. L., Iliopoulos, J., and Maiani, L. (1970). *Phys. Rev. D* **2**, p. 1285.
Glendenning, N. (2000). *Compact Stars: Nuclear Physics and Particle Physics and General Relativity* (Springer (Berlin)).
Gold, T. (1968). *Nature* **218**, p. 731.
Goldhaber, M., Grodzins, L., and Sunyar, A. W. (1958). *Phys. Rev.* **109**, p. 1015.
Goriely, S., Pearson, J. M., and Tondeur, F. (2001). *At. and Nucl. Data Tables* **77**, p. 311.
Hadermann, J. and Alder, K. (1969). *Helvetica Physica Acta* **42**, p. 3.
Haensel, P., Potekhin, A., and Yakovlev, D. (2007). *Neutron Stars 1: Equation of State and Structure* (Springer Verlag (New York)).
Hagedorn, C., Mohapatra, R. N., Molinaro, E., Nishi, C. C., and Petcov, S. T. (2018). *Int. J. Mod. Phys. A* **33**, p. 1842006, arXiv:1711.02866 [hep-ph].
Hajert, H. *et al.* (1973). *Phys. Lett. B.* **45**, p. 138.
Han, M. Y. and Nambu, Y. (1965). *Phys. Rev.* **139**, p. 1006.
Haxel, O., Jensen, J. H. D., and Suess, H. E. (1949). *Phys. Rev.* **75**, p. 1766.

Hayes, A. C. and Vogel, P. (2016). *Ann. Rev. Nucl. Part. Sci.* **66**, p. 219, arXiv:1605.02047 [hep-ph].
Hecht, K. T. (2000). *Quantum Mechanics* (Springer).
Helm, R. H. (1956). *Phys. Rev.* **104**, p. 1466.
Henley, E. M. and Garcia, A. (2007). *Subatomic Physics* (World Scientific Publishing Company).
Henley, E. M., Oka, T., and Vergados, J. D. (1988). *Nuc. Phys.* **476 A**, p. 589.
Hewish, A., Plikington, S. J. B. J.-S., Scott, P., and Collins, R. (1968). *Nature* **219**, p. 25.
Heyde, K. (1999). *Basic Ideas and Concepts in Nuclear Physics* (Institute of Physics Publishing (Bristol and Philadelphia)).
Higgs, P. W. (1964). *Phys. Lett.* **12**, p. 132.
Iguaz, F. *et al.* (2012). *Phys. Procedia* **37**, p. 1079.
Jackson, J. D. (1062). *Classical Electrodynamics* (J. Wiley).
Janka, H. T., Langanke, K., Marek, A., Martinez-Pinedo, G., and Mueller, B. (2007). *Phys. Rep.* **442**, p. 38, arXiv:astro-ph/0612072.
Johnson, K. E. (1986). *Physics Today* **39**, p. 44.
Kamal, A. (2014). *Particle Physics* (Springer), appendix E: Dalitz plots.
Karliner, M. and Rosner, J. (2017). *Phys. Rev. Lett.* **119**, p. 202001, arXiv:1707.07666 [hep-ph].
Khachatryan, V. *et al.* (2017). *J. High Energ. Phys.* **2017**, p. 156, https://doi.org/10.1007/JHEP03(2017)156.
Konopinski, E. J. (1966). *The theory of beta radioactivity* (Oxford University Press).
Kunne *et al.* (2003). *Nucl. Phys. A* **721**, p. 1087.
Landau, L. D. and Lifshitz, E. M. (1977). *Quantum Mechanics, Non-relativistic Theory*, 3rd edn. (Pergamon press).
Lattimer, J. and Prakash, M. (2001). *Astroph. J.* **550**, p. 426.
Lee, T. D. and Yang, C. (1956). *Phys. Rev.* **104**, p. 254.
Lichtenberg, D. B. (1978). *Unitary Symmetry and Elementary Particles* (Academic Press (New York) second edition).
Link, J. *et al.* (2004). *Phys. Lett. B* **585**, p. 200.
Link, J. M. *et al.* (2007). *Phys. Lett. B* **653**, p. 1, the FOCUS Collaboration.
Lipscy, P., Kushida, K., and Incerti, T. (2013). *Environmental Science and Technology*, p. 6082.
Magnon, A. *et al.* (2002). *Nucl. Istr. Meth. A* **478**, p. 210.
Majore, M. (2015). *A Modern Introduction to Quantum Field Theory* (Oxford University Press).
Mayer, M. G. (1948). *Phys. Rev.* **75**, p. 235.
Mayet, F., Guillaudin, O., Grignon, C., Koumeir, C., Santos, D., Colas, P., and Giomataris, I. (2009). *J. Phys. Conf. Ser.* **170**, p. 012011, arXiv:0905.4654,(astro-ph.CO).
Mikheyev, S. and Smirnov, A. (1985). *Sov. J. Nucl. Phys.* **42**, p. 913.
Mohl, D., Petrucci, G., Thorndahl, L., and van-der Meer, S. (1980). *Phys. Rept.* **58**, p. 76.
Mössbauer, R. L. (1958). *Z. Phys.* **151**, p. 124.

Nagado, T. and Nishijima, K. (1953). *Prog. Theor. Phys.* **10**, p. 10.

Oppenheimer, J. and Volkoff, G. (1939). *Phy. Rev.* **55 (4)**, p. 374.

Orrigo, S. L.-A., Alvarez-Ruso, L., and Pena-Garay, C. (2016). *Nuc. Phys. Proc.* **273–275**, p. 414.

Pais, A. (1952). *Phys. Rev.* **86**, p. 663.

Pais, A. and Jost, R. (1952). *Phys. Rev.* **87**, p. 871.

Patrignani, C. *et al.* (2016). *Chin. Phys. C* **40**, p. 100001, particle Data Group.

Patterson, C. (1956). *Geochimica et Cosmochimica Acta* **10 (4)**, p. 230.

Perkins, D. (1972). *Introduction to High Energy Physics*, 3rd edn. (Addison-Wesley).

Perkins, D. H. (1971). *Introduction to High Energy physics* (Addison Wesley).

Pfalzner, S. (2006). *An Introduction to Inertial Confinement Fusion* (Taylor and Francis).

Plastino, W., Lauri, K., Paolo, B., and Francesco, B. (2001). *Radiocarbon* **43 (2A)**, p. 157.

Prakash, M., Ainsworth, T., and Lattimer, J. (1988). *Phys. Rev. Lett.* **61**, p. 2518.

Ring, P. (1996). *Prog. Part. Nucl. Phys.* **37**, p. 193.

Roy, K. K. and Nigam, B. P. (1967). *Nuclear Physics* (John Wiley and Sons (New York)).

Rujula, A. D. (2019). *IJMPA* **34 (32)**, p. 1930015, arXiv:1910.1389 [hep-ph].

Salam, A. (1968). *Elementary Particle Theory - Eighth Nobel Symposium* (Almquvist and Wiksell (Stockholm)).

Salaris, M. and Cassisi, S. (2005). *Evolution of Stars and Stellar Populations* (John Wiley).

Schutz, B. (2009). *A First Course in General Relativity*, 2nd edn. (Cambridge University Press).

Serot, B. and Walecka, J. (1986). *Adv. Nucl. Phys.* **16**, p. 1.

Shapiro, S. and Teukolsky, S. (1983). *Black Holes and White Dwarfs and Neutron Stars* (John Wiley and Sons (New York)).

SO, S. I. and Strottman, D. (1976). *J. Math. Phys.* **20**, p. 153.

Sweig, G. (1964). *CERN Preprint (Th)* **8409**, p. 412.

Swordy, S. P. and Fortson, L. F. (2002). *Astropart. Phys.* **18**, p. 129, astro-ph/0202159.

Tanabashi, M. *et al.* (2018a). *Phys. Rev. D* **98**, p. 030001, the Review of Particle Physics (2018) and 2019 update.

Tanabashi, M. *et al.* (2018b). *Phys. Rev. D* **98**, p. 03001, particle Data Group.

Tolman, R. C. (1939). *Phy. Rev.* **55 (4)**, p. 364.

Tomas, R., Kachellriess, M., Raffelt, A. G., Janka, A., and Schreck, L. (2004). *JCAP* **0409**, p. 015.

Tondeur, F., Goriely, S., and Pearson, J. M. (2000). *Phys. Rev. C* **62**, p. 024308.

Uglietti, C., Zapf, A., Jenk, T. M., Sigl, M., Szidat, S., Salazar, G. and Schwikowski, M. (2016). *Cryosphere* **10**, p. 3091.

Veltman, M. (2003). *Facts and Mysteries of elementary Particle Physics* (World Scientific).

Vergados, J. (2016). *Group and Representation Theory* (World Scientific).

Vergados, J. (2017). *The Standard Model and Beyond* (World Scientific).

Vergados, J., Avignone, F. T., and Giomataris, I. (2009). *Phys. Rev. D* **79**, p. 113001, [arXiv:0902.1055 (hep-ph)].
Vergados, J. D., Ejiri, H., and Simkovic, F. (2012a). *Rep. Prog. Phys.* **75**, p. 106301.
Vergados, J. D., Ejiri, H., and Simkovic, F. (2012b). *Rep. Prog. Phys.* **75**, p. 106301.
Vergados, J. D., Giomataris, Y., and Novikov, Y. N. (2010). *J. Phys. Conf. Ser.* **259**, p. 012100, arXiv:1010.4388 (hep-ph).
Vergados, J. D. and Strottman, D. (2018). Symmetries in subatomic multi-quark systems, *IJMPE* **27 (12)**, p. 1840003.
von Weizsäcker, C. F. (1935). *Z. Physik* **96**, p. 431.
Wapstra, A., Audi, G., and Thibault, C. (2003). *Nuc. Phys. A* **729**, p. 129.
Weinberg, S. (1967). *Phys. Rev. Lett.* **19**, p. 1264.
Weinberg, S. (1980). *Phy. Rev. D* **22)**, p. 1694.
Weinberg, S. (1995). *The Quantum Theory of Fields, Volume 1: Foundations* (Cambridge University Press).
Weinberg, S. (1996). *The Quantum Theory of Fields, Volume 2: Modern applications* (Cambridge University Press).
Weinberg, S. (2008). *Cosmology* (Oxford University Press (Oxford UK)).
Wetherill, G. W. (1975). *Ann. Rev. Nucl. Sci.* **25**, p. 283.
Wigmans, R. (2000). *Calorimetry - Energy Measurement in Particle Physics* (Clarendon Press (Oxford)).
Wolfenstein, L. (1978). *Phys. Rev. D* **17**, p. 2369.
Wotschack, J. (2013). *Mod. Phys. A* **28**, p. 1340020.
Wu, C. S. *et al.* (1957). *Phys. Rev.* **105**, p. 1413.
Zettili, N. (2009). *Quantum Mechanics: Concepts and Applications* (John Wiley and Sons).
Zorich, Z. (2011). *Archaeology* **64 (2)**, p. 39.
Zurlo, N. *et al.* (2006). *Hyperfine Interact.* **97**, p. 105, arXiv:0801.3193 [hep-ex].

Index

www.ingramcontent.com/pod-product-compliance
Ingram Content Group UK Ltd.
Pitfield, Milton Keynes, MK11 3LW, UK
UKHW022106170926
8624IPUK00024B/157